William Coulson

Coulson on the Diseases of the Bladder and Prostate Gland

William Coulson

Coulson on the Diseases of the Bladder and Prostate Gland

ISBN/EAN: 9783337424619

Printed in Europe, USA, Canada, Australia, Japan

Cover: Foto ©berggeist007 / pixelio.de

More available books at **www.hansebooks.com**

[July.]

COULSON

ON

THE DISEASES

OF THE

BLADDER AND PROSTATE GLAND

SIXTH EDITION

REVISED BY

WALTER J. COULSON, F.R.C.S.
SURGEON TO ST. PETER'S HOSPITAL FOR STONE, ETC.
AND SURGEON TO THE LOCK HOSPITAL

NEW YORK:
WILLIAM WOOD & COMPANY
27 GREAT JONES STREET
1881

PREFACE

TO

THE SIXTH EDITION.

TWENTY-THREE YEARS having elapsed since the publication of the fifth edition of this work, numerous alterations have now been found necessary. Nearly all the chapters have been re-written, and several additions have been made to their number. The chapter on the chemistry of the urine, with which former editions commenced, has now been omitted; the subject being too extensive to admit of suitable treatment in a work like the present. In its place will be found chapters on the anatomy and physiology of the bladder and prostate, and on the methods of examining those organs.

The importance of the preventive treatment of calculus being now fully recognized, an attempt has been made to give as complete an account as possible of the causes to which stone-formation is due, and of the conditions under which it takes place. Numerous facts relating to the prevalence of calculus in several parts of the world, and under various conditions of life, etc., will be found in the chapter on the causes of stone.

The invention, by Dr. Bigelow, of a method for the rapid removal of a calculus from the bladder, without the risks attendant upon the use of the knife, bids fair to create a complete revolution in the doctrines hitherto current with regard to lithotrity, and in the mode of performing that operation. The advantages of withdrawing all the fragments of a calculus at one sitting must be obvious to every surgeon who is willing to recognize progress, and to Dr. Bigelow is due the credit of having discovered a method by which those advantages may be obtained without undue risk to the patient. The method, however, is still, to some extent, upon its trial. Its applicability, as a general rule, to all cases suitable for ordinary lithotrity seems, indeed, fully demonstrated; the limits of its range for cases of large calculi have yet to be determined. The chapter on lithotrity in the present work refers to the operation as practised before the introduction of the method in question, to an account of which a separate chapter has been devoted. This arrangement has been adopted in order that the reader may observe the differences between the two operations, and also because many of the details connected with ordinary lithotrity belong, as a matter of course, to Dr. Bigelow's method.

The editor takes this opportunity of thanking his friend, Dr. T. P. Smith, of Reigate, for his aid in the revision of this work, particularly in collecting references from numerous German and French authors. Care has been taken to acknowledge the various sources whence information has been derived, and the editor trusts that the additions and alterations that have been made will enhance the value of the book as a practical guide to those who may consult its pages.

17 HARLEY STREET, CAVENDISH SQUARE:
December, 1880.

CONTENTS.

CHAPTER I.
GENERAL ANATOMICAL AND PHYSIOLOGICAL CONSIDERATIONS.

	PAGE
Functions, shape and size of the bladder	1
Variations in size under certain circumstances	1
Relations of the bladder to neighboring parts	2
Its external surface, divided into regions for purposes of description	2
The anterior region; its relations affected by condition of distention of the bladder. Amount of space uncovered by peritoneum	3
Posterior surface; its relations	3
The summit, apex, or superior fundus of the bladder	3
The base of the bladder; in the male and female	4
The lateral regions of the bladder	5
The neck of the bladder; definition of anatomical and surgical neck	5
Surgical neck equivalent to prostatic portion of the urethra; its length and width; lining membrane; thickness of its walls	6
Internal surface of the bladder	7
Its appearance; orifices of the ureters and urethra; the *trigone*; the *uvula vesicæ*	8
Orifice of the urethra; variations in shape	8
Diameter of the neck of the bladder; its dilatability	8
Walls of the bladder, their composition; the peritoneal coat, the muscular coat; the mucous membrane	9
Absorption by mucous membrane; experiments with reference to this question by Messrs. Jolyet, Alling, Demarquay, Mercier, Mr. Wheeler and Professor Küss. Conclusions to be drawn therefrom	10
Nerves of the bladder; their two sources; explanation of the two forms of paralysis	13
The mechanism by which urine is retained in the bladder, various opinions concerning	14
Normal micturition, parts concerned in	14
The prostate gland; its position, relations, and structure	15
Its measurements and divisions. The prostate largely composed of muscular tissue; its glandular structure	16
Its vessels, nerves, and functions	17
Size of the prostate at different periods of life; Dr. Gross's researches	18

CHAPTER II.
THE METHODS OF EXAMINING THE BLADDER AND PROSTATE GLAND.

Bladder when undistended but little accessible to external examination	19
When distended may be examined by palpation and percussion in hypogastric region	19
Examination *per rectum*, when distended	19
Evidence as to the existence of a tumor; Professor Billroth's case	19
Professor Volkmann's method of " Bimanual " exploration	19
Interior of bladder; methods of examining; catheters and sounds	20
Objects for which catheters and sounds are used	20
Tumors of the bladder, method of examining in suspected cases	20
Catheters, various kinds of; sizes; French and English measurements	21
Introduction of the catheter, method of, and precautions regarding	21
Elastic instruments; French catheters, their advantages	22

	PAGE
Difficulties that may be encountered in passing a catheter	22
Catheters for cases of enlarged prostate	22
Accidents occasionally attendant upon catheterism	23
False passages; cystitis; hæmorrhage; injuries to prostate and neck of bladder; orchitis	23
Constitutional symptoms sometimes following catheterism	24
Rigors, with febrile reaction, occasionally followed by suppression of urine; syncope	24
The endoscope; its invention and uses	26
Rutenberg's endoscope for the female bladder	26
The prostate gland, examination of	26
Examination of the bladder in the female; introduction of the catheter	27
The vesico-vaginal septum; examination of the bladder by means of a sound	27
Introduction of the forefinger into the female bladder, for the purpose of examining the interior	28
Professor Simon's dilators for the female urethra; their advantages	28
The extent to which dilatation may be safely carried	29
Incisions through the vesico-vaginal septum	29
Objects for which dilatation of the female urethra may be practised	29
Indications for incising the vesico-vaginal septum	29
Catheterism of the ureters; Professor Simon's experiments	30

CHAPTER III.

ABNORMITIES OF THE BLADDER—ABSENCE OF THE BLADDER—MULTIPLE BLADDERS—EXTROVERSION—PERVIOUS URACHUS.

Complete absence of the bladder	31
Cases by Professor Mayer and Dr. Gross	31
Cases of atrophy of the bladder by Uytterhoeven and Mr. Vost	32
Multiple bladders	32
Extroversion of the bladder, its nature	33
Its comparative infrequency in females	33
Causes of extroversion; Mr. John Wood's views	33
Varieties of the malformation; the appearances presented	34
Its appearance in children and in adults	34
Attendant complications	35
Case by Mr. Erichsen; post-mortem examination by Mr. McWhinnie	35
Case by Mr. Wiblin	36
Inconveniences arising from malformation of the bladder	37
Treatment—Mr. Earle's Apparatus	37
Messrs. Lloyd and Simon's operations	37
Operations by Professor Pancoast of Philadelphia, Dr. Ayres of Brooklyn, Messrs. Holmes and Wood, and M. Maissonneuve	88
Description of Mr. Wood's operation	88
Dr. Maury's operation	39
Dr. Gross's recommendations	39
The after-treatment	40
Results of the operation; statistics; cause of failure	40
Pervious urachus; the appearances presented	40
Cases by Mr. Paget, Mr. Thomas Smith, and Dr. Charles	41
Treatment of pervious urachus	41
Fissures in the posterior wall of the bladder	41

CHAPTER IV.

HERNIA AND DISPLACEMENTS OF THE BLADDER.

Varieties of the affection	42
Eversion of the bladder; its symptoms; occurs exclusively in females	42
Cases by Dr. Murphy, Mr. Crosse, Dr. Beatty, Mr. Lowe, and Mr. Croft	42
Eversion of the bladder in an adult, Dr. Thompson's case	43
Partial eversion	43
Symptoms of eversion	44
Treatment	44

CONTENTS. vii

	PAGE
Eversion as a complication of vesico-vaginal fistula	44
Hernia of the bladder; its nature, position, and causes	45
Inguinal cystocele	45
Causes of this hernia, its relations and coverings	45
Illustrative cases	46
Complications of inguinal cystocele	48
Symptoms	48
Treatment	49
Prolapsus vesicæ	49
Symptoms	50
Treatment	50
Professor Stolz's operation	51
Mr. Baker Brown's operation	51

CHAPTER V.

WOUNDS AND INJURIES OF THE BLADDER.

Comparative rarity of these accidents	53
Causes various; external violence; rupture from over-distention; as a result of fracture of the pelvis	53
Injuries affecting the peritoneal covering more serious than others	53
Peritonitis and extravasation of urine, the main dangers	53
Gun-shot wounds of the bladder. Incised, lacerated, and punctured wounds. Varieties in the direction of the wounds and in the weapons with which they are inflicted	53
Rarity of injuries to the bladder by swords or other sharp weapons in warfare	54
Symptoms of wounds of the bladder	54
Treatment	54
Gun-shot wounds of the bladder, the most frequent of all	55
Dr. Bartel's tables and statistics	55
Symptoms of gun-shot wounds of the bladder	55
Complications	56
Treatment	56
Rupture or laceration of the bladder without external wound	57
Mode in which the accident may occur	57
Usual seat of the injury	57
Dr. Harrison's views	58
Symptoms and *post mortem* appearances	58
Differences in the intensity of the symptoms and in the time of their appearance	58
Professor Bartel's case	58
Mr. Holmes' case	58
Symptoms of peritonitis and of extravasation of urine	59
Hæmorrhage sometimes absent	59
The diagnosis of laceration of the bladder	59
Cases of the injury; Mr. Hird's case	60
Case in which life was prolonged for seventeen days after receipt of the injury	60
Mr. Prescott Hewett's cases	61
Mr. Syme's case in which recovery took place	62
Dupuytren's case	62
Comparative rarity of the injury in women and children	63
Rupture of the bladder from over-distention	63
Perforation of the bladder in the course of febrile affections	64
Rupture of the bladder during parturition	64
Dr. Bartel's statistics of the various forms of injury to the bladder	64
Medico-legal question with regard to these injuries	65
Treatment of rupture of the bladder	65
Views as to the consequences of the escape of urine into the peritoneal sac	66
Tolerance of the presence of urine by the peritoneum	67
Treatment by incision into the perineum	67
Treatment by abdominal section, and closing the wound in the bladder by sutures	67

CHAPTER VI

ACUTE INFLAMMATION OF THE MUCOUS MEMBRANE OF THE BLADDER.

	PAGE
Forms of the disease	69
Its causes	69
Essential symptoms	70
Symptoms of milder cases	71
Condition of the urine	71
Sources of pus found in the urine	72
Pus and mucus; their microscopical appearances	72
Pathological appearances	73
Ulceration of the bladder	74
Fistulous openings between the bladder and intestines	74
Sloughing of the mucous membrane; diphtheritic cystitis	77
Gangrene of the bladder	78
Distinctive diagnosis of inflammation of the bladder	78
Treatment	78
Venesection, leeches, purgatives, hot baths, morphia	79
Treatment if retention of urine occurs	79
Diet and regimen	80
Dr. George Johnson's recommendation with regard to milk	80
Modifications of the treatment	80
Prognosis	81
Cases of ulceration of the bladder	81
Dr. Prout's commentary on these cases	83
Professor Billroth's remarks	84

CHAPTER VII.

CHRONIC INFLAMMATION OF THE MUCOUS MEMBRANE OF THE BLADDER.

Its causes; rarely an independent affection	85
Symptoms	85
Condition of the urine; causes of its decomposition	86
Duration of the disease	87
Results	87
Morbid appearances	88
Condition of mucous membrane, mucous lining of the ureters and tubular structure of the kidney	88
Hypertrophy of the muscular coat	88
Treatment	89
Stricture as a complication	89
Enlarged prostate, foreign bodies in the bladder	89
The affection occurring in gouty persons	89
Remedies designed to act directly on the mucous membrane; the balsams, uva ursi, pareira brava, and injections	89
Tinct. benzoin. comp.—tinct. ferri perchlorid,—tannin recommended by Niemeyer	90
Injections, objects for which used; method of using	91
Case illustrative of the advantages of injection of acidulated solutions in cases of phosphatic deposit	92
Cauterization of the neck of the bladder	92
Astringent injections; injections of balsam of copaiba	92
Diet and regimen	93

CHAPTER VIII.

ACUTE INFLAMMATION OF THE WALLS OF THE BLADDER—CYSTITIS PARENCHYMATOSA—CYSTITIS TOTALIS—ABSCESS.

The muscular coat of the bladder never inflamed alone, but sometimes especially affected	95
Abscess of walls of bladder	95
Causes of inflammation	95

CONTENTS. ix

	PAGE
Symptoms	95
Anatomical structure as explaining symptoms	96
Ureters enormously distended in some cases	96
Danger of extension of the disease to the peritoneum	97
Progress of the disease	97
Formation of abscess in the coats of the bladder	97
Symptoms	97
Morbid appearances	97
Illustrative case	97
Treatment. Blood-letting, calomel and opium, sedative injections, Aperients, Diet	98
Abscess to be opened, if any external signs	98

CHAPTER IX.

HYPERTROPHY OF THE BLADDER—COLUMNAR AND SACCULATED BLADDER—ATROPHY OF THE BLADDER.

Hypertrophy of the bladder—its causes and varieties	99
Dilatation of the bladder; excentric and concentric	99
Hypertrophy	99
Columnar and fasciculated bladder	99
Hypertrophy of bladder, as a result of want of concord between certain sets of muscles	99
Mr. Guthrie on this condition of the bladder	100
"Stammering bladder," Sir J. Paget's views	100
Appearances on dissection	100
Symptoms	101
Treatment	101
Sacculated bladder	101
Formation and number of sacs	101
Symptoms. The sac sometimes contains a calculus	102
Diagnosis	103
Illustrative case	103
Dr. Murchison's case of an enormous vesical cyst	104
Dr. Warren's case	104
Treatment of sacculated bladder	105
Irregular contraction of the muscular coat, causing divisions of the cavity of the bladder	105
Atrophy of the bladder	106

CHAPTER X.

INFLAMMATION OF THE PERITONEAL COAT OF THE BLADDER, AND OF ITS SUBJACENT CELLULAR TISSUE.

The disease seldom confined to the bladder	107
Morbid appearances	107
Symptoms	107
Peritonitis occurring after lithotomy	107
Inflammation of the peritoneal coat of the bladder generally connected with inflammation of the adjoining lining of the pelvis	107
Dr. Elliotson's case of abscess between the bladder and the symphysis pubis	108
Pseudo-abscess in the peritoneal cavity	108
Symptoms, seldom acute	109
Treatment	109

CHAPTER XI.

TUMORS OF THE BLADDER—TUBERCULOSIS.

The various forms of tumor which are found in the bladder, mucous polypi, cystic growths, papilloma or villous tumor, fibrous tumors, and fibro-myomata; scirrhus, epithelioma and encephaloid	110
Causation	110

CONTENTS.

	PAGE
More common in females ; occur at all ages	110
Mucous polypi and polypoid hypertrophy of the mucous membrane	111
Anatomical appearances	111
Symptoms	111
Case recorded by Mr. Crosse	111
Cases recorded by Mr. Warner and Mr. Savory	113
Cysts	114
The bladder sometimes penetrated by dermoid cysts of the ovary	114
Cases recorded by Dr. Fuller, Dr. Greenhalgh, and Sir H. Thompson	114
Cystic outgrowths from the bladder ; Mr. Erichsen's case	114
Papilloma or villous tumor of the bladder	115
Cases recorded by Dr. Wilks, Sir H. Thompson, and others	115
Villous growths, their nature and structure	115
Symptoms ; hæmaturia the principal one	115
Dr. Hicks' case in which the growth was passed by the urethra	116
Villous growths proper, non-malignant nature of	116
Cancerous growths sometimes throw out villous prolongations	116
Villous growths elsewhere sometimes co-exist with the same disease in the bladder	116
Dr. Ultzmann's cases in which recovery took place	116
Diagnostic symptoms ; condition of the urine ; fibrinuria	117
Fragments of the growth sometimes found in the urine	117
Fibrous tumors and fibro-myomata	118
Dr. Winckel's case	118
Professor Billroth's case, in which the tumor was successfully removed	118
Professor Volkmann's case	119
Sarcoma of the bladder	121
Tuberculosis of the bladder	121
Generally, but not always connected with tuberculosis of other organs	121
Cases by Mr. Prescott Hewett, Dr. Winckel, and Dr. Wilks	122
Recovery possible ; Sir J. Paget and Mr. Thomas Smith's cases ; symptoms	122
Caseous disease of the kidney sometimes produces similar symptoms	122
Malignant disease of the bladder	123
Encephaloid, scirrhus, epithelioma	123
Differences between a malignant and non-malignant growth	123
Epithelioma ; cases by Sir H. Thompson and Dr. H. Fagge	124
Symptoms caused by malignant growths	124
Pain, hæmorrhage, and disturbances of function	124
Condition of the urine	125
Analysis of the symptoms	126
Persistent vesical hæmaturia the most characteristic	126
Examination of the bladder	128
Evidence to be obtained from the urine	129
Case recorded by Dr. Dickinson	129
Cases recorded by Dr. Lankester, Mr. Coulson, and others	129
Symptoms connected with non-malignant growths	134
Simon's method of examining the female bladder	134
Tubercular disease of the bladder	135
Symptoms	135
Treatment ; measures to check hæmorrhage	135
Removal of tumors from the bladder	136
Cases by Civiale, Billroth, Liston, and others	136
Removal of tumors from the female bladder	136
Cases by Dr. Braxton Hicks, Professor Simon, and others	136

CHAPTER XII.

FISTULÆ OF THE BLADDER.

First notice of urinary fistulæ occurring in women	138
Progress of the surgery of these lesions down to modern times	138
Varieties of vesical fistulæ	139
Their causes	139
Two main classes, those connected with the puerperal state, and those having a different origin	139

CONTENTS.

	PAGE
Question as to use of instruments as a cause	139
Dr. Bouqué's statistics	140
Fistulæ of non-puerperal origin	140
Anatomical appearances	140
Varieties; changes in other parts	141
Symptoms and diagnosis; method of examination	141
Prognosis; occasional spontaneous closure	142
Results of operations	143
Treatment, palliative and radical	143
Appliances suggested by Desault, Dupuytren, and others	143
Treatment by caustics	144
Dr. Bouqué's method	145
Treatment by applying sutures	145
Lallemand and Dupuytren's operations	146
Modern method; paring the edges and applying sutures	146
Needles, kinds of and method of using	147
Mr. Baker Brown's suggestions	147
Time at which the operation should be performed	148
Preparation for the operation	149
Mr. Baker Brown and Dr. Sims' operations	149
The clamp-suture, its advantages; the button suture	149
Case by Mr. Baker Brown	150
Professor Simon's suggestions	151
Dieffenbach and Jobert's method for cases in which the loss of substance is large	152
Mr. Lawson Tait's method and cases	152
Fistulæ involving the vagina, uterus, and bladder	153
Fistulæ involving the ureters	154
Method of dealing with various complications	154
Accidents which may attend the operation	154
Recto-vesical fistula	154
Causes and symptoms	154
Treatment	155
Recto-vesical fistula after lithotomy	156
Various other forms of vesical fistula	156

CHAPTER XIII.

NEURALGIA OF THE BLADDER—IRRITABILITY OF THE BLADDER—SPASM OF THE BLADDER.

The so-called neuroses of the bladder, generally connected with structural changes	157
Division into two classes, according as the sensibility and contractility of the organ are increased or diminished	157
Neuralgia of the bladder, a very rare affection	157
Dr. Gross' account of the symptoms	157
Causes and Treatment	157
Irritability of the bladder; the term merely implies a group of symptoms	158
Billroth's view as to possible causation in certain cases	158
Disorders, of which irritability of the bladder is a symptom	158
Diseases of the genito-urinary system	158
Organic renal disease often accompanied by this symptom	159
Irritability of bladder as a consequence of gonorrhœa	159
As a symptom of calculus	159
Occurring after temporary paralysis of the bladder	159
Caused by contracted or adherent prepuce	160
Caused by abnormal conditions of the urine	160
Due to diseases of neighboring organs, and of the brain and spinal cord	160
The complaint connected with some constitutional affection, as gout or rheumatism	161
As a symptom of hysteria, and of oxaluria	162
Its occurrence after exposure to extreme heat or cold	162
Spasm of the bladder; its nature, varieties, and causation	162
Symptoms	163

	PAGE
Treatment of irritability and spasm of the bladder	163
Careful examination necessary in order to detect possible causes	164

CHAPTER XIV.

ATONY, PARESIS, AND PARALYSIS OF THE BLADDER.

Degrees of atony or paralysis	165
Views regarding functions of vesical muscles	165
Differences between atony and paralysis	165
Causes of paralysis	166
Paralysis of the bladder, a symptom of injuries of the head and spine, and of apoplexy	166
Condition of the urine in such cases	166
When occurring from reflex irritation	166
As a symptom of narcotic poisoning	166
Atony from over-distention	167
Its causation and symptoms	167
Case by Mr. Lawrence	167
Temporary paralysis in pregnant women	168
Diagnosis and symptoms of paralysis of the bladder	168
Results and prognosis	169
Morbid appearances	169
Treatment—Relief of distended bladder, in order to prevent distress and extension of mischief; restoration of contractile power	170
Use of the catheter; precautions	170
Remedies for the purpose of restoring the contractile power of the bladder—injections of cold water, electricity	171
Internal remedies; nux vomica, ergot of rye	172
General indications	172
Treatment of cases of paralysis of the sphincter	172

CHAPTER XV.

ENURESIS—INCONTINENCE OF URINE.

A symptom of various affections of the urinary organs	173
Term not to be applied to describe escape of urine from over-distended bladder	173
Enuresis passiva and activa	173
Causes of incontinence—the complaint in females	173
Nocturnal incontinence	174
Conditions under which it occurs	174
Treatment; general and special	175
Mechanical contrivances	175
Special remedies; belladonna, chloral hydrate, iodide of iron, cold sponging and cold baths	175

CHAPTER XVI.

RETENTION OF URINE.

Definition of the term	177
Causes—due either to paralysis or atony, or to obstruction	177
Causes to which obstruction may be due	177
Inflammation of the prostate	178
Treatment of obstruction due to this cause	178
Chronic enlargement of the prostate	178
Malignant disease of the prostate	178
Stricture of the urethra	178
Diagnosis and treatment	179
Puncture of the bladder—various methods	179
Puncture *per rectum*. Supra-pubic puncture	179
Perineal section	180
Tapping the bladder through the pubic symphysis	181

	PAGE
Voillemier's operation	181
The use of a capillary trocar and aspirator	181
Other causes of retention of urine	182
Malignant tumors—abscess *in perineo*—calculus in the bladder or urethra— laceration of urethra	182
Retention of urine in female patients; hysteria	182
Retention due to disease of neighboring organs	183
Retention due to tumors, retained menses, etc.	184

CHAPTER XVII.

HÆMATURIA.

Definition of the term	184
Various alterations in the color of the urine produced by admixture of blood	184
The microscope as a means of diagnosis	184
Blood-corpuscles always present in true hæmaturia	184
Alterations in their size and form	184
Albumen of the blood—coloring matter of the blood, its detection by means of the spectrum-apparatus	185
Red color of urine may be due to the presence of the coloring matter of certain drugs	185
Urine in jaundice; in carbolic acid poisoning	185
Hæmaturia due to hæmorrhage from the urethra, causes and symptoms	185
Hæmorrhage from the prostatic urethra	186
Hæmorrhage from the bladder or kidney	186
Diagnosis—reaction of the urine—pus—color of the urine—presence of coagula	186
Renal hæmaturia—vesical hæmaturia—causes	187
Wounds of the bladder	187
New formations: epithelioma; villous growths—fibrinuria as a symptom; Dr. Ultzmann's views as to its causes	187
Varix of the bladder	189
Hæmaturia due to calculus	189
Parasites in the bladder	189
Hæmaturia as a symptom of inflammation and ulceration of the bladder	190
Renal hæmaturia; causes and symptoms	190
Hæmaturia in certain severe constitutional affections—Hæmatinuria	191
Treatment of hæmaturia	191
Astringents by the mouth and by injection; application of cold	191
Opium—Question as to use of catheter	191
Internal remedies. Styptics	191

CHAPTER XVIII.

THE CHEMISTRY AND STRUCTURE OF URINARY CONCRETIONS.

Historical sketch of the progress of knowledge	193
Discovery of uric acid by Scheele	194
Forms of calculi	194
Their external characters, color, size and weight	194
Number of calculi sometimes present	195
Their internal structure and composition of the nucleus	195
Drs. Rees and Ultzmann's investigations	196
Classification of calculi—Primary and secondary stone-formation	196
The predominance of uric acid as a nucleus	197
Dr. Klien's investigations as to nature of calculi in Russia	197
Dr. Vandyke Carter on calculus in India—predominance of oxalate of lime	197
Disposition of the nucleus	198
Form of calculus, how modified	198
Foreign bodies as nuclei	198
Animal matter in calculi	199
Views of Professor Scharling and Dr. Haskins	199
Classification of calculi	200

	PAGE
Method of examining	201
Uric acid calculus	203
Relative proportions of this species	203
Its varieties	203
Chemical characters of uric acid	205
Urate of ammonia calculus	206
Characters of urate of ammonia	206
Question as to nature of these concretions	207
Xanthine or uric oxide calculus	207
Its rarity	207
Chemical characters and analysis	208
Oxalate of lime calculus	208
Relative frequency with which it occurs	209
Varieties in form	209
Structure; chemical characters and analysis	210
Phosphate of lime calculus	210
Physical characters and chemical analysis	211
Ammoniaco-magnesian phosphate calculus	211
Relative frequency with which it occurs	211
Structure and chemical analysis	212
Fusible calculus	212
Relative proportions	212
Structure	213
Cause of the deposition of the mixed phosphates	213
Carbonate of lime calculus	214
Physical characters and chemical analysis	214
Alternating calculi	215
Various forms	215
Relative proportions	215
Cystine calculus	216
Relative frequency with which it occurs	216
Question as to hereditary tendency	216
Structure and chemical analysis	217
Silicious concretions	218
Fibrinous and blood concretions	218
Urostealith	219
A calculus containing indigo	220

CHAPTER XIX.

THE CAUSES OF STONE.

General principle of the formation of calculus	221
Composition of urea and uric acid	221
Sources of uric acid	222
Causes of its precipitation	222
Relative frequency with which it occurs in calculi	223
Uric acid infarctions in kidneys of children	223
Virchow's views and Ultzmann's investigations	223
Formation of calculus probably due to these infarctions	224
Calculi in adults—primary stone-formation	224
Theories as to diatheses	225
Deposits of uric acid and urates in the kidneys of adults	225
Deposits of oxalates in the kidneys	225
Meckel's theory	226
Heller's views	226
Ultzmann's classification and researches	226
Primary and secondary stone-formation	226
Especial predominance of uric acid	226
Conditions under which stone-formation occurs	227
Metamorphosis of urinary concretions	228
Ultzmann's conclusions with regard to stone-formation	228
Oxalate of lime; its relative frequency and sources	229
Question as to mode of production	229
Drs. Rees and Roberts' views	229

CONTENTS. XV

	PAGE
Cystine	230
Secondary stone-formation, nature of the process, and circumstances under which it occurs	230
Age as connected with the occurrence of calculus	231
Frequency of calculus in children	231
Relative frequency of calculus at different periods of life	231
Tables and statistics	232
Females far less liable than males to vesical calculus	233
Reasons for this difference	233
The influence of certain conditions and habits of life on the production of calculus	234
Frequency of calculus among children of the poor	234
Influence of diet on the production of calculus	234
Influence of water containing lime-salts	235
Calculus, frequency of, in the county of Norfolk	235
Influence of alcoholic liquors	236
Influence of climate	236
Prevalence of calculous affections in certain counties	237
Prevalence of calculous diseases in the United States	237
Calculus, prevalence of in Russia	238
Immunity of the Finlanders	238
Prevalence of calculus in India	239
Calculous disorders in Egypt, their cause	239
Prevalence of calculus in certain parts of China	239
Influence of occupation on the production of calculus	240
Unfrequency of the complaint among sailors	240
Hereditary tendency to calculous disorders	241
Calculi in animals	242

CHAPTER XX.

SYMPTOMS AND DIAGNOSIS OF STONE IN THE BLADDER.

Symptoms of the passage of a renal calculus	243
Early symptoms of stone; peculiarities with regard to children	243
Irritability of bladder; frequent micturition, pain, hæmaturia	243
Circumstances under which hæmaturia occurs	244
Pain as a symptom of calculus	245
Absence of pain under certain circumstances	245
Sudden arrest of the flow of urine	246
Symptoms of irritation of the bladder	246
Condition of the urine	247
Causes of changes in its reaction	247
Spontaneous disruption of calculi within the bladder	248
Mr. Southam and Dr. Ord's cases	248
Rate of increase in size of calculi	248
Production of disease of the kidney; symptoms of this complication	248
Diagnosis of calculus; many of the symptoms met with in other disorders	249
Sounding; best form of instrument	250
Mode of performing the operation	250
Use of the lithotrite to measure the stone	251
Other instruments for the same purpose	251
Information as to the character of the stone	251
The occasional dangers of sounding	251
Auscultation	252
Occasional difficulties in detecting the stone	252
Examples of undetected calculi	252
Effects of position of the stone	253
Calculi concealed in cysts	254
Causes of the difficulty in detecting stone	255
Fungous growths in the bladder; sacculation of the bladder; irregular contraction of the muscular fibres	255
Enlargement of the prostate gland	256
Peculiarities of the calculus	257

xvi   CONTENTS.

	PAGE
Symptoms of stone simulated by those of other diseases	258
Operations, and no stone found	258
Aneurism of the abdominal aorta, giving rise to symptoms resembling those of calculus	259

CHAPTER XXI.

FOREIGN BODIES IN THE BLADDER.

Varieties and manner of introduction	261
Illustrative cases	261
Bullets and other projectiles	263
Symptoms	263
Treatment	263
Instruments adapted for removing the various kinds of foreign bodies	264
Hydatids in the bladder	265
Parasites and worms in the bladder	265
Hairs voided with the urine	265

CHAPTER XXII.

LITHOTRITY.

Object to be fulfilled	267
Preparatory treatment	267
Use of chloroform	268
Instruments employed	269
Preliminary injections unnecessary	270
Position of operator and patient	270
Introduction of the lithotrite	271
Searching for the stone	272
Different methods employed	272
Seizing and crushing the stone	273
Care necessary in withdrawing the lithotrite	273
Amount of work to be done at first sitting	274
Interval between the sittings	275
Symptoms of cystitis, a reason for shortening the interval	275
Question as to removal of débris	275
The after-treatment	276
Necessity that no fragments should remain	276
Hypertrophy of the prostate, operation in cases of	276
Obstacles to the performance of lithotrity	277
Excessive sensitiveness of the parts	277
Great irritability of the bladder	277
Enlargement of the prostate	277
Hypertrophied bladder	278
Size and density of the stone	278
Accidents of the operation	278
Pain excited	279
Hæmorrhage	279
Suppression of urine	279
Acute cystitis	279
Rigors and irritative fevers	280
Retention of urine	281
Lodgment of fragments in the urethra	281
Treatment of this accident	282
Orchitis	283
Condition of the urinary organs requiring modifications of treatment	283
Objections to plan of removal of fragments by scoops	283
Dr. Bigelow's method	284
Administration of alkalies in cases of uric acid calculus	284
Acid injections for phosphatic calculi	284
General indications and contra-indications of lithotrity	284
The size, density, and other peculiarities of the calculus	286
The state of the prostate	287

CONTENTS. xvii

	PAGE
Chronic cystitis	287
Other conditions connected with the bladder	288
Condition of the kidneys; question as to operation when renal disease is known or suspected to exist	289
Indications derived from age	289
Lithotrity in children	290
Recurrence of symptoms of calculus after the operation of lithotrity	290
Causes of such recurrence	290
Prevention	291

CHAPTER XXIII.

LITHOLAPAXY, OR RAPID LITHOTRITY WITH EVACUATION OF THE FRAGMENTS.

Unsatisfactory results of attempts hitherto made at removal of fragments after crushing	292
Professor Bigelow's method; the objects sought to be accomplished	292
Removal of all fragments at one sitting	293
The instruments used in the operation	293
Difference between them and instruments hitherto employed	293
The elastic bulb or bottle—the evacuating tubes—special lithotrite	293
Advantages of the fenestrated lithotrite	295
Method of performing the operation	296
Obstacles, and manner in which these may be overcome	296
Two or more sittings may be necessary when the stone is large	297
Length of time required for the operation	297
Results of the operation as tested by statistics	297
Precautions to be observed in performing the operation	298
Its advantages	298

CHAPTER XXIV.

LITHOTOMY.

The three places at which the bladder may be opened	299
The high or supra-pubic operation	299
General objections to the operation	299
Method of performing it	300
Preference given to this operation by several German surgeons	300
Cases for which it is suitable	301
Danger of wounding the peritoneum	301
The recto-vesical operation	302
Method of performing	302
Objections to the operation	303
The lateral operation	303
Performed on patients of all ages	303
The instruments required	303
Preparatory treatment	303
The operation; necessity of feeling the stone with the staff	303
Method of securing the patient	303
The external and internal incisions	304
Use of button-pointed bistoury for section of the prostate	305
Additional precautions in operating on children	307
The use of the straight staff	307
Dr. Buchanan's rectangular staff	308
Obstacles to lithotomy	308
Enlargement of the prostate	309
Calculus encysted or adherent to the wall of the bladder	309
Calculus concealed behind the prostate	310
Calculus behind the pubes	310
Forcible contractions of the bladder	311
Fracture of the stone	312
Size of the calculus as influencing result of operation	312
Cases of large calculus	313

xviii   CONTENTS.

	PAGE
Mr. Crosse's tables	313
Various methods of dealing with large stones	314
Accidents of lithotomy	314
Hæmorrhage during or after the operation	314
Forms and causes of this bleeding	315
Treatment	315
Questions relating to section of the prostate	316
Cheselden's practice	316
Mr. Martineau's method	317
Comparison between free incisions, and small incisions with risk of laceration	317
Cystitis after lithotomy	318
Peritonitis	318
Urinary infiltration, Mr. Crosse's views	318
Pyræmia	319
Diseases of the kidney	319
After-treatment	319
Retention of urine to be guarded against	320
The bilateral operation	320
Mode of performing it	320
Results of the method	320
The median operation—Allarton's method	321
Mode of performing it	321
Estimate of its value	322

CHAPTER XXV.

CALCULUS IN THE FEMALE.

Women less subject to calculus than men	323
Statistics	323
Symptoms and diagnosis	323
Methods for the extraction of calculi	324
Dilatation of the urethra	324
Advantages of rapid dilatation	324
Instruments. Weiss's dilator and Simon's specula	324
Lithotrity	325
Lithotomy	327
Precautions to be observed	327
Calculus in the bladder of a parturient woman	328
Methods of dealing with such cases	328

CHAPTER XXVI.

STATISTICS OF LITHOTRITY AND LITHOTOMY.

LITHOTRITY.

Civiale's statistics	330
Statistics of Sir Benjamin Brodie	331
Statistics of MM. Leroy, Heurteloup and Amussat	331
Statistics published by Sir W. Fergusson, Professor Keith, and Sir H. Thompson	331

LITHOTOMY.

General mortality of lithotomy	332
Tabular view of 6,505 cases	332
Table illustrating the influence of age on mortality of lithotomy	333
Table of operations at St. Thomas's Hospital	334
Table illustrating the influence of the weight of the calculus on results of lithotomy	334
Tables showing the relative frequency with which lithotomy and lithotrity are performed in London	335
St. Thomas's Hospital	335
Guy's Hospital	335

	PAGE
St. Bartholomew's Hospital	336
St. George's Hospital	336
St. Peter's Hospital	337
General table	337
Mortality of lithotomy according to the different methods	338
The apparatus major	338
The high operation	338
Comparative mortality of the lateral and high operations with calculi of the same weight	338
The bilateral operation	339
The recto-vesical operation	339
The median operation	339
Statistics of lithotomy in India	339

CHAPTER XXVII.

THE SOLVENT TREATMENT OF CALCULUS.

Attempts made at dissolving calculi	340
Mrs. Stephens's solvent	340
Dr. Whytt's remedy	340
Fourcroy and Vauquelin's rules	340
Messrs. Chevallier and Petit's researches	341
Mr. Ure's experiments with borax and phosphate of soda	342
Dr. Roberts's investigations and conclusions	343
Solvent action of alkalies on uric acid	343
Methods of dealing with other calculi	343
Solution of urostealith by carbonate of soda	344
Injection of solvents into the bladder	344
Drs. Hales and Butler's experiments	344
Dr. Roberts's experiments	344
Sir Benjamin Brodie and Mr. Southam's cases of phosphatic calculi treated by injections	345
Dr. Hoskins's experiments with the nitro-saccharate and acetate of lead	345
Experiments with electricity	345
General conclusions	346

CHAPTER XXVIII

THE PREVENTIVE TREATMENT OF CALCULOUS DISEASE.

Premonitory symptoms of calculus	347
Microscopical examination of the urine	348
Premonitory symptoms of calculus in children	348
Indications of treatment directed to prevent the formation of concretions	348
Diminution of the abnormal constituent of the urine	348
Prevention of precipitation of the calculous material	348
Dilution of the urine	348
Uric acid, measures to be taken when deposited	349
Vegetable diet, influence of	349
Alcoholic drinks	349
Exercise	349
Attention to state of secretions and excretions	349
How to prevent precipitation of uric acid	350
Various alkalies considered; potash the most useful	350
Dr. Roberts's experiments	350
Various forms in which potash may be given	350
Summary of medical treatment	350
Method of keeping the urine alkaline	351
Treatment when oxalate of lime is deposited	351
Oxalate of lime usually associated with excess of urea	352
Articles of food to be avoided	352
Condition of the digestion	352
The use of the mineral acids	352
Treatment when calculus is supposed to exist	352

Cystine exceedingly rare ; treatment 352
Phosphatic urine ; ammoniacal decomposition 353
Danger of formation of stone increased by retention of urine . . . 353
Treatment 353
Treatment when calculi have been passed . . , . . . 353
Various results of the presence of calculi in the kidney 353
The symptoms not always characteristic 353
Careful watching of the urine necessary 353
Effects of violent exercise 353
Use of microscope to detect blood and epithelial cells 353
Passage of calculus along the ureter 354
Treatment 354
Calculus retained in the bladder 355
Probable cessation of acute symptoms 355
Treatment 355

CHAPTER XXIX.

ACUTE AND CHRONIC INFLAMMATION OF THE PROSTATE GLAND.

Causes of acute prostatitis 356
Incipient symptoms 356
Progress of the disease 357
Formation of abscess 357
Terminations 358
Chronic prostatitis 358
Treatment of acute prostatitis 358
Treatment of abscess 358
Treatment of chronic prostatitis 359
Periprostatic abscesses 359

CHAPTER XXX.

CHRONIC ENLARGEMENT OF THE PROSTATE GLAND.

Pathology of the affection 360
Condition due to hypertrophy of the elements of the gland or development of tumors 360
Changes in the form, size, and consistence of the gland 361
Effect upon the urethra 361
Effect upon the bladder of enlargement of the middle lobe . . . 362
Vesical calculus as a complication 364
Causes of the affection 365
Incipient symptoms 365
Symptoms of the disease when confirmed 366
Diagnosis 367
Treatment 368
Attention to the diet and state of the bowels 368
Counter-irritation ; iodine, mercury 369
Professor Heine's method of injecting iodine into the gland . . . 369
Use of the catheter 369
Various forms of the instrument 369
Precautions to be observed 370
Hæmorrhage from the prostate 370
Treatment of chronic cystitis 370
Puncture of the bladder in cases of retention of urine 371
Proposed operation for relief of patients suffering from advanced prostatic disease 371
Rupture of the bladder from over-distention 371

CHAPTER XXXI.

MALIGNANT DISEASE OF THE PROSTATE GLAND—TUBERCLE OF THE PROSTATE.

Scirrhus of the prostate very rare 372
Encephaloid, the usual form of malignant disease 372

	PAGE
Symptoms of	373
Cases by Messrs. Stafford, Solly, and Moore	374
Cases by Mr. Haynes Walton and Sir H. Thompson	375
Treatment	376
Tubercular disease of the prostate	376
Pathology; symptoms and treatment	376

CHAPTER XXXII.

PROSTATIC CONCRETIONS AND CALCULI.

Composition of prostatic concretions and calculi	378
Dr. H. Jones, Mr. Quekett, and Sir H. Thompson's descriptions	378
Mode of formation	379
Appearances	379
Effects on the prostate	379
Analysis of the concretions	380
Milky urine	380
Symptoms produced	380
Removal of prostatic calculi	381
Method of crushing	381
Incision into perineum	381
INDEX	383

LIST OF ILLUSTRATIONS.

		PAGE
1.	Interior of a bladder showing hypertrophy of its muscular fibres .	100
2.	A bladder with three cysts	102
3.	A bladder with medullary tumor and two calculi	132
4.	Method of applying sutures in the operation for vesico-vaginal fistula .	151
5.	Figure showing method of holding lithotrite .	273
6.	Stone between the blades of the lithotrite, the screw of the instrument being in action .	274
7.	Bigelow's evacuating apparatus	293
8.	Sir H. Thompson's modification of Bigelow's aspirator .	293
9.	Terminal portion of evacuating tubes, straight and curved .	294
10.	Bigelow's lithotrite	295
11.	Blades of the lithotrite	295
12.	Modern instruments used in lithotomy	304
13.	Bladder showing the calculus struck with the staff	304
14.	Muscles of the peritoneum	305
15.	Vessels of the perineum	305
16.	Position of the patient	306
17.	Section of prostate with button-pointed bistoury	306
18.	Bladder with an encysted calculus	309
19.	Calculus behind prostate	310
20.	Bladder with calculus behind the pubes	310
21.	Enlargement of lateral lobes of prostate	363
22.	Enlargement of middle lobe of prostate	363

DISEASES

OF

THE BLADDER AND PROSTATE GLAND.

CHAPTER I.

GENERAL ANATOMICAL AND PHYSIOLOGICAL CONSIDERATIONS.

THE urinary bladder is the musculo-membranous receptacle which receives the urine as it escapes from the ureters, and from which it is discharged at certain intervals. The bladder, in the adult, is more or less rounded in shape, and is situated in the pelvic cavity, immediately behind the pubes; in the young subject it is pyriform, and lies chiefly in the abdomen. It is the only cavity in the body which fulfils the conditions of a perfect reservoir. It admits of considerable dilatation, thus accommodating itself to the quantity of fluid which is constantly trickling into it, regurgitation being prevented by the peculiar manner in which the ureters terminate. The aperture for the discharge of the fluid is ordinarily closed, and as a general rule the influence of the will is required to open it; but when, under certain circumstances, urine accumulates, and the distention becomes such as to threaten rupture of the coats of the organ, the muscles guarding the aperture of exit usually become relaxed, and an outflow takes place by which the integrity of the receptacle is preserved.

The bladder varies in size and shape according to the degree of distention: when contracted, it is of a triangular form, and is almost hidden in the pelvic cavity. When moderately full it becomes roundish, but when completely distended it assumes an ovoid shape, with the larger part directed towards the rectum in the male and the vagina in the female, and the smaller end towards the abdominal wall, and it is slightly curved over the anterior part of the pelvis as it projects above the bone. A line prolonged through its centre would touch the end of the coccyx, in one direction, and a point on the abdominal wall in the other direction, somewhere between the umbilicus and pubes, according to the degree of distention. In a condition of moderate distention its measurements average four inches and a half in a vertical direction, three and three-quarters from side to side, and three inches from before backwards. Its average capacity is from fifteen to twenty-five ounces. In an exceptional state of distention the bladder may reach the umbilicus, or even extend above it, and may be mistaken for a tumor of another kind.

Various other circumstances affect the size of the bladder—it is large in the fœtus, and relatively larger during early life than in adult age. It is larger in persons who are accustomed to retain their urine for long periods than in those of opposite habits. Hence it happens that the bladder in the female is often larger than that of the male subject, but in the normal condition, according to Luschka, Henle, and Sappey, it is decidedly smaller than that of the male. The size of the bladder is, of course, influenced by various pathological conditions; thus whenever any obstacle to the escape of urine through the urethra has existed for any length of time, the bladder becomes more or less dilated, and may, under such circumstances, contain as much as eight or ten pints of urine, and mechanically impede the respiration, and the circulation through the abdominal veins. On the other hand, the bladder's capacity may undergo diminution, as frequently happens in cases of hypertrophy, chronic catarrh, and in calculous and fungous disease. Its cavity may be also encroached upon by tumors connected either with itself or with other pelvic or abdominal organs. Its walls vary in thickness according to the state of distention, measuring under varying conditions in this respect from 2 to 15 millimeters.

The bladder is in relation anteriorly with the triangular ligament of the urethra, the sub-pubic ligament, and the symphysis and body of the pubis, and, when distended, with the lower part of the anterior wall of the abdomen. It is attached to these parts by loose areolar tissue, and to the bones by two strong bands of the recto-vesical fascia called the *anterior true ligaments*. Posteriorly it is in contact with the rectum in the male, and with the uterus in the female. From its summit there arises the urachus, a remnant of fœtal life, and connecting it with the umbilicus; below and anteriorly the bladder is continuous with the urethra, the commencement of which is encircled by the prostate in the male. The bladder is kept in position by its connections with neighboring parts, by the anterior true ligaments above referred to, by other portions of the same fascia forming the lateral true ligaments, and by folds of peritoneum called *false ligaments*. The obliterated hypogastric arteries, the ureters, and the urachus also assist in supporting the bladder. The anterior and lower portions are those which are most firmly connected with the adjoining parts.

The bladder presents for examination an external and an internal surface. The external surface is convex, but somewhat irregular, the convexity being most marked posteriorly and above; the anterior aspect and base are flattened. For purposes of description the external surface may be divided into an anterior and posterior region, a summit, base, and two lateral regions.

The anterior region is bounded below by the anterior true ligaments and the upper part of the urethral orifice, and above by the place of attachment of the urachus, while at the sides it is continuous with the lateral regions. In the empty state of the bladder this region is altogether free from peritoneum, and corresponds to the symphysis and body of the pubis, and to the fascia covering the obturator muscles on both sides, being attached to these parts by loose cellular tissue. Numerous veins course over it in a direction obliquely downwards and inwards, and form a trunk, which traverses the quadrilateral space bounded by the anterior ligaments, and empties itself into the dorsal vein of the penis. To this region are attached the anterior true ligaments, two bands of the recto-vesical fascia extending between the bladder and the body of the pubis. These ligaments are in relation below with the upper surface of the prostate, and the quadrilateral space of which they form the lateral boundaries is occupied by a thin fibrous membrane which is traversed by the anterior vesical veins. In the operation suggested by M. Voillemier, of puncturing the bladder below the pubis, the trocar would be plunged through this membrane.

When the bladder is distended, the relations of its anterior surface become considerably modified. It rises above the level of the symphysis, and comes into contact with the posterior aspect of the recti muscles covered by the fascia transversalis, and, in an extreme state of distention, the summit of the bladder may reach as high as the umbilicus. Under such circumstances the anterior surface of the bladder becomes much increased in size, and only a portion of it is covered by peritoneum; the bladder can therefore be punctured above the pubis without wounding the serous membrane. The space that is left uncovered by the peritoneum varies considerably, but it may be taken on an average to extend one and a half or two inches above the symphysis pubis, when the bladder is much distended. Formerly it was supposed that the uncovered space far exceeded this measurement, the idea being that the urachus kept back the peritoneum and prevented it from descending into the space between the distended bladder and the abdominal wall. But inasmuch as the urachus possesses a considerable amount of elasticity, it permits the peritoneum to form a cul-de-sac of varying depth in front of the bladder, the summit of which as it rises in the abdomen gradually recedes from the wall of that cavity, being at a distance from it of nearly one inch when the organ is fully distended. The various operations that may be performed on this part of the bladder are described in the chapter on retention of urine.

The posterior surface of the bladder is entirely free, and covered everywhere by the peritoneum, which in the male subject is prolonged also for a short distance upon the base of the bladder. When the bladder is empty this surface is more or less flattened. It rests against the middle piece of the rectum in the male and the uterus in the female, but is generally more or less separated from these parts by convolutions of the small intestine. The distance between the integuments and the cul-de-sac which the peritoneum forms behind the bladder varies according to the state of distention of the latter organ. When the bladder is empty, the border of the cul-de-sac is rather more than two-fifths of an inch from the base of the prostate, and about two inches and a quarter from the anus. Both measurements increase as the bladder becomes distended, but the distance between the anus and the recto-vesical cul-de-sac never exceeds three and a half inches. The sides of the cul-de-sac are bounded by the posterior false ligaments, the folds of peritoneum which pass in the male along the sides of the rectum to the posterior and lateral aspects of the bladder. In the female, these posterior folds pass forwards from the sides of the uterus, and are comparatively small. A few smooth muscular fibres have been discovered in these posterior false ligaments.

The summit, apex, or superior fundus of the bladder is rounded, and, considered as a region, varies in size according to the state of distention of the organ. When the bladder is empty, the summit is represented by the point to which the urachus is attached, and is applied against the posterior aspect of the symphysis pubis. As the bladder becomes distended, its apex gradually leaves the abdominal wall, and comes into contact with the convolutions of the small intestine, and when the distention is very great the summit may rise above the level of the umbilicus. Attached to the summit in the median line is a ligamentous cord, the urachus, which passes upwards between the linea alba and the peritoneum to reach the umbilicus, where it becomes blended with the dense fibrous tissue of that part. The urachus in early fœtal life forms a tubular communication between the bladder and the allantois, and towards the sixth or seventh month its cavity commences to undergo obliteration at its allantoic end. The abdominal portion is often permeable at birth, at least in a portion of its extent, and in adult life it often preserves, according to Luschka, vestiges of its original condition in the form of a long interrupted cavity with irregularities and dilatations lined with epithelium similar to that of the bladder, and sometimes communicating by a fine opening with the vesical cavity. Urinary fistulæ at the umbilicus, caused by the persistent permeability of the urachus, have been observed by Dupuytren, Bréaud, Mr. Paget of Leicester, and others, and calculi have been sometimes found in its canal. The urachus is composed of fibrous and elastic tissue, mixed at its base with some smooth muscular fibres which are prolonged upon it from the bladder.

The base of the bladder is a very important part of the organ from a surgical point of view. Its relations are different in the male and female subject.

In the male it is triangular in shape, and comprises that portion of the organ which is bounded at the sides by the lateral regions, in front by the base of the prostate, and posteriorly by the cul-de-sac formed by the recto-vesical fold of peritoneum, which, however, generally covers a small portion of the margin of the surface. Behind the prostate are the vesiculæ seminales and the vasa deferentia. The former are external to the latter; they are directed inwards and forwards, their greatest distance from each other at their posterior extremity being about two and a half inches, while in front they approach closely towards each other as they enter the prostate. Each has a vas deferens on its inner side, so that a triangular space is formed, its base directed backwards, its apex toward the prostate, and its sides bounded by the vasa deferentia and vesiculæ seminales. This space varies in size according as the bladder is distended or the reverse. When the bladder is empty, the vesiculæ seminales fall back upon the sides of the rectum, and are distant from each other about two inches and a half. When the bladder is distended, they approach each other, and the base of the triangle which they limit does not exceed two inches. The base of the bladder is directed backwards as well as downwards, and rests upon the second portion of the rectum, to which it is adherent by dense fibro-areolar tissue. In the triangular space the bladder may be punctured from the rectum without injuring the peritoneum.

In the female the base of the bladder is of less extent, and does not reach so far back in the pelvis; it is separated from the rectum by the vagina and neck of the uterus. This part of the bladder adheres to the vagina, but its relations differ according to the state of distention. When empty the base rests almost entirely upon the superior portion of the vagina and the neck of the uterus. Above this surface the peritoneum forms numerous folds, more marked in the median line, between the bladder and the uterus. When the bladder becomes distended, its fundus enlarges in the antero-posterior direction, covers the upper portion of the vagina and the neck of the uterus, and even reaches the junction of the neck with the body of the womb. The peritoneum is thus raised and its folds effaced. Thus all that portion of the anterior wall of the uterus which is situated between the reflexion of the vaginal mucous membrane over the neck of the organ and of the peritoneum over its anterior aspect comes into relation with the fundus of the bladder.[1] The parts are connected together by loose cellular tissue. The uterine arteries give off small branches to the bladder which run along the sides of the vesico-vaginal septum. The ureters open into the bladder on each side of this septum. These tubes have a longer course in the pelvis of the female than in that of the male, and run along the sides of the cervix uteri and upper part of the vagina before reaching the bladder. The relations of the base of the bladder in the female are of considerable importance, inasmuch as vesico-vaginal fistulæ involve this part of the organ.

The lateral regions of the bladder, when it is distended, are rounded and prominent, but when the viscus is empty, and flattened from before backwards, the sides are represented by curved edges. They are covered above and posteriorly by peritoneum, but below they are in relation with the levator ani muscle on each side, covered by the pelvic fascia. The bladder is connected with this fascia by means of a quantity of loose cellular tissue, which in the female is continuous with that of the broad ligaments. Each side of the bladder is crossed obliquely along its upper part by the obliterated hypogastric artery, which is connected posteriorly with the superior vesical artery, and passes forwards and upwards to the umbilicus, joining the urachus above the summit of the bladder. Above and behind this cord the side of the bladder is covered with peritoneum. Crossing it lower down, in a direction from before backwards and downwards, is the vas deferens, which turns over the obliterated hypogastric artery and passes upon the inner side of the ureter to the under part of the bladder. The lateral true ligament on each side is a broad expansion from the recto-vesical fascia, and is fixed to the lateral part of the prostate gland at the upper border and to the anterior and lower part of the side of the bladder, where it is prolonged forward on the veins which cover the prostate, and is firmly adherent to the capsule of that organ except at its base, where an angular furrow, occupied by large veins, exists between the prostate and bladder.

The term "neck of the bladder" is one which has been employed in a very indefinite manner by anatomists in general. It is used by some to designate simply the vesical orifice of the urethra, and by others to include also the prostatic portion of the canal. Dr. Mercier describes an anatomical and a surgical neck, and, following his example, the former may be defined as consisting of the orifice leading into the urethra, while the surgical neck of the bladder may be considered to include the prostatic portion of that canal. The urethral orifice has a well-defined border, and is in both sexes the lowest part of the bladder[2] when the body is in the erect position; it lies at the angle of meeting of the base and the anterior surface. It was formerly believed that the base or fundus was the lowest part of the bladder in the adult male, and hence the origin of the term. The inferior position of the urethral orifice was supposed to be peculiar to women and children, while in the male subject the base of the bladder was stated to form a pouch the bottom of which was lower than the neck of the organ. This, however, is not the case, as will appear from a consideration of the following circumstances. "The symphysis pubis is placed very obliquely; the ischial tuberosities are little lower than the inferior margin of the symphysis pubis, and the triangular ligament is therefore almost horizontal; the lower part of the sacrum and

---

[1] Mercier, Anatomie et Physiologie de la Vessie, p. 30.
[2] Quain's Anatomy, 8th edition, vol. ii., p. 422.

the coccyx are nearly vertical, being only slightly curved forward, and the tip of the coccyx is on a somewhat higher level than the inferior margin of the symphysis pubis; the curve and position of the rectum are determined by those of the sacrum and coccyx, until it passes in front of the coccyx, when it turns vertically downwards; the prostate gland, situated entirely on the upper or interior side of the triangular ligament, rests on the last turn of the rectum, and the base of the bladder is in contact with the rectum above that place." (Quain.)

The neck of the bladder is the most muscular part of the organ, and is closely enveloped by the prostate gland, into the base of which it enters at the junction of the anterior one-fourth with the posterior three-fourths of the gland, the larger portion of which is, therefore, situated behind the vesical neck. The so-called third lobe of the prostate intervenes between the neck of the bladder and the spot where the vasa deferentia enter the gland. When enlarged it projects upwards behind the orifice of the urethra, and mechanically impedes the escape of urine. Anteriorly the neck of the bladder corresponds to that portion of the prostate which is situated above the canal of the urethra, to the anterior ligaments of the bladder, and to the posterior aspect of the symphysis pubis. It is covered by numerous veins. Laterally the neck of the bladder corresponds to the lateral portions of the prostate, to the puboprostatic aponeurosis, and to the anterior fibres of the levator ani muscle. In the female, the neck of the bladder is nearer to the symphysis pubis in front, while posteriorly it is applied to the anterior wall of the vagina.

The anatomical neck of the bladder is formed by the circular ring of fibres at the junction of the urethra with the body of the organ. The function of these fibres is to permit the accumulation of urine and to prevent its escape, except under the influence of the will. Very varying accounts are given of the sphincter vesicae by different anatomists. According to Sappey, the sphincter has the form of a large ring which embraces the neck of the bladder and all the posterior half of the prostatic portion to the urethra. Its external surface corresponds below and on each side to the prostate, to which it adheres closely and without any line of demarcation; above, it is covered by the anterior longitudinal fibres of the bladder, which cross it at a right angle, and are closely united to it. By its internal surface, it is in relation with the longitudinal fibres of the urethra and urethral mucous membrane. Above, it is continuous with the muscular fibres of the bladder; below, it reaches the posterior extremity of the verumotanum. It is from ten to twelve millimeters in breadth, and three or four in thickness at the neck of the bladder, but it becomes thinner as it passes forward. On the internal surface, the anatomical neck of the bladder appears as an auricular orifice, bounded by a fold of mucous membrane. This opening will admit the little finger. It becomes much altered in shape by enlargement of the prostate. The vesical orifice is situated about an inch and a quarter behind the symphysis pubis, and four-fifths of an inch above a line drawn from the coccyx to the under part of the symphysis.

The surgical neck of the bladder includes that portion of the urinary passage which extends from the anatomical neck of the organ to the membranous portion of the urethra. Its termination anteriorly corresponds to the apex of the prostate. The surgical neck of the bladder, therefore, thus considered, is the prostatic portion of the urethra. This part of the canal is about fifteen lines in length; it is wider in the middle than at either end; its normal diameter is twelve millimeters, and it may be dilated to fifteen or sixteen millimeters, but it gives way if the dilatation is carried beyond this point. It is, however, more dilatable than any other part of the urethra. Its axis describes a curve with the concavity directed upwards and forwards; this curve is slight in young subjects, but increases with age. The walls of the canal are in close apposition, one being anterior and the other posterior. The canal traverses the upper part of the prostate, so that it has much more of the gland below it than above.

Its lining membrane is continuous with that of the bladder, and is thrown into longitudinal folds. On the floor of the canal is a narrow median ridge, from twelve to fourteen millimeters in length, rounded posteriorly, and more acute at its anterior or lower end; this is the crest of the urethra, and it is also called the *verumontanum* or *caput gallinaginis*. At the fore part of this crest is a depression, within the margins of which are situated the openings of the ejaculatory ducts. On each side of the crest is a longitudinal groove, the prostatic sinus, into which the prostatic ducts open. Behind the crest, a slight elevation is continued backward through the neck of the bladder, into the uvula vesicæ. The mucous membrane is covered by a flat laminated epithelium; it has on its outer side a layer of longitudinal muscular fibres, and these again are covered by circular fibres which are more marked above. A second layer of longitudinal fibres separates these latter from the substance of the prostate gland, which has on its outer side interlacing muscular fibres and a close plexus of veins.

The thickness of the walls of this canal is a point of considerable importance, and it has been carefully studied by Mercier and others. According to Mercier, the median anterior measurement of the structures which form the wall amounts to five millimeters, the median posterior to seventeen, the transverse to fifteen, and the oblique (downwards and outwards), to twenty-three millimeters. Senn, of Geneva, gives nearly the same measurements. Making allowance, therefore, for the size of the instrument employed for extraction, it follows that an incision outwards and downwards through the substance of the prostate will allow a calculus of only two centimeters (or four-fifths of an inch) in thickness to pass without dilatation of the opening. In Dupuytren's bilateral operation, room is gained for the passage of a calculus thirty-six millimeters in diameter.

The internal surface of the bladder is lined by a mucous membrane, the color of which varies with age, being pale bluish in the young subject, grayish red or ashy-gray in the adult, and of a deeper shade in old age. In the child the walls are perfectly smooth, but with age the muscular fibres often become hypertrophied, and project into the bladder, giving its internal surface a reticulated appearance. When this condition is very marked, the bladder is said to be "fasciculated" (*vessie à colonnes* of the French). The mucous membrane is moulded over the muscular ridges, and sinks into the depressions between them, giving rise to a number of more or less distinct sacs. The bladder thus affected is said to be "sacculated" (*vessie à cellules* of the French). When the bladder is empty, the anterior and posterior aspects of its internal surface are in contact; the lateral surfaces, under these circumstances, are simply borders.

The internal surface presents at its lower part three orifices, placed in the form of a triangle—two posterior, those of the ureters, and one anterior, that of the urethra. Immediately behind the latter is the smooth triangular surface, having its apex turned forward, and differing from the rest of the internal surface in presenting no rugæ. This surface is called the *trigone;* at its posterior angles are the orifices of the ureters. The triangle is nearly equilateral, and each of its sides measures about an inch and a half. In some bladders there is a slight elevation of the mucous surface extending from the internal orifice of the urethra to the base of the trigone. This has been termed the *uvula vesicæ* (*luette vésicale* of the French). This elevation exists, though in a slighter degree, in the female, and is therefore not due to the presence of the prostate. The ureters run obliquely through the coats of the bladder for about three-quarters of an inch, and open on its inner surface by two narrow and slit-like openings, distant about an inch and a half from one another. Between these orifices, there extends a curved elevation, having the convexity forwards, and due to a subjacent muscular band, which joins them to each other, and to the neck of the bladder. From opposite the middle of the elevation, the uvula projects forwards towards the neck of the bladder. The uvula is slightly in advance of the middle lobe of the prostate, and may be traced through the urethral orifice, which, when normal, it may assist to close. It is produced by a thickening of submucous tissue. Behind the posterior border of the trigone, the base of the bladder is somewhat depressed, and to this portion the term *bas-fond* has been given by the French. When the median portion of the prostate is enlarged, this depression becomes more marked, and under such circumstances it is the favorite spot for the lodgment of calculi, which often take the shape of the fossa. The *bas-fond* increases in size with the development of the prostatic enlargement and the decreasing energy of the muscular coat. Its presence favors accumulation of urine, which by its decomposition tends to produce cystitis.

The orifice of the urethra is situated at the apex of the triangle. It is, as stated above, the lowest portion of the bladder in both sexes; its shape varies according to age. In children and adults it is circular in form, and has puckered margins; in old age it is often more or less irregular, the alteration being due to distortion of its inferior margin, as a result of prostatic hypertrophy, or of hypertrophy of the muscular tissue of the uvula vesicæ. Under these circumstances the neck of the bladder loses its circular form, it becomes triangular, radiated, crescentic, or semicircular, and at the same time an obstacle is caused to the passage of a catheter. The urethral orifice is habitually closed, and some amount of force is required to open it.

The diameter of the neck of the bladder, and the extent to which it may be dilated are points of considerable importance. Deschamps,[1] quoted by Mercier, gives the diameter of the urethral orifice, when dilated to its fullest extent, at 16 millimeters; the circumference, therefore, is 48 millimeters. This measurement has been confirmed by other anatomists; but in estimating the size of a calculus which could be extracted without laceration, it is necessary to make a subtraction, on account of the instruments employed. According to Deschamps, also, the apex of the prostatic portion of the urethra is less dilatable than the vesical orifice of the urethra. Dolbeau, the advocate of perineal lithotrity, maintains that the dilatability of the neck of the bladder has generally been overestimated; and some experiments have been made, which show that such supposed dilatation always causes more or less

---

[1] Mercier, loc. cit., p. 53.

laceration. Dolbeau, however, affirms that, by means of his dilator, the neck of the bladder can, in the healthy adult, be dilated up to a diameter of twenty millimeters without causing laceration; beyond this, the prostate and urethra are torn. His dilator expands to twenty millimeters, and his largest lithoclast can be passed through an opening having this diameter.

The walls of the bladder are composed of a serous investment, which forms an incomplete covering; of a muscular coat disposed in three layers, and of a mucous membrane separated from the muscular coat by dense and abundant connective tissue. The serous investment facilitates the movements of the bladder upon adjacent organs; the function of the muscular coat is to retain and discharge the contents of the bladder, while the mucous membrane protects the other coats from contact with urine.

The serous or peritoneal coat does not form a complete investment. It covers the whole of the posterior aspect of the organ, a portion of its sides and of its anterior aspect. When the bladder is empty, however, its anterior aspect is uncovered by peritoneum; when the organ is much distended, the serous membrane forms a cul-de-sac in front extending for a variable distance, but always leaving uncovered a certain portion above the pubes.

The muscular coat is composed of fibres uniting to form rounded fasciculi, separated by sheaths of connective tissue. The fasciculi are arranged in tolerably definite layers, the fibres of the most external layer being more or less longitudinal, those of the next being disposed in a circular direction, while still more internally, according to some authors, there is a layer of delicate longitudinal fibres. The muscular arrangements of the bladder have been especially studied by Ellis[1] and Pettigrew;[2] but the variation in the disposition of the fibres is, according to Obersteiner,[3] so great that it is impossible to give minute details respecting the course of the muscles, any minute description being applicable only to the individual specimen. The external fibres, however, which are longitudinal, are well marked towards the vertex of the organ, and taken together, they constitute the *detrusor urinæ* muscle. The circular fibres on the other hand are most marked at the neck of the organ, where they form a complete ring, the *sphincter vesicæ*. Pettigrew describes the fibres of the cervix as traceable into the verumontanum, which he believes to act as a valve to the urethra, falling down into the tube and obstructing it when the muscular coat of the bladder is not acting, and raised up so as to stand erect in the middle line, and thus allowing the urine to flow past it, when the fibres contract. Mercier believes that the longitudinal fibres have a double action: they force the urine towards the neck of the bladder, and open the orifice and keep it dilated while micturition is going on. Certain fibres extending between the opening of the ureters and the neck of the bladder have been called the muscles of the ureters. They have been described as originating behind the orifices of the ureters, and converging at the back part of the prostate, into the middle lobe of which they are inserted by means of a fibrous process. Sir Charles Bell, who described them, supposed that during the contraction of the bladder they served to retain the oblique direction of the ureters, and so to prevent the reflux of urine into them.

[1] Medico-Chirurgical Trans., 1856.   [2] Philosophical Trans., 1866.
[3] Stricker's Manual of Human and Comparative Histology, New Syd. Soc. Translation, vol. ii., p. 127.

The mucous membrane of the bladder is continuous above with that lining the ureters, and in front and below with that of the urethra. It is loosely united to the muscular coat by means of submucous connective tissue, and in the empty state of the bladder is thrown into numerous folds, except at the triangular area, at the base, where it is always more even. The membrane is soft and smooth to the touch; its color, as above mentioned, varies with age. The mucous membrane is provided with numerous small racemose glands, most abundant near the urethral orifice, and lined with columnar epithelium. The epithelium of the mucous membrane is arranged in several layers, the cells of each of which present marked differences. Those of the most internal strata are large and irregularly spheroidal or polyhedric in shape, the most superficial being somewhat flattened. Their under-surface presents a concavity which is, as it were, moulded over the rounded ends of the next layer, the cells of which are pyriform in shape, with the broader portion directed towards the surface. The lower end of these cells, directed towards the deeper tissues, forms a long and unbranched process of varying length, which is received among smaller rounded irregularly shaped cells, forming the deepest layer, and filling up the intervals between the prolongations. The processes of the pyriform cells appear to terminate in the superficial layers of the submucous connective tissue. There are no villi on the surface of the mucous membrane.

One of the most interesting questions connected with the physiology of the bladder is that which refers to the capacity for absorption possessed by its mucous membrane. Sir Henry Thompson[1] states that the mucous membrane in question appears to have no absorbing power, and he cites an experiment of his own in which four drachms of Liq. opii sed. were injected into the bladder of a patient with chronic cystitis, without producing any of the local or general effects of opium. Such an experiment, however, taken by itself, proves only that absorption did not occur in this particular case, a result which might very possibly be due to the advanced condition of disease which affected the mucous membrane. It is, indeed, only on some such hypothesis as this that the varying statements of experimenters can be reconciled; and in estimating the value of the experiments, it is necessary to distinguish between those in which the absorption might have taken place by the mucous membrane of the urethra and those in which such a possibility was excluded. The absorbing power of the urethra appears to be admitted on all hands, and the experiments of Messrs. Jolyet and Alling[2] show that this power may be very great when the bladder possesses little or none. These observers introduced a catheter into a dog's bladder, and then opened the abdomen and placed a ligature round the urethra close to the neck of the bladder. The urine was then drawn off, and the condition of the pupils carefully noticed. Five centigrammes of atropine dissolved in water were then carefully injected into the bladder; an hour was allowed to elapse, during which time the pupils were constantly watched. No dilatation was manifested. The catheter, its end being sealed, was then withdrawn from the bladder, and only its extremity allowed to remain within the urethra. A ligature was placed round the meatus, and pressure made with the fingers on the bladder. The fluid was thus forced into the urethra, and in five minutes the pupils began to dilate. Five minutes afterwards the dilatation was complete. M. Alling also injected a solution containing five centi-

[1] Clinical Lectures on Diseases of the Urinary Organs, 5th ed., p. 302.
[2] Mercier, loc. cit., p. 72.

grammes of hydrochlorate of morphia into his own bladder, and experienced no effect. He subsequently introduced into his urethra a small bougie smeared over with ointment of morphia, and soon experienced the characteristic effects of that drug.

Iodide of potassium has been employed for similar experiments by M. Demarquay. In eight out of sixteen cases, iodine could be detected in the saliva, but not until several hours had elapsed. In the remaining eight there was no evidence of absorption. All his patients were suffering from stricture of the urethra, and as in such cases the mucous membrane of the bladder is generally in a more or less unhealthy condition, Messrs. Jolyet and Alling, commenting on these and similar experiments, advanced the opinion that in the cases in which absorption took place there was erosion of the mucous membrane, and that where there were no signs of absorption the membrane in question was sound, and they believe that in all cases the absorption or non-absorption of substances injected is dependent upon the condition of the mucous membrane.

Another experimenter, Dr. Mercier of Neuchatel, who has paid considerable attention to this subject, believes that in a perfectly sound condition the mucous membrane of the bladder possesses very little absorbing power, and that when absorption takes place it is because the vesical epithelium has undergone more or less desquamation, the degree of absorption being in proportion to the changes that have occurred. In this manner he attempts to explain the conflicting results of various experiments. He also thinks that the urethral mucous membrane, *when intact*, has no more power of absorbing than that of the bladder, and that the reason why absorption appears to occur so readily in the case of the former membrane is because the passage of a catheter and the distention of the urethral canal cannot fail to produce erosion of the epithelium. He cites the case of a woman, suffering from hysterical retention of urine, into whose bladder, apparently quite normal, one centigramme of hydrochlorate of morphia was injected, and produced the usual effects of the drug. A repetition of the experiment led to the same result, and the explanation offered was that the mucous membrane had been stretched and damaged by the distention, and possibly scraped off by the catheter. In another case, in a male patient without any vesical symptoms, the injection of three centigrammes of morphia produced deep sleep. Admitting the correctness of Dr. Mercier's theory, all that can be said is that the vesical mucous membrane, in a perfectly natural condition, has not been shown to possess any power of absorption, but that it acquires this power when its structure has been modified by disease or mechanical agencies. In some cases of chronic cystitis, the general symptoms, such as fever, etc., are doubtless attributable to the absorption of purulent matter, and the experiments which have been referred to show that it is, to say the least, hazardous to inject large quantities of narcotic substances into the bladder in the treatment of such cases.

The question as to whether absorption takes place from the bladder has recently been under discussion at a meeting of the Surgical Society of Ireland,[1] but it can scarcely be said that any fresh light has been thrown upon the subject. The discussion was based on some experiments by Mr. Wheeler on a patient, the subject of extroversion of the bladder. Mr. Wheeler was anxious to test the value of Sir H. Thomp-

---

[1] At a meeting held on January 18th, 1878, for a report of which see Medical Press and Circular, February 6th, 1878.

son's deduction, above referred to. He washed the surface of the extroverted bladder with tepid water, and then applied an ointment containing twenty minims of tincture of iodine. In three minutes iodine could be detected in the urine. He also, on another occasion, placed morphia on the mucous surface when cleansed, and found that sleep was produced. In another case, of ulceration of the bladder, he applied morphia through an endoscope; sleep came on in eighteen minutes, and lasted for four hours. With regard to the first of these experiments, it may be urged that the iodine might have found its way directly into the ureters, and thus become mixed with the urine in which it was afterwards detected, and on the other hand, even allowing that absorption took place through the exposed mucous surface, it by no means follows that the surface of the normally closed reservoir would discharge a similar function. No fresh experiments are required to prove that an ulcerated mucous surface will absorb matters presented to it, but such experiments prove nothing with regard to the normal mucous membrane of the bladder. Professor Kuss[1] states that the epithelium of the bladder is remarkable for its impermeability, and that it absolutely opposes the transmission of liquids. A solution of belladonna may be kept in a perfectly healthy bladder for a long time, and so, also, may solutions of opium, without any risk of poisoning. If, however, the epithelium be diseased, absorption immediately occurs; dilute alcohol injected into a bladder affected with catarrhal inflammation soon produces symptoms of intoxication. Professor Kuss also asserts that the vesical epithelium, even for some hours after death, preserves its vitality, and consequently its impermeability. He bases the statement on an experiment in which a solution of ferrocyanide of potassium is injected through a tube (thus preventing contact with the urethral mucous surface) into the bladder of an animal just killed. If the bladder be then exposed and a salt of iron placed on its walls, no Prussian blue appears. If, however, by means of a wire the epithelial coat of the bladder be scratched, Prussian blue will be immediately found, and Dr. Kuss believes that the opposition to the passage of liquids results solely from the presence of the epithelium.

Dr. Mercier,[2] of Neuchatel, commenting on experiments similar to those that have been referred to, draws the following conclusions:—

(1) It is highly probable that in a healthy bladder absorption does not take place in any sensible degree, but it must be remembered that evidences of slight epithelial desquamation are to be found in the majority of specimens of urine.

(2) The alterations in the epithelium are very frequent and various, and this fact explains the differences in the absorption of certain substances, the absorption being always in direct proportion to the degree in which the mucous membrane is altered.

(3) The variations in the result obtained by experimenters are due to the fact that some of the experiments were performed on healthy bladders, and others on bladders more or less changed by disease.—One thing seems certain, viz., that an experiment such as that performed by Sir H. Thompson is not altogether free from risk, and that when medicinal substances, poisonous in large doses, are injected into the bladder, care should be taken not to exceed the ordinary dose, at least until it has been proved that this is inefficacious. Injections containing anodyne remedies are often indicated in the treatment of certain diseased states, in the majority

---

[1] Manual of Physiology, p. 473. [2] Op. cit., p. 74.

of which the mucous membrane is more or less affected, and that absorption does occasionally take place under such circumstances is a fact which admits of no dispute.

The bladder derives its nerves partly from the hypogastric plexus of the sympathetic, and partly from the sacral plexus. Those from the former source contain a larger proportion of spinal nerves than is found in the branches distributed to the other pelvic viscera. They are said to be chiefly distributed to the upper part of the bladder, while the spinal nerves, especially the visceral branches of the fourth sacral nerve, may be traced more directly to its neck and base. Some of the nerves are distributed to the muscular fibres, and others to the mucous membrane. "According to Kisseleer, the nerves form a network immediately under the epithelium, from which filaments pass amongst the epithelium-cells. Gangliated plexuses of nerves accompany the blood-vessels, and send branches both to these and to the muscular coat of the bladder"[1] (F. Darwin). According to Obersteiner,[2] the nerves may be followed as medullated fibres in the connective tissue layers, especially at the fundus near the urethral orifice, where they are present in great numbers. He has not been able to trace the manner in which they terminate.

The fact that the bladder derives its nerve-supply through a double channel serves to explain the course of the symptoms frequently observed in cases of partial or complete paralysis of the organ. The bladder often suffers in cases of spinal disease or injury, and the most frequent symptom is that of paralysis of the body of the organ, as shown by retention of urine. As the disease advances, the retention may be succeeded by true incontinence, due to a secondary paralysis of the sphincter of the bladder. To explain these symptoms, it has been supposed that the retention in such cases is due to spasmodic action of the sphincter, the detrusor muscle being unaffected but unable to overcome the spasm. This explanation, however, is unsatisfactory, for it is contrary to analogy to suppose that spasm of a muscle like the sphincter can be complete and permanent. It is far more reasonable to assume that the disease affects first one and then the other channel of nervous influence.

These two channels have been demonstrated by Budge: one passes in the anterior roots of the third and fourth sacral nerves, the other is derived from the hypogastric plexus; the former is connected with the brain by a strand of fibres which may be traced from the cerebral peduncle along the anterior columns of the spinal cord. The motor nerves contained in the hypogastric plexus come from the lumbar enlargement of the spinal cord. The fibres, therefore, which transmit to the sphincter the influence of the will come from the encephalon, while those which are distributed to the detrusor of the organ come from the hypogastric plexus, and are concerned in the involuntary action of the bladder. The body of the organ is, however, supplied by some fibres derived from the first-named source. In the majority of nerve-lesions affecting the bladder, the nervous supply of the detrusor muscle is first and especially implicated. The antagonism of the two sets of nerves is proved by the phenomena witnessed in normal micturition. The accumulation of a certain quantity of urine in the bladder gives rise to a peculiar sensation which is transmitted to the spinal cord, and excites the motor

---

[1] Quain's Anatomy, vol. ii., p. 427.
[2] Stricker's Histology, New Syd. Soc. Trans., vol. ii., p. 128.

nerves. The influence of the will is then brought to bear upon the sphincter, and the urine as a consequence is kept in the bladder for a longer or shorter period until its discharge is determined by the volition of the individual. The detrusor and sphincter muscles act therefore independently of and in opposition to each other.

With regard to the mechanism by which urine is retained in the bladder, there appears to be considerable difference of opinion. Dr. Carpenter[1] states that "under ordinary circumstances the escape of urine is prevented by the moderate contraction or tone of the sphincter vesicæ, to overcome which in the rabbit the pressure of a column of water from sixteen to twenty inches in height, and in the dog of about thirty inches, is required. On section of the lower part of the spinal cord or of the vesical nerves, the pressure of a column of water of only five or six inches in height is sufficient to force a passage for the fluid through the sphincter." Kuss[2] points out that in addition to the contraction of the sphincter, the vesical orifice is kept closed by certain mechanical arrangements connected with the bladder, so completely indeed that no physiological act or contraction is required to prevent the exit of the urine. In the first place, distention of the bladder causes the urethra to be compressed between it and the adjacent tissues, owing to the direction which the canal takes, viz., first vertically downwards and then forwards and upwards. Again, the prostatic portion of the urethra is encircled by the gland in such a way as to have its walls kept closely in contact. Kuss believes that the urine is kept in the bladder (in the male subject) mainly by the elasticity of the muscular elements of the prostate gland. He thinks also that the closure of the urethra is assisted by the arrangement of the perineal fasciæ, the fibres of which compress the sides of the urethral canal in their course from the ischium to the pubes. Alluding to the fact that urine is often found in the bladder of the dead body, he thinks that the conditions which restrain its escape during life are simply of a mechanical nature. It would appear, however, that he somewhat under-estimates the functional importance of the sphincter of the bladder. There can be little doubt that in the female the escape of urine is partly at least prevented by the activity of this muscle. The muscles connected with the urethra also render essential aid in closing the canal of exit. Winckel[3] alludes to cases of vesico-vaginal fistula, in which after successful operation it was found that the neck of the bladder and the upper part of the urethra were completely disorganized, but the lower part of the canal, which had not been involved in the fistula, remaining intact, the power of retaining urine was preserved. These cases would appear to prove that even after urine has escaped through the sphincter, its course may be checked by the action of other muscles.

When a certain quantity of urine has accumulated in the bladder, a peculiar sensation is produced, and this leads to a series of expulsive actions. The impression produced by the distention of the bladder is the probable excitant of the desire to pass urine. Kuss, however, thinks that it is the contact of the urine with the prostatic mucous surface which gives rise to the sensation of a necessity or desire for micturition, and that if this desire be unheeded, a reflex irritation is produced, which is followed by the contraction of the constrictor urethræ or urethral sphincter; the urine can then advance no further, and is even obliged to

---

[1] Human Physiology, p. 545.   [2] Manual of Physiology, p. 474.
[3] Handbuch der Frauenkrankheiten, p. 10.

pass backwards, by reason of the contraction of the muscles on the anterior portion of the prostate, and so re-enters the bladder, the contractions of which have ceased. As to whether the bladder is able to empty itself by its own contractions, such at least would appear to be the case from experiments upon the lower animals. After opening the abdomen of a dog, the bladder may be seen to empty itself completely by its own contractions. In man, however, the organ is less muscular, and the contraction of the diaphragm and abdominal muscles is probably necessary in order to empty the bladder completely. The same muscles also assist at the commencement of micturition, but when once the flow is established, the involuntary contraction of the bladder alone is sufficient to make it continuous. The escape may, however, at any time be accelerated by the assistance of the abdominal muscles. Toward the end of the expulsive act the same muscles come into play, and there is an interrupted discharge of the small quantity of urine that remained. The escape of the last drops is promoted by the contraction of the muscular walls of the urethra, and the action of the accelerator urinæ and other perineal muscles. Kuss, as above mentioned, thinks that the reflex movement starts from the mucous membrane of the prostatic urethra, and he explains enuresis nocturna by supposing that it is due to the lack of sensibility of the mucous surface to the contact with urine, and to the consequent absence of a premonitory sensation of the desire to urinate. He also thinks that every time a true resistance is offered to the passage of urine, the opposition is due to the action not of the sphincter of the bladder, but of that of the urethra (the constrictor urethræ muscle), which is the only one of these muscles which is striated or voluntary.

From the above account it will appear that extremely different opinions are held by physiologists as to the manner in which the closure of the bladder is normally effected. Professor Winckel's cases prove that with a defective sphincter the urine may be completely retained in the female bladder by virtue of the muscularity of the urethra.

The frequency with which the urine is discharged depends mainly upon the habits of the individual, the quantity of liquid taken, and the activity of the skin, which again is influenced by the temperature of the atmosphere and other conditions.

The prostate gland has been already in part described in the account given of the surgical neck of the bladder. It remains to add a few words with regard to its position, relations, and structure. The prostate gland is a firm, muscular, and glandular body surrounding the neck of the bladder, with which it is directly connected. It resembles a horse-chestnut in shape and size; it is situated in the pelvic cavity, and its apex reaches in front the posterior layer of the sub-pubic fascia. Its base limits anteriorly the triangular space at the base of the bladder. The body of the organ is directed downwards and forwards, and is situated about half an inch behind and below the symphysis pubis, to which it is connected by the anterior ligaments of the bladder. Its posterior surface is somewhat convex, and is in contact with the rectum, being separated from it by a thin layer of the recto-vesical fascia, which forms a sheath for the gland. It rests against the middle portion of the rectum, the lower part of which gradually recedes from the urethra, leaving an angular interval between the bowel and the apex of the gland. It is this portion of the prostate which can be felt by the finger introduced into the bowel. After passing over the bulb of the urethra and the membranous portion, the apex of the prostate can be made out, and behind this the body of the gland, becoming wider towards the base, which can also be felt if the bladder is empty. This surface is marked by a median hollow, which indicates the division into lateral lobes.

The measurements of the gland in various directions are as follows: From base to apex it measures an inch and two-fifths; transversely, at the base, one inch and three-quarters; and in depth about three-quarters of an inch. Hence it follows that the greatest extent of incision possible, without completely dividing its substance, is obliquely outwards, downwards, and backwards. Such an incision, commencing at the urethra and terminating at the outer border of the organ, will measure seven-eighths of an inch in length. According to Sir H. Thompson's researches, the usual weight of a healthy adult prostate may be estimated at from $4\frac{1}{2}$ to $4\frac{3}{4}$ drachms.

The prostate gland is maintained in its position by its connection with the neck of the bladder; by its sheath of pelvic fascia continued from the posterior surface of the pubic bones; by prolongations of the same fascia forming the anterior true ligaments of the bladder; and by the anterior portion of the levator ani muscle on each side. This portion, which has been named the *levator seu compressor prostatae*, descends by the side of the prostate, and unites with its fellow below the membranous portion of the urethra, thus supporting that canal as in a sling.

The prostate consists of two lateral lobes, and a portion at the base situated between the neck of the bladder and the ejaculatory ducts, and called the "middle" or "third" lobe. The lateral lobes are distinct and symmetrical; they exist as independent portions up to the fourth month of fœtal life. In adult age the division is indicated by a median hollow on the under surface, and sometimes by a slight groove along the middle of the upper surface. The so-called "middle" lobe has no claim to be regarded as an independent portion of the normal gland. It is a small piece of the gland uniting the posterior border of the lateral lobes. When enlarged, it projects upwards into the bladder, and may interfere considerably with the flow of urine and the passage of a catheter, and, as Sir H. Thompson points out, any appearance of a lobe in this situation must be regarded as belonging not to normal but to morbid anatomy.

The prostate is traversed by the first portion of the urethra, which runs nearer to the anterior than to the posterior surface of the gland. The larger portion of the prostate is therefore behind the urethra, but the measurements of the anterior and posterior portions are by no means constant. It sometimes, though rarely, happens that the amount of glandular tissue in front of the urethra equals or even exceeds that behind the canal. On either side are the masses of the lateral lobes. The prostatic urethra, as forming part of the neck of the bladder, has been already described. The lower part of the prostate is traversed by the common ejaculatory ducts, one on each side; these enter at the base of the organ, and are inclosed in a special hollow part of the gland; they open into the vesicula prostatica of the urethra.

The structure of the prostate gland consists of glandular substance and muscular tissue, enveloped in a fibrous coat. On section the gland appears reddish in color, and is firm to the touch, but it can be torn without difficulty. The proper fibrous capsule is distinct from the investment derived from the pelvic fascia, and is separated from it by a plexus of veins. The proper stroma of the prostate is made up of muscular tissue; the connective tissue is scanty, and serves to support the vessels and nerves, and to separate the muscular bundles. The glandular substance constitutes only about one-third of the entire mass. According to Professor Ellis's investigations,[1] the results of which agree with those of Kölliker, the prostate is essentially a muscular body, consisting of circular or orbicular involuntary fibres directly continuous behind with the circular fibres of the bladder. "In front a thin stratum, about one-thirtieth of an inch thick, is prolonged forwards around the membranous portion of the urethra, so as to separate this tube from the surrounding voluntary constrictor muscle." The tube of the urethra is encased by its own layer of longitudinal fibres, and is quite distinct from the circular fibres. On the upper surface of the prostate are some longitudinal fibres derived from those of the bladder. From the cortical muscular layer strong bands of muscular fibres run towards the centre, interlacing freely in their course, and forming meshes in which the gland substance is imbedded. The portion of prostate anterior to the urethra consists almost exclusively of muscular tissue. The glandular substance, on the other hand, is most abundant in the portion lying behind the urethra. Dr. Klein[2] states that "transversely striated muscular tissue also occurs in the prostate in the form of continuous bands, internal to the transversely striated fibres of the sphincter urethræ. Henle describes similar circular bands existing in the uppermost of those portions of the prostate lying in front of the urethra." Fasciculi of fibres of the same kind are found also in the cortical layer of the segment situated behind the urethra, especially in the upper part, where, in company with trabeculæ of smooth muscular tissue, they penetrate into and divide the glandular substance.

The glandular structure presents the characters of the so-called acinous glands. It consists of numerous small saccules, of various sizes, spherical or oval in shape, opening into elongated tubes which unite to form excretory ducts. These latter open, by from twelve to twenty orifices, on the floor of the urethra on each side of the verumontanum; a few openings are occasionally found behind that structure. The saccules and ducts are lined by columnar epithelium.

The prostate gland is supplied with blood by branches of the vesical, hæmorrhoidal, and pudic arteries. The veins form a plexus round the gland, and communicate with the dorsal vein of the penis, and with veins at the base of the bladder. This plexus is often much developed in elderly subjects, and may give rise to considerable hæmorrhage when cut in the operation of lithotomy. The nerves are derived from the hypogastric plexus, and consist of medullated and non-medullated fibres; in the cortical portion they exhibit numerous large ganglion cells. Pacinian bodies, according to Dr. Klein, are also found in the cortex of the gland. The prostate secretes a milky fluid, having an acid reaction, and containing epithelial particles, numerous molecules, and granular nuclei. Its use in the economy is unknown. By some it is supposed to act as a diluent for the semen. Mr. Ellis thinks that as only so small a portion of the prostate is glandular, the propriety of calling that body a gland is rendered doubtful. He thinks that the small secretory glands are but appendages of the mucous membrane, which project among the muscular tissue in the same way as the other glands of the urethra extend into the subjacent tissues. In advanced age the ducts of the prostate often contain concretions, varying in size, and composed in varying degrees of animal

---

[1] Trans. Med. Chir. Soc., vol. xxxix., p. 330.
[2] Stricker's Handbook of Human and Comp. Histology, New Syd. Soc. Trans., vol. ii., p. 297.

and earthy matters. These will be found described in a subsequent chapter.

The size of the prostate in early life is a subject of considerable importance. Dr. Gross[1] has given engravings exhibiting the size and form of the prostate in four young subjects. The results of his researches are as follows:—

Prostate at birth. Width at base, 4 lines; a little above middle, 5 lines; at apex, 2 lines; length along the middle, 4 lines, and at the edge, $4\frac{3}{4}$; thickness at base, 2 lines; at middle, $3\frac{1}{4}$, and at apex, $1\frac{1}{4}$. Weight, 13 grains.

Prostate at 4 years. Breadth at base, 6 lines; just above middle, 7; at apex, $2\frac{1}{2}$; length along the middle, 6 lines, and at margin, 7 lines; thickness at base, $2\frac{3}{4}$ lines; at the middle, 4, and at apex, 2. Weight, 23 grains.

Prostate at 12 years. Width at base, $8\frac{1}{2}$ lines; above the middle, $9\frac{1}{2}$, and 3 at apex; length along the middle, 8 lines, and $8\frac{1}{2}$ at the edge; thickness at base, 3; middle, $4\frac{1}{2}$; apex, $2\frac{3}{4}$. Weight, 43 grains.

Prostate at 14 years. Width at base, 11 lines; at middle, $9\frac{1}{2}$; at apex, 4; length along the middle, 8 lines, and 10 at margin; thickness, $3\frac{1}{2}$ at base, 5 at middle, and 3 at apex. Weight, 58 grains.

The prostate appears to increase in size up to the age of 25; it then, according to Dr. Gross, measures 18 lines at the base, and 15 along the middle, while in its thickest part it measures 11 lines; its weight at that age is $4\frac{1}{2}$ drachms. In childhood its consistence is less firm than in after-life, and its shape is more globular; its capsule is also more readily lacerable. It has a more vertical position in the pelvis, and its base more nearly approaches the peritoneum than it does in the adult. These last-mentioned differences are due to the position of the bladder, which in the young subject is pyriform in shape, and is rather an abdominal than a pelvic organ.

---

[1] Diseases of the Urinary Organs, 1st edition, p. 71.

CHAPTER II.

THE METHODS OF EXAMINING THE BLADDER AND PROSTATE GLAND.

In consequence of the manner in which it is surrounded by other organs, the bladder, undistended by urine and free from disease, is but little accessible to external examination. When empty, lying deeply in the pelvis, and with its apex not projecting above the symphysis pubis, it cannot be detected on percussion, while its base, which rests upon the second portion of the rectum, may be reached by the finger, but forms no prominence in the bowel. In proportion, however, as the bladder becomes distended, and projects above the symphysis, and into the rectum, it becomes accessible to examination in those regions.

Its anterior surface, as the bladder rises in the pelvis, is at first closely applied to the posterior surface of the abdominal wall. As the size of the organ increases, the apex recedes somewhat, and as it approaches the umbilicus it is found to be from half an inch to an inch from the wall of the abdomen. Long, however, before this condition is reached, the bladder may be detected by palpation and percussion, and when the distention is very great, the tumor becomes visible to the eye. On examining the lower part of the abdomen, the bladder will be found to present itself as a tense, hard, elastic tumor, rounded or pyriform in shape, and extending above the pubes towards the epigastrium. On percussion, the swelling will yield a dull note, while the bowel around it is clear and distinct. On passing a finger into the rectum, the base of the bladder will be felt there as a rounded prominence, giving the sensation of fluctuation or pressure. Also, if the other hand be placed above the pubes and pressure be made, a wave will be communicated to the index finger in the bowel. These symptoms, then, are those of distention, and it is this condition of the bladder which makes it accessible to examination. It sometimes happens that the existence of a vesical tumor can be thus positively diagnosticated, and a large growth may so distend the bladder that the organ may project above the pubes and yield a dull note on percussion. In a case under the care of Professor Billroth, a vesical tumor, of a myomatous nature and as large as a man's fist, could be felt through the abdominal wall of a boy; it was discovered to be firm, tolerably movable, and somewhat sensitive to pressure. It could also be felt through the rectum, and by making pressure at the same time above the symphysis pubis its consistence, size, and locality could be made out with such clearness that an operation was undertaken for its removal and attended with a perfectly successful result. The existence of a small soft tumor could not, however, be thus demonstrated.

Professor Volkmann[1] lays great stress upon the advantages of a "bimanual" exploration of the bladder. This is effected by passing two fingers of the left hand as far as possible into the rectum, the patient being under the influence of chloroform. An assistant places both hands

---

[1] Langenbeck's Archiv, Bd. xix., 681.

above the symphysis, and makes pressure downwards and backwards towards the rectum. When the adipose tissue is not very abundant and the bladder nearly empty, the superior fundus of the organ is brought near to the fingers. If anything abnormal is felt, the surgeon passes his right hand carefully under the hand of the assistant, and endeavors to ascertain more closely the nature of the object. By this method, Professor Volkmann asserts, he has been able to detect the presence of a calculus no larger than a bean. The manipulation must, however, be conducted as gently as possible. In one case it was evidently the cause of ecchymoses, found after death, in the coats of the bladder.

The interior of the bladder can be examined, in the male subject, only with the aid of instruments passed down the urethra. Such an instrument may be either hollow or solid, and may be composed of materials of various kinds, some being hard and resisting, others soft and flexible. The hollow instruments are termed catheters; the solid ones, sounds or bougies. Catheters are used for removing the urine from the bladder, or for testing the capacity of the organ; sounds are employed to detect the presence of calculi or other foreign bodies, and to examine the interior surface of the bladder. Sounds are sometimes made hollow, so that during an examination fluids may be injected into the bladder, or urine withdrawn from it.

By passing a catheter or sound into the bladder we gain evidence as to the capacity of the organ; the hardness and tenderness, or the opposite conditions, of its coats; the elasticity of its walls; the smoothness or roughness of its internal surface; the presence or absence of cysts and of tumors, calculi, and foreign bodies in general. With regard to its capacity, this may be tested by noticing the quantity of urine which is withdrawn from it after retention for a definite period, and by observing the length of catheter required. The urethra, however, is by no means uniform in length; it varies from six to eight inches, and, after the bladder is reached, it should be possible to pass the catheter for three or four inches further. In a healthy bladder the examination, if properly conducted, gives rise to no pain. The bladder, however, if inflamed, becomes exquisitely tender. The softness, or the reverse, of its walls is the next point to be ascertained. After the cathether has been felt to touch the wall of the healthy bladder, it can be pushed in for at least an inch further, and is ejected to that extent by the elasticity of the organ when the pressure is taken off. When inflamed, the least pressure causes acute pain, and when the organ is hypertrophied and its cavity contracted, the diminution in space will be appreciable and the walls will evince a total absence of elasticity. In the normal condition the inner surface will appear smooth and even, but in cases of hypertrophy the projecting muscular columns communicate a decided sensation of roughness to the hand in contact with the catheter, when the point of the latter is moved over the internal surface of the bladder. If the organ be dilated, the increase in capacity will be appreciable, and when sacculated, it sometimes happens that the point of the catheter passes into the abnormal pouch and draws off a large quantity of urine. When atony co-exists, the flow from the catheter is slow and feeble, and pressure on the abdominal wall expedites the escape of urine, and causes more to flow, even after the stream has ceased.

The fact has been already alluded to, that tumors in the bladder have occasionally been felt through the abdominal wall, but in cases of this description the diagnosis will be facilitated by the introduction of a sound.

By moving it from side to side, the presence of an obstacle connected with the walls of the bladder will be detected, and with a finger in the rectum knowledge will be gained as to the thickness of the structures at the base of the bladder. The existence of induration at any spot is the point to be determined. Malignant tumors of the bladder generally occupy the base of the organ, and form more or less irregular, hard prominences, which can in part at least, especially when the growth has had time to develop, be felt from the rectum. The surface of such a tumor is uneven, and the mass itself differs from an enlarged prostate in being irregular and unsymmetrical in form. Such an examination by sound and finger causes great pain. Villous growths cannot be detected in this manner. The softness and yielding nature of such growths prevent them from offering any resistance to the movements of a sound.

Catheters, as above stated, are made of materials which differ very widely as regards flexibility. Metallic instruments,—silver being the metal usually selected,—are hard and inflexible; catheters made of vulcanized india-rubber are perfectly soft and pliant, and between these extremes we have instruments of various materials and possessing more or less flexibility and softness. The English gum-elastic catheter may be rendered soft and flexible by dipping it in warm water, it can then be made to assume any curve that is desired, and when dipped in cold water the curve becomes permanent. This property can be utilized in cases where it is desirable to use a catheter with a considerable curve, and one which, although not so hard as a metallic instrument, yet offers considerable resistance to any force that would change its shape. Catheters are of various sizes; the English instruments are (or were until recently) made in twelve sizes only, No. 1 being the smallest, No. 12 the largest. The French, on the other hand, have thirty sizes. No. 1 is one millimeter in circumference, No. 2 is two millimeters, and the increase is uniform throughout the series. The English No. 1 is about equal to the French No. 3, while No. 12 is equal to No. 21 of the French scale. The advantages of the greater range in size of the French instruments are obvious.

When a catheter has to be introduced the patient may be either standing, with his back against a wall so as to afford him support, or lying on his back. The latter position is the more advisable when the instrument is about to be passed for the first time, and when the bladder is suspected to contain a large quantity of urine. Under ordinary circumstances, and with a healthy condition of the bladder and other parts, the passage of a catheter gives rise to little or no pain, but in some patients the neck of the bladder is peculiarly sensitive, and when the instrument reaches that part, severe pain, faintness, and other nervous symptoms may come on. Also if the bladder contains a large quantity of urine, and the patient be in the erect position, the sudden withdrawal of the fluid may cause great faintness, or even fatal syncope. In ordinary cases, and when the patient becomes habituated to the use of the catheter, the erect position will probably be found the most convenient. The instrument should be warmed (especially if a metallic one be used) and well oiled. For the latter purpose, vaseline will be found a very convenient material. It is highly lubricating and cleanly, inasmuch as it never becomes rancid. The patient stands facing the surgeon and with his back against a wall. The surgeon, taking the penis between the middle and ring fingers of his left hand, opens the meatus with the thumb and forefinger, and passes the catheter gently down the canal, keeping the penis slightly on the stretch, and the shaft of the instrument close to the body in the line of Poupart's ligament on the left side. The catheter is carried steadily and slowly in as far as the triangular ligament, a distance of from four to five inches. When this is reached, the shaft of the instrument should be brought across the abdomen to the median line of the body, and then downwards in a horizontal direction, and gradually depressed as the point passes under the triangular ligament, and along the prostatic part of the urethra, upwards into the bladder. No force should be used, but the instrument should appear to find its own way into the bladder. When the patient is recumbent, the surgeon stands on the left side, and introduces the instrument as before. When the point reaches the triangular ligament, the shaft is brought to a right angle with the patient's body, and depressed between the thighs as the point traverses the prostatic portion of the urethra. During this latter part of its course, its passage will often be much assisted by introducing the finger into the rectum, so as to tilt up the point. The escape of urine and the ease with which the point of the catheter can be moved, will indicate that the bladder has been reached.

With regard to elastic instruments, the proceedings are the same; generally speaking, if there is no obstruction in the urethra, these instruments can be passed with ease. This is especially the case with regard to the French catheters *à boule*, the use of which has now become common in this country. These instruments are extremely flexible, and have a long tapering extremity terminating in a rounded bulbous point. No damage can be done to the urethra by an instrument of this kind, and a patient may safely be trusted to use it himself. It is the best form of instrument that can be used to relieve retention of urine in cases of atony or paralysis of the bladder. Another form of elastic instrument, in common use in France, is that which is called *coudée* or elbowed. This has a soft and flexible stem, but towards the point, where it is turned up so as to form a short beak, the material of which it is composed is made more firm and stiff. As it passes down the urethra, the beak keeps close to the roof and slides over any obstruction which may exist in the floor of that canal.

Certain difficulties may be encountered in the attempt to introduce a catheter, even in a normal condition of the urethra. In the first place, the point of the catheter may lodge in the *lacuna magna*, a cul-de-sac in the roof of the urethra, about an inch from the meatus. If an obstacle be met with here, the instrument should be withdrawn slightly, and its point kept on the floor of the canal and then pushed forward. The next obstacle is the bulb of the urethra, a spot where false passages are apt to be made. This obstacle is avoided by keeping the point of the instrument against the roof the canal. The last obstacle is at the prostatic portion of the urethra, near the neck of the bladder, and this is also avoided by keeping the point of the instrument against the upper wall as before.

For examining the bladder in cases of enlarged prostate, instruments possessing a different shape are necessary. The catheter must be longer, and have a larger curve. If a gum-elastic instrument be selected, it should have been kept on a well curved stylet, so as to retain its curve when the latter is withdrawn. Sir H. Thompson recommends that the instrument should be well over-curved, by keeping it on a stylet for a month or two, and that the shaft should be turned back immediately before using, so as to undo the extreme curve and produce an ordinary one. The catheter has then a tendency to become more curved as it passes down the passage, and its point will the more readily pass over any obstruction in the floor of the canal. The *coudée*, or elbowed, catheter is also useful for such cases.

The introduction of an instrument into the urethra for the purpose of exploring the bladder may be followed by various troublesome, or even dangerous consequences. These accidents may be divided into two classes, according as their effects are of a local or general character.

The local accidents which may be caused by catheterism or sounding are, false passages; inflammation of the bladder, hæmorrhage, injuries to the prostate and neck of the bladder, and orchitis. False passages are caused by the instrument passing through the walls of the urethra into the surrounding tissues. They may be easily made when the urethra is affected with stricture, and they occur most frequently in the bulbous and membranous portions of the canal. Their direction is usually downwards and to one side of the urethra. In addition to these spots, false passages are occasionally made through the prostate, in cases of chronic inflammation of that gland. The false passage may be complete, that is, after leaving the urethra and tunnelling under its walls, the point of the catheter may again pass into the right passage. It may, in like manner, perforate a portion of the prostate, and, in these cases, serious consequences are not likely to result, but a false passage sometimes extends between the bladder and the rectum, allowing escape of urine, and infiltration of the connective tissue about the neck of the organ. The least serious, and the most common form of false passage is made when the point of the catheter simply passes through the wall of the urethra for a few lines, and then on being withdrawn from the opening thus made, finds its way into the natural channel. When a false passage is made, the point of the instrument is felt to slip towards one side of the urethra; the patient complains of sudden pain, and blood escapes by the side of the catheter. A rough sensation is communicated to the hand of the surgeon, who finds that although the catheter has passed deeply, its point is not in the bladder. If the instrument has passed between the rectum and bladder, its point will be felt by the finger introduced into the bowel to be much nearer than it ought to be. In order to avoid making a false passage, the point of the catheter or sound, after it has passed beyond the fossa navicularis, should be kept against the roof of the urethra. No force should be used, but the instrument should be allowed to find its own way into the bladder. When a false passage has been made, a large catheter should, if possible, be passed into the bladder, and retained there for some days until the laceration has healed. The patient must be kept at rest and placed on low diet. When a false passage already exists, especial care is needed in passing a catheter. The point of the instrument must be kept away from the side of the canal on which the opening is known to be, lest it should become entangled in the valve-like portion of the mucous membrane which forms part of the wall of the false passage. If an obstruction be met with at this spot, the instrument must be withdrawn, and by changing its direction the opening into the false passage may perhaps be avoided.

Cystitis of a severe and even fatal kind has been known to follow the passage of a catheter or sound. It is only reasonable to assume that in the majority of such cases an undue amount of violence was the cause of the mischief. The symptoms are those of acute inflammation of the bladder; they must be treated according to the directions to be found in the chapter devoted to that subject. Hæmorrhage may be caused by catheterism, and the blood may come from the urethra, prostate gland, or bladder itself. Hæmorrhage is a constant symptom of a false passage and of injury to the prostate, especially in cases where the gland is enlarged. The bladder itself may also be injured by a catheter or sound roughly manipulated. Such accidents can generally be avoided; when they occur, they must be treated in the manner laid down in the chapter on Hæmaturia. As a general rule, it is unadvisable to attempt to pass a catheter when the penis is much congested or in a state approaching priapism. Hæmorrhage is almost certain to be caused, and it will be almost impossible to find the passage into the bladder. False passages, rupture of the urethra, and extravasation of urine are the probable consequences of attempts at catheterism under such circumstances.[1]

Orchitis is an occasional consequence of the passage of a catheter or sound. It is probably produced by passing too large an instrument and over-distending the canal of the urethra.

The constitutional symptoms which sometimes follow the passage of a catheter are (1) rigors with febrile reaction, occasionally followed by suppression of urine, (2) syncope. The rigors, with their subsequent train of symptoms, have been described as urinary or urethral fever. Such symptoms are by no means infrequent in patients suffering from various affections of the urinary organs, especially of the kidney, and they are very apt to be excited in such cases by the passage of a catheter, or operations upon the deeper parts of the urethra. They are sometimes, however, produced by the introduction of instruments even where there are no indications of any kidney affection. The symptoms vary much in intensity; when marked, they closely resemble those of a severe attack of intermittent fever. A distinct rigor is followed by febrile reaction, and this by profuse sweating. The rigor is generally accompanied by nausea and vomiting, pain in the back and headache, and may supervene almost immediately, or not until some hours after the introduction of the instrument. In milder forms the symptoms are less severe. There is perhaps a little headache, nausea and chilliness, followed by slight feverishness and perspiration. On the other hand, in the severest forms, the attack may resemble remittent fever of the gravest type, and may prove fatal in a few hours. Between the most severe and the slight types the symptoms are extremely variable as regards duration and severity. The graver forms have been generally noticed to occur in patients suffering from stricture or other form of urinary obstruction, with some affection of the kidneys as a complication.

The cause of the symptoms of urethral fever is very obscure, but it seems probable that by reflex action, the passage of an instrument along the urethra induces, under certain circumstances, a disturbance in the circulation of the kidney, or interferes in some manner with its function. It was formerly supposed that the symptoms were due to absorption of urine from a wound or excoriation in the urinary passages, but on the one hand, the rigor often occurs before any urine is passed, and on the other, it is not found that rigors invariably follow lithotomy or internal or external urethrotomy, after all of which operations urine is more or less frequently in contact with a wounded surface of the urethra.

---

[1] For an account of cases in which these accidents occurred, generally as a consequence of the patient's own efforts, see Fleming's Injuries and Diseases of the Genito-Urinary Organs, p. 112.

In the majority of instances the symptoms pass off, leaving the patient more or less exhausted. They may, however, recur at uncertain intervals. In other cases fatal symptoms supervene within a few hours. Mr. Banks,[1] of Liverpool, has recorded several cases of this kind. In one case, that of a man suffering from stricture, the passage of the catheter was followed by a severe rigor, the patient rapidly passed into a state of profound syncope and died in a few minutes. In a second case of a similar affection, death occurred in six and a half hours after catheterism. On examination it was found that no injury had been done to the urethra, and the only morbid appearance was slight congestion of the kidneys. Sir H. Thompson[2] mentions a case of the same kind. A man with an old-standing and narrow stricture died fifty-four hours after the passage of an instrument which had been habitually used at least a hundred times before. No damage whatever was found to have been inflicted on the urethra, rigors and vomiting commenced about an hour after catheterism, and not another ounce of urine was secreted from that time until death. On post-mortem examination, the kidneys were found to be congested to an extraordinary degree, and their substance was so soft and pliable as to give way under very gentle pressure. No sign of inflammation existed in any part of the excretory urinary apparatus. Dr. Roberts[3] records another case in which death in twenty hours, with total suppression, followed catheterism in an old case of stricture, when instruments had been repeatedly used before without any ill effects. The only post-mortem appearance of consequence was intense congestion of the kidneys. To this congestion Dr. Roberts attributes the suppression of urine. He thinks that disturbance of the innervation of the organs is probably the primary cause, and possibly also in many cases the direct cause of the suspension of the secretion. The fatal issue is not due to the non-elimination of the urinary excreta, for death is sometimes too rapid to be thus caused. Mr. Banks also points out that even when no urine has been secreted and time has been given for the non-eliminated urea to act noxiously, the symptoms have not been those of uræmic poisoning.

When the symptoms of urethral fever are but slight, very little is required in the way of treatment. The patient should be kept warm in bed, warm bottles applied to the feet, and hot tea or coffee given to drink. When diaphoresis appears, a full dose of quinine should be administered, and it may be repeated after a few hours' interval. Quinine may also be given before catheterism, should rigors have followed a previous operation. It is worthy of notice that patients who have suffered from intermittent fever are especially liable to rigors after operations on the urethra, and in these cases quinine is especially indicated both before and after catheterism. Aconite has also been highly recommended for preventing and treating rigors. Its value for this purpose was first brought under the notice of the profession by Mr. Long, of Liverpool, who used to give two minim doses of Fleming's tincture every few hours. It may be combined with quinine. Opium has been recommended, but should be given with caution, on account of the possibility of a renal complication.

When symptoms of suppression of urine occur, the case is far more serious. A hot bath would appear to be the best treatment, and this

---

[1] In a paper "On Certain Rapidly Fatal Cases of Urethral Fever after Catheterism." Edinburgh Medical Journal, June, 1871.
[2] The Pathology and Treatment of Stricture of the Urethra, 3d edition, p. 94.
[3] Urinary and Renal Diseases, p. 25.

may be supplemented by hot fomentations and poultices to the loins. A hot-air bath, if it can be obtained, may be tried instead of the hot water. If there should be any tenderness in the loins, dry cupping will probably be serviceable as an adjunct to the warmth. A saline purge is also indicated, and its action may be assisted by warm enemata. These remedies will tend to relieve the kidneys. Diuretics are inadmissible, and opium in any form is strongly contra-indicated.

Some years have elapsed since the first attempt was made to gain a view of the interior of the bladder, by passing into it a tube illuminated by a lamp at its outer extremity, and closed by a piece of glass. This instrument, called an endoscope, was first used in England for the purpose of examining the urethra, by Mr. Avery,[1] of the Charing Cross Hospital. M. Desormeaux, of Paris, however, some years afterwards, devised an improved instrument, which has been still further modified by Dr. Cruise, of Dublin, and Mr. Warwick. At present, the endoscope would appear to have fallen into disuse. Probably, in this country at least, it has been submitted only to a very partial trial. Sir H. Thompson appears to regard it as an instrument of but little value. The urethra may be explored by its means, and changes of color and texture of the mucous membrane can be easily recognized. Sir H. Thompson states that no one has yet been able by its means to identify the verumontanum, but, on the other hand, Dr. Grünfeld,[2] of Vienna, declares that he has repeatedly succeeded in discovering this portion of the urethra. With regard to the bladder, very little can be seen on looking down the endoscope. Only a very small portion of the posterior wall of the cavity is immediately opposite to the opening of the tube, and the presence of urine would prevent the mucous membrane from being clearly seen. A portion of a calculus or other foreign body might become visible, but far more reliable evidence of its presence would be given by a sound.

A form of endoscope for the female bladder has been devised by Rutenberg.[3] In order to obviate the difficulty caused by the presence of urine, he conceived the idea of emptying the bladder, and then, by means of a contrivance connected with the endoscope, of distending the organ with air. Chloroform is required, for the operation causes some amount of pain. It is said, however, to be quite free from danger. By means of this instrument, the posterior and lower part of the bladder can be very clearly seen. Rutenberg describes it as of a dusky grayish-red color before air has been injected. After the insufflation, the color becomes more decidedly red, and the numerous plexuses of blood-vessels are clearly seen, and also the muscular fasciculi. Towards the base, the color is somewhat dull, the blood-vessels more abundant, and the muscular layer more distinct. Dr. Winckel regards this instrument as a valuable aid in diagnosis: he states that by its means it is easy to recognize hyperæmia, ulcers, sloughs, etc., and to apply medicaments with preciseness to any spot that may be desired.

The prostate gland may be examined by the finger introduced into the rectum, and by a sound or catheter passed along the urethra. Its posterior or rectal surface is smooth or slightly convex; the two lateral lobes, with a shallow depression between them, can be usually distinguished by making slight pressure with the finger. On passing the

---
[1] About the year 1850, according to Sir H. Thompson. Diseases of the Urinary Organs, 5th edition, p. 16.
[2] Der Harnröhrenspiegel. Wiener Klinik, Nos. II. and III., 1877, p. 60.
[3] Winckel, Handbuch der Frauenkrankheiten, p. 18.

finger through the sphincter ani, the palm of the hand being upwards and the patient lying on his back, the posterior extremity of the bulb and the membranous urethra are first felt, then the apex of the prostate, widening out into the body of the organ. The base of the prostate may be felt by the finger, and sometimes a portion of the base of the bladder, if the latter organ be distended with urine. The examination will be facilitated by making pressure with the other hand above the pubes. In its normal condition the prostate is firm and fleshy to the touch, and its examination causes no pain.

When the prostate is enlarged, a long catheter with a large curve will be required, as with an ordinary instrument it will be found difficult, if not impossible, to reach the bladder. It is also especially necessary that the point of the catheter should be kept against the upper wall of the urethra, and its passage along the prostatic portion of the canal will be much facilitated by the finger introduced into the rectum. The bladder in these cases will usually be found to contain more or less urine, even after the patient has done his utmost to empty it.

The bladder in the female admits of much more minute examination than is possible in the male subject, for, owing to the shortness and dilatability of the urethra, almost the whole of the internal surface of the bladder can be explored with the finger. Owing to the same causes, there is seldom any difficulty in passing a catheter or sound. The patient is placed on her left side, near the edge of the bed, and with her knees flexed. The surgeon, holding the catheter between the right index finger and thumb, feels for the orifice of the urethra which is below the clitoris and just above the entrance into the vagina. The catheter, guided by the finger passed a little way into the vagina, is then introduced. If necessary, the left forefinger may be used as a guide for the introduction for the catheter.

By means of a vaginal examination, various morbid conditions of the bladder may be detected. Deficiencies in the vesico-vaginal septum, and prolapse of the bladder may be thus recognized. A calculus also, especially if large, may often be readily felt, and the various tumors to which the bladder is liable are more or less accessible to examination through the vagina. Such an examination, a sound being also passed along the urethra, is sufficient for the diagnosis of prolapsus vesicæ and vesical fistulæ, opening into the vagina. This examination by means of a sound enables us also to ascertain various points connected with the condition of the bladder; its size, the softness or hardness of its walls, and its sensitiveness. In a healthy bladder, the sound may be introduced for four or five inches, measuring from the external orifice of the urethra, without causing any pain, and after bringing the instrument into contact with the coats of the organ, it can be pushed in for about an inch further, owing to the elasticity of the parts. When the pressure is taken off, the same force causes the instrument to be ejected to the same extent. When the bladder is inflamed, it resists any degree of pressure with the sound, and the examination is very painful. In the normal condition, the introduction of a sound, if accomplished with care, causes no pain, and the instrument can be freely moved about in all directions. If, however, there be even a slight degree of inflammation, the mucous membrane becomes very sensitive. As a matter of course, in nervous hysterical women the parts may be exceedingly sensitive, and yet free from any structural change.

The presence and size of calculi and tumors may also be determined by means of a sound, but for establishing the diagnosis in these cases the introduction of the forefinger into the bladder is the most trustworthy means. In the female the canal of the urethra is capable of considerable dilatation, which can be effected in various ways. The late Professor Simon,[1] of Heidelberg, perfected a method by which the interior of the bladder can be explored with little, if any, risk of producing incontinence of urine or other mischief. His system comprehends rapid dilatation of the urethra by means of a series of plugs or specula made of hard india-rubber. These plugs are of seven different sizes; the smallest is three-quarters of a centimeter in diameter, the largest, two centimeters. The end of the plug or dilator can be removed, and the remainder of the instrument then forms a small speculum.

Before the dilators are introduced, the orifice of the urethra should be slightly divided, as this opening is the narrowest and most unyielding portion of the canal. Dr. Simon's directions are that two lateral incisions, each a quarter of a centimeter deep, should be made in the upper border, and another, half a centimeter deep, into the urethro-vaginal septum. These incisions much facilitate the introduction of the instruments and subsequently of the finger, and prevent laceration; they are best performed with a pair of scissors. By their means also the urethra will not only be enlarged, but also slightly shortened, and these objects are of considerable importance, inasmuch as the thickest part of the finger occupies the urethra during the examination. There is no risk of producing incontinence, as very few, if any, of the circular fibres of the urethra are divided. The incisions soon cicatrize, and the enlargement of the urethral orifice remains permanent.

Dilatation of the canal of the urethra forms the second portion of the operation. This may of course be effected by means of the finger, the little finger being introduced first, and afterwards the index-finger. Dr. Simon's plugs, however, being perfectly smooth, are much easier of introduction, and dilate the canal in a much more uniform manner. Mr. Christopher Heath,[2] who recommends dilatation to be effected by means of the finger, confesses that in all cases the mucous membrane is split beneath the pubes, and there is usually some degree of incontinence for twenty-four hours. The use of Dr. Simon's plugs is not attended with this inconvenience; the patients, with very few exceptions, are able to retain their urine immediately after the operation. The dilatation is effected in from five to seven minutes, and after the largest plug has been introduced, there is no difficulty in passing the finger into the bladder.

In the third stage of the manipulative proceedings, the forefinger is introduced into the bladder. The examination will be facilitated by passing the middle finger into the vagina; in this way the forefinger is enabled to penetrate further into the organ. With the left hand, pressure should at the same time be made over the hypogastrium so as to bring the bladder towards the finger. Every part of the organ is thus made accessible to examination; the upper lateral portions alone are somewhat difficult to reach. As a matter of course, incontinence of urine may follow this process for dilating the urethra, but if the directions are properly carried out, it is of very rare occurrence. It is important to ascertain the extent to which dilatation may be safely carried. This has

---

[1] Ueber die Methoden, die weibliche Harnblase zugängig zu machen, Volkmann's Sammlung klin. Vorträge, No. 88.
[2] Med. Times and Gazette, April 11th, 1874, p. 391.

also been determined by Professor Simon. He states that in adult females the urethra may be dilated by means of plugs having a diameter not exceeding two centimeters. In two cases, in which the removal of papillomata was the object to be attained, he carried the dilatation one centimeter further. Incontinence of urine, though not permanent, was the result in both cases. In girls from 11 to 15, the dilatation may be made with plugs having an extreme diameter of 1.5 to 1.8 centimeters, and from 15 to 20, the diameter must not exceed 1.8 to 2 centimeters. These measurements are equal to that of an ordinary forefinger together with that of the handle of an instrument. Chloroform is generally necessary for the first operation, but afterwards the bladder can sometimes be examined without the aid of an anæsthetic. The patient is to be lying on her back.

Where dilatation of the urethra is not sufficient for the purpose to be attained, and where a tumor or large calculus is suspected, the bladder may be made accessible both to eye and finger, by incising the vesico-vaginal septum and inverting the organ through the opening thus made, which should be T-shaped. The horizontal incision should be about half a centimeter in front of the os uteri, and from the middle of this another incision is carried, at a right angle, as far as the upper end of the urethra. Through an opening of this size a large calculus may be removed, or the bladder may be inverted by means of hooked forceps, when the incisions are made for the purpose of removing tumors from its interior.

According to Dr. Simon, dilatation of the female urethra, as above described, may be practised for the following objects:
1. The diagnosis of diseases of the mucous membrane of the bladder, and of calculus and other foreign bodies. The advantage of being able thus to examine the interior of the organ is obvious. Mr. Heath records a case in which nothing could be detected by the sound, but on introducing the finger a tiny calculus was at once felt, and removed with the aid of a scoop. 2. The removal of calculi and foreign bodies. The operations necessary in these cases will be much facilitated by the introduction of the finger. 3. For applying various remedies to the internal surface of the bladder, and for treating fissures of the urethra. 4. For the diagnosis of the position and attachments of tumors in the vesico-vaginal septum, and for the removal of tumors, especially papillary growths, from the walls of the bladder. 5. For the discovery and removal of calculi from the vesical extremity of the ureters. 6. For the opening of hæmatometra, the evacuation of which between the bladder and rectum is impossible or dangerous, as, for example, in cases of congenital absence, either partial or complete, of the vagina. 7. For the treatment of vesico-intestinal fistula.

The indications for incising the vesico-vaginal septum are less numerous, but the operation may be advisable or necessary: 1. For the removal of very large stones and tumors which cannot be dealt with by simply dilating the urethra. 2. To provide for the escape of urine in very obstinate catarrhal affections of the bladder, accompanied by ulceration. 3. For dealing with vesico-intestinal fistulæ, which cannot be treated in any other way.

It has even been proposed to carry the examination further, and to attempt the relief of obstruction in the ureters by passing a catheter along those canals. It has also been proposed to compress, by means of an instrument remembling a lithotrite, the vesical opening of the ureter on one side, so as to prevent the escape of urine from it, and in this way to collect the secretion from one kidney only. Were such a proceeding readily feasible, valuable information might be obtained in cases of disease affecting one kidney: such an experiment, however, is almost impossible on the male subject. In the female, however, it is easy to recognize with the finger the openings of the ureters. The urethra is first dilated, the finger is then introduced and the raised band which connects the orifices of the ureters is sought for. Guided by this elevation, which runs across the base of the bladder at a distance of one inch from the internal aperture of the urethra, a catheter of special construction can be passed into the opening of the ureter, in a direction upwards and outwards. The handle of the catheter is brought across to the opposite side and raised towards the pubic arch. Its point may easily become entangled in the mucous membrane of the bladder, instead of passing into the ureter, but in the former case it will be readily distinguished by the finger. The catheter is to be passed onwards towards the kidney in a direction upwards and outwards, until it touches the bony margin of the pelvis, at a height of seven or eight centimeters. The handle must now be brought down close to the inner side of the thigh on the same side, while the catheter is directed upwards in a line with the vertebral column, but with its point slightly towards the wall of the abdomen. With very slight force, the catheter will pass up the ureter and into the pelvis of the kidney.

This experiment has been performed seventeen times by Professor Simon, and in all but two instances the pelvis of the kidney was reached without difficulty. In no case was the experiment attended with any bad consequences. The instruments used are a sound and catheter, each twenty-five centimeters long, and passing through a handle furnished with a screw, by means of which the length of the instrument to be passed can be regulated. The instruments are made of metal and are only very slightly flexible. They have a slight curve near the point.

An examination, made as described, would be useful for the diagnosis of stone in the ureter or kidney. A concretion lying near the commencement of the ureters could be pushed back so as to occupy a more advantageous position. Strictures of the ureters could also be dealt with, and an attempt could be made to relieve some forms of hydronephrosis by drawing off the fluid. It is, however, very difficult to insert the point of the catheter into the vesical opening of the ureters; Professor Winckel[1] states that he has made several attempts, but has never been successful.

---

[1] Op. cit., p. 16.

CHAPTER III.

ABNORMITIES OF THE BLADDER.—ABSENCE OF THE BLADDER.—MULTIPLE BLADDERS.—EXTROVERSION.—PERVIOUS URACHUS.

THOSE organs which are placed in the median line of the body are more subject than the limbs to certain deviations from their normal development. The bladder, though so important an organ, does not escape the influence of this law. It is subject to several abnormities, some of which are irremediable, while others admit of relief by surgical means.

In some cases of congenital malformation the bladder is completely wanting. This anomaly is always attended by more or less malformation of other organs. In the more complex examples, the kidneys and ureters are likewise absent, and the external organs of generation imperfectly developed. When the kidneys exist the ureters are also present, and may be either incomplete, forming canals with closed extremities, or may open into the urethra, the vagina, or the rectum. The abnormal opening of the ureters does not, however, necessarily imply complete absence of the bladder, which may be imperfectly developed and in a rudimentary state. In a case of complete absence of the bladder, lately recorded by M. Fleury,[1] of Clermont, the ureters opened into the urethra. The patient was a girl who had menstruated regularly for two years, and had suffered from incontinence of urine from birth. This latter symptom had become very distressing, and the external genital organs and the thighs were covered with severe erythema. A catheter could be introduced for an inch and a half only, and caused great pain, but no urine came away. The girl died from peritonitis, and on post-mortem examination no trace of a bladder could be found. The urethra was an inch and a half long, and terminated posteriorly in a cul-de-sac into which the ureters opened. Professor Mayer, of Bonn,[2] has published a very remarkable example of absence of the bladder which he observed in the case of a new-born child. The external organs of generation were replaced by a small pedunculated sac, filled with cellular tissue. There was no trace of an anus, the intestinal canal terminated abruptly in the descending portion of the colon. The kidneys, ureters, and bladder were completely absent, but the supra-renal capsules were twice as large as natural. There was no trace of the prostate gland or seminal vesicles. A case somewhat similar is recorded by Dr. Gross.[3] Nearly the whole mass of the small intestines was contained in a tumor attached close to the umbilicus, in which they terminated. The entire colon was wanting. No trace of the bladder could be detected anywhere. Both ureters terminated in the sac which contained the intestines; the pubic bones were absent, and the anus was imperforate.

---
[1] New Syd. Soc. Biennial Retrospect, 1873–74, quoted from Gaz. Hebd. No. 6, 1874.
[2] Tiedemann's Zeit. für Physiol., Bd. 2, s. 1.
[3] Gross, A Practical Treatise on the Diseases, etc., of the Urinary Bladder, 3d edition, p. 356.

The case recorded by Uytterhoeven [1] as one of absence of the bladder is really an instance of atrophy of the organ. A woman aged forty had suffered from incontinence of urine for nearly thirty years. She was found to be the subject of urethro-vaginal fistula, through which the urine passed immediately it escaped from the ureters. The bladder had become atrophied from disease, and all that remained of it was a minute sac not larger than an ordinary pea. The woman stated that in early childhood she had been able to retain and pass her urine naturally. In a case recorded under the same heading by Mr. Vost [2] there was extroversion with atrophy of the bladder. The patch of mucous membrane which represented the bladder was as large as a florin.

The sacculated condition of the bladder, described by some anatomists under the term bilobed, is probably rather an acquired than a congenital state, the septum being produced by elongation of the mucous membrane; but, from arrest of development at different stages of intra-uterine life, the bladder may appear to be double, or it may be fissured along the middle line.

Cases have been reported in which two or more bladders co-existed in the same individual, but such abnormities are exceedingly rare. A double or triple bladder can be said to exist only when each division has a ureter of its own, and all the natural coats of the bladder enter into its structure. A case of this kind has been described by Blasius. The bladder was completely divided into two sacs, opening into a common urethra, and each having its own ureter. Scibelli, of Naples, has described a triple bladder, and Molinetti has recorded the case of a woman who had five bladders, six ureters, and five kidneys. It is probable that numerous other cases which have been referred to this category were instances of sacculation, the result of various morbid conditions, such as stricture of the urethra, prostatic enlargement, and other obstacles to the escape of urine. A curious case reported by Heyfelder is not an instance of double bladder, though the separation was almost complete. In the bladder of a boy, six years of age, on whom lithotomy had been performed, no stone could be found, although previous to the operation its presence had been detected by the sound. The case terminated fatally, and on examination it was discovered that the bladder was divided into two parts communicating by a small opening through which a sound could be passed. The ureters opened into the lower part of the anterior portion. Mr. Holmes, in his "Treatise on Surgery," p. 733, describes a case of malformation in which a large congenital cyst, or false bladder, communicated with the true bladder. The walls of this secondary cyst were very thin, and destitute of muscular fibres. Cases of this kind are interesting from a surgical point of view, on account of the difficulty which would be experienced in extracting calculi which are very liable to become formed in such cysts.

The bladder has in a few instances been found imperforate, but such an anomaly is extremely rare, and, like those just described, admits of no interference on the part of the surgeon. It remains to describe a much more common and interesting malformation, viz., extroversion of the bladder. This anomaly depends on imperfect development, whereby the union between the lateral moieties of certain organs remains incomplete in the median line. In the cases now alluded to, there is a deficiency of the anterior wall of the bladder, and as this deficiency is accompanied by

---

[1] Presse Méd. Belge, 1860, No. 29.   [2] Lancet, 1875, vol. ii., p. 265.

imperfect union of the abdominal walls, the result is the abnormity denominated fissure, or, more commonly, extroversion of the bladder. The latter term, however, does not sufficiently express the nature of the affection, for the bladder is deficient as regards its walls, as well as misplaced. The malformation occurs much more frequently in males than in females. Of sixty-eight cases collected by Mr. Earle, sixty had occurred in males, only eight in females.

M. Geoffrey St. Hilaire found about one-fourth of the cases appertaining to females, and the experience of subsequent observers has been to the same effect. Mr. M'Whinnie states, in an interesting paper in the *Medical Gazette*, that out of nine cases which had fallen under his observation, two only were females. Of eight cases examined by Sir H. Thompson, six belonged to the male sex. Indeed, the fact that females are much less liable than males to this malformation, had been long before pointed out by Dr. Duncan. No attempt has been made to explain why the male should be more subject than the female to this abnormity. It must, however, be noticed that the defect is seldom confined to the bladder alone, but extends to other parts of the genito-urinary system; and the more complex or complete a system of organs, the more it is subject to deviations from the normal type.

With regard to the cause of this abnormity very different views have been held by various writers from time to time.¹ Bursting of the fœtal bladder from over-distention at an early period, the abdominal walls being still incomplete, has been assigned as a cause. Others have suggested that it may be due to ulceration of the abdominal walls and bladder, between the second and third months of intra-uterine life. M. Vrolik (*Archiv für Anat.*, 1868, p. 165) supports the theory that extroversion is due to an arrest of development of the bladder and urachus from the fœtal allantois, and he points out the analogy which exists between this abnormity and other fissures of a congenital character in the median line. As the allantois expands into the urinary bladder, the urogenital sinus, or cloaca, which is the termination of the common intestinal cavity, becomes divided into two by a septum, so as to form the bladder in front and the rectum behind. Arrest of development of this septum produces those malformations in which the bowels and bladder open into each other, or the ureters open into the rectum. "If, about this period of gestation, by inflammatory change, or adhesion, or some degenerative process arising from one of those specific diseases, such as syphilis, to which we know the fœtus *in utero* to be liable, the normal progress is arrested in the outer abdominal or amnionic layer of the allantois, without affecting the inner or mucous layer, the result at time of birth would be the formation of the deformity termed by Vrolik '*ectopia vesicæ*'; i. e., a fissure of the hypogastric wall only, with a completed condition of the walls of the bladder itself. But if the morbid process extends also to the subjacent portions of the internal cylinder or mucous layer of the *allantois*, then, according to the position or extent of the abnormal change, the result would be a more or less extreme degree of *epispadius*, an open or imperfect urachus, or a completely fissured and extroverted bladder, with separation of the pubic bones and the other changes associated with that deformity."

---

¹ For an excellent account of extroversion of the bladder, with cases treated by plastic operations, see Mr. John Wood's paper in the Medico-Chirurgical Transactions, 1869, p. 85 et seq. The *résumé*, given in the text, of the causes of extroversion has been mainly obtained from this source.

Extroversion of the bladder presents several varieties of form and degree, depending on the amount of deficiency in the vesical or abdominal walls, and on the abnormities of other organs by which it may be accompanied. It should also be remarked that the deficiency in the anterior walls of the bladder and abdomen is attended by more or less displacement of the urinary reservoir; its posterior wall is pushed forward by the intestines, becomes everted, and presents itself in the form of a round red tumor, at the lower part of the abdomen. In the more simple form of extroversion there is merely a deficiency of the vesical and abdominal walls; in the more complicated form, which is likewise the more frequent, the pubic bones are separated or deficient at the symphysis, and other anomalies occur which will be presently noticed.

The tumor formed by the everted bladder is characteristic, and cannot readily be mistaken for any other organ. It exists at the lower part of the abdomen, on the median line, or more commonly in front of and just above the pubes.

In this situation it forms an oval or rounded swelling, the surface of which is red, vascular, and in most cases can be seen to be composed of mucous membrane. From the lower part of the tumor the urine dribbles away almost constantly: and on examining the swelling at this part, we discover two small eminences, perforated by openings which are the orifices of the ureters. The tumor is often increased by coughing, straining, etc., during which a portion of the abdominal viscera may become engaged in the sac formed by the everted walls of the bladder. In some cases, by gradual and continuous manipulation, the tumor may be partially returned into the abdominal cavity, through the fissure between the recti muscles. Hernia of the bowel is often present at the sides of and below the extroverted bladder.

The size of the tumor varies at different times in the same subject, but chiefly according to the patient's age, or, in other words, according to the duration of what, practically speaking, may be called his disease.

In new-born children the swelling does not project much, and rarely exceeds the size of a walnut. In the adult, as can be readily understood, the tumor gradually attains a greater volume, and in many cases becomes as large as the fist, being two or three inches in its vertical, and four or five inches in its transverse diameter. Although exposed for years to external influences, the mucous membrane of the everted bladder undergoes much less change than might be supposed. At first it is somewhat tender and bleeds readily, the hæmorrhage being in some cases profuse, but it soon gets accustomed to external impressions, the vascularity diminishes, the surface is covered with a coat of mucus, and in old cases the membrane may become quite smooth. The gait and posture of the patients are somewhat characteristic. The thighs are kept wider apart than usual, and the body leans forward so that the affected part may be relieved from the pressure of the clothes.

In a few rare cases, as above mentioned, the malformation is confined to the anterior walls of the bladder and abdomen.[1]

---

[1] A case of ectopia, in which only the abdominal walls were involved, is quoted in New Syd. Soc. Biennial Retrospect, 1873-74, p. 323, from Lichtheim, Archiv für Klin. Chir., xv. The bladder lay in a depression of the abdominal wall, was covered externally by mucous membrane, was closed, and could be replaced. The boy, aged 8, could retain his urine perfectly, and passed from six to seven ounces at a time. The mucous membrane covering the bladder had several layers of flattened epithelium resting on a soft basis of connective tissue. The ends of the horizontal rami of the pubic bones were united only by bands of ligament.

In the great majority of cases, however, the abnormity is of a complex kind, involving the pubic bones and external organs of generation. The parts appear to follow the same law, under the influence of which the bladder and recti muscles remain separate; their deviations consisting mainly in a separation of the lateral moieties, on the median line. In such cases the pubic bones are separated from each other at the symphysis to the extent of several inches, or rather the body of the pubis is absent on each side; the deficiency may even amount to five inches, but during life the vesical tumor and hernia accompanying it often prevent us from determining the state of the pelvic bones in a satisfactory manner. The extensive separation of the pelvic bones necessarily involves some anomaly of the male organs of generation, which are so peculiarly connected with them. The penis is usually in a rudimentary state; and as the urethra is not required for the emission of urine, it opens in various directions, in the male above the penis, in the female above the clitoris. Epispadias, more or less complete, usually co-exists with this abnormal situation of the urethra. The testicles are often retained in the abdomen, or they may occupy an imperfect scrotum. In some cases, the external organs of generation are completely wanting, and the imperfect tegumentary sac which represents the scrotum, often contains a portion of the intestinal canal.

Extroversion of the bladder is not only less frequent in the female, but it is less apt to be attended by remarkable deviations of the genital system. The clitoris and labia are usually in a state of imperfect development, and prolapsus of the uterus is very liable to occur from the separation which exists between the pubic bones.

The records of surgery contain a great number of cases of this affection. In one of these, related by Mr. Erichsen, the penis was rudimentary, with complete epispadias; the urine constantly distilled from the open mouths of the ureters at the lower part of the swelling, and fell, during the erect posture, in part upon the open groove of the urethra, which served imperfectly as a spout. In each groin was a projection which became more marked as the growth of the body advanced, corresponding with the widely separated pubic bones. From the want of the symphysis and consequent support in front of the pelvis, there resulted considerable weakness in the part, and much awkwardness in motion. These inconveniences, as well as the general deformity, were further increased by the distance to which the thighs were thrown apart from one another. This case is particularly interesting from the fact that an account of the post-mortem examination has been published by Mr. M'Whinnie in the *Medical Gazette*, the patient having died of phthisis. "In the groin," he observes, "on each side of the malformed genito-urinary organs, is seen the prominence corresponding with the body of pubis. The general aspect of the body, as well as the signs manifested during life, showed that the period of puberty had not been retarded. There was no appearance of umbilical cicatrix; the recti abdominis, in the upper two-thirds of their extent, were about an inch apart; below this a triangular space was left by the divergence of each muscle to the corresponding pubic bone, which was separated from its fellow to the distance of about two inches and a half. Through this fissure the irregularly nodulated mucous surface of the posterior wall of the bladder protruded; its muscular fasciculi had acquired considerable strength. The mouths of the ureters, directed towards each other, were surrounded by prominent papillæ. On looking into the pelvis, the vacuity there was very striking; the space between the anterior wall of the abdomen and the rectum, ordinarily filled by the bladder, being here entirely unoccupied.

"At the upper margin of the vesical tumor, the obliterated umbilical arteries converged to come into contact with the remains of the umbilical vein, which took a longer course than usual to reach the liver.

"The pubic bones, often connected by strong ligaments in cases where they are disjoined, and which in some measure counterbalance the want of osseous union, had merely a few thin fibrous bands passing between them, quite inadequate to give fixity to the parts, or to support the viscera.

"The rudimentary penis measured about one inch and a half in length; the glans, cleft at the upper part, had appended to it below an imperfect preputial covering. The epispadias was complete; the urethra presenting a simple open groove or furrow, upon the surface of which several follicles opened. The prostate, deficient above, was of tolerable size beneath the canal. In this part of the urethra the seminal ducts terminated in the usual manner.

"The testes, though diminutive, were, as in most analogous cases, natural in structure. They had descended into a small contracted scrotum. The vasa deferentia had their proper relation to the vesiculæ seminales, which were below the natural size. The ureters, sometimes so enormously dilated, as if designed almost to compensate for the want of the proper reservoir, were quite natural. Each corpus cavernosum diverged from its fellow; and, taking a more transverse direction than usual, was connected by its crus with the corresponding ischium, where it was covered by a few faint fibres of an erector muscle. Some slender elongated fibres occupied the situation of the transversus perinei. As in many other instances, the anus was immediately behind the scrotum."

Mr. Wiblin, of Southampton, has recorded, in the *Medical Times*, an interesting case of this deformity, illustrated by a drawing, exhibiting the appearance presented upon the front surface of the abdomen. The patient, twenty-eight years of age, of robust and masculine appearance, was employed as a coal-lifter on the quay. At a very early period of his life he had been examined by several medical men, who conveyed to his relations the impression that he was an hermaphrodite; and under this appellation he had passed the latter years of his life, subjected to the most painful annoyances. His dress consisted of a very long smock-frock, but in consequence of the great inconveniences occasioned by the dress being sometimes found frozen to his person, this attire was abandoned for high leggings. The pubic bones were separated to the extent of at least three or four inches, and a distinct cartilaginous band passed across from one angle of the pubic symphysis to that of the other, and in a measure tended to compensate for the weakened condition of the parts which such a separation must necessarily produce. The rudimentary penis was about one inch and a half in length, and had a perfectly formed corona glandis; the prepuce was bifid, and had the appearance of two wart-like appendages. Upon the upper surface of the organ was a groove, in which were seen numerous lacunæ, and at its base the termination of the ejaculatory ducts, on either side of the caput gallinaginis. Through the extensive triangular opening in the inferior parietes of the abdomen the posterior wall of the bladder protruded. When the patient assumed the erect posture the swelling was about the size of a very large orange. The mucous membrane of the upper portion of the tumor was highly vascular and shining. On the inferior and lateral portions of the protrusion there existed an exceedingly sensitive and painful strawberry-like pulpy surface, and below this on either side, the external openings of the ureters, from which the urine dribbled away as fast as it became secreted. Above the parts thus described, and immediately under the apex of the triangle formed by the separation of the abdominal parietes, was a small hernial protrusion. In this case, as in those recorded by the late Mr. Earle and Mr. M'Whinnie, there was no umbilicus.

In cases of extroversion of the bladder the principal inconvenience arises from the constant dripping of the urine, which produces excoriation of the parts around and not only renders the patient very uncomfortable, but seriously interferes with his ordinary avocations. Relief may sometimes be afforded by the application of instruments to catch the urine as it escapes from the ureters. An apparatus similar to the one recommended by the late Mr. Earle may occasionally be useful. It consists of a silver bowl which covers and protects the tumor and external organs, and is retained *in situ* by a double truss or appropriate bandage. The collected urine flows through a funnel at its lower part into an india-rubber bottle placed between the knees.

Such an apparatus, however, would be of little service unless it could be fitted very accurately to the parts and retained in exact position at all times, and these conditions are very difficult of fulfilment. Numerous attempts have therefore been made to afford relief by operative proceedings, and two classes of operations have been performed. The establishment of a communication between the bladder and rectum was the object aimed at in the first attempts of this kind, while more recently several surgeons, notably Messrs. Holmes and Wood, have with considerable success attempted to cover the extroverted bladder with flaps of integument taken from the adjacent parts of the abdomen. The inconvenience of passing the urine through the rectum is, of course, very considerable ; moreover, the presence of urine in the bowels is apt to cause much irritation and diarrhœa, and the results of the operation performed by Messrs. Lloyd and Simon are not such as to encourage the belief that the methods they adopted are worthy of a farther trial. In one case Mr. Lloyd passed a seton through the posterior wall of the bladder from the rectum, with the view of obtaining a fistulous communication, and he intended subsequently to close the abdominal opening by paring its edges and uniting them by suture. The patient, however, died from peritonitis on the fifth day after the operation, and on post-mortem examination it was found that the recto-vesical fold of peritoneum descended much lower than usual towards the neck of the bladder and had been perforated by the seton. The case was one in which a plastic operation would have been likely to prove successful. Mr. Simon performed a somewhat different operation on a boy thirteen years of age. He passed canulæ armed with threads through the openings of the ureters and brought the threads out through the rectum. Serious constitutional disturbance ensued, but the patient was eventually able to bear the pressure of a pad in front, by means of which the opening in the abdomen was closed and all the urine forced into the rectum. Peritonitis however set in and became chronic, and the boy died from exhaustion twelve months after the operation. Calculous concretions had formed in the ureters, and had led to great obstruction. In a case of this malformation, treated by Breschet, small elastic tubes were passed into the ureters, and their other ends connected with a receptacle for the urine. The tubes, however, set up so much irritation that they had to be withdrawn.

In the second class of operations, performed for the relief of extroversion, a different object has been aimed at, viz., the partial restoration of the abdominal wall, so that a convenient receptacle may be applied, in order to catch the urine. Dr. Pancoast of Philadelphia, in 1859, was the first to perform a successful operation of this kind, and he was followed by Dr. Ayres, of Brooklyn, and soon afterwards by Messrs. Holmes and Wood, of London. A flap is taken from one side of the abdomen or groin, dissected up, and turned over like the leaf of a book so that the epidermis comes in contact with the mucous membrane. A second flap is then taken from the groin on the opposite side, and its raw surface applied to the raw surface of the former flap. The upper and lower edges of the flaps are then sewn together, and thus a thick bridge is formed over the cleft. A certain time is allowed to elapse, so that the parts may return to their state of healthy nutrition. The upper border of the flap is then implanted into the neighboring edge of the abdominal parietes, the lower border being left free. Below this latter a small opening remains to which a receptacle can be fitted to catch the urine. A similar operation, performed with successful results by M. Maisonneuve at the Hôtel Dieu, is reported by Mercier. The patient was a boy fourteen years of age, and the opening measured two and a half inches in the vertical direction, and one and a half from side to side. When the parts had healed, a small fistulous opening remained above in addition to the aperture left below for the escape of the urine. About a month after the operation, an india-rubber urinal could be conveniently worn, pressure being made upon the upper fistulous opening by means of a small pad.

Mr. John Wood's method [1] is somewhat different. A large flap, nearly square in shape, and embracing the whole width of the exposed bladder, is dissected from the central abdominal region above the umbilical cicatrix and reversed upon the exposed mucous membrane. The thickness of the flap should include all the superficial structures, and its length should be sufficient to cover the whole of the extroverted surface. A lateral, pyriform flap is taken from each groin, with the base turned towards the scrotum or thigh, so as to receive blood from the external pudic and superficial epigastric arteries. These flaps are brought together over the first flap, and their adjacent borders are united in the median line by hare-lip pins which are made to transfix the greater or entire thickness of the reversed flap. The margins of the wounds caused by dissecting out the flaps are then brought together with hare-lip pins and wire sutures, and broad straps of adhesive plaster are placed transversely and obliquely across the abdomen and groins, to draw the integuments towards the median line and lessen the strain upon the sutures. By thus taking a flap from the umbilicus the occurrence of a fistulous opening above is prevented. If the healing process has gone on satisfactorily, all pins and sutures may be removed about the sixth or eighth day, and cicatrization will probably be complete about three weeks afterwards. If the umbilical flap be too short, fistulous openings are likely to remain at the angles, and, as the cicatrices go on contracting for some time, the opening left at the root of the penis may gradually increase in size. Under such circumstances the mucous membrane will again protrude, and there will be a return of the troublesome symptoms. This condition

---

[1] Lancet, 1869, vol. i., p. 259, and 1874, vol. i., p. 198. Med. Chir. Transactions, vol. lii., p. 85 et seq.

of things may be prevented by taking care to have the umbilical flap sufficiently long, and in some cases a further closure may be effected by a subsequent operation which has for its object the formation of a covering for the fissured penis and urethra. To effect this, a flap of skin, like a bridge, is raised from the front part of the scrotum and lower surface of the penis. This flap includes the *dartos*, and is connected with the groin on each side. It is lifted over the penis and placed upon a raw surface prepared by turning down a fold of skin from the lower arched border of the superficial flaps covering the bladder. Another fold of skin may, if necessary, be dissected up from each side of the penis, and turned over with its skin-surface towards the mucous membrane covering the urethral groove. A continuous wire suture is applied to keep the reversed flaps in place, and the transplanted scrotal structures are held in their new position by points of interrupted wire suture pretty closely applied. The sides of the scrotal wound are drawn together vertically by the same means. In one of Mr. Wood's cases the results of the two operations were very satisfactory. As the cicatrices contracted, the orifice of the artificial urethra became more and more tightened and braced up, and urine was found to accumulate in the bladder, while the patient was lying down, in quantity sufficient to be expelled in a stream on rising, to a distance of a few inches from the penis.

Dr. F. F. Maury has suggested and performed a somewhat different plastic operation for the relief of extroversion in the male subject. He takes a flap from the perineum and scrotum by carrying a curvilinear incision from the outer third of Poupart's ligament across the middle of the perineum to a corresponding point on the opposite side. The flap is carefully dissected up until the root of the penis is reached, and the latter organ is slipped through a small opening made in the centre of the flap, so that the urine flows away without coming in contact with the wound. A curvilinear incision is then carried across the abdomen and a short flap dissected up for about an inch, and under this the scrotal flap is passed and attached by suture. In two of Dr. Maury's successful cases a double inguinal hernia was cured by the contraction of the exposed granulating surfaces.

Dr. Gross,[1] from whom the above description of Dr. Maury's operation is taken, considers that, for extroversion occurring in the male, this method is superior to all others. The continuous extent of raw surface, however, which its adoption involves is very great, and the after-treatment must be long and tedious. As a general rule, the size and character of the opening and the condition of the adjacent parts determine the kind of operation to be selected. During the subsequent treatment the patient must be kept almost in a sitting position, with the knees drawn up and supported by pillows so as to secure complete relaxation of the pelvic muscles and abdominal walls, and the parts should be protected as far as possible from contact with urine. Carbolic acid putty has been recommended by Professor Lister, but Mr. Wood has found pledgets of lint dipped in a solution of the sulpho-carbolate of zinc (grs. 3–6 to an ounce of water) to be a more satisfactory application. When the flaps have completely united, a suitable urinal should be adjusted, and the newly formed cavity should be frequently washed out with an injection of dilute nitric acid, in the proportion of two drachms to the half pint of water, in order to prevent the formation of phosphatic concretions.

---

[1] Gross, loc. cit., p. 365.

Mercier suggests that when the cicatrices have acquired a proper amount of consolidation, a pad should occasionally be placed over the artificial opening, in order to produce slight distention of the walls of the bladder and thus to increase its capacity.

The hairs which grow on the reversed surface of the flap may give rise to some inconvenience. They may be detached from time to time with a pair of forceps, but a better plan would be to remove them with some depilatory application some days before the operation. In Mr. Wood's method the small hairs growing on the umbilical flap are not likely to cause much irritation, and they are soon destroyed by contact with the urine.

The operations as above described have now been repeatedly performed, and, in the majority of cases, the results have been very satisfactory. The condition of persons suffering from extroversion of the bladder is such an extremely miserable one that any alleviation of their distress is well worthy of surgical efforts. Of fifty-three cases operated upon by various surgeons, thirty-nine were successful: in eight the operation failed, and in six it led to fatal results. Sloughing of the flaps from defective nutrition is an occasional cause of failure, but the breaking up of the adhesions by the protrusion of the abdominal viscera has more frequently rendered the operation unavailing. In the fatal cases death has been sometimes due to peritonitis, which is likely to supervene if the flaps are dissected too deeply from parts of the abdomen where the walls are very thin.

The last class of abnormities of the bladder consists in perviousness of the urachus, and this condition may be either partial or entire. In the latter case the tube forms an outlet for the urine at the umbilicus. The integuments in front of the abdomen are excoriated, the umbilicus itself is red, and presents at one spot a prominent red papilla or elevation, in the centre of which is an opening through which the urine is from time to time discharged: a probe can be readily passed downwards in the direction of the bladder. This affection is often associated with some congenital obstacle to the passage of urine. A contracted and adherent prepuce has been found to be the exciting cause.[1]

Mr. Paget, of Leicester, has recorded, in the "Transactions of the Royal Medical and Chirurgical Society,"[2] the case of a man, 40 years of age, in whom a ring-shaped calculus, formed upon a hair in the bladder, was extracted through a patent urachus at the umbilicus. The bladder being distended, Mr. Paget introduced his finger through the duct, and removed the calculus without trouble. In this case it was obvious that the bladder and the urachus formed one urinary receptacle which in shape might be compared to a cupping-glass with a curved neck. The calculus was removed in August, 1844, and the patient declined to submit to any operation for the cure of the fistulous opening, which was about an inch in diameter, and surrounded by a thick and almost cartilaginous margin. The nose of a catheter, passed into the bladder, could be easily made to appear at the umbilical opening. The patient wore a girdle, with a thick flannel pad to catch the jets of urine which were apt to occur when he was at work. The man continued in this state until 1860, when, having heard that Mr. Paget had performed a successful operation upon a child with a similar affection, he was anxious that the

---

[1] A case of this kind is recorded by Dr. Charles of Belfast. Brit. Med. Journal, Oct. 16th, 1875.   [2] Vol. xxxiii., p. 293.

experiment should be tried upon himself.' Accordingly, the edge of the opening was pared to an extent of one-third of an inch in thickness and two-thirds of an inch in breadth. The wound was brought together by means of three common curved needles, which were preferred to hare-lip pins, as taking a deeper hold of the parts. Strips of lint were wound round these needles, first in a figure of 8 form, then elliptically, until sufficient breadth of pressure was given, and a catheter was introduced into the bladder. The needles were removed five days afterwards, when the opening was found to be closed. The patient was discharged cured in the fourth week after the operation, and when seen some months afterwards the closure was perfect and complete. In Mr. Paget's other case, which occurred in a girl four months old, the opening would admit an ordinary cedar pencil. The edge was pared and the wound closed with a suture pin. In a few weeks there was complete union of the parts, and two years afterwards the result was found to be perfectly satisfactory.

The application of a ligature to the protrusion has also sufficed to produce permanent closure of the opening. Mr. Thomas Smith[2] records a case treated in this manner. The patient was a boy two years of age.

In those cases in which an obstacle to the flow of urine through the urethra co-exists with the fistulous opening at the umbilicus, the surgeon's efforts must of course first be directed towards the removal of the impediment. In Dr. Charles's case, occurring in a boy a year old, the prepuce was long, contracted and adherent to the glans, and urine was passed with difficulty by the urethra, but flowed freely from the umbilicus during micturition. The phimosis was remedied by slitting up the prepuce, and the operation was followed by considerable improvement as regards the umbilical fistula, to which tincture of iron was subsequently applied. Cruveilhier was of opinion that there is always some obstruction of the urethra in cases of pervious urachus, but Mr. Paget's case sufficiently proves that the malformation is not necessarily connected with any impediment to the natural flow of the urine.

Instead of solely occupying the anterior wall of the bladder, the fissure or deficiency may be seated in the posterior wall of the organ. In all such cases the anterior wall of the rectum is likewise deficient, and a kind of cloaca is formed for the common reception of urine and feces —a malformation representing the normal structure in certain animals. It is almost unnecessary to say that such an abnormity does not admit of operative interference. The case resembles, in one respect, recto-vesical fistula, but the loss of substance is too great to be remedied by plastic surgery. All that we can do is to devise means for alleviating the effects of the malformation.

---

[1] Med. Chir. Trans., vol. xliv., p. 13.
[2] Medical Times and Gazette, March 28th, 1863.

CHAPTER IV.

HERNIA AND DISPLACEMENTS OF THE BLADDER.

DISPLACEMENTS of the bladder are not of frequent occurrence. They consist of perineal displacements, and that very rare form which some writers call "inversion," but which would be more correctly designated by the term "eversion of the bladder." In this latter affection the organ is turned inside out, in a more or less complete manner, and protrudes either through the dilated urethra, or through an accidental opening in some neighboring septum.

Eversion of the bladder with protrusion through the urethra, is an affection which occurs exclusively in females, and is almost peculiar to infants. The displacement is sometimes congenital. The symptoms are usually well-marked and distinct. A small pyriform, elastic tumor, of a deep-red color, and prone to bleed when touched, is observed projecting between the labia, below the clitoris and in the direction of the orifice of the urethra. When the child cries, or when pressure is made on the abdomen, the size of the tumor increases. The orifices of the ureters, perhaps distilling urine, may be discovered on careful examination. The urethra is not distinguishable. More or less incontinence of urine will be described as present. Reduction of the tumor can be effected by manual pressure, but when this is taken off the swelling is prone to reappear.

Several cases of this affection have been placed on record. In the earliest, reported by Dr. Murphy,[1] the subject was a female child, four years of age. The tumor was pyriform, depending by its apex from the upper angle of the labia, of a dark mahogany color, and about the size of a small hen's egg. The orifice of the meatus urinarius could not be detected; but on making slight traction of the tumor, the eversion was rendered complete, and the orifices of the ureters were brought into view. Its return was readily effected by steadying the neck of the bladder with the thumb and forefinger, while the fundus was pushed upwards by the end of a gum-elastic catheter.

In a case recorded by Mr. Crosse,[2] the tumor was about the size and shape of a walnut, and projected visibly between the external labia. It was of a florid red color and somewhat granulated on the surface, so as to resemble a large strawberry. Towards the posterior part of the tumor, and on its sacral aspect, there was an aperture which was supposed to be the entrance into the displaced urethra; a small female catheter could be passed through this aperture into what seemed to be the bladder. Near the posterior junction of the labia, and concealed in a fold of the tumor, were two orifices, not far asunder from each other, and from which the urine oozed. The tumor was easily returned by pressure, and the canal through which it retired was found to be the urethra, dilated sufficiently to admit the little finger, although the patient was only

---

[1] Med. Gazette, vol. xi., p. 112, Jan. 19th, 1833.
[2] Trans. of Prov. Med. and Surg. Assoc., vol. xiv. p. 185.

between two and three years of age. It is worthy of notice that, but for Mr. Crosse's expostulations, a ligature would have been applied, the inverted viscus having been mistaken for a vascular tumor by the surgeon in immediate charge of the patient. The results of the treatment were satisfactory and permanent.

A case of complete eversion of the bladder has also been recorded by Dr. Beatty.[1] The patient was a girl aged 22 months, and the bladder protruded between the labia in the form of a scarlet tumor, the size of a chestnut. The tumor could be forced back and replaced by the finger introduced through the urethra. The eversion followed a fit of crying when the child was twelve months old. There was also prolapse of the rectum. Unfortunately the child died from croup while under observation in hospital, and before any operative measure could be undertaken. In a case of congenital eversion of the bladder recorded by Dr. Lowe,[2] the patient was a girl aged two and a half years, and the bladder formed a vascular tumor as large as a walnut, between the labia. The protrusion became increased when the child cried, and the urine flowed freely away over the thighs, causing much soreness and distress.

In a case which occurred to Mr. Croft[3] there were several additional symptoms of interest. The patient was fourteen months old, and the red, vascular, pear-shaped protrusion had been observed a few hours before the child was brought to the hospital. (It had, however, been noticed more than once before.) The tumor was tense with fluid, and its surface seemed formed by mucous membrane. During the examination, the swelling burst and gave exit to a clear straw-colored fluid, and then partly collapsed. A very small clot of blood formed at the site of the rupture. The collapsed sac again refilled, and the fluid was ejected. Under chloroform the tumor collapsed, and, as insensibility became complete, the little finger easily returned the remainder through the meatus urinarius and urethra. An examination of the fluid confirmed the opinion that the bladder had given away, and that the fluid came from the peritoneum. No bad symptom followed, and on the third day the child left the hospital. A year afterwards the parents reported that there had been no relapse, but an occasional dribbling and a gush of urine on coughing or sneezing. With regard to the fluid which escaped, chemical examination proved that it was not urine. The spot, moreover, bled and the rupture was in front, while the ureters were behind and below. It appeared that the bladder, having become everted through the urethra and meatus urinarius, had carried its partial peritoneal coat with it; serous fluid had been poured out, and had accumulated so as to distend the sac, which ultimately gave way.

Eversion of the bladder has also been observed in an adult. Dr. Thompson[4] has reported a case occurring in a woman, aged 40, who was the subject of chronic cystitis. The protrusion was caused by the woman's efforts at micturition; it formed a small, hard, nodulated tumor the size of half a walnut-shell, and was covered by phosphatic incrustations. In the case communicated by Mr. Percy to Chopart, it does not seem certain whether the tumor was formed by the mucous membrane of the bladder, or by the everted walls of the organ. Percy, however, considered it as

---

[1] Dublin Quarterly Journal of Medical Science, August, 1862.
[2] Lancet, March 8th, 1862.
[3] St. Thomas's Hospital Reports, vol., ii. 1871, p. 195.
[4] Lancet, 1875, Jan. 9th, p. 47.

a true example of eversion. The patient was a very fat abbess, the subject of severe and chronic bronchitis.

Protrusion of the mucous membrane has been described as a partial form of eversion of the bladder. In a case, recorded by Noël,[1] the urine, owing to obstruction of the ureters, had found its way between the muscular and mucous coats of the bladder, and the separated portion of the latter coat protruded from the urethra in the form of a transparent bag filled with clear fluid.

The diagnosis of eversion of the bladder presents, as a rule, no particular difficulties. The diseases with which the affection is most likely to be confounded are, vascular growths from the mucous surface of the bladder or urethra, or protrusion of the relaxed lining membrane of this latter canal beyond the meatus urinarius. The peculiar feel of the tumor, its position, and the manner in which it yields under pressure, the oozing of urine from its surface, the way in which it is girt by the urethra, and, above all, the existence of the ureters at its lower part, will suffice to indicate the true nature of the affection. In prolapsus of the urethral mucous membrane, of which complaint there are a few cases on record, the opening of the canal will be found in the centre of, or above, the tumor; in eversion of the bladder the urethra surrounds the prolapsed portion, the pedicle of which is also thicker and firmer than that of a tumor consisting exclusively of mucous membrane.

Replacement of the tumor is, of course, the first object to be aimed at in the treatment of eversion of the bladder, and this can generally be effected without difficulty. The patient being placed on her back, with the thighs flexed on the pelvis, the tumor should be gently but firmly pressed with the fingers and made to pass back through the urethra. The administration of chloroform may facilitate the replacement, and, if necessary, the point of a gum-elastic catheter may be used to assist. The patient should be kept for some time in the recumbent position, and the urine should be drawn off with the catheter so as to prevent any efforts which might bring back the protrusion; constipation should be prevented by enemata or castor oil. Faradization and astringent injections are likely to aid in restoring the tone of the urethra and reducing its calibre, and a compress may be placed over the orifice, and retained in position by a bandage. In Dr. Lowe's case, the actual cautery was applied several times to the canal of the urethra with satisfactory effects. Some degree of incontinence remained, but no protrusion took place. If these measures should not suffice to prevent a return of the protrusion, the canal of the urethra should be diminished by removing a small strip of mucous membrane, the edges of the wound thus made being brought together by a few sutures. During the healing of the wound the urine should be drawn off by a catheter. The patient should be instructed to avoid all violent efforts at defecation.

Eversion of the bladder may also take place through an artificial opening in its own wall and in that of the vagina. In most cases of vesico-vaginal fistula, accompanied by extensive destruction of the soft parts, the mucous lining of the bladder projects more or less into the cavity of the vagina—a plug is, in fact, formed by the partially everted bladder. The eversion may be even complete; Mr. Crosse saw a case in which the bladder, everted through a fistulous opening, had escaped between the external labia, forming there a tumor from which the urine continually

---

[1] Mémoires de l'Acad. Roy. de Chir., t. iv., p. 17.

dribbled at points found to be constituted by the termination of the ureters. A similar case was communicated by Dr. Howard, of Columbus, to Dr. Gross. The patient, during her first labor, five years previously, had received an extensive laceration of the vesico-vaginal septum and perineum. Four years afterwards she gave birth to another child, and some months after that event she observed a tumor in the vagina. On examination, Dr. Howard discovered complete eversion of the bladder, which hung through the vulva in the form of a red mass, as large as an orange, and of a globular shape; its rounded and rather narrow pedicle could be traced to the edges of the fistula, and the orifices of the ureters were seen at its posterior extremity within the vagina. After two unsuccessful attempts, Dr. Howard returned the tumor through the fistulous opening.[1] The treatment of this form of eversion of the bladder will be that of the fistula on which it depends.

A portion, or even the whole of the bladder, may be displaced in such a way as to produce a hernial tumor. In the male, the protrusion usually takes place through the inguinal canal into the groin or scrotum; in the female, the displaced organ descends into the vagina, or towards the floor of the pelvis, giving rise to the accident which is called vaginal cystocele, or prolapsus vesicæ.

Hernia of the bladder is said to occur more frequently amongst males, but if we include cases of vaginal cystocele amongst herniæ, the proportion must be reversed; it will be found that displacement of the bladder happens much more frequently in the female. The relative disposition of parts in the male and female pelvis readily explains why this difference should exist.

Inguinal cystocele exists when the bladder protrudes through the inguinal canal into the groin or scrotum. The tumor may contain a portion, or even the whole, of the bladder. Beaumont relates a case of the latter kind, where the scrotum contained the whole of the urinary bladder, in which a large calculus was also lodged. The hernia may be single or double; it presents great varieties in size, external appearance, etc., and it may be complicated either with protrusion of the bowels, or formation of calculus within the displaced portion of the bladder.

The causes of inguinal cystocele have not been clearly explained. Violent efforts at passing water in cases where obstruction exists, difficult labors, great lateral distention of the bladder occurring during pregnancy, have been mentioned as exciting causes. There must, however, exist some peculiarity, either congenital or acquired, in the abdominal walls, or in the shape and size of the bladder, which brings it nearer than natural to the opening through which it protrudes. Sir A. Cooper attributes the cause to a relaxed state of the bladder, and suspects that the protrusion may also be promoted by neglect to empty the bladder at the time required. This probably holds good in many cases. It is necessary to the occurrence of a cystocele, that the bladder should be placed immediately behind, or very close to the ring, and that it should hold that situation when empty; for the distended condition of the organ is obviously so unfavorable to a protrusion that it can hardly be deemed possible in that state.

It is probable that, in many cases, the inguinal canal may have been prepared for the reception of the bladder by the previous descent of intestine; the hernial tumor will then contain both bladder and intestine; or the bowel may have returned into the abdomen, and left the sac free for

---

[1] Dr. Gross, loc. cit., p. 327.

the reception of the bladder. However this may be, it is certain that, as a cystocele may give rise to a protrusion of omentum, so an enterocele or epiplocele may cause a descent of the bladder. The symptoms of the latter occurrence have not been observed in many instances until long after the patients have been incommoded by an intestinal or omental hernia, and it has even been suggested that hernia of the bladder is always preceded by one or other of these affections; that is, however, contrary to experience, which has shown that a protrusion of the bladder may exist alone.

The manner in which an ordinary omental or intestinal rupture may become complicated by the addition of a cystocele, can be easily understood when we consider that the peritoneum forming the sac is placed immediately behind the ring, and continued over the superior fundus of the bladder. If the original hernia be neglected, its increase elongates the hernial sac, gradually drawing into the ring that portion of the peritoneum which is attached to the bladder, and the bladder itself, if it be disposed to yield to this force. Thus a portion of the organ becomes situated behind the cavity of the first rupture, and it passes through the ring just as the fixed portion of the large intestines sometimes does in the gradual increase of a common scrotal rupture. The anatomical description is the same in this as in the preceding case. The protruded portion of the bladder is here interposed between the original hernia and the spermatic cord. The posterior surface of the sac, at its upper part at least, consists of the peritoneum covering the fundus and back of the bladder, and the proportion of the sac formed in this way depends on the extent of the protrusion.

A bubonocele taking place through the abdominal canal gradually brings the upper opening behind the lower one, so that we can conceive the possibility of the bladder being drawn through the ring in the subsequent increase of the swelling. But the relative positions of the opening and the bladder render the occurrence of cystocele more probable as a consequence of direct inguinal rupture.

The investments of the hernial mass differ in different cases. There may be either one or two sacs, or the protrusion may be destitute of an investment of this kind. Thus, if we conceive the bladder to elevate the peritoneum above the level of the internal abdominal ring, and then to protrude through the opening with that portion of its wall which is uncovered by peritoneum, it is evident that we should find a cystocele without any sac; such an occurrence, however, must be very rare; more commonly the superior fundus, covered by its serous investment, protrudes through the ring; or a portion of the bladder, covered or not by peritoneum, may be found in the sac of an intestinal rupture. In old cases, the walls of the bladder are frequently adherent to the adjacent intestine or sides of the sac; and the viscus may become greatly expanded as in the following case, the history of which is given by Mr. Clement. In the patient referred to, the entire bladder had passed through the inguinal canal into the scrotum, where it had become enlarged by distention, so as to form a tumor of enormous magnitude. The patient, a very corpulent man, about sixty years of age, had been troubled with the swelling for twenty-five years. It was small at first, and slowly increased to its immense size, having varied but little during the greater part of that time. The patient could not make water without first raising the rupture towards the belly, and then rolling it about for a short time, when the urine would pass in a full stream, though he was unable to make much at one time. He had repeatedly suffered from constipation and slight attacks of hemiplegia; but his health had been good in other respects, and he had been accustomed to take regular, and occasionally laborious exercise. The swelling had produced no alarming or dangerous symptoms until about a fortnight before death, when obstinate constipation came on, with paralysis of the left side. To these affections was joined stillicidium urinæ, for which the catheter was used several times; but not more than a tea-cupful of urine was drawn off on each occasion. Neither purgatives nor enemata produced any relief of the bowels, but the patient's symptoms were due more to retention of urine than to constipation. The pain and other symptoms increased, and delirium and death supervened. The true nature of the hernial tumor does not seem to have been suspected during life. The circumference of the swelling was twenty-nine inches; its length fourteen inches and three-quarters. The whole penis was retracted within the integuments, and the urine had been discharged through an opening resembling the navel. One of the testicles could be distinctly felt at the middle of the tumor; but the other was not discoverable before the parts were dissected. Although the rupture was so large as to extend generally over the pubes, and occupy both inguinal regions, the protruded parts obviously came through the left abdominal ring. When the inguinal canal had been exposed and opened, a portion of the colon was seen traversing it, distended with feces, but not inflamed, nor compressed, as the entire hand could be passed through the opening. On continuing the dissection, and after removing the testicle, the tumor was found to consist of a sac, resembling an enormous hydrocele, from which two quarts of fetid urine escaped by the rupture of a part which had become red and pointed before death. It was now discovered that the urinary bladder had passed through the abdominal ring, so as to form the immense scrotal tumor already described.

Another case of hernia of the bladder into the scrotum is recorded in the "Transactions of the Pathological Society."[1] The patient was a man, aged 80, who had suffered from a hernia in the left groin since he was sixteen years of age. He had enjoyed excellent health, though he had never worn a truss. When he was 70 years old the hernia appeared to increase in size, and a swelling formed also in the right groin. This gradually became larger, but evidently diminished after some urine had been passed. Symptoms of retention occurred, a catheter was introduced, and two quarts of urine were drawn off. The patient died, and on post-mortem examination it was found that the displaced portion comprised two-thirds of the bladder, and would hold fifty ounces of urine. The pelvic portion of the viscus had a capacity of twenty ounces. The communication, in the inguinal canal, between the two parts measured one inch in diameter. The epigastric artery was at the inner side of the neck of the protrusion.

The part undergoes further changes after it has passed through the ring. It becomes contracted in the opening, and expands again below. Mr. Keate found it contracted at the ring, dilating itself again in the abdomen and pelvis, and forming a kind of double bag divided by the ring. And the same change had occurred, to a still greater extent, in a case operated on by Mr. Pott. He discovered a membranous bag, growing narrower as it proceeded upwards; and a membranous duct, about

Vol. iv., p. 187.

the size of a large wheat-straw, was continued from its upper end through the ring. The urine flowing through this, when it was divided, proved the case to be a hernia of the bladder.

The cystocele is occasionally complicated with the formation of calculi in the protruded part of the bladder; and it is, indeed, to be wondered that this does not more frequently occur, when we remember the irritation to which the bladder is exposed, and the manner in which the urine stagnates in the protruding sac. Many cases of the complication now spoken of are on record; and it led to some curious mistakes formerly, when the nature of the complaint was less known. Thus Petit felt several calculi in a hernial tumor, but they subsequently disappeared, for they had been discharged through the urethra. Sala had a patient who presented the rational symptoms of calculus, but the stone could never be felt with a sound; after death the calculus was discovered in a portion of the bladder engaged in the inguinal canal. Vidier relates a case where a calculus contained in a cystocele was mistaken for bubo, and caustic applied, with the effect of establishing urinary fistula. Even Pott himself once cut into a cystocele through mistake.

Calculi may, likewise, be found in cases of cystocele in the female. In a case related by Hartmann, a pudendal cystocele contained a stone weighing three ounces. Cloquet and other writers mention several examples. Cruveilhier, however, found a calculus in the undisplaced portion of the bladder.[1]

The symptoms indicating this displacement, are—difficulty in making water, with occasional retention; sudden escape of a large quantity of urine upon introducing a gum-elastic catheter; enlargement of the hernial swelling after the urine has been retained; and the escape of urine from the urethra upon pressing the tumor.

The symptoms of cystocele will be different according as the protruded portion is full or empty; confined to the groin, or continued into the scrotum; and simple, or combined with intestinal or omental rupture. When the part is empty, its volume is not considerable, the sides collapse, and examination discovers nothing but a soft membranous substance rolling under the fingers. The swelling is sometimes round and uniform, sometimes elongated and irregular; when compressed, it causes a desire to make water, and sometimes the involuntary flow of urine; and it increases and diminishes in size according as the bladder is full or empty. There is often a constant pain in the region of the kidneys and perineum. But the most characteristic circumstance is the state of the urinary evacuation. When there is frequent desire to expel the urine with occasional retention; when the tumor increases after retaining the water for some time, and is diminished, or entirely disappears on voiding the urine, the case must be a cystocele. As an additional means of diagnosis, warm water may be injected into the bladder. This will be followed by an increase of the swelling, which will again subside when the water is allowed to escape through the catheter. The patient sometimes feels unable to expel the urine without elevating and compressing the tumor, but he can accomplish it easily by these means. After voiding all that he can, a further desire to make water is excited by pressing the swelling.

Should retention of urine have occurred, it will be observed that the tense tumor usually formed above the pubes by the distended bladder

---

[1] Anat. Path., liv. xxvi., p. 3.

is absent; but we must not forget that only a small portion of the bladder may be contained in the groin. M. Civiale¹ insists on the necessity of our studying the manner in which the urine is excreted while a catheter remains in the bladder. Thus, if the body of the organ has been emptied by the catheter, and if a fresh quantity of fluid flows through the instrument when pressure is made on the tumor, we have convincing proof that some portion of the urinary reservoir is contained in the hernial sac. Unless the catheter be employed, the urine which stagnates in the tumor might be merely forced into the bladder, and a useful means of diagnosis would be omitted. The urine obtained by compressing the tumor is usually more or less turbid, and deposits a copious precipitate. When the bladder has descended into the scrotum and is full of urine, it might be mistaken for hydrocele. The dysury, the power of diminishing the swelling by pressure, and desire of making water consequent on this, sufficiently distinguish the case. To the peculiar symptoms of cystocele will be added those of an intestinal or omental rupture, when the affection is complicated. The diagnosis would be rendered difficult if the opening leading into the displaced portion were blocked up by a calculus. It might in such a case be impossible to reduce the size of the tumor by compression.

In some cases the protrusion of the bladder has been attended with no symptoms. Its existence was not known until after death in Mr. Keate's case, where the greater part of the viscus had passed into the scrotum; and the same observation may be made concerning a case related by Arnaud.

If the hernia be reducible, it should be replaced within the abdomen and a suitable truss applied. If reduction be impossible, the tumor should be supported by means of a suspensory bandage, or a truss with a hollow pad. It seems to be hardly susceptible of strangulation; but if such a circumstance should occur, the operation must be resorted to, as in other cases of hernia. If a stone be discovered, we ought to remove it by an incision. No ill consequences followed in two instances where openings were made in the protruded portion of the bladder.²

In all cases great care should be taken to prevent the urine from becoming stagnant in the protruded sac. The catheter should be frequently used. If retention should occur and the symptoms are not relieved by the catheter, the tumor should be punctured.

Inguinal cystocele may likewise occur in the female, especially during pregnancy; but in a less rare form of displacement, the bladder descends behind the pubes into the vagina, forming a tumor which surgeons call vaginal cystocele, while accoucheurs generally denominate the complaint prolapsus of the bladder. The causes of the accident are well understood, and in most cases sufficiently obvious. Relaxation of the walls of the vagina affords a predisposing cause, joined to which any unnatural pressure on the bladder may force it downwards and inwards on the vagina, where the least support is afforded. Hence the complaint is often the result of difficult and protracted labors, especially when the urine has been suffered to accumulate, and the displacement is particularly liable to occur in those unfortunate cases where the support, which should be afforded by the pelvic floor, has been lost through laceration of

---

¹ Civiale, Traité Pratique sur les Maladies des Organes Génito-Urinaires, vol. iii., p. 37.
² Mém. de l'Acad. Roy. de Chir., tom. ii., pp. 11, 13.

the perineum. In many other cases the bladder is mechanically brought down by prolapsus of the vagina and uterus.

Inconvenient as this accident is to unmarried females, it has occasioned the greatest embarrassment during delivery; impeding the passage of the head, or misleading the accoucheur. Dr. Hamilton mentions a case in which the prolapsed bladder was mistaken for the membranes, and punctured; and Dr. Merriman records one, in which a distended bladder, pushed before the head of the child during labor, was actually opened under the notion that it was a case of hydrocephalus. The obstacles which the accoucheur may have to encounter in cases of this kind, are well illustrated by an interesting case, related by Dr. M'Cormack, of Portadown. Although vesico-vaginal fistula had existed for seven years, the bladder formed a large tumor at the entrance of the vagina, and delivery was effected only by perforation of the child's head.[1]

Cases of prolapsus of the bladder seldom come under the notice of the surgeon until the ordinary means of relief have failed. The symptoms are usually well-marked. The patient experiences a sensation of weight and bearing-down with frequent desire to pass urine; there is also a peculiar pain referred to the navel with a sense of tightness or dragging. This pain is greatest when the bladder is distended, and diminishes after the passage of urine. The bladder is never completely emptied, for its lower and posterior part,which projects into the vagina soon loses its power of contractibility, and retains a quantity of urine in spite of the patient's efforts at expulsion. In some cases the patients are obliged to replace the protrusion before they can pass water. The bladder becomes very irritable, and the urine often contains much ropy mucus and phosphatic deposits.[2]

On examination, a soft, elastic, fluctuating tumor will be felt at the orifice of the vagina. If the displacement be but slight, it will be advisable that the patient should retain her urine for an hour or two so that the protrusion may be well-marked. In severe cases, however, this will be unnecessary, for the tumor constantly remains full, notwithstanding the patient's attempts to empty the bladder. The swelling varies in size, generally according to the duration of the case, from a slight fulness extending backwards from the urethra, to a considerable projection which fills up the vagina, or even projects beyond it. On introducing a catheter, its point can be felt in the tumor, which diminishes in size as the urine is withdrawn. The finger can be passed into the vagina below the tumor, and the os uteri can be felt behind it nearly in its natural situation. When the bladder is distended, the surface of the tumor is smooth, moist, and shining, but thrown into transverse folds when the bladder is empty. There is usually more or less leucorrhœa, caused in part by the relaxed condition of the vagina, and partly by the irritation produced by the size and friction of the tumor. Prolapsus of the bladder can easily be distinguished from prolapsus uteri. In the latter affection the tumor is firm, the os uteri lower than natural, and the vagina shortened. Prolapsus of the bladder, however, may accompany various forms of displacement of the uterus. Calculi have been known to form in the prolapsed portion, as in a case reported by Gendron.[1]

The nature of the treatment to be adopted will depend upon the extent of the protrusion and the condition of the parts. If the patient

---
[1] Lancet, Feb., 1850, p. 204.
[2] Dr. Golding Bird, Med. Times and Gazette, 1853, Jan. 1st.
[3] Bull. de l'Académie, tom. xxiv., p. 47.

be young, and the protrusion recent and moderate in extent, the bladder should be kept empty by the frequent use of the catheter, the recumbent position should be enjoined, and astringent injections containing alum, sulphate of iron, or decoction of oak bark may be employed. A sponge pessary, soaked in a decoction of oak bark with alum, forms a convenient method of applying an astringent to the relaxed parts. The bowels should be kept open by mild purgatives, such as small doses of sulphate of magnesia with a grain or two of sulphate of iron, and the patient should be directed to avoid all efforts at straining. If the prolapsus be of long standing and of considerable extent, not much benefit can be expected from local astringent applications. Various operations have been performed with the object of contracting the calibre of the vagina. The anterior wall of the relaxed canal may be touched with the actual cautery, so that a slough is produced. When this comes away, the contraction may suffice to prevent the formation of the vesical pouch. The same object may be attained by removing a triangular slip of mucous membrane, the base of the slip being turned towards the orifice of the vagina. The edges of the wound are brought together by suture. Another operation, suggested by Prof. Stolz,[1] of Nancy, consists in removing a circular piece of mucous membrane, about the size of a five-shilling piece, from the surface of the protruded portion. When the bleeding has been checked, a silk suture is passed round the bared surface about an inch from its margin, in the same manner as is done in making an ordinary calico bag. On drawing the two ends together, the edges of the wound are approximated, and the canal of the vagina is much diminished in length and circumference. The cervix of the uterus is retained in a pouch above the narrowed vaginal surface.

In the more severe forms of the affection occurring in women beyond the period of child-bearing, the operation proposed and practised by the late Mr. Baker Brown[2] has been attended in several cases by very satisfactory results. The contraction of the calibre of the vagina is the object sought to be attained. The following is the mode of operating:—The bowels having been previously opened by enema, the patient is placed in the lithotomy position, and an assistant holds up the tumor with Jobert's bent speculum, and presses it under the pubes in its natural position. A piece of mucous membrane, about an inch and a quarter long, and three-quarters of an inch broad, is then dissected off longitudinally, just within the lips of the vagina. The upper edge of the denuded part being on a level with the meatus urinarius, the edges are drawn together by three interrupted sutures; the operation is then repeated on the other side of the vagina. The next step consists in dissecting off the mucous membrane laterally and posteriorly in the shape of a horse-shoe, the upper edge of the shoe commencing half an inch below the lateral points of denudation, taking care to remove all the mucous membrane up to the edge of the vagina where the skin joins it. Two deep sutures of twine are then introduced about an inch from the margin of the left side of the vagina, and brought out at the inner edge of the denuded surface of the same side, and again introduced at the inner edge of the pared surface of the right side, and brought out an inch from the margin of the right side of the vagina, thus bringing the two vascular surfaces together, which are then kept so by means of quills, as in the operation for rup-

---

[1] Brit. and For. Med. Chir. Review, 1875, vol. ii., p. 225.
[2] Med. Times and Gazette, 1858, vol. i., p. 395.

tured perineum. The edges of the new perineum are lastly united by interrupted sutures, and the patient placed in bed on a water-cushion. Two grains of opium are given directly, and one grain every six hours; simple water-dressing applied to the parts; beef-tea and wine for diet. A flexible catheter, to which is attached an elastic bag to catch the urine, is introduced into the bladder; by this means the bladder is constantly kept empty.

The object sought in this operation is the contraction of the calibre of the vagina, which, as may be imagined, is enlarged and flabby. The first step of the operation is directed to the contraction of the vagina laterally, so as to prevent the tumor from falling down from above; the second step is for the purpose of contracting the vagina posteriorly; and thus in the end, the orifice of the vagina being diminished by at least two-thirds, and the perineum increased in extent, should the prolapsus not be restrained by the lateral contractions, it cannot extrude beyond the orifice of the vagina, but must necessarily fall upon the new perineum. The after treatment in most particulars resembles that pursued after the operation for ruptured perineum. Opium is given to allay irritation and pain, and to prevent defecation; the strength is supported by nourishing diet and wine; water-dressings are applied, and perfect repose is enjoined. The use of injections is, however, contra-indicated; for the sutured parts must not be interfered with in any way. It is of the greatest importance to keep the bladder emptied; and this point is best secured by retaining a catheter in the bladder, with a bag to receive the urine as it escapes. After the seventh or eighth day, according to the integrity of the union of the parts, the patient may pass the urine resting on her hands and knees.

The time for the removal of the sutures must be regulated by the circumstances of each case; but, in general, the deep ones may be withdrawn from the third to the fifth day, the others a few days afterwards.

CHAPTER V.

WOUNDS AND INJURIES OF THE BLADDER.

ALTHOUGH carefully protected, by the bony framework of the pelvis and adjacent soft parts, from the effects of injury, the bladder occasionally suffers from external violence.[1] Various kinds of weapons and missiles may reach the bladder through the walls of the abdomen or pelvis, the perineum, rectum or vagina, and the resulting wounds will be either incised, lacerated, punctured or gunshot, according to the weapon employed. In this first class of cases there is an external wound communicating with the wound in the bladder.

The bladder may also be injured in a different manner, viz., by pressure from within outwards, as when the organ gives way from over-distention. Such an accident, however, from this cause alone, is extremely rare, but it not unfrequently arises when the bladder, in a state of distention, is subjected to direct or indirect violence. The bladder may also be contused or lacerated by fragments of bone, or ruptured from pressure, without perforation, in cases of fracture of the pelvis. In the female, the bladder is liable to injury during parturition, either from pressure of the head of the child, or from unskilfully used instruments. The bladder may also be injured by a catheter or lithotrite roughly manipulated.

In all injuries of the bladder the locality and extent of the mischief are the main points of importance. As a matter of course, an injury affecting a part uncovered by peritoneum is less serious than one which involves any portion of that membrane. Wounds of the anterior wall, neck and base of the bladder are therefore, *cæteris paribus*, less liable to be followed by fatal consequences than those of other parts of the organ. Freedom of escape for the urine by the wound is another point of importance, extravasation into the cellular tissue being generally attended by fatal results. Profuse hæmorrhages, usually venous in character, sometimes occur as a result of external violence, but peritonitis and extravasation of urine into the cellular tissue are the main dangers to be dreaded in all forms of injury to which the bladder is liable.

The first class of cases, which includes wounds of the bladder caused by weapons and missiles of various kinds, embraces gunshot wounds, but these latter may be with advantage considered separately, inasmuch as they form a perfectly distinct class.

Incised, lacerated, or punctured wounds of the bladder, caused either by sharp instruments, such as swords, knives or lances, or by blunt weapons, as stakes, sticks, and the like, are of comparatively rare occurrence. Out of 504[2] cases of injury to the bladder collected by Prof.

---

[1] The comparative rarity of injuries to the bladder is shown by the fact that only two instances occurred among 16,711 surgical cases treated at St. Bartholomew's Hospital, 1869-75.
[2] Langenbeck's Archiv für klin. Chir., Bd. xxii. In the revision of this chapter much assistance has been derived from Dr. Bartels' article.

Max Bartels, only 50 belonged to this class. The weapons by which such injuries are caused are very various, and the bladder may be reached through the perineum or rectum, the obturator foramina, or, above the pubes, through the walls of the abdomen. The injuries are more frequently caused by the patient falling upon or against a comparatively blunt object, such as the leg of a chair, the handle of a broom, a wooden stake, etc. In a few cases the bladder has been penetrated by the horn of a bullock. In the remaining instances belonging to this category the organ has been wounded by a lance, sword, dagger, or other pointed instrument.

In 22 of the 50 cases above alluded to, the perineal region was the seat of injury, and only in five of these cases was the wound inflicted with a sharp weapon. In the remaining seventeen the patients fell from a height upon a more or less blunt extremity of a stick, handle of a broom, etc. In thirteen instances, the point of the object passed through the bowel into the bladder; ten of these patients recovered.

Only one case of injury to the bladder through the obturator foramina is placed on record by Dr. Bartels. The injury was inflicted on a cavalry soldier by the lance of a Cossack.

The third channel by which the bladder may be reached by a weapon is through the abdominal wall, in the hypogastric region. Injuries of this kind (gunshot wounds, of course, excepted) are extremely rare. Dr. Bartels has found only twenty-six cases of wounds of the bladder from injuries inflicted in this region, and only ten of these cases belong to military surgery. There is not a single case on record of a bayonet wound of the bladder. It would seem to be a well-authenticated fact that the bladder is rarely injured in military warfare by swords or other sharp weapons. Thrusts with the bayonet are directed at the thorax and upper part of the abdomen, and a sword cut could scarcely be so made as to reach the bladder. It appears from the medical reports of the war of the American Secession, that no case of injury to the bladder by lance, sword, or bayonet, came under the notice of the surgeons.

Out of Dr. Bartels' fifty cases only eleven proved fatal. In all of these the peritoneum was opened, in some the intestines also were injured, and death usually occurred with great rapidity, the patient never surviving beyond the third day, unless there was a second opening through which the urine could escape. In many of the cases, the cure was prolonged by the occurrence of a fistula, while in some instances fragments of the patient's clothes, or of the offending object, were retained in the bladder and gave rise to various complications, such as the formation of urinary concretions, etc.

The symptoms of wounds of the bladder are usually sufficiently clear to render the diagnosis easy. The direction of the wound, the escape of urine from it, the passage of bloody urine through the urethra, are the characteristic symptoms of the injury. If, however, the wound be small and its direction very oblique, little or no urine may escape from it. The subjective symptoms are not necessarily characteristic; they are those of severe injuries in general.

In the treatment, the main indication to be fulfilled is to prevent infiltration of urine and to combat any symptoms of peritonitis. It is better not to introduce a catheter if the urine escapes freely from the wound. Hæmorrhage is to be arrested in the usual manner, and foreign bodies removed when within reach. The pain and vesical tenesmus are to be alleviated by opium in some form, and any symptoms of peritonitis must be treated by calomel and opium, leeches, fomentations, etc.

Gunshot wounds of the bladder form the majority of all cases of injury of this organ. Out of Dr. Bartels' 504 cases 285 were instances of this nature. In 131 cases the injury to the bladder was complicated by a gunshot fracture of one or more bones. This complication does not appear to influence the rate of mortality; only 38 deaths occurred among these last mentioned cases. Fracture of bone, however, causes the treatment to be much more prolonged, owing to the time required for the formation and separation of sequestra. Tedious suppuration is also a frequent consequence, and in some cases fragments of bone are retained in the bladder, and form the nuclei of urinary concretions.

The course taken by the projectiles, in injuries which involve the bladder, is very various. In the majority of cases the aperture of entrance is in front. This was the case in 126 of Dr. Bartels' cases. In 100 instances the bullet entered posteriorly. The mortality was the same in each class, viz. 26. In 136 instances the bullet traversed the body; the number of deaths was 30. In 92 cases the bullet lodged; of those, 23 terminated fatally. The conclusion to be drawn is that a double wound is less dangerous than a single one. With regard to the missiles employed, these, in Dr. Bartels' tables, are bullets of all kinds, shot of various sizes, fragments of shells, etc. In some cases, the injury to the bladder is not caused by the bullet itself, but by fragments of bone which are forced through the coats of the organ. In other cases the bullet lodges close to the bladder, and subsequently enters its cavity by a process of ulceration. As a matter of course the injury is often complicated by wound of the bowel and peritoneum. When the latter structure is penetrated the result is always fatal, and the same may be said when the opening into the peritoneum is caused by a secondary process of ulceration. In the fatal cases with uninjured peritoneum, the cause of death is infiltration of urine, with acute septicæmia, or profuse suppuration with chronic septicæmia. Professor Pirogoff states that in three-fourths of the fatal cases, death is due to urinous acute purulent œdema of the fascia of the pelvis. When the peritoneum is likewise involved, it would appear that death is often due to septicæmia.

The symptoms of a gunshot injury of the bladder are generally very marked. The patient usually falls prostrate to the ground; only in rare instances is he able to walk after receipt of the injury. Pain in the wound and abdomen generally, retraction of the testicle with great pain in that organ, vomiting, urgent desire to make water, escape of bloody urine or pure blood from the urethra, are the other prominent symptoms. The escape of urine from the wound almost establishes the diagnosis, but such a symptom may occur (though much more rarely) after a similar injury involving the kidney, the ureters, or the prostatic portion of the urethra. If there are two wounds, the urine generally escapes from both. If the intestines likewise are injured, feces and flatus will probably escape from the wounds. After a time the escape of urine from the wounds ceases, owing to swelling of the parts, and painful desire to pass water then becomes the most prominent symptom. In the course of the second or third day, the sloughs commence to separate, and suppuration, accompanied by fever, sets in. It is at this period that infiltration of urine is likely to occur, a symptom marked by the occurrence of rigors and fever. When the sloughs are detached, the track of the wounds being covered by granulations, infiltration of urine is no longer a possible danger. Where the peritoneum has been wounded, acute inflammation of that membrane is the prominent symptom. The abdomen becomes painful and tumid, the patient is uneasy and complains of thirst, and is further distressed by vomiting and hiccup. Delirium soon sets in, the skin is cold and covered with perspiration, and death speedily ensues. In non-fatal cases, when there are two wounds, one of them usually heals long before the other, which may remain open for a considerable period and form a fistula, continuing sometimes for months or years. The anterior wound is more likely to take this latter course. These fistulæ have their external opening in various parts, such as the rectum, the vagina, the abdominal wall, the groin, perineum, scrotum, the gluteal region or upper part of the thigh. In some cases the fistulæ are due to the suppuration, and to the pus making its way through the skin, or to artificial openings for its evacuation.

With regard to the complications of gunshot wounds of the bladder, Dr. Bartels gives a summary of his cases, from which it appears that the bones of the pelvis often suffer, the thigh-bone and joints being rarely touched. In 18 cases, vessels of some size were injured; in 28 the peritoneum was involved. The ureters, the genital organ and the kidneys occasionally suffer, and likewise various portions of the bowel. The cases in which the colon and ileum were injured all ended fatally; in 60 instances the rectum was wounded; 17 of these patients died.

Another complication of gunshot wounds of the bladder is the retention, within the organ, of the missile itself, or of various other foreign bodies, such as portions of clothes, pieces of bone, fragments of wood, coagula, etc. Small articles belonging to these classes have been spontaneously expelled by the urethra; in other cases the foreign bodies have remained behind, and given rise to the formation of urinary concretions. In 46 of Dr. Bartels' cases of gunshot wounds of the bladder, the operation for lithotomy was eventually performed for the removal of the foreign body, and in all but two instances with a successful result. The ammoniaco-magnesian phosphate forms the usual incrustation of the foreign body; the urates, phosphate of lime, and organic matters are less frequently met with. Iron projectiles become incrusted with greater rapidity than leaden ones.

In the treatment of gunshot wounds of the bladder, it is a point of the greatest importance to provide as much as possible for the free escape of the urine, so as to lessen the risk of infiltration. The wound or wounds should be covered with some light material and kept as clean as possible. If the urine escapes freely, it is unnecessary to introduce a catheter, but if retention occurs, a gum-elastic instrument must be used. It is, generally speaking, unadvisable to keep a catheter in the bladder. If the symptoms of retention be urgent and a catheter cannot be passed, owing to the neck of the bladder being implicated, a free opening must be made upon a catheter or staff passed down the urethra as far as it will go. Mr. Guthrie records a case in which the attempt to pass a catheter failed and no other treatment was attempted. Extravasation of urine took place, and the patient died.[1] Mr. Guthrie also records several other cases of gunshot injury to the bladder, terminating favorably.

Urinary infiltration must be treated by incisions. Pain is to be relieved by opium in the form of a suppository, or morphia subcutaneously, and by warm applications to the hypogastrium and perineum. Dr. Bar-

---

[1] Commentaries on Surgery, p. 611.

tels thus summarizes the conclusions to be drawn from the histories of the cases of gunshot wounds of the bladder: These wounds are certainly fatal when they involve the peritoneum, large vessels, or the hip-joint. Under all other circumstances the prognosis is favorable rather than otherwise. It is unfavorable when there is any amount of urinary infiltration. In favorable cases the wounds heal in from three to six weeks, but progress is often much delayed, especially when a fistula forms. The possible formation of a urinary concretion is a point always to be borne in mind. Even in favorable cases the patients often suffer for life from various disturbances in the function of the bladder.

It now remains to consider the third class of injuries to which the bladder is liable: this includes cases of rupture or laceration of the organ without external wound. An injury of this nature is almost always the result of some great external violence, the bladder at the same time being full of urine. Sudden blows over the hypogastric regions, falls against hard projecting bodies, and the pressure of a heavy weight on the lower part of the abdomen, are the most frequent causes of the accident. When arising from the last-mentioned cause, the injury to the bladder is often complicated by fractures of the pelvis, and it may be sometimes difficult to determine whether the laceration was the direct result of the injury or was caused by the penetration of splinters of bone. As illustrating the manner in which rupture of the bladder may be caused, Mr. Coulson has described a case in which it occurred between two powerful men, after drinking; the one fell with his knee bent on the abdomen of the other, and caused a laceration of the bladder, which was speedily followed by fatal peritonitis. In another case, the injury was caused by a fall from the top of a coach. A gentleman was riding on the box of his carriage with his coachman, when the vehicle was upset; the coachman fell on him, and the master's bladder was lacerated. The patient died on the third day after the accident.

In a few cases the bladder gives way, not under direct violence, but from general concussion of the frame, as when a man falls on his feet from a height. Mr. Cusack, of Dublin, has related a case of this kind.[1]

Professor Bartels, in the paper before referred to, has collected 169 cases of rupture of the bladder. In 57 of these the injury was caused by the patient falling against some hard projecting object; in 51 instances the injury was the result of blows or kicks upon the abdomen, and in 52, the pressure of heavy weights, e.g., the wheel of a wagon passing over the hypogastrium, was the cause of the accident. In the first subdivision, the majority of the patients were in a state of intoxication at the time of the accident; in the second, the injury was generally the result of a fight, while in the third, the wheel of a carriage passing over a body, or the fall of a wall was the most frequent cause of the accident. Dr. Bartels believes that the excessive use of alcoholic liquors tends to diminish the tone and elasticity of the muscular fibres of the bladder, and in this way he accounts for the frequency with which laceration of the organ occurs when persons in a state of intoxication receive a blow on the lower part of the abdomen. Chronic inflammation of the bladder would likewise render its coats more fragile.

The part of the bladder which gives way under sudden violence is generally that portion of the posterior wall which is covered by the peritoneum. Dr. Harrison explains this fact in the following manner:

---

[1] Dublin Hospital Reports, vol. ii., p. 310.

"The several tunics of the bladder allow of considerable distention; but least of all, the peritoneal; when, therefore, the bladder becomes fully distended, and is subjected to sudden or violent compressing force, this tunic, which is then tense, and comparatively unyielding, will crack, while the subjacent tunics, which are connected to it, will be torn along with it; whereas in other situations, where cellular tissue occupies the place of serous membrane, the coats of the bladder will yield considerably before they give way or admit of laceration." In Dr. Bartels' cases the laceration involved the peritoneum in 93 instances, and in 39 of these the injury was at the posterior part of the bladder. In 40 cases of extra-peritoneal injury, the laceration was at the anterior part of the bladder in 18 instances. Sub-peritoneal rupture occurred nine times. In a few instances the coats of the bladder were lacerated at two or more places.

The extent and appearances of the laceration vary in almost every case. It is sometimes a mere fissure; sometimes it extends to several inches, the edges of the lacerated surface being ragged and everted. The appearances will also vary somewhat, according as the injury may have occurred in the peritoneal or extra-peritoneal portions of the bladder; but when the patient has survived for a few days, the inflammation which supervenes will mask the appearances belonging to the original lesion. In the third class of cases, those in which the laceration is complicated or caused by fracture of the pelvic bones, various other injuries will be found after death.

The symptoms of laceration and rupture of the bladder vary much in kind and intensity in different cases. The early symptoms are probably connected with the accident itself; those which set in later depend on the inflammation excited by the effused urine. In some cases the patient seems to experience little inconvenience from the serious injury inflicted on him. A case of this kind is quoted by Prof. Bartels. A man received an injury in a drunken brawl. On the following day, being unable to pass water, he walked three miles to have a catheter passed. He continued at his work till the third day, when he sought admission into the hospital on account of abdominal pain. He died on the eighth day from peritonitis. On post-mortem examination, a rent, one and a half inches long, was found in the posterior wall of the bladder. The pelvic cavity contained a quart of colorless urine. A somewhat similar case is cited by Mr. Holmes in his "Treatise on Surgery," p. 215. A man was admitted into St. George's Hospital, complaining of pain in the hypogastric region, the result of a fall or blow received about thirty-six hours previously. He had walked to the hospital, and displayed no distress of any kind. The case terminated fatally, and, on post-mortem examination, the bladder was found extensively lacerated and communicating freely with the peritoneal cavity. In cases such as these, the early symptoms proper to laceration are absent, and the injury may not be suspected until inflammation begins to set in; more commonly, however, the patient experiences severe pain at the lower part of the abdomen immediately after the accident, and then exhibits the ordinary signs of violent shock. He becomes weak and faint; the pulse is feeble, the countenance anxious, the extremities cold; he is unable to expel the urine, and sometimes loses all desire to make the attempt. The little urine that escapes, or may be drawn off by the catheter, is tinged with blood; nausea or vomiting next ensue; the patient becomes restless, and the abdomen is somewhat tender to the touch, or may be prominent just above the pubes. These are the precursory signs of inflammation, which sets in at a very uncertain period after the accident; sometimes after the lapse of several days.

This inflammation and the symptoms which accompany it will be modified according as the injury has taken place in the part of the bladder covered or not by peritoneum; in other words, according as the urine may be effused into the peritoneal cavity or into the cellular tissue of the pelvis. Dulness on percussion, extending symmetrically over the abdomen, would indicate that the urine is effused into the peritoneal cavity. Circumscribed patches of dulness, on the other hand, are signs that the effusion is extra-peritoneal.

The occurrence of peritonitis is indicated by the febrile reaction, and the condition of the abdomen: as the symptoms of collapse from shock pass away, or after a certain interval of ease, the patient begins to complain of uneasiness and pain in the lower part of the abdomen; these quickly spread until the whole belly becomes extremely tender to the touch. At first the lower part of the abdomen is somewhat full; it now becomes tympanitic, the febrile symptoms keep pace with the local signs, delirium ensues; the well-known symptoms of peritonitis are developed, and the patient dies.

When the urine has been effused into the cellular tissue of the pelvis, the symptoms and progress of the case very much resemble those in certain fatal cases of lithotomy. The shock produced by the accident is the same, but the inflammatory symptoms which supervene are not so well marked; there is more prostration, and as the peritoneum is involved in a secondary manner only, the symptoms of peritonitis occur at a later period, or may be absent altogether. This chiefly happens when the bladder has given way low down, at the back part; if the laceration has taken place above and in front, the peritoneum is more rapidly affected.

Hæmorrhage, as before mentioned, is another source of danger, but in some cases of rupture of the bladder the urine contains only slight traces of blood. Mr. Fleming[1] mentions the case of a boy who was run over by a cart, and sustained an extensive fracture of the pelvis. The urine, withdrawn by a catheter, appeared to the naked eye to be free from blood, but some blood corpuscles were visible under the microscope. Death ensued on the seventh day, the bladder acting voluntarily for two days previously. The pelvis was found to be extensively fractured, and the bladder perforated anteriorly in a portion destitute of peritoneum. In another case, in which the anterior part of the bladder was lacerated below the peritoneal coat, hæmaturia was absent.

The diagnosis of laceration of the bladder is seldom difficult. The history of the case and the condition of the patient clearly point to the nature of the accident. It arises from external violence or general concussion acting on a full bladder; and when these injuries are immediately followed by severe pain in the region of the bladder, with constant desire but inability to pass water, and the peculiar symptoms of sinking which indicate a violent shock to the system, we have a train of circumstances which show not only that some viscus has been ruptured, but that the viscus is the bladder. Other symptoms will also assist us in doubtful cases. There may be a tumor at the lower part of the abdomen; but as the swelling depends on effused urine, it is rather a diffused ten-

---

[1] Diseases and Injuries of the Urinary Organs, pp. 39 and 58.

sion, unlike the circumscribed tumor resulting from retention of urine. On introducing a catheter, a small quantity of bloody urine or even blood may come away; and on observing the manner in which the urine flows, it will be seen that it comes away irregularly and in small quantities at a time, not like the natural flow from an organ desirous to relieve itself of its contents. As occurred in Baron Dupuytren's case, the end of the catheter may become engaged between the edges of the opening, and the diagnosis is thus rendered certain; but such a chance cannot be reckoned on. In cases of rupture the manner in which the effused urine makes its way into the cellular tissue above the pubes or at the floor of the pelvis, is an additional sign of great value. It is worthy of notice that, in cases of rupture of the bladder from external violence, there may be no marks on the surface of the skin. This point may become of importance in medico-legal cases.

The following cases of rupture of the bladder from external violence exhibit several points of interest. The first was related at the Medical Society of London by Mr. Hird :—A bricklayer, after drinking with some friends at Kensington, was knocked down and run over by a cab; he was taken to the house of a surgeon in Lambeth, who dressed a slight wound on the head. After this he walked home to Pimlico, a distance of nearly two miles, and about three hours after, at the request of the gentleman who ran over him, Mr. Hird saw the patient. He found him in bed, and complaining of severe pain in the hypogastric region; but having drunk to excess, the patient was unable to give an account of his sensations at the time of the accident. He had vomited two or three times, and had felt great desire to make water; but not more than two or three table-spoonfuls, slightly tinged with blood, had dribbled away. He was cold, and occasionally shivered, and was covered with a clammy perspiration. The pulse was full, and 120. The abdomen was distended, but did not evince, on examination, the circumscribed form observable in cases of distended bladder from retention. The patient died three days afterwards. On opening the abdomen the bladder was found small and contracted, with an oblique fissure, an inch and a half long, in the posterior inferior part, and opening into the peritoneal cavity. The mucous membrane of the bladder was healthy, and slightly everted at the aperture.

The above case illustrates the fact that a long interval may take place between the receipt of the blow and the manifestation of symptoms connected with the bladder. The following example is also instructive in this respect :—A horse-breaker, aged thirty-eight, while driving a restive horse in a gig, was thrown out, and the vehicle passed over the lower part of his abdomen. He was immediately carried to the Chester Infirmary: no marks of injury were displayed externally, and after rallying from the effects of the fall, he seemed pretty well. The following afternoon, however, it was ascertained that he had not made water since the occurrence of the accident, and the catheter being introduced, about eight ounces of fluid, deeply tinged with blood, were withdrawn. He then also began to experience considerable pain in the hypogastric region, and there was much tenderness on pressure, for which he was bled and leeched. The catheter was required to be introduced at intervals for several days, after which he regained the power of voiding water; the urine assumed its natural appearance, and the febrile symptoms, to a considerable extent, subsided. He, however, declined in strength, typhoid symptoms came on, and he sank on the seventeenth day from the receipt of the injury; two days before his death he again had retention of urine, and the fluid removed by the catheter was very dark-colored and fetid.

The body was examined twenty-four hours after death. No fluid was contained in the peritoneal cavity. The omentum was firmly attached, by old adhesions, to the abdominal parietes in front and to the intestines below. The small intestines were also firmly united together, and had become adherent to the edges of the pelvic cavity. The bladder displayed, at the posterior part of the fundus, a lacerated opening capable of admitting of the passage of two fingers, which led into a cavity situated posteriorly or inferiorly to it, and which was bounded by the adherent intestines above, by the sides of the pelvic cavity laterally, and by the rectum behind and below. The cavity contained fetid urine, mixed with shreds of lymph and gangrenous cellular tissue. On the side where it was bounded by the rectum there was an ulcerated aperture, by which it communicated with the bowel. This aperture was nearly the size of half-a-crown, and was situated at about four inches from the verge of the anus. The adhesions by which the omentum and intestines were united together and to the edges of the pelvic cavity, were evidently of old date, and long anterior to the accident. The coats of the bladder were much thickened, and its lining membrane was covered with tough mucus mixed with lymph. The kidneys were much enlarged; they were easily deprived of their proper tunic, and the surface exposed was of a deep purple color. Their texture was infiltrated with small purulent deposits, and the organs were generally very soft and readily tearable.

The prolongation of life in this case was evidently due to the peculiar situation of the laceration of the bladder, in consequence of which the urine had escaped into the recto-vesical cellular tissue. The extension of inflammation to the peritoneum had also been prevented by the old adhesions between the omentum and the intestines in the hypogastric region.

In the "'Transactions of the Pathological Society," Mr. Prescott Hewett has related 10 cases of laceration of the bladder. In 2 of these 10 cases there was no injury of the bones; in 8 the pelvis was extensively fractured. In 2 cases the bladder was ruptured into the peritoneum (one being with fracture of the pelvis, the other without). In 8 cases the rupture was into the cellular tissue of the pelvis; in these 8 cases, the bladder was ruptured in its fore part in 5, in its lateral part in 2, at its neck in 1.

"Both the cases into the peritoneum were most interesting: the one from its rarity, the injury having occurred in a woman, an accident of which there are very few cases on record; the other from there being so small a quantity of fluid found in the peritoneum, merely an ounce, and yet the rent in the bladder was large enough to admit of the passage of the first two fingers, as well as there being no trace of inflammatory action above the serous membrane.

"The variety of points at which urine extravasated into the subperitoneal cellular tissue may show itself, was also well illustrated in some of these cases. In one, the patient living twenty-three days, large abscesses made their appearance above the pubes and in both iliac fossæ. In another, in addition to these regions, the scrotum and perineum became extensively infiltrated. In a third, the right side of the scrotum was the part principally affected, the urine having passed through the right internal ring and down the inguinal canal. In a fourth, the upper part of both thighs was affected, the urine having passed through the obturator foramina.

"Lastly, the reparative efforts sometimes made by nature in accidents of this kind were also well shown. In a patient who lived five days after the accident, the cellular tissue in the immediate neighborhood was condensed by lymph, forming a species of pouch connected with the margins of the rupture. In another case, living twenty-three days after the injury, the surrounding cellular tissue was so condensed, and so firmly attached to the margins of the rupture, except at a small point, that the secondary cavity thus formed presented the appearance of a sacculus of the bladder which had given way."

Mr. Syme[1] has published a very interesting case of rupture of the bladder from violence, which terminated in recovery. The patient, a youth 17 years of age, received a violent blow on the lower part of the abdomen, in attempting to leap over a wooden paling. He immediately experienced intense pain, and a feeling as if his bowels had protruded; as soon as he was brought home a catheter was introduced, and four ounces of bloody urine were drawn off. The patient complained of severe pain in the belly, which was distended; his look was sunken, anxious, and he presented all the usual signs of ruptured bladder. On the fourth day after the accident the abdomen had become much swollen, with evident signs of fluctuation, and it was thought that an incision through the linea alba, above the pubes, might be of use. This was made, and a quantity of urine escaped through the wound, the patient experiencing much relief from its discharge. Two days afterward, the urine ceased to flow through the wound, and unfavorable symptoms set in. The incision was therefore enlarged with a bistoury, and from that time no dangerous symptoms appeared. On the 20th, Mr. Syme, while removing a portion of sloughy membrane, passed his finger into the bladder through a rent more than an inch long on the pubic side of the reflexion of the peritoneum.

On the thirtieth day, seven ounces of urine came away through the urethra, and in a fortnight afterwards the patient was well. The fortunate result in this case evidently depended on the rupture having occurred in a part of the bladder not covered by peritoneum.

Another remarkable case, which occurred in the practice of Dupuytren,[2] illustrates the manner in which nature makes an effort to heal the wound. A powerful man, 30 years of age, after having drunk a considerable quantity of wine, became engaged in a quarrel with his neighbor, and received a violent kick on the lower part of the abdomen. He instantly experienced most severe pain in the part, and was carried to the Hôtel Dieu. Blood was drawn from the arm, and leeches were applied over the pubes, but the febrile symptoms were not relieved. Although the bladder did not appear to be distended M. Dupuytren introduced a catheter, and drew off a small quantity of turbid reddish urine. While directing the point of the catheter towards different parts of the bladder, Dupuytren remarked that the instrument penetrated much further than it should at a particular point near the upper and anterior wall. On pushing the instrument it penetrated still further, and then more urine came away. It was now suspected that effusion of urine had taken place into the abdominal cavity. The patient died on the seventh day.

---

[1] Path. and Pract. of Surgery, p. 332.
[2] Bull. de Thérap., 1832, tom. iii., p. 349.

On examining the abdomen, some strong cellular adhesions were found between the walls of the abdomen, the sides of the bladder, and the viscera contained in the pelvis. These adhesions formed a kind of sac behind the bladder, containing a turbid fluid which had a urinous smell and was mixed with shreds of lymph. In the anterior wall of the bladder were found two lacerated openings, one of them about two inches long. The abdominal viscera were not inflamed, but a few points of the mucous membrane in the small intestines and stomach were slightly injected.

Laceration and rupture of the bladder occur more frequently in males than in females; in adults than in children. Dr. Stephen Smith, of New York, who has published a valuable monograph on this subject,[1] gives an analysis of 78 cases, reported by various authorities. Of the 78 patients, 67 were males, and 11 females; only 3 of the patients were under ten years of age. Among Dr. Bartels' 169 cases, only 16 occurred in women and girls. Dr. Harrison ascribes the greater rarity of the accident in females to the greater size of the pelvis, the cavity of which is not so extensively occupied by the bladder, when this organ is full of urine.[2] We should also bear in mind, that in five cases out of six the accident arises from external violence, to which women are much less exposed than men.

The comparative exemption of children from injuries of this nature may be explained by the fact that they rarely suffer the organ to become much distended with urine. In children also the sacral promontory is much smaller than in adults, and the bladder, inasmuch as it lies higher in the abdomen, is less liable to be injured in accidents involving the bones of the pelvis. The elasticity of the surrounding parts also helps to prevent laceration.

In addition to violence as a cause of laceration of the bladder, the injury has been known to occur as a result of over-distention alone. The accident occurring under such circumstances is probably always preceded by softening or some other change in the vesical walls; it generally takes place at the neck or at the base of the bladder, below the reflected fold of peritoneum.

Retention of urine within the vesical cavity is the determining cause of rupture in most of these cases, but the condition of the mucous membrane and other coats is usually such as facilitates laceration. In cases where any obstacle to the natural escape of urine has existed for some time, chronic cystitis is very apt to be set up, and may result in softening or ulceration. The state of distention being more or less permanent, the coats of the bladder in their diseased condition will readily give way, and the urine may be forced into the peritoneum, or cellular tissue of the pelvis.

The retention of urine which leads to rupture of the bladder, may depend on several causes. Impermeable stricture of the urethra, enlargement of the prostate, any lesion, in a word, which effectually opposes the discharge of urine, may lead to retention, over-distention of the bladder, and rupture. Sir E. Home and Sir B. Brodie have recorded cases of rupture of the bladder from stricture. M. Amussat met with a case in which the accident arose from enlargement of the prostate. It may even take place without the existence of a mechanical obstacle.

---

[1] New York Journal of Medicine and Surgery, new series, vol. vi., p. 374.
[2] Dublin Journal of Medical Science, vol. ix., p. 372.

Thus, Hauff relates the case of a person who was seized at the dinner-table, after having drunk a quantity of wine, with violent pain in the abdomen; the abdomen, however, was not tense, nor was there any tumefaction over the region of the bladder; a small quantity of urine only was discharged through the urethra. The patient died in four days, with symptoms of violent peritonitis; the bladder was small, thickened and perforated; urine was effused into the cavity of the abdomen.[1]

In cases of this kind, on post-mortem examination, the peritoneal cavity, or the cellular tissue of the pelvis, according to the seat of the perforation, is found in a highly inflamed state. Near the fundus of the bladder there are one or more perforations, with black and ragged edges; while the viscus is relaxed and empty, and the abdomen or pelvic tissues contain a quantity of urine.

After typhus fever and some other debilitating affections, perforation of the bladder occasionally occurs; a ragged irregular opening, through which the urine has escaped, being found after death, and generally in the posterior part of the organ. Such a condition, however, could not occur without great neglect on the part of the practitioner. He should not be satisfied with the nurse's report that the patient is passing water freely, for this may be quite correct and yet the bladder may remain enormously distended, but should examine the hypogastrium at least twice daily, and, if the expulsive power of the bladder seems to be in abeyance, a catheter should be regularly introduced.

Rupture of the bladder occasionally takes place during parturition. The mechanism of the accident in such cases is easily understood. A calculus in the bladder of a parturient woman has been known to cause rupture of the organ during the progress of labor.

In this last class of cases, retention of the urine arises in consequence either of neglect or of external pressure; the head of the child compresses the distended organ: and rupture of its walls may ensue directly in consequence of this pressure or from inflammation and sloughing. In a case of this kind, related by Mr. Hey,[2] the symptoms attendant on retention of the urine did not take place till the second day. The patient, who was thirsty and vomited, had frequent desire to void urine, which she did very suddenly, but not in greater quantity than a teaspoonful at a time. The pulse was quick, the belly swelled, and pressure gave her pain. She died about the eighth day, and the bladder was found to be ruptured at its upper part. A case is also related by Mr. Bedingfield,[3] in which the bladder seems to have given way during a very easy labor of only two hours' duration. The patient died of peritoneal inflammation.

Before passing on to a description of the treatment of rupture of the bladder, Dr. Bartels' statistics may be again referred to. His account includes 169 cases, 152 (90 per cent.) of which terminated fatally. In 94 instances the peritoneum was injured, and only one of these patients survived. Of the other 75 cases in which the peritoneum was not involved, 46 patients died.

The following table taken from Dr. Bartels' paper shows the relative frequency of the various forms of injury to the bladder, the mortality among each class, and the frequency with which the accident is complicated with injuries of other parts.

---

[1] Civiale, Traité Pratique, t. iii., p. 84.
[2] Medical Observ. and Inquir., vol. iv.
[3] Lancet, June, 1837.

WOUNDS AND INJURIES OF THE BLADDER.

Class of Injury	Total	Deaths	Mortality per cent.	Intra-peritoneal		Extra-peritoneal		Rectum injured	Bones injured
				Deaths	Cases	Deaths	Cases		
Incised wounds.	50	11	22.	10	10	1	40	13	.....
Gunshot wounds	285	65	24.5	27	27	38	258	60	131
Laceration (rupture)	169	152	90.	93	94	46	75	1	65
Total	504	228	45.	130	131	85	373	74	196

A question of considerable importance in a medico-legal point of view may arise in connection with wounds of the bladder. A man receives several injuries from the effect of which he dies; on post-mortem examination his bladder is found ruptured or lacerated, and it may be important to determine the manner in which the injury was caused. Dr. Taylor refers to such a case, quoting the details from Professor Syme. A man received a kick on the abdomen during a brawl. He went home, and an hour after the injury fell with his abdomen against the doorstep. He died in four days, with symptoms of laceration of the bladder, and the diagnosis was confirmed by the autopsy. The question arose as to how the injury to the bladder occurred. The surgeons attributed it to the fall. Dr. Bartels points out that it was more probable that the injury was caused by the kick. A partial or incomplete rupture may then have taken place, involving perhaps only the muscular coat. Subsequently the other coats gave way, possibly after the fall, and the fatal symptoms became developed. Such cases, however, can very rarely admit of any positive opinion being given.

In the treatment of rupture of the bladder, the main indications to be fulfilled are simple and sufficiently manifest : a highly irritating fluid has been effused into the cavity of the peritoneum, or the cellular tissue which envelops the bladder at those points uncovered by serous membrane : we cannot withdraw the effused urine from the cavity of the abdomen, nor in the first instance from the pelvic cellular tissue, while the irritating fluid is almost sure to produce violent inflammation of the textures with which it comes in contact. Our first object, then, must be to oviate the risk of further effusion by keeping the bladder as empty as possible ; our next will be to combat any inflammatory action that may arise ; and in case of pelvic infiltration our final object will be to limit the effect of the effused fluid by giving egress to it whenever circumstances will permit. The means by which these several indications may be fulfilled are so obvious that it seems almost unnecessary to dwell on them. The patient should be kept perfectly quiet in the recumbent posture, and the depressing effects of the shock obviated by the usual remedies. To prevent any accumulation of fluid in the bladder, and the consequent risk of further extravasation, the catheter should be passed frequently, or, if possible, a full-sized catheter should be kept constantly in the bladder. Pain should be combated by opiates in full doses, and warm applications to the abdomen.

As soon as the symptoms of shock have been relieved, the antiphlogistic treatment should be commenced, if the patient be young or strong, and vigorously pursued as the symptoms of peritonitis become developed. No internal remedies can be of any avail, if the urine has been effused into the cavity of the abdomen; but the effusion may be extra-peritoneal, and the inflammation of the serous membrane may be secondary. Under these circumstances, the serous inflammation may be, perhaps, confined to that portion of the peritoneum which adjoins the mass of effused urine, as was seen in the case quoted from Dupuytren, and there may be a faint chance of saving the patient's life. The application of leeches to the abdomen, and the administration of calomel with opium, are the most powerful means of arresting peritoneal inflammation; but, if it be evident that the effusion has taken place into the extra-peritoneal cellular tissue, we must be cautious how we push debilitating measures to excess. The ill-conditioned inflammation, caused by the contact of urine with cellular tissue, is little amenable to blood-letting or any internal remedies; on the other hand, every means must be employed to prevent extension of the inflammatory action to the peritoneum, or to limit its progress when it does ensue.

The treatment in cases of this kind, that is, when the urine has been effused into the cellular tissue, should be the same as that adopted for infiltration of urine after the lateral operation for stone. The two cases are nearly identical, the only difference, so far as the urinary organs are concerned, being that in one case we have a lacerated, in the other an incised wound of the bladder. Hence, while the bladder is kept empty by the introduction of the catheter, the signs of urinary infiltration should be carefully watched. The urine may make its way above the pubes, and become infiltrated beneath the abdominal muscles, but more frequently it occupies the cul-de-sac, comprised between the peritoneum above, the rectum and bladder on the sides, and the perineum inferiorly. From the original source it may pass in various directions, as the cases of Mr. Hewett show, into the iliac fossæ, the scrotum, or even to the thighs, through the obturator foramina. An important point will be to discover the presence of infiltrated urine at the very earliest period possible, and then to provide for its evacuation by free incisions; a perusal of Mr. Syme's valuable case will at once show the importance of this precept. Harrison has even proposed that an incision or puncture should be made, with a curved bistoury or a trocar, through the rectum, into the cul-de-sac already spoken of.

During the last few years important modifications in the treatment of these injuries have been suggested and adopted by several surgeons, and the reports of cases in which recovery has taken place tend to show that rupture of the bladder is not necessarily so fatal an injury as was formerly supposed. Peritonitis and extravasation of urine are, as above described, the main sources of danger, but the former need not occur if the injury affects a part uncovered by serous membrane, and extravasation of urine into the cellular tissue may be combated by prompt treatment. Even if the peritoneum be involved, the injury is by no means necessarily fatal, for the contact of urine with the serous membrane does not always produce inflammation. Mr. le Gros Clark[1] states that "in some instances the presence of the urine seems to be tolerated almost passively by the serous membrane," and that "it seems not improbable that urine

---

[1] Lectures on the Principles of Surgical Diagnosis, p. 341.

may be absorbed by the peritoneum." Several cases have been published which appear to support this latter view. In a case, reported by Sigallas,[1] of rupture of the bladder in consequence of prostatic abscess, the symptoms and post-mortem appearances led to the belief that most of the urine originally effused into the peritoneum had become absorbed. Peritonitis, however, had been set up and had caused death. The tolerance of the presence of urine by the peritoneum is shown by a somewhat remarkable case recorded by Mr. Thorp,[2] and which he treated by washing out the bladder and peritoneal sac with warm water from time to time, a catheter being kept in the urethra, with its extremity projecting into the bladder. Calomel and opium were freely administered, and the patient made a good recovery. Mr. Bryant proposes to make a free incision through the perineum into the bladder, so as to provide an outlet for the urine as it reaches the organ. In this suggestion he has been anticipated by Dr. W. J. Walker, of Boston, U.S., who, more than thirty years ago, thus treated with success a case of rupture of the bladder.[3] Another American surgeon, Dr. Erskine Mason, of New York, has more recently adopted the same mode of treatment. In his patient, the rupture involved the posterior wall of the bladder, and peritonitis had supervened. Sixty-two hours after the accident, a free incision was made into the perineum, as in the lateral operation of lithotomy. The symptoms of peritonitis subsided, and the man was discharged cured on the thirty-seventh day after the injury.

Rupture of the bladder has been treated still more heroically, but with equally successful results, by another American surgeon, Dr. Walter, of Pittsburg, U.S.[4] The patient was a man, twenty-six years of age, whose bladder had been ruptured by a blow on the abdomen. All the symptoms of such an injury were present; a little bloody urine was drawn off by means of a catheter, but no relief was obtained. Ten hours after the accident, the abdomen was opened by an incision commencing one inch below the umbilicus and terminating one inch above the pubes. A pint of bloody urine was removed by sponging out the peritoneal cavity, and a rent two inches long was found in the base of the bladder. The abdominal wound was closed up by pin suture, opium was given in full doses, and a catheter retained in the bladder. During the third week the catheter was used every four hours, and the patient made a good recovery. Mr. Holmes[5] thinks that such an operation is justifiable when the laceration appears to involve the peritoneum, and suggests that the wound in the bladder should be closed by means of silver, or carbolized gut sutures. This additional procedure, however, was adopted in another case reported by Mr. Alfred Willett.[6] After removing the urine and blood from the peritoneal sac, the wound in the bladder, three and a half inches long, was united by means of eight silk sutures. The abdominal wound was closed, a drainage-tube inserted, and a catheter retained in the bladder. Some relief to the pain and other symptoms followed the operation, but the patient died twenty-two hours afterwards, or fifty-one hours after receipt of the injury. On post-mortem examination, the wound in the

---

[1] New Syd. Soc. Retrospect, 1873–74, p. 323, from Marseille Médical, 1874, x.
[2] Dublin Quarterly Jour. of Med. Science, Nov., 1868.
[3] Gross, loc. cit., p. 323.
[4] Philadelphia Med. and Surg. Reporter, Feb., 1862.
[5] Treatise on Surgery, 1st edition, p. 216.
[6] St. Barth. Hosp. Reports, 1876.

bladder was found to be partly closed, but at one spot there was an opening through which fluids passed.

A consideration of the cases thus briefly summarized will suffice to show that the prognosis in cases of rupture of the bladder is more hopeful than has been hitherto supposed, and that there are various operative measures which the surgeon would be justified in adopting. A free incision into the perineum would certainly not increase the gravity of the case, but the same remark can scarcely be applied to the operation of abdominal section. Strict antiseptic precautions, however, would doubtless greatly lessen the dangers of this latter procedure. The highly satisfactory results of ovariotomy combined with the antiseptic method show that the mortality after injuries to the peritoneum may be very considerably reduced. For the injuries under discussion the operation in question offers the only prospect of success in cases otherwise hopeless, and, with the additional and valuable aid of the antiseptic method, it would seem that it is not only justifiable, but that in the interests of the patient its performance is urgently demanded. Further experience alone can show whether sutures should or should not be applied to the wound in the bladder.

CHAPTER VI.

ACUTE INFLAMMATION OF THE MUCOUS MEMBRANE OF THE BLADDER.

The bladder is liable to be attacked by various forms of the inflammatory process, which may involve one or more of the structures of which it is composed. The mucous membrane is the part most commonly affected, and inflammation of this tissue may be either acute or chronic, and of the catarrhal, croupous or diphtheritic kind. The peritoneal covering and the sub-serous cellular tissue may also be the seat of inflammation, while in other cases the muscular coat of the organ is the part most extensively involved. Inflammation of these tunics, however, almost invariably results from extension of the disease from the mucous membrane. Cystitis, in one or other of these forms, is a common disease, and it may occur alone, or may, as is often the case, complicate other affections of the urinary organs.

Acute inflammation of the mucous membrane of the bladder is always a severe disease, and often attended with danger. It occurs under a variety of circumstances, to which attention should be paid, for not only the appearance and progress of the malady, but its termination also, will be modified by the causes which give rise to the attack, and the circumstances attending its development.

The disease may be confined to a small part of the mucous membrane, or spread over a great portion ; it may be idiopathic, but this form is comparatively rare ; the bladder resists many influences which would inevitably excite acute inflammation in other mucous membranes. More frequently by far the disease is produced under the influences of causes which are sufficiently obvious. In many cases the acute attack supervenes on chronic inflammation of the mucous membrane which has already existed for a considerable time ; in others, the inflammation extends from the urethra, or from some of the pelvic viscera.

Direct irritation, of a mechanical kind, is a frequent cause of cystitis, as is likewise the use of certain chemical substances which have a specific action on the genito-urinary mucous membrane. Thus the disease may be caused by operations on the bladder, external violence, wounds, the unskilful introduction of catheters and similar instruments, or the retention of such instruments for too long a period ; by injuries sustained during difficult labors, or by the presence of calculi or other foreign bodies. Among the chemical irritants may be mentioned turpentine, cantharides, and the balsams. Alkaline decomposition of urine, under circumstances of partial retention, is another exciting cause, and this decomposition has been attributed to the presence of bacteria [1] introduced

---

[1] Traube, referred to by Niemeyer, Lehrbuch der spec. Pathologie und Therapie. Aufl. ix., Bd. 2, s. 80 ; see also English translation of same text-book, 8th edition, vol. ii., p. 67. Niemeyer alludes to a case, occurring in his clinic, in which alkaline fermentation of the urine appeared to have been caused by the introduction of dirty catheters into the bladder. The urine contained many vibriones and fungi, but neither cellular elements nor large quantities of mucus.

upon a catheter, or gaining access to the bladder in some other way. The extension of inflammation from the urethra, as may occur in gonorrhœa, gives rise to a very acute form of the disease. Inflammatory action may also extend from the prostate, rectum, vagina, or uterus. Cystitis is also sometimes due to chilling of the general surface of the body, or exposure of the lower part of the abdomen and feet to cold and moisture. A rheumatic or gouty habit of body is said to predispose to the disease.

Croupous or diphtheritic inflammation of the mucous membrane of the bladder is an occasional symptom during the course of certain severe febrile affections, such as pyæmia, typhoid and scarlet fevers, other mucous membranes being at the same time similarly affected. This form is that which occurs as a result of poisoning by cantharides, and it has also been noticed as a consequence of retention of urine with marked alkaline decomposition.

The severity and extent of the inflammation arising from wounds and other external injuries will mainly depend on the nature of the wound and the circumstances under which it has been inflicted. The health of the patient at the time of attack will also exercise considerable influence in determining whether the cystitis shall be partial, or diffused over the whole mucous membrane. It is on this account, perhaps, that the inflammation which succeeds lithotomy so often proves fatal; but this form should be referred rather to general inflammation of all the coats of the bladder than to inflammation of its mucous membrane alone.

It has been mentioned above that inflammation of the bladder may be connected with, or dependent on, several other morbid states of that organ. It is, therefore, necessary to distinguish the essential symptoms of the disease from those which may be called accidental, inasmuch as the accidental symptoms arise from the morbid states which contributed to the production of the complaint.

Unless it supervenes on the chronic form, acute inflammation of the bladder generally commences in a sudden manner, and runs a rapid course. The patient first experiences some pain in the region of the bladder; this is quickly followed by frequent and irresistible desire to make water, which is voided at short intervals, and in small quantities; these two symptoms rapidly increase until they acquire a most distressing degree of intensity. The pain is first experienced above the pubes, and may be dull for a short time; but it soon becomes violent and extends along the urethra, shooting into the perineum and down the thighs. Pressure over the pubes or on the perineum greatly increases the pain, and if the posterior wall of the bladder be examined by introducing the finger into the rectum, it will be found that this part of the organ also is extremely sensitive. The slightest movement of the body increases the pain; the patient lies with the limbs drawn up, so as to relax the abdominal and pelvic muscles.

As the disease advances, the discharge of urine takes place very frequently, for the desire to make water becomes more and more urgent, there is a sense of heat and burning along the urethra, and the pain felt in passing a few drops of urine is often compared by the patient to the passing of molten lead. At the early stage, the pain subsides after the urine has ceased to flow, but it returns as soon as a small quantity of fluid collects in the bladder. The mucous membrane, in fact, soon becomes altogether intolerant of the contact of the urine; and the agony thus produced, together with the incessant and irresistible calls to evacuate the organ, are valuable diagnostic signs of acute inflammation. The irritability of the bladder is communicated to the rectum, and tenesmus is, consequently, a frequent symptom.

The condition of the urine varies with the stage of the disease; it is at first mucous, then tinged with blood during the height of the disease, and, in many cases, purulent towards the end. These conditions will be subsequently referred to.

Whenever the inflammation is severe, and occupies any considerable portion of the mucous membrane, constitutional symptoms, as might be expected, quickly set in. Rigors occur, and symptomatic fever is developed, severe in degree, but often of a nervous character and attended by great disturbance of the digestive organs. Hence, vomiting frequently occurs, and the patient is a prey to nervous symptoms of a very distressing nature.

After a few days, unless the disease be arrested, the local symptoms acquire an extraordinary degree of severity; the pain is constant and of a most distressing kind, being accompanied by incessant but unavailing efforts to empty the bladder; the urine comes away by drops, and the bladder gradually loses its expulsive power, though aided by violent contraction of the abdominal muscles. A firm, painful tumor may now often be felt above the pubes, for retention is succeeding the incontinence produced by irritation; the sufferings of the patient, carried to their highest pitch, may appear to subside a little; but this is deceptive; the character of the fever changes; the countenance becomes anxious, and the features sunken; general prostration increases rapidly; low delirium sets in; there is often hiccough; the pulse becomes weak, irregular, and then intermitting; the skin is covered with a clammy perspiration; and the patient finally sinks about the eighth or tenth day in a state of prostration or absolute coma.

Such is the general history of acute inflammation of the bladder of a severe and fatal type. This form is that which sometimes occurs after operations on the bladder, or external violence of various kinds. There is, however, another and a more common form of acute cystitis of a less serious nature. This form is most commonly met with as a sequela of gonorrhœa, but it may arise from any of the causes previously mentioned. The symptoms are milder in character, and the neck of the bladder would appear to be the part mainly affected. The constitutional symptoms are either slight or altogether absent. This form of the disease usually terminates in resolution during the second week. The symptoms gradually subside; the pain abates; the calls to pass water become less frequent, and the urine gradually assumes its normal characters. Generally speaking, however, some amount of irritability of the bladder remains for a considerable period, and any indiscretion on the part of the patient is liable to be followed by a return of pain and other symptoms.

The condition of the urine requires some further notice. In the milder forms of cystitis the color of the urine is normal, but the fluid itself is cloudy when passed. The reaction is acid, but is apt to become alkaline after the urine has stood for a few hours. The sediment is considerable in quantity, cloudy and easily miscible with the urine on stirring. It consists mainly of mucus which has been poured out in increased quantity. Under the microscope numerous epithelial cells from the vesical mucous membrane and smaller granular corpuscles can be abundantly demonstrated. After the urine has stood a few hours, crystals of the ammoniaco-magnesian phosphate make their appearance.

In the severer forms of cystitis the urine is at first cloudy and dusky yellow in color. The cloudiness is due to the presence of pus and mucus. The sediment is greenish yellow in color, and adheres to the containing vessel. The reaction of the urine is alkaline. Albumen is present in a proportion corresponding to the pus present. The urine contains a considerable quantity of carbonate of ammonia. The sediment consists of pus mixed with crystals of the ammoniaco-magnesian phosphate, and amorphous phosphate of lime. Blood-corpuscles, urate of ammonia, and numerous epithelial cells from the bladder may be likewise present. In the advanced stages of severe cystitis, the urine becomes much darker in color, owing to the disintegration of the blood-corpuscles by the carbonate of ammonia. The urine has a strongly ammoniacal and fœtid odor, and sometimes contains sulphuretted hydrogen or sulphide of ammonium in sufficient quantity to blacken a catheter. Albumen, the coloring matter of the blood, and carbonate of ammonia are present in considerable quantity.

It is often a point of some difficulty to decide whether the albumen in purulent urine is derived solely from the pus, or is a symptom of co-existing disease of the kidneys. Such a question would assume considerable importance in cases of stone in the bladder, with acute or chronic cystitis as an accompaniment, and symptoms pointing to some affection of the kidneys. The quantity of albumen, however, derived from liquor puris is comparatively small; the presence of a large quantity would indicate serious disease of the kidneys. The presence or absence of tube-casts would assist the diagnosis. Should the inflammation extend along the ureters to the kidneys, pus will be found in increased quantity, and the deposit will also contain the irregularly shaped epithelial cells of the pelvis and infundibula. These and other varieties in the appearance of the urine depend on the condition of the kidneys and mucous membrane of the bladder, and the exciting causes of the attack. Though the urine is discharged in very small quantities at a time, the total in twenty-four hours may equal the normal amount.

When pus and mucus exist in the urine in small quantities, it is difficult with the unassisted eye to distinguish one from the other. Pus, however, may be distinguished from mucus, when diffused through a fluid, by rendering the latter opaque. Urine containing pus is more or less turbid and milky when passed, but after it has stood for some time the pus subsides to the bottom of the vessel, and the fluid resumes its transparent character. If mucus be present likewise, the pus is found lying on the mucus, and presents a much yellower tint; it is also quite opaque, whereas mucus is more or less transparent.

The microscope affords at once the most ready and certain mode of detecting the presence of pus in the urine. The pus-corpuscle is spherical, granular on the surface, and yellowish in color; its diameter is $\frac{1}{3500}$th of an inch. It is about one-third larger than a red blood-disk. If the urine be acid, the nucleus of the pus-corpuscle may be seen without the aid of reagents. Should, however, the peculiar shaped nucleus not be clearly detected before the addition of acetic acid, a drop of that fluid, by rendering the envelope of the corpuscle transparent, will instantly bring it into view. The characteristic nucleus of the pus-globule is ordinarily tripartite, sometimes quadripartite, more rarely only bipartite; each division of the nucleus is biconcave, and measures about $\frac{1}{8000}$th of an inch in diameter.

The only bodies with which the pus-corpuscles, when passed in considerable quantity with the urine, can be confounded, are young epithelium scales. These, however, are readily distinguished by their solitary nucleus; although a pus-corpuscle may occasionally be found in which only one undivided nucleus is to be seen, it will be lost amid the numbers which present their normal tripartite nuclei. It is probable, however, that the epithelial cells become transformed into pus-corpuscles.[1] There is no strict line of demarcation between the corpuscles of mucus and those of pus; they pass, by insensible gradations, into each other. The corpuscles of mucus, however, usually contain only a single nucleus, and they are somewhat larger and more regular in outline than those of pus. When solution of potash or ammonia is added to pus, a viscid mass is formed by the mixture.

A copious sediment, apparently resembling pus, is sometimes produced by the presence of triple phosphate in the urine. The nature of the deposit is shown by the rapidity with which it is dissolved in nitric acid, and by the shape of the crystals when a little of the fluid is placed beneath the microscope.

Urine containing pus may have an acid or an alkaline reaction. In the former case the pus ordinarily owes its origin to pyelitis, in the latter to cystitis. If urine, very turbid from the presence of pus, be allowed to stand for a while, it becomes clear, and throws down at the same time a more or less abundant sediment; the supernatant fluid frequently has a peculiar greenish yellow tint, and contains albumen. The importance of determining whether the albumen in purulent urine is derived from the liquor puris or directly from the blood, has been already alluded to.

The morbid appearances of acute catarrhal inflammation of the bladder are the same as those produced by the same disease in other mucous membranes. We find, as in cases of enteritis, hyperæmia, swelling and softening of the mucous membrane; the redness being more or less uniformly distributed, or appearing in streaks or patches. The surface of the membrane is covered by a copious muco-purulent secretion; it readily bleeds on being scraped, and it often exhibits minute erosions. The inflammation may extend over the whole inner surface of the bladder, or be limited to some particular part of it; that part which adjoins the neck of the organ is most prone to be affected. When the inflammation has been very severe and has lasted for some time, the mucous membrane becomes still more swollen, and its folds project considerably; it loses almost all its epithelial covering, and exhibits numerous patches of extravasated blood. The muscular coat is also seen to be thickened, and abscesses are occasionally found between it and the mucous membrane. Some of these may have burst into the bladder, or externally beneath the peritoneum. In the croupous and diphtheritic forms of cystitis, layers of false membrane are found adhering to the surface, or the deeper layers of the mucous membrane are infiltrated with plastic material, and in addition to detachment of epithelial cells, necrosis of large portions of the mucous membrane has been occasionally observed. The layers of false membrane are often coated by phosphatic deposits. They may be expelled during life,[2] and be mistaken for structures of another kind. These flakes of false membrane may be found to extend for some distance into the urethra or even into the ureters.

---

[1] For a short account of the most recent views with reference to the pathology of catarrhal inflammation of mucous membranes, see Billroth's Surgery, New Syd. Soc. Trans., vol. i., 372.

[2] See Discussion at Path. Soc., London, reported in Medical Times and Gazette, 1863, vol. ii., pp. 522, 678.

It is frequently difficult to decide as to the existence of ulceration when resulting from inflammation; but there is always reason to suspect it when disease of the bladder has been of long continuance, when blood is found in the urine, and the pain is extensive and increasing. The ulcerative process is attended with constant pain and irritation, keeping up the desire to void the urine, which is never suffered to accumulate; while, at the same time, increased difficulty and pain generally attend micturition.

The neck and base of the bladder are the regions in which ulceration most commonly occurs, and the process may spread over a greater or less extent of the mucous membrane. The ulcers vary in shape; sometimes they are circular with a well-defined edge, sometimes they are irregular in outline and present ragged and puckered borders. The walls of the organ generally become thickened, and the muscular fibres, denuded of their mucous covering, stand out in strong relief. Cases are recorded in which ulceration has extended so deeply into the substance of the bladder as to cause perforation of its parietes and extravasation of urine into the abdomen. In these cases, life is soon destroyed by peritoneal and cellular inflammation. But when the ulcerative process extends to the other tunics of the bladder, it usually happens that it is accompanied by effusion of lymph, from adhesive inflammation exterior to the ulcer, by which process the bladder becomes united to the neighboring parts, and the escape of urine is prevented. In these instances a communication is sometimes formed between the fundus of the bladder and the ileum or the sigmoid flexure of the colon, or between the under surface of the bladder and the rectum; in the former of which cases, the feces pass into the bladder, and through the urethra; while in the latter the urine generally passes into the rectum, and is voided with the feces.

Mr. Wilson[1] says, "I have preserved the bladder and ileum of a person, which had adhered fifteen years before the death of the patient. Ulceration to a large extent had taken place in this connected part: and for the whole of the above-mentioned period, the feces readily passed from the ileum into the bladder. The patient died when sixty-eight years of age. Being a female, the shortness of the urethra had allowed the substances which passed into the bladder a tolerably free escape, and no calculus formed." Dr. Gross records the case under his care, of a clergyman, eighty-five years of age, from whose bladder fecal matter has been discharged for upwards of a quarter of a century. His health, with the exception of an occasional attack of colic, has been excellent during all this period. The passage of feces by this route was at first intermittent, but of late it has been almost constant.

Ulceration between the bladder and rectum is a somewhat rare occurrence. Mr. Coulson met with a case of a gentlemen, æt. sixty-two, in whom the feces found their way into the bladder, and were passed by the urethra. He complained of frequent desire to make water, which was attended with smarting and scalding during its passage. He had also great distention about the lower part of the abdomen. Micturition was usually followed by severe darting pains at the end of the penis, and by the escape of flatus through the urethra. The urine, which was acid and albuminous, contained much feculent matter, with a good deal of mucus. It was clear that in this case there existed a communication between some portion of the intestinal canal and the bladder, through

---

[1] Lectures on the Urinary Organs, p. 317.

which the feces passed; and, during their evacuation, the pain was intense. The bowels were pretty regular; the motions were often solid; and their passage was not attended with pain. The treatment which was adopted consisted in the use of suppositories and other sedatives, so as to keep the rectum and bladder in as quiet a state as possible. Some time after, when the fistulous opening seemed to be closed, and the general health was much improved, hæmoptysis occurred. During the treatment pursued for the relief of this attack, there were constant feculent discharges from the bladder attended with great exhaustion.

After this attack, the balance of the circulation was never restored; tympanites occurred, with distention of the colon to an enormous magnitude; ascites followed the inflation of the colon, serous diarrhœa and anasarca supervened, and in a few months afterwards the patient died. For some weeks prior to death the escape of feces through the bladder had altogether ceased.[1]

On dissecting back the integuments, and laying them open, the colon presented the appearance of a large tromboon, amazingly distended, and stretching across the umbilical region, with its sigmoid flexure, equally large, in the left iliac fossa. The space above its line was occupied by an enlarged liver, and that below by the small intestines. The omentum was shrivelled, and drawn across to the right iliac fossa, where it was adherent to the parietes above the internal abdominal ring. The peritoneum was universally opaque and injected. The sigmoid flexure of the colon, the ileum, and the cæcum, with its appendix, were each and all adherent, *en masse*, to the fundus of the bladder, and involved in a general thickening of the surrounding textures. The bladder, with the adherent intestines, was removed, and a particular dissection revealed the following appearances.

The colon hypertrophied, singularly muscular, and in circumference about the size of a man's arm, was, together with a convolution of the ileum, and the appendix cæci, adherent to the fundus of the bladder. The natural calibre of the gut was impeded by a contraction or stricture, which commenced inferiorly in the rectum, about a forefinger's length from the anus (just at the base of the triangle formed by the vesiculæ seminales), and extended upwards for about two inches, barely admitting the entrance of the little finger. A section of the gut in this part resembled scirrhus: and the glands were, with the surrounding tissues, thickened immediately above this stricture; the coats of the bowel were studded with ulcerations and openings leading into a channel which separated the bladder from the intestine. This channel was, in fact, a fecal abcess, situated beneath the reflected portion of the peritoneum, between the bladder and bowel. It was degenerate in structure, lined with a dark membrane, and filled with a muco-purulent secretion. It opened anteriorly into the fundus of the bladder above, into the colon; below, into the rectum: and posteriorly, through the colon into the ileum; so that there was a false passage, by which the natural course of the colon was diverted, and forced between the bladder and strictured part of the intestine down into the rectum below, and at the same place, by means of a fistulous opening into the bladder, in front. The orifice of this fistulous opening within the bladder was curtained by a fungous growth or thickening which overhung it like a valve. Thus, exactly at this point of the fecal abscess, there was this strange lesion, the

---

[1] Vide Guy's Hospital Reports, No. xiii., p. 405.

colon, the rectum, and the ileum—each, conjointly with the fecal abscess, possessing one common entrance into the bladder itself. Within the bladder, the rugæ of the mucous membrane were vascular, and the muscular coat was considerably hypertrophied.

In another case under Mr. Coulson's care, a communication existed between the fundus of the bladder and the colon, just above a stricture in the sigmoid flexure. The patient complained of a deep-seated dull pain in the lower part of the abdomen, extending from the pubes to the sacroiliac symphysis on the left side; he said he had been affected with pain in the bowels for nearly twelve months: and he also complained of considerable uneasiness in the urethra, and of a burning sensation towards the extremity of the penis. The urine was high colored. On pressing the hypogastric region with firmness, he suffered considerable uneasiness; his tongue was dry, a little furred, and of a brownish hue ; his pulse was quick, about 110, and rather firm. He was bled from the arm to eight ounces; a dozen leeches were applied to the perineum, and a blister to the lower part of the abdomen; a purgative was administered. On the following day, the pulse was softer, and the urine high-colored, with a very considerable deposit of a purulent kind, and of a very fetid smell. Liquor potassæ and extract of hyoscyamus were then given. On the following day, the urine contained a considerable quantity of feculent matter, which led to the conclusion that there was ulceration of the intestines. Flatus was also passed through the urethra; stools, voided per anum, were always loose, and the patient said he had not passed a solid stool for a considerable time. On examination per rectum, the finger could not reach any ulceration or stricture; and it appeared that no urine had ever passed per rectum. The pain and uneasiness increased; and the treatment was confined to giving various forms of opium. The strength of the patient gradually decreased, and he died worn out by his sufferings. On examining the body a stricture was found in the sigmoid flexure of the colon, and above the stricture there existed slight adhesion with the fundus of the bladder, toward the left side. There was an opening through the stricture, about the size of a goose-quill, through which some feces passed into the rectum ; but from the narrowness of this orifice it was impossible that any solid feces could pass. Two inches of the intestine above the stricture, as well as part of the fundus of the bladder, appeared gangrenous, and in a state approaching to sloughing; the coats of the bladder were thickened; the bladder itself was not bigger than a large orange ; and from being much thickened, it appeared incapable of further distention.

This case is a good illustration of the effort which the system at times makes to evacuate what is extraneous by the most direct passage, when there exists an obstacle to its expulsion in the natural way. The more usual course, however, is, that ulceration gradually extends over a considerable portion of the mucous membrane, and then the muscular structure is shown more clearly than any dissection can exhibit it. In the progress of the ulceration, disease commonly manifests itself in one of the kidneys; and more frequently in the left. This is indicated by pain in the loins on pressure (for the pain is never severe), by shiverings, sickness, and the albuminous state of the urine. At this stage, large quantities of pus are voided with the urine, and the latter is tinged with blood.

With ulceration of the mucous coat there is generally combined hypertrophy of the muscular coat, and a contracted state of the bladder.

The urine, no longer contained in a viscus capable of dilatation, partly escapes involuntarily by the urethra, and is partly thrown back upon the ureters, which become dilated and tortuous. The mucous membrane, extending upwards along the ureters to the pelvis and the infundibula of the kidneys, becomes inflamed and rough, and pours forth a quantity of pus. The glandular structure of the kidney undergoes slow absorption from pressure. The capsule adheres with preternatural firmness to the exterior of the gland. Upon making a section of a kidney so diseased, we find that, although apparently of large size, it consists in great part of dilated tubes, and that the true vascular or secreting part is in smaller proportion than usual.

In another case, where the symptoms had been those of inflammation, the whole mucous membrane was found detached in the form of a smooth, grayish-white, continuous layer. It appeared to have sloughed away during life, and to have separated from the surrounding muscular tunic. At first it might have been taken for a false membrane of organized lymph; but upon closer inspection it was seen to possess traces of the normal structure. The bladder, when first divided, seemed to contain a large cyst or bag of hydatids; for the yellow membrane, partly distended with urine, protruded through the cut surface. It was attached to the muscular coat both at the neck of the organ and at the entrance of the ureters, where the mucous membrane was red and congested, and more loosely connected than natural to the adjacent structures, but not deprived of vitality. Under the microscope the well marked fibres of areolar tissue were seen, in which, however, the yellow elastic tissue prevailed; the white fibres having in great part undergone decomposition.

Several similar cases have been reported by various authors. They would appear to be instances of the diphtheritic form of the disease. The Museum of the Royal College of Surgeons of England contains a preparation of a saccular membrane, six inches by four, which was removed by the supra-pubic operation from the bladder of a patient suffering from retention of urine. The outer surface of this membrane is flocculent and in parts distinctly fibrous, the inner surface granular and reticulated, like superficially ulcerated mucous membrane. Its shape indicates that it lined the whole interior of the bladder, and was thrown off from it in one piece. This, and other cases sufficiently prove that exfoliation of the mucous membrane of the bladder really occurs, though the fact has been doubted. Mr. Spencer Wells has met with a case in which nearly the whole of the mucous membrane of the bladder was passed through the urethra by a woman, six weeks after a tedious labor. The patient recovered, and could retain her urine for some time, but lost the power of expelling it. The exfoliation was attributed to the caustic action of the urine, which had become ammoniacal from retention. A similar case was communicated to the Obstetrical Society (July 1st, 1863) by Dr. Martyn. In another case, which occurred to Mr. Maunder,[1] doubts were expressed as to the real nature of the membrane, indeed, the examiners appointed by the Pathological Society stated that the specimen formed no part of the bladder of a woman. This decision was, however, subsequently reversed. The circumstances of the case were all favorable to the woman's statement. She had suffered from retention of urine caused by retroversion of the uterus. A catheter,

---

[1] For report of discussion at Path. Society, see Medical Times and Gazette 1863, vol. ii., pp. 522, 678. Conf. Path. Trans., 1862, vol. xiii.

introduced to relieve her, had slipped into the bladder and remained there three days. Acute cystitis followed, and had led to the formation of the cast. As Mr. Spencer Wells remarked, the possibility of the existence of these casts should be borne in mind, and it is of great importance to understand their real nature, so that retention may be guarded against. When suspected to exist they should be sought for, and if found, carefully removed.

If the inflammation reach a high degree, the muscular coat is also attacked, presenting here and there gangrenous spots, or even being completely destroyed by it. But as the muscular coat is connected but loosely to the inner membrane of the bladder, the inflammation does not easily pass from one to the other.

Gangrene is a very rare termination of acute cystitis, but it may occur in traumatic cases of the disease. Besides the general symptoms of gangrene, the blackish or brownish color of the urine, and its fetid cadaveric odor will sufficiently indicate the state of the case.

Acute inflammation of the mucous membrane of the bladder cannot easily be confounded with any other affection of that organ. This applies especially to idiopathic cases; for in others, connected with the presence of foreign bodies, etc., in the bladder, the symptoms arising from the cause may, in some degree, mask those which depend on the inflammation thus excited. It may be more difficult to distinguish inflammation of the mucous membrane from inflammation of the deeper seated tissues. Many surgeons deny the possibility of our drawing such a distinction, and, in a practical point of view, it is not of very great importance; but when the muscular tissue is chiefly involved, it generally happens that the power of passing urine does not exist, and the desire to void it is less frequent, as it is not experienced until a good deal of urine is accumulated in the bladder, and then comes on in violent paroxysms. Neither is there the burning sensation along the urethra which is felt when the mucous membrane alone is affected.

The disease may sometimes be mistaken for stone. The uneasiness in the bladder, the frequent desire to make water, and the passage of blood with the urine, are symptoms of stone as well as of this affection. But in cases of stone the pain is principally experienced after the bladder has been emptied, whereas, in acute inflammation of the mucous membrane of the bladder, the pain is most intense when the bladder contains urine, and it subsides when that viscus is empty; in stone, also, larger quantities of blood are passed than in this disease, and the urethra is seldom so irritable. The diagnosis will of course be aided by the history of the case, and if a calculus be suspected the introduction of a sound will clear off the difficulty.

Acute inflammation of the mucous membrane of the bladder is a formidable disease, though if taken in time, and treated with energy, it often yields under appropriate treatment. The main indications to be fulfilled are:—the removal, if possible, of the cause, and the relief of the pain and symptoms of irritation. In cases where the cause cannot be ascertained, or where either it cannot be removed, or its effects continue after removal, the main object should be to prevent the inflammatory process from extending to the deeper seated tissues or to the ureters; and this can be attained only by the adoption of energetic measures. Venesection, as recommended by Dr. Gross,[1] is seldom necessary, but in

---

[1] Gross, loc. cit., p. 22.

severe attacks leeches may be applied to the perineum, or over the pubes, according to the especial seat of the pain. A brisk saline cathartic should be administered. After the bowels have been freely opened, the patient should be placed in a hot hip-bath for ten minutes or a quarter of an hour, twice or three times a day. During the intervals, hot fomentations or linseed meal poultices should be applied to the hypogastric region.

At an early stage, morphia is a most valuable remedy, given in sufficient doses to allay the pain about the bladder and along the urethra, as well as the frequent desire to pass water. These are the symptoms which particularly distress the patient, and if not mitigated they soon wear out his strength. When the pain and spasm are very distressing, and the nervous symptoms predominate, morphia administered by subcutaneous injection will often afford more relief than if given in any other way. The same remedy may also be used in the form of injection or suppository. In milder cases, extract of henbane may be given in full doses.

The bowels must be kept gently open by means of the mildest laxatives. The effect of any irritation of the lower intestines on the bladder points to the expediency of using castor oil, or any other gentle evacuant. Stimulant injections into the rectum must for the same reason be avoided, but an enema of simple water may be used to keep the rectum empty, and it will also relieve the tenesmus.

M. Civiale justly remarks that one of the most important points in the treatment of acute inflammation of the bladder is to keep the organ in a state of as perfect repose as possible. If the functions of the bladder could be completely suspended, and the contractions necessary for the expulsion of urine avoided, the disease would lose a great part of the danger which accompanies it. Hence he recommended that we should carefully watch the condition of the bladder, and as soon as urine begins to accumulate, even though the patient experience no great desire to make water, an instrument should be introduced. This is excellent advice, and should be followed whenever any symptoms of retention of urine manifest themselves; but the pain and irritation produced by the passage of any instrument along the urethra are so severe, that it would be unadvisable to employ the catheter with the sole object of keeping the bladder empty. Should, however, retention occur and no relief be obtained by the use of the warm bath, warm fomentations and narcotic remedies, a catheter should at once be introduced, and an elastic instrument is to be preferred. This should be well oiled and passed along the urethra with the utmost gentleness, and only a small portion of its extremity introduced into the bladder. Should the attempt fail, owing to spasm at the neck of the bladder, a silver catheter, well oiled and warmed, should next be tried. The previous subcutaneous injection of morphia will, in such a case, much facilitate the introduction of the instrument. The catheter should not be retained in the bladder, but should be re-introduced from time to time if necessary. The bladder must not be allowed to become distended, for not only would such distention aggravate all the symptoms, but decomposition of the retained urine would be very likely to ensue, and thus another source of serious mischief would be added to those already in existence. When the active symptoms have abated, it will be advantageous to apply counter-irritation above the pubes.

In the treatment of these cases, however, we find that no remedy, opium or morphia perhaps excepted, long retains any influence over the disease. The practitioner must be provided with a variety of agents, so as to be ready to substitute one for another when it loses its effect. Infusion of diosma in the proportion of an ounce to a pint of water, small doses of copaiba and essential oil of cubebs, infusion of hops, and the alkalies, will, all in their turn, be found useful.

To the diet of the patient the greatest attention should paid. Animal food, wine, spirits, and acid drinks, should be interdicted; the diet should be light, consisting of bland, farinaceous food; and the drink should be water, toast-and-water, linseed tea, and milk; but not to such an extent as to increase, in any very considerable degree, the secretion of urine. The patient should also be kept as quiet as possible, and in a rather warm temperature.

Dr. George Johnson[1] has lately recommended an exclusively milk diet for patients suffering from all forms of cystitis, and the cases which he has cited prove that very satisfactory results have been obtained by the adoption of this regimen. It is, however, adapted rather for the subacute and chronic forms of cystitis than for the acute disease, in treating which we should, for obvious reasons, abstain from administering large quantities of fluids. Diminution of the acidity of the urine is, of course, an important indication to be fulfilled, but an undue increase in quantity cannot but aggravate the distress of the patient. Milk, however, will be found very suitable as the patient's ordinary drink, but he should not be encouraged to take more than is absolutely required in order to assuage his thirst. With the milk may be advantageously combined an equal part of seltzer or other alkaline mineral water. As the more acute symptoms gradually subside, the diet of the patient may be improved, but all stimulants should be carefully avoided. When the patient is convalescent and able to return to his ordinary avocations, he should be cautioned against exposure to cold and moisture, rough or violent exercise, or any influence likely to cause a return of the disease.

In dealing with cases of cystitis, certain modifications, varying with the exciting cause, may be required in the treatment. When, for instance, the disease is due to the presence of a calculus, we must endeavor to relieve the acute symptoms before attempting any operative procedure for the removal of the stone. Rest in bed is especially indicated. In that form which is caused by cantharides or turpentine, the warm bath, warm fomentations, morphia by the mouth or in a suppository, and demulcent drinks, will usually suffice for the relief of the symptoms. Camphor also has been especially recommended; it may be combined with the morphia. When the inflammation occurs in a gouty or rheumatic subject, colchicum will be likely to prove advantageous, and if in such a case the symptoms be very severe, small doses of calomel combined with opium may be given every few hours. When the inflammatory action has extended from the urethra or prostate gland, leeches and hot fomentations to the perineum are especially indicated. When the disease has spread from the rectum, vagina, or other organs, the treatment must be directed to the parts primarily affected. When the case is complicated by the presence of a stricture, we must wait until the acute symptoms have in some measure subsided, and then treat the obstruction either by dilatation or internal urethrotomy. The presence of a stricture, however, would be an additional reason for the prompt introduction of a catheter whenever symptoms occurred indicative of retention of urine. In all

---

[1] Lancet, vol. ii., 1870, p. 847.

cases of cystitis, inquiry should be made as to the possible presence of a foreign body in the bladder.

The prognosis of the disease is favorable in the majority of cases, but very unfavorable if the ulcerative stage once sets in. This is usually indicated by the continuance of the pain of the bladder on motion, and in making water. If, therefore, the pain be not early subdued, little hope can be indulged of a successful termination. By judicious management, life may be prolonged for some time, but the patients are seldom cured, the bladder being in a state unfit to perform its functions, and the morbid changes having extended along the ureters to the kidneys. When the inflammation spreads to the mucous lining of the ureters, the pelvis, and tubular structure of the kidney, it is attended by a profuse discharge of mucus, purulent matter, and epithelium.

The symptoms of ulceration of the bladder, accompanied by an affection of the kidney, are well illustrated by the following cases which occurred in Mr. Coulson's practice:

CASE I.—Mrs. M., æt. 36; was supposed to be laboring under symptoms of stone. She had frequent desire to make water, and shooting pains in the region of the bladder, much increased by walking and exercise of every kind. The urine was acid, and contained some shreds of lymph or mucus. No blood or gravel had been passed. On sounding the patient, the instant the instrument was introduced into the urethra, the pain became much aggravated. No stone could be felt. The pulse was small and quick, the skin dry and rough, the tongue white, the countenance anxious, and indicative of much suffering. An unfavorable opinion of the case was expressed under the belief that there was ulceration of the bladder. She had been in this state two months, and various remedies had been tried. The use of pareira brava was suggested, first in the form of infusion, then in that of decoction. For the first six weeks leeches were occasionally applied to the hypogastric region, tartar emetic ointment rubbed in, and at night a thin starch injection, with twenty minims of Battley's sedative solution, was given. Mrs. M. then tried the decoction of pareira brava, which she continued till about two months prior to her death, when, perceiving more mucus in the urine than usual, and occasionally blood, a small quantity of balsam of copaiba was added to the mixture. This brought on sickness, and deranged the stomach so much that thenceforward she was obliged to desist from the decoction. She now had sickness and nausea; pus was voided with the urine; there were complete loss of strength, emaciation of body, and hectic flushes: and six months from the date of the first visit death put an end to her sufferings. It should be observed that, for a few days prior to death, no pus had been voided with the urine, and the pain and frequent desire to make water, for the only time during her long illness, had almost subsided.

The body was examined within forty-eight hours after death. The bladder was not thickened or contracted, but completely divested of its mucus membrane. No dissection could exhibit the arrangement of the muscular structure so well as it was seen in this case. One spot, the size of a shilling, towards the fundus, was black and almost gangrenous. The ulceration had not extended to the urethra; but its lining membrane was highly inflamed. The right kidney was in its natural state, but the left kidney was ulcerated, and its interior filled with pus. The renal extremity of the left ureter was blocked up by a detached portion of the substance of the kidney.

CASE II.—Deborah Mulloday, æt. 46, complained of great uneasiness, pain in the lower part of the belly, and frequent desire to void urine. After the bladder was emptied, the pain and uneasiness usually subsided. These symptoms were at first relieved by the use of the decoction of the pareira brava. The pain became very acute, the desire to make water more frequent; the urine contained a good deal of pus; and on two or three occasions, it was tinged with blood. The pulse was small and quick; the countenance pale and sallow; and there was emaciation of the body, with occasional shiverings and cramps. She had no pain in the loins, and there was no sickness until about ten days prior to her death, when it was very distressing, and of some days' continuance. On Friday, July 26th, she was seized with paralysis, and expired on the following Tuesday. Twenty-four hours after death the body was examined. The mucous membrane of the bladder was ulcerated in several spots, but was not so extensively destroyed as in the preceding case. The bladder, which was thickened and contracted, contained a good deal of pus. The right kidney was in a state of atrophy, and chalky matter was deposited in its interior. The urethra was inflamed, but not ulcerated.

CASE III.—George Scandreth, æt. 56, tailor, complained of frequent desire to pass urine, of great pain in passing it, a scalding sensation in the urethra, and of pain above the pubes and across the loins. The urine was high-colored, and contained numerous fine shreds of lymph; it was also acid, scanty, and albuminous. Some soothing remedies were ordered, and for a time diminished the pain, but had no other effect upon the disease. The patient became gradually worse, and in two months he died. On examination, the mucous membrane of the bladder was found almost completely destroyed, minute filaments hanging loosely from its inner surface; the bladder was contracted in size, and a little thickened; the lining membrane of the urethra highly inflamed and ulcerated about the membranous portion; the prostate not enlarged; the left kidney in a state of ulceration, and containing pus.

CASE IV.—William David Sadler, æt. 17, of a delicate constitution, and light complexion, applied at the General Dispensary, laboring under an affection of the bladder. He said that about seven months previously, in making water, he first felt pain, which lasted for some minutes, and then disappeared. Since that time, this symptom had never left him. Occasionally he was suddenly seized with a darting pain near the neck of the bladder, accompanied with an irresistible desire to make water, after which the pain subsided, and he felt easier. These attacks varied in frequency—sometimes occurring every hour or oftener, at other times at longer intervals. The urine was turbid, voided in small quantities, and extremely acid. The appetite was good; the general health not much deranged; the pulse quick; the tongue white and dry. Two or three times a sound was introduced, with great pain, but no stone could be felt. Alkalies and hyoscyamus; small doses of cubebs, with carbonate of potash; the decoction of pareira brava alone, and in combination with small doses of copaiba, mucilage, and hyoscyamus, were tried, but without benefit.

The patient ceased to attend; but he died a short time afterwards, and the medical man who examined the body informed Mr. Coulson that there was ulceration of the bladder and disease of the left kidney.

CASE V.—Charlotte Mason, æt. 56, a married woman without family; much subject to pains in the limbs, was seized in the month of March with frequent desire to void urine, always attended with great pain, and a burning heat in the urethra; the pain continued for some time after the urine was passed, so that she could scarcely stand or sit down; she had great pain in her back and the lower part of the abdomen, headache, and at times pains in the left hip and leg, great weakness, trembling, and general debility. During the night she was generally obliged to micturate four or five times; the urine was alkaline, of a light straw color, and latterly contained a good deal of pus. The disease resisted all the remedies which were used. Morphia alone gave her relief. She gradually lost flesh and strength, and died on the 30th of the following December. On post-mortem examination the whole of the mucous membrane of the bladder was found to have been completely destroyed by ulceration. The left kidney was also in a state of ulceration, and the lining membrane of the left ureter, to the extent of two or three inches, close to its vesical extremity, was thickened and lined with coagulable lymph.

In these cases there was evidently something more than simple ulcerative inflammation of the mucous membrane; but they are related because they appear to be instructive.

Dr. Prout, in alluding to these cases, considers them nearly allied to irritable bladder dependent on organic mischief in the kidney. "The urine in these cases," says Dr. Prout, "is generally acid, of a pale greenish wheylike color, opalescent, from the presence of minute epithelium or mucus; of low specific gravity (that is, generally below 1.020); often serous, but rarely bloody. Sometimes, on being heated, it deposits the phosphates; but the urate of ammonia is seldom so abundant as to be spontaneously separated on the cooling of the urine; and when this circumstance does take place, the color of the sediment, instead of being yellow or red, is usually of a grayish ashy tint. After standing for some time, the urine becomes clearer, but seldom acquires perfect transparency, even by filtering; and the peculiar sediment in general is very easily remixed on shaking.

"In conjunction with these appearances of the urine, the patient usually complains of the following symptoms: there is a frequent and urgent desire to pass water, the period varying from one to three hours, and the quantity from one to two or three ounces, both by night and by day. At the moment of passing water, and for some time afterwards, there is an uneasiness, sometimes amounting to severe pain, felt along the whole of the urethra, but particularly just behind the scrotum; and of this uneasiness or pain, a sense of burning or scalding is one of the elements. There is no mechanical impediment to passing the water; and in the earlier stages, after a short time, the whole uneasiness subsides, and the patient remains quite well till the period arrives when he is called upon, as before, to empty the bladdr. As the disease advances, all these symptoms become augmented. The unnatural properties of the urine, and of the mucous deposit, increase; the symptoms, and particularly the calls to pass the urine, are more urgent and frequent; the general health and strength, which from the commencement had been disordered and enfeebled, now daily decline, and the patient becomes emaciated, weak, and irritable, and more than ever susceptible of the influence of atmospheric changes. During the whole of this period, there is but little uneasiness felt in the region of the kidneys, and what little there may be is usually referred to weakness. On minute inquiry, patients will sometimes admit the existence of a dull aching sensation in the loins, and occasional darting pains down the course of the ureter, and even to the testicles, etc.; but these are so trifling compared with the bladder sensations, that they are seldom complained of, unless particularly inquired after. The termination of the complaint is various; most frequently, perhaps, as follows:

"The pulse gradually becomes more weak and feeble, and the stomach, from being weak and disordered, often rejects what is taken, so that the patient is very apt to be sick after eating. At the same time the urine, though not improved in quality, is diminished in quantity, and the calls to pass it are in consequence less frequent. The patient complains of nothing, but he daily becomes more indifferent and drowsy; as the sickness increases, the urine is still further diminished in quantity; at length everything is rejected, the secretion of the urine ceases altogether, and the patient expires, generally in a comatose state. Occasionally the termination is more sudden and unexpected; and in such instances, inflammatory symptoms have been generally superinduced from exposure to cold or some other exciting cause. Now and then the patient becomes phthisical—in short, the fatal termination, though always certain, may be various, and depends upon the peculiarities of the patient's constitution, and accidental circumstances.

"Several of the cases described by Mr. Coulson, under the denomination of acute inflammation of the mucous membrane of the bladder, seem to me to be nearly allied to this form of disease. I have reason to believe that the present disease often exists for years in a chronic form, and confined chiefly to the kidney; that when the degenerating process reaches and attacks the bladder, it sometimes assumes, either spontaneously or from accidental circumstances, a more acute form, and terminates fatally, like those described by Mr. Coulson, with complete destruction of the mucous membrane of the bladder, etc."

These remarks are worthy of consideration, but there is one mark of distinction sufficiently prominent to enable a diagnosis to be made between the affection above described and irritable bladder dependent on organic disease of the kidney; and that is the intense pain which attends inflammation and ulceration of the mucous membrane of the bladder, and soon exhausts the strength of the patient. Now in irritable bladder dependent on organic affection of the kidney, there is sometimes, but not always, pain in passing the urine; the frequency of making water is the most distressing symptom; and even when pain exists, it is never so severe as to wear the patient out, but may be, and frequently is, endured for years. The intensity of the pain and the rapid exhaustion of the vital powers, mark the distinction between the two classes of cases.[1]

---

[1] Dr. Podrazki (Handbuch der allg. und spec. Chir. redigirt von v. Pitha und Billroth, p. 65) in alluding to these remarks, and the cases on which they are founded, says:—"Billroth believes that, under such circumstances, the vesical catarrh is dependent upon constitutional disturbance, and in the majority of cases is simply the most prominent symptom of catarrh of the pelvis of the kidney, with or without tuberculosis of that organ (chronic caseous nephritis). The catarrh manifests itself in the course of time by its well-known symptoms, especially when the pyelitis has become fully developed and the urine is acid and contains a copious purulent deposit. The connection between these affections has been pointed out by Coulson."—Archiv für Klin, Chirurg., Bd. x., s. 526.

CHAPTER VII.

CHRONIC INFLAMMATION OF THE MUCOUS MEMBRANE OF THE BLADDER.

This is a frequent disease, and in many respects worthy of serious attention. It may arise from a great variety of causes, and presents itself in different degrees of intensity. The disease may succeed acute inflammation of the mucous membrane. In such cases it is sometimes accompanied by ulceration, as has been stated in the preceding chapter, and constitutes a dangerous affection; but generally speaking the inflammation is subacute or chronic from the commencement, and is characterized by an abundant discharge of mucus with the urine, whence the term *vesical catarrh*. The disease is rarely, if ever, an independent affection of the mucous membrane. In almost every case which comes under the notice of the surgeon, he will, on due examination and inquiry, be able to trace the origin of the catarrh to some co-existing malady, or to the forced retention of urine, such retention being due either to some obstruction to the natural flow of urine, or to atony of the coats of the bladder.

The most common exciting causes, therefore, are stricture, stone, and enlargement of the prostate; after these come exposure to cold, indulgence in ardent spirits, diuretic and irritating remedies such as cantharides, violent exercise on horseback, and venereal excesses; the disease also exists as a symptom in connection with many organic diseases of the bladder and kidneys, and with hæmorrhoids and other diseases of the rectum. In cases of injury and disease of the spine, this state of the bladder is by no means unfrequent. Men are more subject to this complaint than women, and elderly persons more so than the young. It would appear to be uncommon in certain countries; while in others it occurs more frequently, and, according to some authors, occasionally assumes an epidemic character. The disease prevails in this form in Egypt, and is due to the presence of a parasite, the Bilharzia hæmatobia. (See chapter on Hæmaturia.)

Some persons are apparently more predisposed to catarrh of the bladder than others; such are those of irritable scrofulous temperament with fair skins, and a tendency to cutaneous affections, more especially if they have been accustomed to live freely, or are given to venereal excesses, or have suffered from syphilitic affections, or gout. In such individuals, exposure to cold is the most frequent exciting cause of this affection, and those who actually labor under it generally suffer much more severely in cold weather. Gouty persons are very subject to this affection.

The symptoms of the disease may be divided into two classes: those which belong to the inflammatory element, and those connected with the state of the urine, the chief of which latter class is the presence of vesical mucus in superabundant quantity.

The inflammation itself seldom gives rise to any general symptoms, as fever, etc., while the local signs are not very well marked, except in old standing and severe cases, in which the inflammation may assume at intervals a subacute character.

In many cases the symptoms are mild, and the patient experiences little inconvenience; there is no pain in the region of the bladder; but the urine is voided more frequently than is natural, and the passage of that fluid is accompanied by a sensation of heat which extends along the urethra, or shooting pains towards the anus, with a sense of weight in the perineum.

Cases of this character have been observed to terminate in a short time, or to assume an intermittent form, especially when associated with hæmorrhoids, or certain petechial affections; but the duration of the complaint is uncertain. Old persons mostly retain it as long as they live.

At other times the disease assumes a serious character and may prove fatal, especially in old and weak persons. The obstinacy and danger of the complaint mainly depend on the causes which have produced the inflammation, and upon the extent to which the kidneys are involved.

In these severer cases the functions of the urinary apparatus are seriously impeded. The bladder is never emptied in a complete manner. The expulsion of the urine, and particularly of the last few ounces, is more or less painful, according to the violence with which the abdominal muscles and bladder contract. The sense of heat in the bladder and urethra is converted into scalding; the desire to make water becomes more frequent, and is attended by violent straining efforts; and retention sometimes takes place from obstruction of the urethra by clots of inspissated mucus.

These symptoms are relieved by drawing off the urine with the catheter; but they return as the organ becomes filled with fluid.

The patient is very restless and uneasy, and complains of thirst; the bowels are irregular, either constipated or relaxed; there exists pain at the extremity of the penis, round the anus, and in the region of the loins. Great prostration of strength and wasting of flesh are present.

The condition of the urine varies with the duration of the symptoms and the causes which have produced the disease. In slight and recent cases of chronic catarrh the urine is more or less turbid, and contains the same cellular elements as are found in the acute form, viz., mucus, epithelium and pus-corpuscles, together with an amount of albumen corresponding to the quantity of pus. The reaction of the fluid is acid or feebly alkaline, and, after the urine has been standing for some hours, the cellular elements form a more or less copious, loose deposit at the bottom of the glass. In more severe cases, and as the disease advances, the changes in the urine become more manifest; while still in the bladder the urea undergoes conversion into carbonate of ammonia; the urine therefore becomes decidedly alkaline in reaction, and emits an ammoniacal and offensive odor. Various theories have been advanced to explain this metamorphosis of the urea. It was formerly supposed that the mucus secreted by the inflamed membrane acted as a ferment, but the more recent view is that the presence of a peculiar ferment or excitant of putrefaction (either in the form of organized bodies such as bacteria, or of a non-organized material such as a particle of putrid matter) is a necessary condition for the change in question. The fact, alluded to by Niemeyer, that the decomposed and altered state of the urine has been known to follow the introduction into the bladder of a dirty catheter, appears to indicate that something more than vesical mucus is required to produce the change. Dr. Owen Rees has suggested that alkalinity of the urine may be sometimes due to the secretion of an alkaline mucus by the vesical mucous membrane, but Dr. Roberts, having had under observation a patient with extroversion of the bladder, was not able to satisfy himself that the alkalinity of the exposed mucous membrane was not owing to the blood-serum, which oozed from the raw surface, rather than to any mucous secretion which might be yielded by an inflamed mucous membrane.¹

When the ammoniacal decomposition has fairly set in, other changes also take place in the urine. It becomes muddy, and often more or less dark from the presence of the granular pigment-matter of disintegrated blood-corpuscles; the purulent deposit becomes more copious, and the pus-corpuscles cohere into a gelatinous tenacious mass, which has occasionally been found so glutinous, that on pouring it from one vessel into another it was drawn out above a foot in length without rending. Enormous quantities of this deposit are sometimes passed in the course of twenty-four hours, and it contains, in addition to cellular elements, the ammoniaco-magnesian phosphate, phosphate of lime, and bacteria. When there is ulceration of the mucous membrane, red blood-corpuscles will also be found. When this glutinous deposit comes away in large quantities, it is discharged with effort and may occasion retention of urine. After micturition, the burning sensation in the region of the bladder ceases, but gradually returns as the mucus again collects. If the secretion be very copious, symptoms of hectic may supervene and the patient dies from exhaustion.

Chronic cystitis may last for several years, the symptoms varying in urgency from time to time. When the urine has become decidedly ammoniacal, another cause of irritation is added to those which already exist. The acrid fluid irritates the mucous membrane and induces fresh inflammation; the purulent secretion becomes augmented and promotes the decomposition. There are therefore two sources of mischief, each tending to aggravate and perpetuate the other, and thus it happens that a case of chronic cystitis, if left to itself, invariably goes from bad to worse. Ulceration of the mucous membrane is a not unfrequent consequence, and when that occurs the local symptoms become more marked. When the disease is about to terminate in death, the patient usually falls into a low febrile state; the tongue becomes dry and the stomach irritable; prostration increases, and death is ushered in by delirium and coma. Various views have obtained from time to time with regard to the causation of these symptoms, and some have attributed them to the absorption into the blood of the carbonate of ammonia. Rosenstein, however, found that the injection of this substance into the veins of animals always excited violent muscular convulsions, a symptom which, though characteristic of uræmia, is not present in the cases under consideration. He attributes the final symptoms of this so-called "urinous fever" to the presence of bacteria in the blood.

Pyelitis is a frequent consequence of chronic inflammation of the bladder. The morbid process extends up the ureters to the pelvis of the kidney, producing suppuration of the infundibula and straight tubes, and the formation of disseminated abscesses with ultimate destruction of the renal tissue. Other complications and consequences of chronic cystitis are, thickening and induration of the walls of the bladder and the

---

¹ Roberts, Urinary and Renal Diseases, p. 57.

formation of cysts and diverticuli. Attacks of the acute form of the disease are also of frequent occurrence in the course of chronic cystitis, and may be produced by various causes, such as errors in diet, exposure to cold, and anything which disturbs the general health. When such an attack supervenes, the urine becomes comparatively clear, but retention, accompanied by tenesmus, rigors, and delirium, is apt to occur. These symptoms may terminate fatally, or may pass off after a few days; in the latter case, the improvement is contemporaneous with a return of the deposit and other appearances in the urine.

The morbid appearances found after death are those of chronic inflammation.

In the commencement, they are usually confined to the neck and posterior part of the bladder; the mucous membrane, usually pale, becomes dotted and streaked with blood, which in part is contained in dilated blood-vessels, and in part is extravasated. These spots are generally black, the blood having lost its normal color. As the disease advances, the discoloration becomes deeper and more general; the membrane is thickened, softened, and flocculent; it tears readily from the muscular coat, and is found abraded, especially in the neighborhood of some large extravasation. The surface is covered with a muco-purulent layer, and the contained urine is dark-colored, turbid and strongly ammoniacal. In a few cases of old standing the mucous surface is pale, and its appearance would never lead us to infer the existence of inflammation in any degree. If the disease has spread along the mucous lining of the ureters to the tubular structure of the kidneys, those canals will appear filled with a muco-purulent fluid, and the kidneys will present the morbid appearances characteristic of pyelitis. The walls of the bladder become thickened from effusion into its cellular tissue; ulceration often takes place in the mucous membrane, which, as in acute inflammation, may be entirely removed, leaving exposed the hypertrophied muscular fibres. Ulceration, however, is more frequently observed whenever, from some occasional cause, the inflammation assumes an acute character. Perforation of the bladder, with suppurative peritonitis, may be found as a consequence of the ulceration.

The most prominent portions of these muscular columns are usually of a bluish-red or purplish color; while between them, the membrane is pale, swollen, soft, and offers little resistance; occasionally small ulcerations are found. But what is very remarkable, between the hypertrophied columns, pouches or sacs generally co-exist with dilated ureters, both states being produced by the same physical cause. These pouches often contain calculous concretions.

In extreme cases of this kind, the secreting structure of the kidney becomes reduced to a thin layer, covering the widely dilated pelvis and infundibula. The ureters, in such instances, are both dilated and tortuous, and the lining membrane is rough and granular, and in some instances is covered by flakes of lymph.

In cases where the obstacle to the escape of urine has existed for a considerable period, the walls of the bladder, and particularly the muscular coat, will be found enormously hypertrophied, and such hypertrophy may be either concentric or excentric. In the former case the capacity of the organ may be much diminished, but in excentric hypertrophy, which is much more common, the bladder may be so much dilated as to contain several pints of urine and to reach as high as the umbilicus. In addition to the morbid appearances presented by the bladder itself, those of the various conditions upon which the disease depends will also be found. Among those may be mentioned strictures of the urethra, prostatic enlargement, calculi, etc.

It has been said that when vesical mucus is passed in small quantity, the disorder may be mistaken for an involuntary discharge of semen, which accompanies in some persons the escape of the urine and feces. These two fluids are somewhat analogous in their appearance, but may easily be distinguished by the aid of the microscope.

The urine in this disorder may be also distinguished from chylous urine, because the latter, immediately it is passed, presents a whitish milky appearance or opaline tint, due to the presence of fatty matter which forms a creamy layer on the surface after the urine has stood for some hours; on the contrary, the urine in vesical catarrh is at first turbid; on standing, the sediment becomes viscid, ropy, and flocculent, or united into one clot.

From the preceding observations it will be evident that in the treatment of vesical catarrh, we must direct our attention not only to the chronic inflammation of the bladder, but to the original disease which has been its exciting cause, and to which its persistence is due.

When this disease co-exists with, or is produced by stricture of the urethra, it is often extremely difficult of treatment. In some cases, it suffices to dilate the stricture to obtain a cure of the catarrh; in others, the pain and irritation along the urethra are often so great as to render the use of the catheters and bougies impracticable; but unless the condition of the urethra is improved, no material benefit can be expected from internal remedies. Under these circumstance, free use must at first be made of sedatives; and when the pain and irritation of the urethra have subsided, this canal must be dilated with bougies or the gum-elastic catheter. If this method fails, recourse must be had to external or internal urethrotomy. The cure of the stricture is indispensable for the cure of the cystitis.

Chronic inflammation of the bladder connected with disease of the prostate, or with lesions seated about the neck of the organ, is always more severe and difficult of cure than when dependent on stricture of the urethra. The difference between the exciting causes readily accounts for this difference in the affection which they produce. It is slow and insidious in its progress, and is liable to vary much in intensity at different periods of the disease. A careful examination of the bladder will alone enable the surgeon to ascertain the particular nature of the co-existing lesion, and determine the method of treatment required for it.

The presence of a foreign body in the bladder is a frequent exciting cause of catarrh; and its connection with vesical calculus may be easily understood. In cases of this kind we possess in the sound a certain means of diagnosis, and the treatment must be guided by the principles laid down under the head of urinary calculi.

The same remark will apply to the various other exciting causes of this affection; and to avoid repetitions, reference must be made to what is said in different parts of this work concerning the treatment of enlarged prostate, atony of the bladder, organic disease of that organ, etc.

If any constitutional tendency to gout exist, colchicum should be administered. The form in which it may be given is the acetous extract, in the dose of one or two grains at bed-time. In these cases, small doses of copaiba, or of the essential oil of cubebs, with hyoscyamus, will often do great good, and they may either be added to the infusion of buchu, or decoction of pareira brava, or be given alone.

But it will not be sufficient to get rid of any constitutional tendency, or confine our treatment to the disease of which the catarrh appears to be an effect. The internal surface of the bladder is in a state of chronic inflammation, and to this, likewise, the attention of the practitioner must be directed; while endeavoring to remove or palliate the exciting causes, he must also employ measures calculated to diminish in a direct manner the inflamed state of the mucous membrane of the bladder.

To attain this object, antiphlogistic treatment must be employed, cautiously, and with due attention to the circumstances of each case. It is seldom desirable to have recourse to the local abstraction of blood, except at an early period of the disease, in healthy persons, and when the severity of the symptoms indicates that the inflammation is one of a sub-acute character.

Of the remedies destined to act either directly or indirectly on the bladder, those from which the greatest benefit has been derived are the balsams, opiates, uva ursi, the pareira brava, and medicated or simple injections.

In many cases the catarrhal state of the bladder seems to be kept up rather by irritation than actual inflammation, and here opiates are obviously indicated; they may be administered by the mouth, or, better still, in the form of enema, care being taken to obviate, by mild laxatives, any constipation which they may have a tendency to produce.

Sir Astley Cooper and Baron Dupuytren placed great confidence in the balsams as a remedy for catarrh of the bladder. Sir A. Cooper says: "The best remedy that can possibly be taken is the balsam of copaiba; no medicine so completely robs the urine of mucus as this. Eight or ten drops three times a day will usually be found quite sufficient; it may be given in conjunction with sweet spirits of nitre and camphor mixture, or in ℨ ij mucilag. gum. acaciæ et ℨ x aq. font." Dupuytren employed Venice turpentine in the form of pills, instead of copaiba.

This latter remedy may be given with advantage in combination with small doses of zinc, Chian turpentine, or sulphate of iron; but whatever other remedies may be used, the surgeon must not omit the use of morphia or opium once, or oftener, in the twenty-four hours. Cubebs and copaiba must, however, be administered with care; for, after the long-continued use of these medicines, chronic inflammation of the bladder sometimes comes on. Mr. Coulson frequently prescribed the tinctura benzoini composita (in the dose of a teaspoonful) three times a day, with considerable relief to the patient. When the urine is alkaline, and contains a good deal of mucus with the phosphates, the alchemilla arvensis may be administered with advantage. An ounce of the dried plant is to be infused in a pint of boiling water for three or four hours; and two ounces of the infusion are to be taken three times a day.

Other medicines found most serviceable where there is much secretion of mucus, are the decoction of uva ursi, with the tincture of the perchloride of iron, and small doses of powdered galls and nitre. Niemeyer recommends the continuous administration of tannin, a remedy which often proved very efficacious in his hands. If there be much pain and irritability of the bladder, the decoction of pareira brava is an excellent medicine; and it may be combined with nitric or nitro-muriatic acid, or dilute phosphoric acid, to lessen the secretion of mucus. If there be much pain and restlessness, morphia or opium ought on no account to be omitted. Barthez mentions a case in which an almost incredible quantity of mucus was passed in thirty-six hours, and which was cured solely by the exhibition of large doses of opium internally, and in the form of clyster.

The bladder, when affected by chronic inflammation, is susceptible of being acted on directly by various remedies introduced through the urethra. The attention of surgeons has been much directed to the influence of injections, the use of which is indicated in almost all forms of the disease. It rarely happens that patients suffering from chronic cystitis are able to empty the bladder completely. After each effort a certain quantity of urine always remains behind and undergoes decomposition, and thus adds to the irritation. It is obvious that so long as this condition exists, no real benefit can be expected from the exhibition of remedies by the mouth. The condition is a mechanical one, and must be treated by mechanical means. The catheter should be introduced at regular intervals, in order to empty the bladder as far as it is possible to do so, but inasmuch as pouches and diverticuli of varying form and size are frequent accompaniments of chronic cystitis, it follows that more or less urine often remains behind when the flow from the catheter ceases. The same thing happens as before, viz., decomposition of the urine, and irritation as a result. The use of the catheter is therefore insufficient for the removal of the mischief, which can be dealt with only by washing out the bladder by means of injections. M. Civiale used to prefer simple to medicated injections, and their well-established influence over irritability of the bladder naturally leads to the inference that they may be found useful in a complaint where to irritation is added a considerable change in the qualities of the urine. According to M. Civiale, the objects we should endeavor to attain by injection of the bladder are—to remove any deposits or offending matter thrown down by the urine, to modify the sensibility of the organ, and restore its natural contractile power, which has been impaired by the chronic inflammation. For the success of this practice it is essential that the introduction of an instrument be not productive of much irritation; for if it be, the pain and irritation caused by the manipulations will exceed any benefit which the injections can produce. An elastic catheter is introduced and the urine withdrawn, and then a small quantity (not more than two ounces) of water, at a temperature of 100°, is injected by means of an india-rubber bag having a stop-cock and tapering nozzle to fit the catheter. After a short interval the water is allowed to escape, and the operation should be repeated several times. When the water comes away tolerably clear, it may be concluded that the bladder has been thoroughly washed out. The water should be introduced gently and slowly, so as not to cause rapid distention of the bladder. The india-rubber bag and catheter are all the apparatus required; a contrivance, however, suggested by Hegar is a very convenient instrument. It consists of an elastic tube attached to a catheter at one end, and at the other to a funnel. The catheter is introduced into the bladder and water is poured into the funnel. The latter is then raised a foot or two above the level of the bladder, into which the water flows; when the funnel is lowered, the water escapes. Dr. Keyes, of New York, has devised a somewhat similar, but more complicated apparatus. The operation should be repeated once a day, or oftener if the last portions of the urine continue charged with mucus. In a few days the patient will become accustomed to the injections, and will pass water more freely, while the quality of the urine will be evidently improved.

Nitric acid injections are of great use in cases where the effects of the disease are local; the general health not being much impaired, and the patient feeling comparatively well when the scalding and pain after passing urine have subsided. In cases also where the internal administration of acids appears to have no effect on the secretion of mucus, the injection of acid is of great service. The following case strikingly illustrates the beneficial effects of this method:

Mr. F., æt. 67, slightly rheumatic, consulted Mr. Coulson, April 7th, on account of an affection of the bladder. He complained of frequent desire to make water, with severe pain and scalding along the urethra, just before and during its passage, and of uneasiness or sense of weight above the pubes. He had occasional retention, and was frequently compelled to pass a catheter to draw off the urine. The disease was brought on by retaining the urine on one occasion, after the desire to void it had come on. There was no enlargement of the prostate, nor any stricture. The urine was alkaline, and contained a considerable quantity of mucus, which was of a dirty white color, and sank to the bottom of the vessel; it was stringy, and on being shaken with the urine did not mix: it was free from blood. The urine, on being passed, had a slightly ammoniacal smell; but this became very strong on standing twenty-four hours. Its specific gravity was 1.015. His general health was not impaired; and he felt pretty well when the burning pain which attended the passing of urine had subsided. The decoction of pareira brava, with the dilute nitric acid and sedatives, afforded no relief.

May 8th.—The quantity of mucus with the phosphatic deposits increasing, the injection of the bladder was commenced, first throwing in, by means of a gum-elastic syringe and catheter, four ounces of decoction of poppy, and, on withdrawing this, four ounces of water, containing two minims of strong nitric acid. This was kept in for two or three minutes, and then withdrawn. He took half a grain of morphia immediately after the injection. The symptoms were a little, but not materially, improved by the injection.

May 10th.—Mr. Coulson injected ten ounces of distilled water with one minim of the strong nitric acid to the ounce of water (five ounces at each injection), and caused each quantity to be retained for two or three minutes, until slight pain or uneasiness was complained of.

May 12th.—The symptoms and condition of the urine were slightly improved. Ten ounces of distilled water, with two minims of the strong nitric acid to each ounce of water, were injected in the same manner as before. This injection was followed by considerable pain, which continued several hours; but on the following day the urine was much changed in appearance, contained little or no mucus, and had lost its dark-brown color. The irritability of the bladder was considerably lessened. The injection was used twice after this, but with a smaller proportion of acid; the reaction of the urine, which had been alkaline, became acid, and the patient lost all his bad symptoms.

When the chronic inflammation does not yield to simple injections, or when it appears to be connected with irritation about the neck of the bladder, some practitioners have had recourse to the injection of certain remedies known to influence the sensibility of mucous surfaces. Nitrate of silver is the one generally preferred. M. Lallemand used to employ this remedy in solution, and also in substance, with the object of effecting temporary cauterization.

The application of the solid nitrate of silver is made by means of a *porte caustique* attached to a stylet and passed through a catheter open at both ends. This method of treatment is applicable for cases in which the neck of the bladder is the part mainly affected, and where irritability and hyperæsthesia are marked symptoms.

When it is desired to apply an astringent and sedative to the mucous surface of the bladder, solutions of nitrate of silver, commencing with a quarter of a grain to the ounce, or acetate of lead in equal quantities, may be employed. The bladder must first be washed out with two or three ounces of warm water as above described, and afterwards the same quantity of the solution injected and allowed to remain a few minutes. If there be much pain, anodyne injections, containing two grains of extract of opium or belladonna, may be used, but a better effect will be obtained from the use of suppositories containing a smaller quantity of these remedies. When the urine is highly alkaline and fetid, solutions of permanganate of potash or carbolic acid may be used as injections.

Dr. Devergie has recorded eight cases of chronic catarrh, some of long standing, which were cured by injecting balsam of copaiba into the bladder. Some of these cases had been developed gradually, and maintained throughout their chronic character. A moderate quantity of an emollient fluid must first be injected, to ascertain the capacity of the bladder, but not in sufficient quantity to irritate it. General measures must be resorted to, to calm the inflammatory action and local pain. Narcotics must next be added to the emollient injections, and these may be repeated three or four times a day. When the irritation of the bladder and neighboring parts is allayed, the copaiba should be injected. A dose of uniform strength is not suited to every case. A drachm of balsam of copaiba to an ounce of barley-water is sufficiently strong for the first injection; but the quantity of balsam must be regulated according to its effects. The combination of narcotics with copaiba renders the latter less exciting. The balsamic injections may be allowed to remain in the bladder for a period of from ten to twenty minutes. The quantity of copaiba is to be gradually augmented; and it should not be injected more frequently than once daily, nor intermitted more than two days. The injection is to be continued until the muco-purulent secretion has quite ceased.

In very severe and obstinate forms of vesical catarrh, when all other methods of treatment have been found unavailing, it has been proposed to open the bladder by an incision similar to that made in the operation of lithotomy. The urine then drains away as fast as it reaches the bladder, and the organ remains at rest. This method was originally suggested by Mr. Guthrie, and it has been adopted by several surgeons. A similar operation has been performed for the relief of chronic cystitis in the female, the incision being made through the vesico-vaginal septum.

In the severer forms of vesical catarrh, where there is great depression of the vital powers, the patient should be sustained with light nourishment, and small quantities of wine given from time to time. In the milder forms, patients may be allowed to take animal food, but beer, wine, and spirits must be strictly prohibited. Milk is especially indicated: an exclusively milk diet produced the happiest results in some cases thus treated by Dr. George Johnson.[1]

The functions of the skin should be attended to, and for this purpose

---

[1] Lancet, vol. ii., 1870, p. 847.

the frequent use of the warm or tepid bath will be found very serviceable. Patients must also be cautioned against irregularities of every kind, and must be told the consequences of neglecting the advice given them. From want of care on the patient's part, the mild has often assumed the severe form, and an attack of acute inflammation has come on and destroyed life.

CHAPTER VIII.

ACUTE INFLAMMATION OF THE WALLS OF THE BLADDER—CYSTITIS PARENCHYMATOSA—CYSTITIS TOTALIS—ABSCESS.

INFLAMMATION of the bladder almost always commences in the mucous lining of the organ, but when very severe often extends to the deeper seated tissues. Some authors suppose that the muscular structure alone may be the seat of inflammation; but analogy and experience fully support the opinion of Dr. Prout, who stated that he was quite unacquainted with any facts indicating that inflammation ever originates in the muscular structure, or is ever limited to this portion of the organ. Inflammation may confine itself to the mucous tunic, or it may limit its action to the peritoneal covering of the bladder; but with respect to the muscular coat of this viscus, it is seldom, perhaps never, exclusively the seat of inflammation.

In some cases, however, abscesses form between the serous and musculars coverings, or in the sub-mucous connective tissue. Morbid anatomy proves to us that inflammation may arise in these situations, that the walls of the organ become thickened, and pus is secreted in one or more cysts, which in course of time burst through the mucous membrane or open in other directions. The muscular structure in these spots is wasted and in great part destroyed, only a few stray fibres being found about the walls of the abscess. Such cases have been regarded as examples of inflammation of the muscular coat; but it would be more correct to regard the connective tissue, and not the muscular fibres, as the seat of inflammation; in short, to class the affection as we should class an abscess occurring in any other situation. From this connective tissue the inflammation has a great tendency to spread in every direction; and hence it is that we never observe it to be much inflamed, without finding the other tissues of the bladder more or less involved in the morbid action. This disease more frequently attacks adults than the young or old, and strong than delicate persons. It is also more common in males than females.

The causes of inflammation occupying the deep cellular tissue of the bladder are nearly the same as those which excite acute inflammation of the mucous membrane. It may be caused by retention of urine from stricture, enlarged prostate, etc.; calculus or the use of caustics may give rise to it; and in some cases the disease can evidently be traced to external injury, or the violent use of instruments. It is sometimes caused by exposure to cold, and indulgence in spirituous liquors; or it may occur as a result of extension of gonorrhœal inflammation. In these cases the disease usually extends from the mucous membrane.

The symptoms are obscure, and are seldom sufficiently well marked to lead to a positive diagnosis, unless the existence of abscess can be ascertained by manual exploration.

The patient first complains of a dull aching pain in the region of the bladder; this soon becomes more violent, and extends to the neighboring organs. The pain is increased by pressure, and is attended by desire to pass urine, without the power to accomplish it. The desire comes on in paroxysms, attended with pain; the urine is at first evacuated in small quantities; and the attempt to pass it causes great distress. The thickening and induration of its walls prevent the normal contraction and dilatation of the bladder, and hence the difficulty in passing water. The straining is due to the pressure exercised by the urine which is continually trickling into the organ. In some cases, however, where the inflammation is limited to small portions of the bladder, the patient has never complained of any painful sensations. The small quantity of urine which escapes is of a dark color, sometimes not unlike coffee in appearance; at other times it is of a deep red, and even blood color; and at last complete retention occurs. There is a sense of fulness in the lower part of the abdomen, and pains in the lumbar region, in the groin, and down the thighs; but there is not the burning sensation along the urethra and in the perineum which exists in inflammation of the mucous membrane.

The disease is ushered in by rigors, which are soon succeeded by great constitutional disturbance. The pulse is full and hard, the thirst great, and the skin hot, with general uneasiness and sickness. If the inflammation increase, pains are felt in the intestines, particularly in the rectum, combined with tenesmus; delirium comes on; the pulse rapidly sinks; and the patient soon dies.

If the inflammation should be seated in the neck of the bladder, as is frequently the case, the urine is retained by the tumefaction ensuing from the inflammatory action; the bladder soon becomes distended, and projects above the pubes. There is a sense of weight in the perineum; there are often painful erections of the penis; and examination by the rectum gives great pain.

The anatomical structure of this part readily accounts for these symptoms. The triangular space is at once very vascular and highly sensitive;[1] its nerves, arising from the third and fourth sacral pairs, as well as from the great sympathetic, descend on each side through the inferior mesenteric and the hypogastric plexus, and communicate more particularly upon this space. We have, therefore, only to recollect the relative connections and ramifications of these different nerves to be able to explain, not merely the strong and constant desire to evacuate the bladder that prevails when this part of the bladder is irritated or inflamed, but also the remote symptoms by which this inflammation is liable to be accompanied.

If the inflammation exists somewhat higher up in the bladder, where the ureters enter, the orifices of the latter become contracted, and the bladder being closed against the influx of the urine, the ureters become enormously distended. The orifice of the ureters is surrounded by a dense elastic substance, which lies between the muscular and the mucous coats of the bladder. Beginning at the base of the triangular space, this substance inclines inwards as it advances towards the neck, forming in a great measure the orifice, and appearing to be continued through the urethral passage as its elastic membrane. This elastic triangular substance yields, in some measure, to the pressure of the urine when impelled by the detrusor, and returns to its original situation when the

---

[1] With regard to the urethral opening of the bladder, Mr. Guthrie observes, "that fibres have been described surrounding this part, though no anatomist has demonstrated them so as to warrant their being called a sphincter muscle; that this part may be both muscular and elastic, but the older anatomists supposed the power which prevented the flow of urine to reside in other muscles surrounding the membranous part of the urethra."

pressure is removed. If the inflammation be situated more in the upper part of the bladder, there is danger of its extension to the peritoneum, and the pain is greater on pressure; but the desire to pass urine is not so frequent, nor the difficulty so great.

The progress of the disease keeps pace with the severity of the symptoms. Very severe cases sometimes terminate fatally within a short time from the commencement of the attack; but, in ordinary cases, if active measures be employed, the pain, after two or three days, begins to subside, and the water flows with greater facility, is less acid, and of lighter color. The febrile symptoms, and at the same time the local uneasiness, gradually lessen.

This kind of inflammation sometimes terminates in the formation of abscess between the coats of the bladder; in such a case the symptoms will present a formidable character, depending upon the size and situation of the abscess. Percussion and palpation over the bladder will cause intense pain, but it will rarely be possible to diagnose with accuracy the seat of the mischief. In some cases, however, a swelling intensely painful on pressure will be felt above the symphysis pubis, or on examination by the vagina or rectum, and the pain and swelling will be found to continue after the bladder has been emptied by means of a catheter. Inflammation of the peritoneal coat will, however, give rise to the same symptoms. The abscess may open into the cavity of the bladder, and in this case pus is evacuated with the urine, and the patient experiences great relief: it may open into the cavity of the abdomen, or into the rectum or vagina, or the matter may extend into the cellular tissue of the pelvis, and make its way either through the intestines, probably the rectum, to the perineum, or even to the groin, and in these cases the result is most frequently fatal. Mr. Wilson mentions an interesting case in which extensive suppuration had taken place in the coats of the bladder, from the prostrate even to the fundus, the matter being lodged everywhere between the tunics; while near the fundus several ulcerations had penetrated the mucous membrane, by which the matter had passed into the cavity of the bladder.

After the formation of an abscess the urgent symptoms subside; but dull pain in the region of the bladder, occasional rigors with febrile excitement, and uneasiness in passing urine and feces still continue.

This disease may easily be mistaken for acute inflammation of the prostate. The uneasiness and pain in the region of the bladder and perineum, the occasional but strong desire to pass urine, every effort being attended with great pain and retention, are symptoms common to both diseases. In inflammation of the prostate, however, there is more fulness with tenderness on pressure in the perineum, and on examination *per rectum*, the prostate is found exceedingly sensitive, painful, and swollen.

In recent cases, the morbid appearances in acute inflammation of the walls of the bladder are thickening and great vascularity, the deep-seated tissues being throughout injected with blood, and of dark red color. Sometimes the walls are found even to be gangrenous; instances are recorded where they had given way, and urine had escaped into the pelvis. The mucous membrane is also found of a dark red color. In other cases the muscular tissue is thickened, and the bladder itself contracted. Pus is often infiltrated through the tunics, or else circumscribed in the form of abscess.

The following case well illustrates this form of inflammation of the bladder. Richard Sergiter, æt. 68, watchmaker, after a slight attack of gout, was seized with rigors, which were succeeded by fever and great constitutional irritation. The pulse was extremely quick; the skin was hot and dry; and there were great thirst and sickness. He had a strong desire, coming on in paroxysms, to void urine; but he could pass only a few drops at a time. There was pain in the region of the bladder and loins. Colchicum and saline aperients were administered; leeches and warm fomentations were applied to the region of the bladder, and the warm bath was tried. The bladder being distended, a catheter was introduced; and between two and three pints of very dark urine were drawn off. None of these measures gave relief; the constitutional disturbance increased, delirium came on, and the man died within forty-eight hours from the commencement of the attack. On the day after his decease an examination was made; the bladder was found in a state of intense inflammation; in its structure there was no organic change; but the tunics, particularly the muscular, were of a very deep red color.

The symptoms of this disease are so severe, and its progress so rapid, that prompt and decisive measures must be adopted. If the patient be strong and robust, general blood-letting must be first employed; or if he be delicate and of spare habit, local bleeding, by leeches to the pubes, or cupping in the perineum, may be substituted; or both may be put in requisition as auxiliaries to general blood-letting.

Hot fomentations should be constantly applied to the pubes; and, after the bleeding, the patient should be placed in a hot bath. As already observed, there is in these cases retention; and the urine must, from time to time, be drawn off by the catheter. Internally, our main reliance must be placed on the use of calomel and opium, which must be given every three or four hours; this medicine affords the most speedy relief. As tenesmus often exists, the proportion of opium should be large; and sedative injections should, at the same time, be administered. After the urgency of the symptoms has subsided, saline aperients, combined with the vin. semin. colch., will be found beneficial, especially if the attack has occurred in a gouty subject. The diet must consist entirely of lukewarm mucilaginous drinks.

If these means be employed early, the patient may experience a diminution of pain; urine is passed in greater quantities and with less suffering; the constitutional symptoms improve; and the patient falls into sound and refreshing sleep. If, on the contrary, these measures be delayed, the symptoms before described become aggravated, and death ensues.

Where there are any external signs of the formation of an abscess, it should be opened as promptly as possible, but with the greatest care. If fluctuation can be felt through the rectum or vagina, a curved trocar should be plunged into the swelling and the contents evacuated. In doubtful cases, as, for example, where there is indistinct fluctuation above the symphysis pubis, a very fine trocar should be employed after the superficial structures have been carefully divided.

CHAPTER IX.

HYPERTROPHY OF THE BLADDER—COLUMNAR AND SACCULATED BLADDER—ATROPHY OF THE BLADDER.

HYPERTROPHY of the bladder is a very common result of obstruction to the escape of urine, and therefore frequently accompanies strictures of the urethra, prostatic enlargements, and calculus in the bladder. The muscular coat of the organ, stimulated to increased functional activity, is the part in which the main increase takes place, but inasmuch as a chronic catarrhal state of the mucous membrane is a constant accompaniment in such cases, this latter coat also becomes increased in thickness from interstitial deposit. Hypertrophy of the muscular coat may be either excentric or concentric, that is, may be associated either with enlargement or diminution of the cavity of the bladder. The former condition is produced when the muscular fibres, contracting but feebly, gradually yield to the pressure exercised by the urine. The resulting dilatation varies in degree; the cavity of the bladder may become so enlarged as to contain several pints of urine, and the organ appears simply distended and its walls much thinner than natural, the muscular tissue being often in a state of fatty degeneration. In other cases the increased capacity of the bladder is accompanied by increased thickness of its walls. In a third class of cases, the hypertrophy is concentric; the muscular fibres, stimulated to increased efforts in order to overcome the obstruction, become enlarged in proportion to the call made upon them, and, as a result of this change, the cavity of the organ becomes diminished in size. When the hypertrophy is considerable, the muscular fibres, which in their normal condition are scarcely perceptible, become much thickened and cord-like, and project into the cavity of the bladder, crossing each other in various directions and giving to the interior of the organ that peculiar appearance which has been termed *vessie à colonnes*, or "columnar and fasciculated bladder." The hypertrophy may involve mainly the fibres of the detrusor muscle, or may extend to the whole muscular tissue of the organ, the walls of which may measure half an inch or even an inch in thickness. The formation of sacculi, which will be presently alluded to, often forms an additional complication.

As a matter of course hypertrophy of the bladder is far more common in men than in women, and in adults than in young subjects. The disease has, however, been found to occur as a result of phimosis in young children.

Hypertrophy of the bladder has been said to arise, in certain cases, from a want of concord between the muscles which must contract for the expulsion of the urine and those which must at the same time relax to permit the urine to escape. The neck of the bladder and the urethra are surrounded by the levator prostatæ (anterior portion of the levator ani), the compressor urethræ and the accelerator urinæ muscles. All these muscles must relax before a drop of urine can pass; and any irregular action in them directly obstructs the passage, whilst it tends to develop hypertrophy of the bladder by rendering its contractions more violent and long-continued. The contraction, indeed, of one set of these fibres, and the relaxation of the other, belong to the same act, like the condition of opposing muscles in the motions of the limbs; so that the bladder is not in a state to execute its functions if these muscles are not in a condition to relax. "The elasticity of the neck of the bladder," says Mr. Guthrie,[1] "is impaired; it will not dilate with the ordinary action of the detrusor muscle; and this action is therefore augmented. A sensibly increased delay is experienced before the water begins to flow, the patient is obliged to expel it; and he becomes conscious of the augmented effort made by the bladder. The desire in which this originates soon amounts to uneasiness, and rapidly afterwards to pain—relieved indeed on evacuating a little water, but too soon to return; for now the bladder is never completely emptied, and the urine which remains is a source of great irritation, although the quantity be really inconsiderable."

Such symptoms have been classed by Sir James Paget[2] as instances

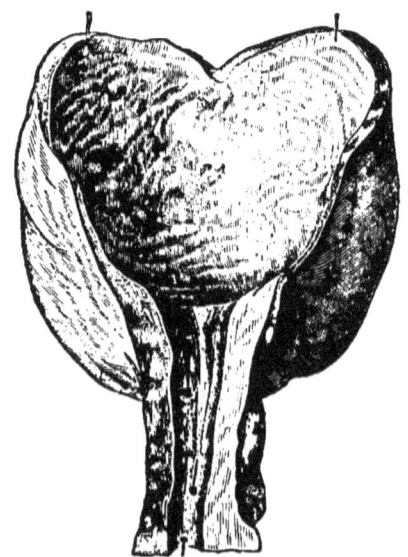

FIG. 1.—Interior of a bladder, showing hypertrophy of its muscular fibres.

of "urinary stammering," and he compares them with similar difficulties connected with the organs of speech. He believes, however, that this "stammering" does not produce structural disease of the urinary organs. "In cases of long-continued urinary stammering, some of which began in very early life, and some of which I have known for many years, I have seen no indication of supervening organic disease."

Besides the increase in thickness due to the hypertrophy, the walls of the bladder are often still further thickened by interstitial deposit, the result of chronic inflammation of the connective tissue. This is shown to be the case by the manner in which the muscular fibres are sometimes

---

[1] On Diseases of the Bladder, Lect. XV., p. 256.
[2] Clinical Lectures and Essays, p. 80.

matted together, and by the uniform way in which the thickening of the walls extends over the whole bladder.

The symptoms of the disease are those of chronic cystitis and of the obstruction to which the hypertrophy is due. There is uneasiness about the region of the bladder, frequent desire to pass water both night and day, but especially by night, and the urine does not flow so readily as usual; there is also occasional prolapsus of the rectum, and these symptoms are increased by exercise. On examination with a sound, the cavity of the bladder is found to be contracted, while the inner surface is rough and uneven, and the elasticity of the coats is more or less diminished. In concentric hypertrophy, the tumor formed by the bladder can be felt through the rectum or vagina.

When this disease is produced by stricture or any local cause, it is clear that the primary affection should especially engage our attention. In gouty, rheumatic, or plethoric persons, colchicum given at night, in the dose of one or two grains of the acetous extract, will be found of great service; and as, in these cases, the urine is acid, and often scanty, the alkalies should also be given. These latter should be taken after meals; and a mixture of bicarbonate and nitrate of potash forms a very suitable combination. In addition to these medicines, great benefit will be derived from pareira brava, a very useful medicine, which may be given in ten-grain doses of the extract, three times a day, or in the form of decoction.

The diosma, in the form of infusion, combined with bicarbonate of potash and tincture of hyoscyamus, will also be found of great service. If the urine is not acid, or, as is not unfrequently the case, should the alkalies produce headache and restlessness, or uneasiness about the region of the stomach, their use must be discontinued, and recourse had only to sedatives, as extract of hop, and of uva ursi, or nitric ether, with tinctur. camph. comp., and the occasional exhibition of suppositories. The diet should be plain, but nutritious; beer, wine, and spirits should be prohibited; and exposure to wet and cold, which invariably aggravates the symptoms, carefully avoided.

As in these cases the bladder is seldom completely emptied by its own efforts, a catheter should be introduced from time to time, and the patient be instructed to do this for himself. Unless this direction be strictly attended to, the patient will become worse, and serious consequences ensue.

Hypertrophy of the bladder is often accompanied by another state, depending on the same causes, and denominated "sacculated bladder." Whenever the bladder is excited to violent contraction, while there exists at the same time some obstruction to the flow of urine, the mucous membrane has a tendency to protrude between the hypertrophied muscular fibres which are often separated by considerable intervals. At these spots the mucous membrane, having lost its natural support, yields to the pressure of the urine, and is forced between the muscular bundles. Such protrusions vary much in size; some are so small as scarcely to project beyond the peritoneal coat, others form tumors larger than a normal bladder. The protrusions constantly tend to increase in size, and they sometimes become adherent to neighboring organs; as a general rule, the size of the cysts is inversely proportionate to their number, but one or more large cysts have been occasionally found co-existing with numerous smaller ones.

There is also a great variety in the number and form of the cysts.

Sometimes there is only a single cyst of moderate size, but two or three are frequently found to co-exist. On the other hand, Civiale met with a bladder almost entirely covered with small cysts so as to resemble a bunch of grapes. Only a small part of the anterior aspect of the bladder, near the neck, remained normal. Platner (quoted by Civiale) saw a bladder which had thirty-nine sacs, each one containing a calculus. When the sac is small, the orifice through which it communicates with the bladder corresponds in size; as the sac enlarges under the influence of pressure, the orifice becomes proportionately wide; though in some cases the entrance into a large sac has been found comparatively narrow. When the orifice of communication is small, it is sometimes more or less occluded by bands of hypertrophied muscular fibre or a fold of mucous membrane. At an early stage of their formation, the walls of these cysts are composed entirely of the mucous membrane, which has been driven out between the muscular fibres, carrying the peritoneum before it. This disposition of the cysts should not be lost sight of in reference to the operative proceedings connected with them. In some old cases, however,

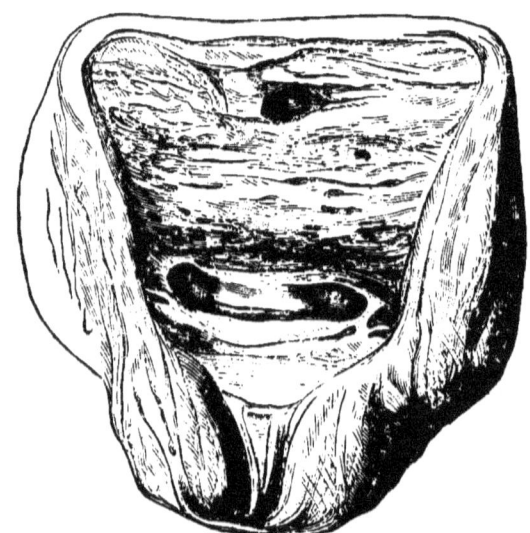

Fig. 2.—This figure represents a bladder with three cysts.

a few thin muscular fibres may often be traced over those parts of the sac in the vicinity of its orifice, and it sometimes happens that muscular fibres enter into the formation of the walls of the cyst. Under such circumstances, as in an instance to be presently referred to, the walls of the cyst possess a considerable amount of contractile power. The cysts generally contain urine, which is often fetid or mixed with pus; M. Andral mentions tubercular deposit, but this must be extremely rare. Phosphatic deposits and calculi, on the other hand, have often been discovered in them.

Every part of the bladder is liable to the formation of these sacs; they have occasionally been observed at the upper part, but they most frequently exist near the fundus or at the posterior surface. On examining the preparations preserved in our anatomical museums, we rarely find cysts in the neck and anterior part of the bladder. When, as sometimes happens, a large sac springs from the posterior part of the bladder, near the base, the transverse fold which bounds the latter posteriorly, becomes relaxed and appears to cover the orifice.

The presence of these sacs often gives rise to serious symptoms. The urine in them cannot be expelled, owing to the absence generally of a muscular covering, and their orifices may be contracted by the formation of a band or valve. The urine may deposit phosphates, become fetid, and, as noticed by Chopart, excite inflammation, ulceration, or abscess of the walls of the sac. Calculi may drop into or be formed in them, and are then said to be "encysted." If the sac be large and distended with urine, it may excite many of the symptoms of retention; and the lining membrane may undergo a process of softening and disorganization attended with general disturbance of the constitution. It has been observed that the sac often contains pus, while no trace of inflammation can be found in any other portion of the bladder. Sometimes an opposite condition exists. Dr. Wilks mentions a case in which the bladder was severely inflamed, while a sac from its side, of greater size than the bladder itself, was quite free from inflammation.[1]

The diagnosis of this condition is always attended with difficulty. The distended sac may sometimes be felt through the abdominal parietes, especially in thin subjects; and after the bladder has been emptied by the catheter, the tumor and the desire to pass water remain, while pressure over the pubes may cause the evacuation of an additional quantity of urine of a fetid character. This circumstance is a strong indication of the nature of the case. In other instances, after the bladder has been apparently emptied, the point of the catheter accidentally enters the sac, and a flow of urine unexpectedly takes place. The distance to which the instrument penetrates will here furnish a useful indication; but great care must be taken, while we employ it, lest the point should penetrate through the thin walls of the cyst. The following cases illustrate this condition of the bladder.

A man, æt. 51, was admitted into St. Thomas's Hospital, April 22d, 1851, with retention of urine. He had suffered from stricture for three years. The stricture gradually became worse, and at last ended in complete retention, for which he had been admitted, under Mr. South, in January. He stated that he had never had a catheter passed before this, but that he had several times experienced a discharge of blood with urine. On this occasion all attempts to introduce a catheter into the cavity of the bladder failed; the instrument, though passing far enough to reach that organ, not moving freely, and no urine coming away through it. The retention, however, was relieved, and the urine, at first thick and alkaline, was rendered clear by the administration of dilute sulphuric acid. He went out, passing his urine in a very good stream, the latter end of February. Soon, however, the stream lessened, and for ten days previous to admission the urine only dribbled away. Several ineffectual attempts had been made to introduce the catheter, and these had caused profuse hæmorrhage.

On re-admission, a large-sized catheter was introduced sufficiently far to reach the bladder, but no urine followed. On examination per rectum, the prostate was found of natural size. Next morning, Mr. Solly suc-

[1] Pathological Anatomy, p. 523.

ceeded in introducing a full-sized catheter, and drew off a quantity of thick, very offensive urine, apparently containing a good deal of pus. There remained an irregularly lobulated tumor on the left side of the hypogastric region, painful, and tender on pressure, and apparently solid.

On the 27th, he complained of severe pain in the tumor; he had passed a considerable quantity of urine, so that it was thought needless to introduce a catheter. The countenance became anxious, the tongue furred, and the pulse quick and weak. The urine was dark-colored, very thick, extremely offensive, and showed, under the microscope, many crystals of phosphate of lime, a few blood-corpuscles, and a quantity of granular matter aggregated in irregular masses. Next day he became comatose, and died the following morning.

Upon examination after death, a stricture was found about the membranous portion of the urethra; there were also several false passages. There was no disease or enlargement of the prostate. The muscular coat of the bladder was very much thickened, and in several places large herniæ of the mucous coat had escaped between its fibres, and formed capacious cysts full of urine, and communicating by narrow orifices with the cavity of the bladder. It was these cysts which had formed the tumor in the left hypogastrium, noticed during life. Both ureters were very much distended throughout, and at intervals swollen into large cysts. The mucous coat of the bladder was very much thickened, inflamed and ulcerated. Both kidneys were larger than usual and pale.

The enormous size to which a cyst of the bladder may attain is shown by a case reported by Dr. Murchison to the Pathological Society.[1] A large abdominal tumor, the exact nature of which was at first very difficult to diagnose, was caused by a distended sacculus of the urinary bladder. The tumor extended as far as the spine of the ilium, and its nature had been determined by puncturing it with a fine trocar. The pressure of the sacculus had caused obstruction and inflammation of the femoral vein. After death the sacculus was found to be much contracted, and the orifice leading into the bladder was just large enough to admit the point of the finger. The contracted state of the sacculus was due to certain bands of muscular fibre being extended over it from the muscular coat of the bladder. Fifty-four ounces of urine had been obtained, within half an hour, by the puncture and by a catheter introduced into the bladder. Another instance of the kind is recorded by Dr. Warren (*American Medical Times*, N.S. IV. 13). A man, aged 85, who had suffered from dysuria for several years, met with an accident which caused his symptoms to increase. His abdomen gradually became enlarged as though he were suffering from ascites, and only a small quantity of urine could be obtained by the catheter. On examination after death, the tumor, which reached as high as the stomach, was found to have been caused by an enormous diverticulum from the bladder, the walls of which were much hypertrophied. On the left side, a little above the neck, was a round opening leading into the larger tumor, which was formed of the peritoneal and mucous coats, and held a gallon of urine. From the history of the case it was evident that the prostatic enlargement and consequent dysuria had led to hypertrophy of the muscular coat. At the time of the injury this tunic had given way, and, as time went on, the mucous and peritoneal coverings had become distended and formed the enormous cyst.

---

[1] Path. Soc. Trans., 1864.

For the treatment of sacculated bladder little can be done beyond obviating the effects of retention of urine. The morbid condition of the bladder does not admit of any remedy, and we must confine ourselves to attempts at the removal of any obstacles which may exist to the evacuation of the bladder, while we endeavor, by the occasional use of the catheter, to draw off any fluid which may have accumulated in the cysts. We may also attempt to wash out the cysts by injecting the bladder, and then exercising pressure over it in different directions. The patient should also be directed to change his position from time to time when passing his urine, and while the injections are being used. In some cases a well-fitting abdominal belt may be found useful.

A morbid condition of the bladder, somewhat analogous to the above, may be produced by irregular contraction of the muscular fibres, whereby the bladder appears to be divided into two cavities.

Dr. Baillie remarks: "I have not had an opportunity myself of examining this singular disease, but I have received an account of such a case from Dr. Ash, which had many years ago fallen under his observation. The upper chamber of the bladder in this case was generally much distended with urine, so that a round tumor could be easily distinguished by the touch above the pubes. When a catheter was introduced into the bladder, only a few ounces of urine came away, and the tumor above the pubes remained as before. When the patient stood up, a quart of water sometimes passed away involuntarily, the tumor very much subsided, and the complaint was relieved for the time. After the death of the patient, the bladder upon examination was found to be divided into two chambers by a strong membranous substance, and the aperture of communication was almost obliterated."

There would seem to be only two ways in which a division of the bladder into two chambers can happen. One is by a morbid growth of the inner membrane, forming a ridge at some particular part, and at length, by a continuation of this process, making a septum more or less complete in the bladder. Mr. Coulson once saw a case in which the calibre of the œsophagus was very much narrowed at one part by a permanent ridge being formed in its inner membrane. Something of the same kind has also been seen in a part of the small intestines. We may therefore admit the possibility of a similar process taking place in the inner membrane of the bladder. Another way in which the bladder may be supposed capable of being divided into two chambers, is by a very strong contraction of its transverse muscular fibres at some particular part. This will be analogous to the hour-glass contraction of the uterus, which is known occasionally to take place. When a complaint of the bladder depending upon its being divided into two chambers has been temporary, it is reasonable to suppose that it has arisen from the last cause; when it has been permanent, it is more likely to have arisen from the first. It is most probable, however, that in the majority of cases, this so-called "division of the bladder" has been caused by the formation of a cyst, as above described.

When two chambers are just beginning to be formed in the bladder, very little inconvenience is probably felt, because the communication between them at this time is very large. Under such circumstances, it seems hardly possible to detect the nature of the disease in the living body; but when the disease has made considerable progress, and the communication between the two chambers has become very narrow, it may be ascertained, or at least presumed with some probability, to exist from the following circumstances: There will be a considerable circumscribed tumor above the pubes in the situation of the bladder when distended; much less urine will be passed than the natural quantity, and the tumor will not be sensibly lessened by it; or if a catheter be introduced, little urine will be evacuated, and the tumor above the pubes will still remain the same. But it may occasionally happen, in some particular attitude of the body, that the urine passes from the upper chamber of the bladder into the lower, and from this it is evacuated by the urethra: under such circumstances a much larger quantity of urine will be passed than usual, the tumor above the pubes disappears, and the patient receives immediate relief, to continue till another accumulation of fluid takes place in the bladder.

A condition the reverse of that of hypertrophy, viz., atrophy of the bladder, has been known to occur in cases where the organ has been prevented from discharging its function as a reservoir for the urine. In a case already alluded to (see page 32), an unrelieved urethro-vaginal fistula resulted in complete atrophy of the bladder.

CHAPTER X.

INFLAMMATION OF THE PERITONEAL COAT OF THE BLADDER, AND OF ITS SUBJACENT CELLULAR TISSUE.

THERE is one other variety of acute inflammation of the bladder that demands our attention, viz., inflammation of its peritoneal coat, and subjacent cellular tissue. This is seldom confined to the bladder; it generally extends over the whole of the membrane, and is often the close of a fatal disease of this organ. That it takes place, however, under other circumstances, without any dangerous consequences, is sufficiently proved by the old adhesions occasionally found connecting this part to the omentum, to portions of the intestine, to the uterus, or to the rectum.

On dissection, we not unfrequently find the abdominal or peritoneal coat of the bladder inflamed, as well as the mucous and deeper tissues. Inflammation sometimes attacks the peritoneal covering, having spread to it from another part of the membrane. Nevertheless, though the circumstance is rare, cases have occurred in which acute inflammation was limited to the peritoneal tunic of this organ. As a reason for limitation to this particular membrane, Dr. Baillie suggested the quantity of cellular tissue interposed between the serous and muscular coats, and the laxity of their connection.

The pain and its aggravation on pressure, the state of the pulse, the countenance, and the position of the body, clearly indicate the nature of the disease. The same treatment which is employed in general peritonitis must be adopted in inflammation of the peritoneal covering of the bladder. Leeches, calomel and opium, and warm applications, must be vigorously employed. The disease rarely, if ever, proceeds to suppuration; but coagulable lymph is sometimes thrown out on the inflamed surface, forming adhesions with some other part of the peritoneum, where it covers other viscera, or lines the cavity of the abdomen.

Peritonitis is an occasional consequence of the operation of lithotomy, generally as a result of the extension of inflammation from the mucous membrane of the bladder. In these cases the disease often comes on slowly and insidiously, the local symptoms not attaining a high degree of severity. In children, peritonitis is a somewhat more common consequence of lithotomy. It results from the extension of inflammation from the mucous lining, or it may be caused by the peritoneum having been implicated in the perineal wound.

The inflammation of the external covering of the bladder is, moreover, generally connected with inflammation of the adjoining lining of the pelvis. In these cases there occurs a peculiar train of symptoms, requiring a different plan of treatment from that suited to inflammation of the peritoneal coat of the bladder.

The pulse is rapid, varying from 100 to 140 in the minute; the skin is hot, the tongue dry, and the countenance anxious. The patient complains of pain and tenderness upon pressure in the hypogastric region.

The abdomen next becomes tympanitic, the pulse intermits, and hiccough occurs. In many cases delirium ensues; but the mind sometimes retains its clearness to the last. Rigors sometimes precede death. The morbid appearances observed are—infiltration of serous fluid around the neck of the bladder, and, between this organ and the rectum, sloughing of the areolar tissue, discoloration of the peritoneum covering the bladder and rectum, with occasional effusion of a small quantity of lymph upon its surface. This, however, is not a case of uncomplicated peritonitis. The disease spreads from the areolar tissue to the serous membrane, and remedies directed to the removal of the morbid condition alone would prove of little avail. Abstraction of blood hastens the fatal termination; while stimulants may prolong life, if they do not save it. Sometimes the cellular tissue around the bladder is the seat of chronic disease; and abscesses may form in different parts of it, without the bladder being affected. These cases are always involved in great obscurity, and often terminate fatally.

Dr. Elliotson[1] relates the case of a female, 42 years of age, in whom an abscess formed between the bladder and the symphysis pubis, which terminated fatally in about two months from the commencement of the disease. She complained of violent pain in the hypogastrium and over the pubes, shooting back to the loins, and frequently attended with a sensation of numbness and tingling in the right thigh, extending to the toes. The hypogastrium and pubes were very tender to the touch; there was, occasionally, a copious discharge of puriform matter from the vagina, tinged with blood; and the urine generally passed freely, but had sometimes been retained for two days together. There was much tenesmus with pain in the rectum; the tongue was white; there was no appetite, but frequent nausea; the bowels were usually opened twice a day, and the pulse was weak and small.

On examination, an abscess was found in front and deeply behind the symphysis of the pubis, extending laterally beyond the abdominal rings, so that the round ligaments passed through it. The surface of the bone was rough, blackened, and denuded of its periosteum, but not carious. The abscess contained dark-colored and very fetid pus, which had free exit through the urethra, a portion of the whole circumference of which was here destroyed. The suppuration was entirely anterior to the bladder, which was rather turned to the right side, but, with the uterus and vagina, was perfectly healthy. The pelvis was large and well-proportioned.

Sometimes, however, there is found, posterior to the bladder, a pseudo-abscess, which is really seated within the cavity of the peritoneum, within the pouch of that membrane, lying between the posterior wall of the bladder and the rectum. The pus in these cases arises from partial peritonitis; lymph glues together some folds of the intestines, the fundus of the bladder, and the lower part of the sigmoid flexure of the colon; and into the closed sac, of which these parts thus connected form the roof, and the peritoneal pouch the floor and lateral walls, the pus is poured. The contents of this pseudo-abscess may, if small in quantity, be absorbed, and old adhesions alone remain to indicate the pre-existence of the disease. More commonly the pus finds its way into the peritoneal cavity, and death from general peritonitis follows. The inflammation on which, in these cases, the formation of pus depends, assumes a chronic

[1] Med. Gaz., vol. i., p. 130.

character. After a short but sharp febrile attack, attended with hot skin, quick pulse, thirst, white tongue, loss of appetite, confined bowels, with pain in passing solid stools, high-colored urine, the evacuation of which is attended with pain in the hypogastric region, and tenderness on deep pressure immediately above the pubes, the patient becomes hectic. The febrile symptoms cease during the day, return with violence at night, and in the morning terminate with sweating. Sometimes it is the night fever of which the patient most complains, and sometimes it is the night sweats. He still has uneasiness in the pelvis, not, however, amounting to pain. If the case goes on to a fatal termination, the patient either dies worn out by the hectic, or he falls a victim to general peritonitis, induced, as before observed, by the escape of the purulent fluid into the general peritoneal cavity. In the latter case the symptoms resemble those of perforation of the intestine, but ordinarily are somewhat less violent. The treatment consists, at first, in local depletion by leeches, applied immediately above the pubes, or by cupping-glasses placed over the sacrum; calomel and opium administered every few hours, so as gently to affect the gums; hot fomentations, or bran poultices to the hypogastric region; mild aperients and simple enemata; the diet being strictly antiphlogistic.

When hectic symptoms show themselves in such cases, the treatment must be decidedly restorative. The patient must be supported with strong broths, jelly, etc. In most cases a little wine must be allowed. At the same time vegetable bitters and mineral acids will be found of considerable advantage. Quinine or decoction of bark, and dilute sulphuric acid, are the best of their class.

CHAPTER XI.

TUMORS OF THE BLADDER—TUBERCULOSIS.

THE bladder, like other organs of an analogous structure, is the occasional seat of various morbid growths, which may be either of a malignant or non-malignant character. So manifold are the alterations which have been made during the last few years in the nomenclature of tumors in general, that it is somewhat difficult satisfactorily to classify the various morbid growths which have been described as occurring in the organ in question. All such affections are comparatively rarely met with, and of certain kinds of vesical tumors the instances on record are very few indeed. A consideration of the cases which have been more or less minutely described shows that the following morbid growths have been found in the bladder: mucous polypi and polypoid hypertrophy of the mucous membrane, cysts, papilloma or villous tumor; fibrous tumors and fibro-myomata. One case of sarcoma[1] has been placed on record, while tuberculosis of the bladder is not a very unfrequent affection. Examples of schirrhus, of epithelical cancer and of encephaloid also occur in the bladder.

Little or nothing is known as to the causation of these various tumors of the bladder. Contrary to what obtains with regard to other affections of this organ, morbid growths are more common in the female than in the male bladder. They occur at all periods of life; Dr. Winckel[2] found two pedunculated polypi in the bladder of a new-born child; in Mr. Coulson's[3] case of soft cancer complicated with calcareous deposit, the patient was sixty-three years of age. Tuberculosis of the bladder is usually a secondary affection, but it may be primary and confined to this organ, as in a case described before the Clinical Society of London by Mr. Prescott Hewitt.[4] Of four cases of this affection recorded by Dr. Winckel, the patients all being females, and their ages varying from 29 to 66, the disease was of a secondary nature in each instance, in three the disease being secondary to tuberculosis of the lung, and in the other case to the same affection of the kidneys and ureters. With regard to the other morbid growths, injuries of various kinds, the irritation caused by foreign bodies, and displacements of the bladder have been mentioned as exciting causes. The cases on record are, however, too few, and their history too obscure, for any general deductions to be made with reference to causation. In a case recorded by Mr. Warner,[5] the symptoms of a papilloma appeared after lifting a heavy weight, and in a most interesting case of the same affection, successfully treated by Dr. Winckel,[6] the symptoms manifested themselves after an injury received during a fall. In these instances it is possible that an extravasation of blood into the

---

[1] Senfleben, Langenbeck's Archiv, Bd. i., s. 128.
[2] Loc. cit., p. 168. [3] Lancet, 1860, vol. ii., p. 187.
[4] Clinical Society, Nov. 27th, 1874. [6] Loc. cit., p. 177.
[5] Referred to by Mr. Hutchinson, Med. Times and Gazette, 1857, vol. i., p. 434.

tissues of the bladder was the starting-point of the affection. In the female, malignant disease sometimes spreads from the uterus to the bladder.

We will first consider separately the various forms of non-malignant growths as above enumerated.

I. *Mucous polypi, and polypoid hypertrophy of the mucous membrane.*—These are usually pedunculated growths, but they are sometimes broad excrescences of the mucous membrane. In the latter case, the base is thickened and spongy in texture, and often covered with phosphatic incrustation. On section the surface has a gelatinous appearance, and the blood-vessels are enlarged and abundant. There is also thickening of the muscular and serous coats, and hypertrophy of the former layer. Mucous polypi of the bladder are sometimes congenital. Dr. Winckel found two such growths in the bladder of a female child who died thirty-two hours after its birth. One of these was situated in the superior fundus, and was rather larger than a pea; the other, somewhat less in size, was situated in the lower part of the bladder. Both tumors were pedunculated, semi-globular in form, and very vascular. Single polypi have been found as large as a hen's egg or even larger.[1]

The symptoms to which these tumors give rise resemble those of calculus, but hæmorrhage rarely occurs. The patient passes water frequently, and with pain and straining. The urine contains more or less pus and epithelial debris. On examination with a sound a resisting object may be felt, or by passing the finger into the rectum the tissues between the finger and the sound may be found abnormally thick. If the tumor occupies the upper part of the bladder, it may, in a thin subject, be felt through the abdominal parietes.

The late Mr. Crosse, of Norwich, has related a case of polypus of the bladder which is worth recording. A child, æt. one year and a half, was troubled with frequent inclination to make water, attended with great straining and much pain. Medicine failing to relieve, M. Crosse at length sounded the bladder, but felt no stone. Alkalies, opiates, and the warm bath were employed; the patient always resting better on the night when the bath was taken, although his symptoms increased in severity. He was continually wet with urine, which was passed in drops, an effort being made, with violent straining, at various intervals, from a few minutes to half an hour. The nights were passed in the same manner as the days, except that during the former the little patient was said to scream more, and strain rather less violently; unless laudanum were given to him he never got any rest till the morning, when he would occasionally sleep for an hour or two. At almost every attempt to pass urine feces were passed, but the rectum never prolapsed. About every two or three weeks, Mr. Crosse used the sound gently, and twice thought he felt a stone, but not satisfactorily; the impression was conveyed on passing the sound towards the left side of the bladder. No bleeding had followed the different soundings performed during three months, and the urine was generally voided more easily afterwards, and sometimes as much as a table-spoonful at a time.

Dec. 28th.—Mr. Crosse again used the sound, which was resisted by something unusual towards the left side of the bladder; for several days after this examination the urine was tinged with blood. The patient had become gradually emaciated by this time, the skin hanging flabbily about

---

[1] Hutchinson, Medical Times and Gazette, May 2d and 9th, 1857.

him, and the countenance presenting as expressive a picture of suffering and grief as could be witnessed. Soon afterwards another opinion was taken, but the evidence respecting the existence of a stone was unsatisfactory. Various ideas passed through Mr. Crosse's mind; he thought that a stone might be encysted at the termination of the left ureter, as it was thereabouts that the resistance was always felt. While sounding he had frequently introduced the finger into the rectum without gaining information, except that when the child screamed, and the muscular coat of the bladder contracted upon its contents, the organ felt as a firm tense ball, about twice the size of a walnut.

The little boy was evidently sinking under his painful disease; and when two years old it was determined that an operation should be attempted for his relief. Mr. Crosse introduced the curved staff without difficulty; the anus prolapsed from violent straining. The surgeon who assisted at the operation stated his opinion that he felt a stone; Mr. Crosse, on the other hand, could only feel a resisting body at the left side of the bladder, about the termination of the left ureter. After a few minutes' delay it was determined to cut into the bladder. The staff having been withdrawn during the consultation, it was re-introduced, and held in a proper position, when Mr. Crosse observed a great fulness of the perineum. As soon as he had cut down to the staff, and opened the membranous portion of the urethra, a semi-transparent substance appeared in the wound, resembling the mucus which had been passed from the rectum by the child's straining when first placed upon the table, and Mr. Crosse feared he had, for the first time in his life, wounded the rectum, an impression which momentarily invaded the minds of the bystanders. With the assistance of the left fore-finger, guided by the staff, he carried the scalpel forward to the neck of the bladder, and on withdrawing the knife, he observed that the wound became instantly filled with a mass resembling what one would have expected to see had the peritoneum been opened, and the vermiform process, and several folds of small intestines had protruded. The protruded parts were pushed back; and on carrying the left little finger into the bladder, Mr. Crosse felt no stone, but found the cavity filled with soft tumors, and with a firmer substance near the orifice of the left ureter.

"If I betrayed," says the operator, "any judgment in this case, it was undoubtly in this stage of the business, by relinquishing the forceps held in readiness for me, and in withdrawing my finger from the bladder." The same parts then protruded, and they proved on inspection to be tumors, connected together like a cluster of grapes, some more, some less transparent; resembling in firmness, appearance, and structure, the mild polypus nasi; the membrane by which they were connected with each other, and with the inner surface of the bladder, was long and loose enough to allow some of the tumors to hang externally dependent at the wound: they had entered the urethra by the violent straining of the child when first placed upon the table. The nature of these tumors being now understood—namely, that they were mild polypous masses, growing from the inner surface of the bladder, it became obvious, that the only chance for the patient's recovery must be sought by removing them. All that were in sight were cut off with the scissors, but the violent straining efforts, which the child had kept up during the operation, brought several more tumors, as big as grapes, sufficiently low down to admit of their being cut off. Upon introducing the forefinger into the bladder, Mr. Crosse ascertained that more of the diseased structure remained behind than had been removed; but, as many of the remaining tumors were attached to the bladder by a broad basis, it was deemed advisable to make no further attempts for their removal.

Notwithstanding the administration of a powerful opiate, the child continued to have violent fits of tenesmus, and could scarcely be prevented from placing himself in his usual posture, resting upon his knees and elbows, to give full effect to his exertions. At the end of four hours, Mr. Crosse placed him again upon the table to ascertain whether these efforts had caused any fresh tumors to protrude. The wound was plugged up with a coagulum, under which was a tumor as big as a nut, and of purple color; this was easily brought down and cut away with scissors. Much of the diseased structure remained which could not be meddled with safely. After forty-four hours of incessant suffering the little patient died.

Examination of the body.—The peritoneum was free from signs of inflammation; the rectum was untouched; the ureters were much enlarged and contorted; the pelvis of each kidney so increased in size, that between one and two ounces of urine could be contained in it. The muscular coat of the bladder was much thickened; at its fundus there was a convex prominence, covered by peritoneum, which, when cut into, proved to be a firm mass of thickened cellular substance, situated external to the muscular coat, and containing a small central cavity filled with pus. The lining or mucous membrane of the bladder, of gelatinous appearance, but loosely connected, and in folds, gave origin to the disease. From the interior part near the neck numerous tumors sprung, occupying the cavity of the bladder; one, of large size with broad basis, of firmer consistence than the rest, was situated near the termination of the left ureter; this must have been the resisting body felt on sounding. Several small detached tumors, from the size of a pea to that of a bean, were loose in the bladder: towards the neck the tumors presented a wart-like surface: but all were covered with their proper membrane continuous with the inner coat of the bladder, which was uninjured, except in three or four spots, where the tumors had been cut off with scissors.

The neck of the bladder and the prostatic portion of the urethra were much dilated, and the tumors could readily drop into this part of the canal, a fact which accounted for the fulness of the perineum.

Cases of polypus vesicæ have been recorded also by Collison, Petet, and Corvillard. Mr. Warner[1] relates the case of a tumor growing from the inside of the bladder of a female patient: he extirpated it successfully, but in this case the tumor projected far into the urethra, and could be felt with the finger.

A very remarkable case of polypus of the bladder has been recorded by Mr. Savory.[2] The patient was a female child, aged thirteen months, who had suffered for some weeks from symptoms resembling those of stone. A swelling had also formed in the umbilical region, from which pus, and afterwards urine, had escaped. The child died, and on examination the bladder was found to contain a pedunculated growth, on its inner surface, stretching transversely across the fundus, immediately behind the apertures of the ureters, which were much dilated. The mass was attached on either side, but free in the centre, and so situated that it might lie forward over the urethral orifice, or be propelled in that

---

[1] Cases in Surgery, p. 303.
[2] Medical Times and Gazette, 1852, vol. ii., p. 106.

direction when attempts were made to pass urine. On microscopic examination, the tumor was found to be composed of very fine filamentous fibrocellular tissue, and in much greater part of granular or dim homogeneous substance, with imbedded nuclei. Over these was an immense quantity of tessellated epithelium, with well-formed and large scales like those of the mouth. This epithelium was the most abundant constituent of the small lobes of the polypus. Its form was remarkable, as differing from that of the ordinary epithelium of the bladder. The case was in other respects very interesting. The urachus had undergone considerable dilatation, and was pervious to urine.

II. *Cysts.*—These are of very rare occurrence as independent affections, but the cavity of the bladder is sometimes penetrated by dermoid cysts of the ovary. Such cases have been placed on record by Drs. Fuller[1] and Greenhalgh, Sir H. Thompson and others. In Dr. Fuller's case, the patient was a lady aged 50, and from the history it appeared that an injury had led to a rupture of a dermoid cyst of the ovary. Suppuration subsequently occurred, and the pus forced its way between the vagina and urethra, and ultimately into the bladder. The matters voided were masses about the size of a hazel-nut, composed of a coil of short hairs, entangling in their meshes yellowish matter resembling cheesy tubercle. In Dr. Greenhalgh's[2] case, an ovarian cyst, containing a quantity of light-brown hair, with sebaceous and fatty matter, was found after death, communicating with the rectum, bladder and umbilicus. Sir H. Thompson's[3] case was one of extra-uterine fœtation. It occurred in the practice of Mr. Joseph Thompson, of Nottingham. The patient had all the symptoms of a foreign body in the bladder, and had passed by the urethra two substances which were found to be portions of a fœtal vertebra. From the history of the case, and the condition of the patient, it was evident that the cyst containing the child had suppurated and burst into the bladder. There was no enlargement of the abdomen, but on examining *per vaginam*, a tumor could be felt towards the left side of the pelvis, opposite the sacro-iliac junction, extending along the left side of the pelvis towards the bladder. Mr. Thompson introduced a staff and made an incision through the urethra, downwards, outwards, and to the right, and through the opening thus made he removed the fœtus piecemeal. The patient made a good recovery. In the case reported by Mr. Wagstaffe,[4] a cyst was found connected with the bladder of a still-born child. The cyst was an inch and three-quarters in diameter and contained a milky fluid. There were, in addition, other malformations of the genito-urinary organs.

Sacculi or cysts are not unfrequently formed by protrusions of the mucous membrane between the muscular fasciculi. Morbid conditions of this kind generally depend upon the existence of mechanical obstacles to the passage of urine; they will be found described in the chapter on "Hypertrophy of the Bladder." Distinct outgrowths of this character may, however, occur in the absence of any obstruction. One instance of the kind is recorded by Mr. Erichsen.[5] The patient was a man 35 years of age; a tense, elastic tumor, smooth and rounded, was found to occupy the whole abdomen, and to extend into the pelvis, so as to be felt through

---

[1] Trans. Path. Society, xxi., p. 273.
[2] Lancet, 1870, vol. ii., 741.
[3] Path. Soc. Trans., xv., 156.
[4] Path. Soc. Trans., xviii., 201.
[5] Science and Art of Surgery, vol. ii., p. 849.

the rectum. There was no difficulty in passing urine or feces, and the patient complained only of uneasiness due to the pressure of the tumor. A puncture was made, and seven pints of clear urine drawn off by means of the aspirator. The patient died from syncope, and on examination, "two enormous cysts were found connected with the bladder, one on each side, by a rounded opening that would admit the little finger." The orifices were about an inch and a half above each ureter. The pelvis of the kidneys, the ureter, and the bladder were all dilated. There was, however, no kind of mechanical obstruction; the urethra and prostate were healthy. "It looked as if the whole of the urinary apparatus between the calyces of the kidneys and the neck of the bladder had taken on a dilating outgrowth."

III. *Papilloma or Villous Tumor of the Bladder.*—There are now many instances of this affection on record. Dr. Wilks[1] has given a description of two cases which he examined after death. One of the patients was a young and strong man, who died from hæmaturia in the course of a few weeks. All that was found in the bladder were two tufts, scarcely larger than peas, and from these the fatal hæmorrhage had occurred. They appeared like little tufts of moss growing from the mucous membrane; and, when examined by the microscope, presented villous processes, "covered by columnar epithelium, long battle-dore-shaped nucleated cells, very different from the ordinary epithelium of the bladder. Each villus contained loops of blood-vessels." A typical case was reported to the Pathological Society some years ago by Sir H. Thompson.[2] The patient had suffered from urinary troubles, occasional hæmaturia, and slight uneasiness, for about nine years. Retention of urine had also occurred, and latterly the catheter had been constantly required. Occasionally small flocculent masses were found in the urine, and these, when examined under the microscope, were seen to be made up of full-sized spindle-shaped cells, with very distinct nuclei, and arranged in layers. The absence of pain was a well-marked feature throughout the whole duration of the case. The man died from exhaustion, and on post-mortem examination, a villous mass was found placed transversely above and close to the urethra, in the front part of the neck of the bladder. The mass was as large as a strawberry, and there were also other smaller growths. There were no glandular or secondary deposits, and the duration of the symptoms (nine years) as well as the microscopic appearances precluded the idea of malignancy.

Cases of this kind were formerly described as "villous cancer," but there are solid reasons why the term "cancer" should not be applied to such growths. There is no evidence of malignant cachexia being associated with these tumors; they do not invade other organs, and neither ulcerate nor slough. They are purely local growths of a peculiar kind, an extreme tendency to bleed being the most marked feature which they present. Some resemble a solid polypus, with its surface covered with long vascular projections. Others consist of a loose shaggy mass of tissue with but little solid material, and attached by a pedicle to a healthy surface of mucous membrane. This is the most typical form. In other cases the mucous membrane is dotted over with patches of villous outgrowths, or mamillary projections. The symptoms, which may last for many years, are hæmaturia and derangement of the functions of the blad-

---

[1] Wilks and Moxon, Pathological Anatomy, p. 527.
[2] Path. Soc. Trans., xviii., 176.

der. Retention of urine is very apt to occur from occlusion of the internal orifice of the urethra, as well as from impaction of masses of tissue in the course of the tube. The urine sometimes contains soft flocculent masses, made up of spindle-shaped or other cells, with very distinct nuclei. Pain is by no means a constant symptom, and when death occurs, it is caused by the exhaustion due to the continuous and profuse hæmorrhage. In addition to the cases referred to, other instances of villous tumor have been described by Mr. Sibley,[1] Sir H. Thompson,[2] and Dr. Braxton Hicks,[3] and a collection of cases was published in the *Medical Times and Gazette*, 1857, by Mr. Jonathan Hutchinson. In Dr. Hicks' case, the tumor was passed by the patient, a woman aged 35, during life. She had suffered for about two years from pain, straining and difficulty in micturition, and occasional hæmaturia. A calculus or polypus was suspected, but neither could be discovered on examination. After some straining and passage of blood, she passed a soft mass, which on being minutely examined was found to consist of a central part, more solid and dense, and of villous growths springing from its surface. "The villi were composed of branching tufts of capillary loops, each branch containing in nearly every instance the trunks of three blood-vessels, joined together by numerous capillaries. A very delicate membrane (probably the representative of the basement membrane) surrounded the network of capillaries, and a few delicate ill-defined cells existed in the spaces between them. The central denser portion consisted of an informal assemblage of cells, varying in all degrees from the globular to the elongated spindle-shaped nucleated fibre, some possessing two or three processes. All had very large nuclei, which were invariably single." The passage of such a growth during the life-time of the patient is the interesting feature in the case. In all these instances of villous tumor, no doubt could well be entertained as to the innocent nature of the growth, and it may be remarked that similar growths have been found in the colon and rectum, with an utter absence of any symptoms of malignancy. On the other hand, however, there can be no doubt that malignant growths occur, springing from mucous membranes and having their surface covered with shaggy projections; these may with justice be called "villous cancers." In some examples of this form of the disease occurring in the bladder, there can be no doubt as to the cancerous nature of the tumor which in such cases has a solid base, and infiltrates the coats of the bladder. Constitutional symptoms invariably make their appearance, and often at an early period. Deposits of a like nature, but not covered with villi, may be found on other portions of the mucous membrane. Preparations of malignant disease of the bladder, showing shaggy prolongations, are to be found in many museums. In connection with villous growths, it is interesting to notice that this disease in the bladder has been found co-existing with a similar affection in the kidney.[4] In the instance referred to there was no tumor of the bladder, but the mucous membrane was studded over with numerous long, delicate, villous processes, the mucous lining of the pelvis and calices of both kidneys presenting similar outgrowths.

Dr. Ultzmann[5] has recorded two cases of villous tumor of the bladder, which serve to show that recovery may take place. One of these

---

[1] Trans. Path. Soc., vii., 256.   [2] Ibid., viii., 262.   [3] Ibid., xi., 153.
[4] As in a case recorded by Dr. Murchison, Path. Soc. Trans., xxi., 241.
[5] Ueber Hematurie, Wiener Klinik, Hefte iv. und v., p. 135.

patients was an old man who had suffered for some time from hæmaturia. His urine contained small reddish flesh-like masses, which, under the microscope, were found to consist of recent, well-preserved villous tissue. Various medicines were given, and the patient finally underwent a course of treatment at a hydropathic establishment, after which the hæmorrhage ceased and never returned. The patient's health rapidly improved, and during an interval of three years there was no return of the previous symptoms. In the other case, the patient suffered from great irritation of the bladder accompanied by retention of urine. A metallic catheter was passed for the relief of the latter symptoms, and on withdrawing the instrument its eye was seen to be occluded by a fragment of tissue, which on being examined was found to be "villous" in character. The symptoms never reappeared, and the patient died, some time afterwards, from an affection of the stomach.

Dr. Ultzmann points out that innocent villous growths are confined to the mucous membrane. They do not give rise to a thickening of the coat of the bladder, that is, there is no infiltration of the tissues. Villous cancers, on the other hand, give rise to tumors of the bladder, or to thickenings which may be felt through the rectum or the abdominal walls. They have also an abundant epithelial investment, whereas non-malignant villous growths are but scantily furnished in this respect. If a fragment of the growths can be obtained, and is found to be composed of capillary loops covered by a scanty epithelial investment, and if the patient be young and in good general health, it is probable that the growth in the bladder is of the simple, villous kind. If, however, the capillary loops are indistinct, or almost invisible through a thick epithelial covering, if the patient be advanced in years and in reduced health, the tumor in the bladder, even though it cannot be felt through the rectum, is probably of a malignant character.

Among the symptoms of villous growths, and in addition to hæmaturia, Dr. Ultzmann has noticed that a peculiar sensation of discomfort in the perineum is often complained of by the patients at an early period. Others suffer from irritation of the glans penis, sometimes accompanied by violent priapism. This pain is much more violent than in cases of stone in the bladder; it is sometimes so severe as to cause the patients to protect the part with the hand, and even to make contact with their clothes unbearable. The uneasiness in the perineum is less marked when the patient lies down, and is aggravated by riding in a vehicle, or even by the pressure of a cushioned chair.

With regard to the diagnosis between villous growths and calculus, irrespective of the information to be obtained by sounding, there are some symptoms which are common to both affections. The stoppage of the stream of urine, in the case of villous growth, is due to the occlusion of the urethra by coagula, or by fragments of villous tissue. When these are removed by the stream of urine, there is no longer any impediment to micturition. The pain in calculus is most severe after the urine has been passed, but in villous tumors the discomfort is aggravated by fulness of the bladder and relieved by its evacuation. The pain in calculus is relieved by rest, which has little or no effect upon the symptoms of tumor of the bladder. The hæmorrhage also in the latter affection is neither decidedly aggravated by movement nor relieved by rest. In villous growth, the blood is usually pure; in hæmaturia due to calculus, there is generally more or less pus mixed with the blood. Examination by the rectum, or with a sound in the bladder, causes pain in cases of villous cancer and increases the hæmaturia, whereas the symptoms of calculus are not necessarily aggravated by these manipulations.

The condition of the urine in cases of villous growth will be found described in the chapter on "Hæmaturia." The occurrence of fibrine, and its coagulation in urine of a reddish yellow color, are regarded by Dr. Ultzmann as pathognomonic of villous growths. If fragments of tissue are passed, and these, when examined under the microscope, are found to consist of capillary loops covered by a thin layer of epithelium, the nature of the affection will be sufficiently manifest; but, as Dr. Ultzmann points out, inasmuch as such fragments have been detached by a process of partial mortification, they will not present such a definite structure as is sometimes depicted, and as may really be traced when fragments are accidentally removed in the eye of the catheter. In the case of malignant growths, it is generally difficult to trace with any distinctness the villous structure. The fragment consists rather of a confused mass of disintegrated epithelial cells, pus, and blood-corpuscles, bacteria and shreds of tissue. Rhombic plates of hæmatoidin may sometimes be discovered in these shreds. This appearance indicates a rupture of capillaries and disintegration of blood-corpuscles. Sometimes also the fragments of tissue contain spherules of oxalate of lime; they are also, in cases of long standing, frequently coated with phosphatic deposits.

The duration of the affection varies according to the nature of the growth and other circumstances. In the simple villous disease, the patient may last for several years, and finally succumb to the repeated losses of blood. Malignant villous growths are more rapidly fatal, and may cause death by exhaustion, or may interfere with the escape of urine from the ureters, or may in other ways set up disease of the kidney, and so produce a fatal uræmia.

IV. *Fibrous tumors and fibro-myomata.*—These have been occasionally found in the bladder, but they are of very rare occurrence. Dr. Winckel[1] alludes to one case of fibro-myoma occurring in a woman. The tumor was found to be between the anterior wall of the bladder and the fascia transversalis, and so intimately connected with the former structure that it appeared to originate therein. The tumor had existed for many years; it was lobulated and as large as a man's fist; it occupied a large space in the pelvic cavity, and was prolonged into the urethra. It contained numerous muscular elements, and was covered by a firm fibrous capsule. A few years ago (1874) Billroth removed a myomatous tumor from the bladder of a boy aged twelve years.[2] The operation was a remarkably bold one, and was perfectly successful. The patient had suffered, for about ten months, from pain after passing water, chiefly felt in the glans penis and over the region of the bladder. After a while micturition became very painful and frequent, and the urine escaped involuntarily, notwithstanding the utmost efforts of the patient. The symptoms were supposed to be due to the presence of a calculus, and he was placed under Professor Billroth's care in order that it might be removed. On examination of the abdomen, a tumor was found in the region of the bladder, to the left of the median line. It could be felt through the abdominal walls; it was apparently about as large as a man's fist, firm and somewhat sensitive to pressure, tolerably movable, and apparently originating from the bladder. It could also be felt through the rectum,

---

[1] Loc. cit., p. 169.
[2] Langenbeck's Archiv für klin. Chir., xviii., 411.

but no further information could there be obtained as to its nature. It was not as hard as a calculus, and from its consistence it appeared to be of a fibrous nature. The sound was felt to glide over an uneven rough surface, and it was noticed that the beak of the instrument was directed forward immediately it entered the bladder; on attempting to move it from side to side, it rubbed against a rough surface on the posterior wall of the bladder. Combined examination with the sound in the bladder and the finger in the rectum demonstrated the presence of a tumor at the back and upper part of the bladder. By passing two fingers into the rectum, and making pressure above the symphysis pubis, the consistence of the tumor was ascertained to be that of a fibroma, and its size that of a man's fist; the insertion into the wall of the bladder could not be made out with certainty. The examination with the sound caused the escape of a little blood, but no fragments of tissue were at any time passed. An attack of cystitis, due to the examination, having passed off, the question of an operation was discussed. Removal with a lithotrite was out of the question, by reason of the size of the tumor, and Professor Billroth decided upon opening the bladder above the pubes. He first, however, performed the lateral operation for lithotomy, and explored the growth with the finger. It was impossible to extract the tumor through the opening thus made, and he next performed epicystotomy, dividing both recti muscles at their insertion in order to give sufficient room. He next tore away as much as he could of the tumor with his finger, and then dissected away the remainder, turning the bladder partly inside out. It then appeared that the tumor had originated from the muscular coat, and that it did not involve either the peritoneum (which was not interfered with) or the fibrous coat. Two arteries were tied, and the ligatures brought out through the upper incision which was not closed, but a drainage tube was drawn through it and brought out through the opening in the perineum. Five days afterwards the upper wound was granulating freely, and, as there was no reason to fear any infiltration of urine, the tube was removed. On the twelfth day urine began to pass through the urethra. No unfavorable symptom of any kind occurred, and the patient was discharged with the wound completely healed, 34 days after the operation. The tumor, on examination, was found to be 8 centimeters in length, and 4 centimeters broad; its circumference in one direction was 18 centimeters, in the other 13 centimeters, while that of its base was 7 centimeters. The surface was nodulated, but smooth, and presented no ulcerated spots. In its general appearance it resembled a soft fibroma, but differed in the ease with which it could be split in various directions (*ausgezeichnete Spaltbarkeit*). On microscopical examination, numerous spindle-cells were discovered in all parts of the tumor, not such, however, as are found in sarcomas, but cells which could not be distinguished morphologically from those of organic muscular fibres. There was no doubt as to the myomatous nature of the tumor. Portions of it, however, were rich in cells characteristic of sarcoma, while in other portions, clusters of cells resembling those of epithelium, and suggestive of carcinoma, were very abundant.

In another case [1] of myoma of the bladder, Professor Volkmann removed the tumor, but the patient, a man 54 years of age, survived the operation only a few days. The symptoms were comparatively of recent standing, and consisted in pain and burning sensations in the glans penis,

---

[1] Langenbeck, Archiv, Bd. xix., 681.

and difficulty of micturition, amounting sometimes to complete ischuria. The urine often contained blood, sometimes fluid, sometimes in the form of large coagula. A feeling of great discomfort and weight in the perineum was added to the other symptoms, and afterwards, masses of fibrillated shreds were discharged with great difficulty and straining by the urethra. The largest of these was as long as the little finger, and almost as thick as the thumb, resembling in appearance swollen macerated tendinous tissue, and of a whitish yellow color. When admitted into the hospital (six months after the appearance of the symptoms) the patient appeared very anæmic and much reduced in strength. Nothing abnormal could be felt by means of a catheter alone, but some shreds of tissue were found in the eye of the instrument. Nothing abnormal could be detected by passing only one finger into the rectum, but, on bimanual exploration, the upper part of the bladder was found to be occupied by a hard, movable tumor. Professor Volkmann's method of examining the bladder has been already described. (See page 19.)

In the case in question, the tumor, thus discovered and examined, was found to be globular, smooth, firm, elastic and very movable. It continually slipped away from the finger, and was therefore attached by a pedicle, which seemed to be fixed high up in the bladder and to the anterior wall. Its size appeared to be that of a hen's egg, and, from its consistence and smoothness, it was assumed to be of a myomatous nature. This opinion was confirmed by a microscopical examination of the fragments that had come away. These were found to consist of broad bands of large organic muscular fibre-cells, with characteristic rod-shaped nuclei, and attached to each other by means of a small amount of loose connective tissue. Many of the fibres were in a state of fatty degeneration.

The removal of the tumor by operation having been decided upon, it was thought advisable first to open the bladder from the perineum. An incision was accordingly made in the middle line, and the membranous part of the urethra was opened for about three-quarters of an inch. The tumor could then be distinctly felt by the index finger introduced through this opening, and it was certain that the pedicle was near the summit of the bladder; epicystotomy was then performed. The bladder being opened to an extent of one and a half inches, the spot to which the pedicle was attached was readily found; it was, as supposed, at the summit of the bladder, somewhat anteriorly and to the left side. The pedicle was about half an inch long, soft, and scarcely so thick as the little finger; it was easily torn through with the finger nail. The tumor was then slowly and carefully extracted through the upper opening. It measured $8\frac{1}{4}$ centimeters in length and $6\frac{1}{4}$ centimeters in breadth; its circumference in the longest direction was $21\frac{1}{2}$ centimeters. The operation was accomplished with very little hæmorrhage; the edges of the wound were washed with a solution of carbolic acid, and carbolized dressing was applied; a drainage tube was inserted into the bladder from the wound in the perineum. During two days after the operation the patient appeared to be doing very well, but on the third day, symptoms of collapse supervened and terminated in death. The post-mortem examination revealed evidences of peritonitis, and diffuse purulent infiltration of the connective tissue between the peritoneum and abdominal wall in the neighborhood of the wound. The bladder was somewhat hypertrophied; the mucous membrane almost normal, the remains of the pedicle distinct. No sign of inflammation of the structures of the bladder, but evidences of two circumscribed patches of effused blood in the posterior wall of the organ.

These were considered to have been caused by the examination. On microscopical examination the tumor was found to be of a purely myomatous nature.

Professor Volkmann draws attention to the remarkable fact that the presence of the tumor failed to set up catarrhal irritation of the bladder, and that violent hæmaturia and retention of urine were the only manifestations. Commenting on the treatment after operation, the Professor thinks that the partial use of antiseptics was of little or no advantage. He believes that a drainage tube should have been passed through the bladder, and dilute solutions of salicylic acid frequently injected.

The only case on record of sarcoma of the bladder has been published by Senfleben.[1] The patient was a woman twenty-nine years of age, who had suffered from irritability of bladder and incontinence of urine for about a year. Before her sixth confinement, the patient noticed a fleshy mass protruding from the urethra; portions of this she herself removed on several occasions. After her confinement, small masses of tissue were spontaneously expelled with the urine, and without much hæmorrhage. When she applied for treatment, the urethra was found to be much dilated, large enough to admit the index finger. The hypogastric region was painful on pressure, and an inguinal gland on the right side was as large as a walnut. On passing the finger into the bladder, an elastic soft lobular tumor, having a broad pedicle, was discovered at the posterior and upper part of the organ, on the right side. Pressue on the tumor caused great pain. A fragment was torn away, and examined with the microscope and found to present the appearance of sarcoma. At the request of the patient, an operation was undertaken for the removal of the tumor. This was accomplished by tearing away portions with a polypus forceps, introduced through the urethra, after failing in the attempt to get hold of the pedicle. There was very little hæmorrhage, but the patient died of purulent peritonitis four days after the operation. On examination it was found that the bladder had been perforated.

The bladder is sometimes the seat of tuberculosis, which, however, when it occurs, is generally connected with manifestations of the same affection in other organs. Like other new formations, the tubercles most frequently attack the neck and base of the bladder. The mucous membrane first becomes thickened and infiltrated with gray and caseous miliary tubercle; superficial roundish ulcers are produced by the detachment of the softened caseous masses; these ulcers may become so large as to occupy the greater part of the surface of the membrane, and their floor and edges are more or less thickened owing to extension of the inflammation to the submucous connective tissue. Perforation rarely occurs, though it has been occasionally noticed. The affection is more frequent in children before eight years of age, than in after-life.[2] A very interesting case of the kind was described by Mr. Prescott Hewett before the Clinical Society.[3] The patient was a girl, aged nine, who had complained for some time of symptoms closely resembling those of stone in the bladder. On examination after death, the bladder was found extensively invaded by tubercular deposit, and the mucous membrane in a state of ulceration; perforation had taken place anteriorly, and also inferiorly into the rectum. The absence of tubercular disease in other

---

[1] Langenbeck, Archiv. i., p. 128.
[2] Paget, Medical Times and Gazette, 1858, vol. i., p. 368.
[3] Clinical Society, 1874, Nov. 27th.

parts was a very remarkable feature in the case. Dr. Winckel[1] states that in 2,505 post-mortem examinations of female subjects, tubercular disease of the bladder was found four times. The youngest patient was thirty-one, and the oldest sixty-six, years of age. In all the cases there were wide-spread indications of tubercular disease in other organs of the body. Dr. Burdon-Sanderson believes that tuberculous processes may originate from catarrhal inflammation of the genito-urinary mucous membrane; thus gonorrhœa may give rise to prostatitis, prostatitis to scrofulous catarrh of bladder, etc. The disease, however, usually begins above in the kidneys, and spreads downwards, for the kidneys are generally found more diseased than the ureters, and these latter than the bladder. Dr. Wilks' has recorded a case which strikingly illustrates this mode of extension. The patient was a man, aged thirty-two, who died from advanced phthisis. His urine had been purulent for three years before death. Tubercles were found in the lungs, intestines, and kidneys. The disease had extended downwards to the bladder and vesiculæ seminales. The whole of the *trigonum vesicæ* was affected, but the mucous membrane above the ureters was healthy. The ulceration extended over the surface of the prostate, and along the membranous and spongy portions of the urethra.

Urinary tuberculosis is very variable in its duration, and it would appear that it may even be completely recovered from. Sir James Paget has met with two children in whom recovery took place. A case of partial recovery in an adult has been recorded by Mr. Thomas Smith.[3] The patient was a man, aged twenty-nine, and was under treatment in St. Bartholomew's Hospital for symptoms resembling those of stone. He was also the subject of strumous abscesses over the sternum and in the left upper arm. His other symptoms were cough, swelled testicles, frequent micturition, arrest of stream, hæmaturia, and pain in left loin. The hæmaturia was aggravated by movements, and the pain was increased on emptying the bladder. No calculus could be detected. The symptoms were relieved by steel and cod-liver oil, opium and morphia. The hæmaturia yielded to confection of black pepper.

The early symptoms of tuberculosis of the bladder much resemble those of calculus, and very careful sounding is necessary, especially as the mucous membrane sometimes becomes coated with phosphatic deposits. The urine may contain, in addition to pus and blood, elastic fibres, débris and shreds of tissue. These denote deep-seated loss of substance of the mucous membrane, but are not absolutely pathognomonic of tuberculosis. When, however, cheesy-looking deposits, visible to the naked eye, occur with granular débris and elastic fibres, the existence of tuberculosis is rendered very probable.

In a few rare cases the symptoms resemble those above described, but the bladder has been found uninvaded by the disease. A remarkable instance of this kind has been recorded by Huber.[4] The patient had suffered for some time from pain in the neck of the bladder and urethra, and frequent micturition. The urine was at first pale and contained no albumen. There was no change detectible in the bladder, but the introduction of a catheter caused pain. Some time afterwards pus appeared

---

[1] Loc. cit.
[2] Trans. of the Path. Society, vol. xi., p. 138.
[3] St. Bartholomew's Hospital Reports, vol. viii., p. 95.
[4] Deutsches Archiv für klin. Med., iv., 609.

in the urine; micturition became extremely frequent, perhaps every quarter of an hour, and the urine occasionally contained a little blood. There was no pain over the symphysis pubis. The patient died in a condition of exhaustion, and on post-mortem examination, cavities and caseous masses were found in both lungs. The kidneys, the left one especially, were much enlarged. The structure of the right kidney appeared normal, but the left one contained numerous ulcerating cavities, opening into the pelvis of the organ, and much caseous matter. The bladder was contracted; its coats hypertrophied; the mucous membrane and that of the urethra injected, but presenting no trace of caseous matter. The sexual organs were intact. In this case, all the symptoms were referred to the bladder, which was comparatively normal. The hyperæmia of the mucous membrane was probably due to the constant straining provoked by the irritation of the altered urine. This latter was always acid, and it probably contained irritating materials in the form of fatty acids. It is just possible that the irritation was due to the cystitis, which in such a case might be supposed to bear the same relation to the kidney-affection as simple bronchitis does to more specific processes in the lungs. There was never any caseous débris detectible in the urine. Both kidneys were sensitive to pressure, but there was no tumor. The patient was the subject of spinal curvature.

The most common new formations found in the bladder are malignant growths, several varieties of which are met with. Encephaloid is the most usual form, schirrhus occurs less frequently, and only a few cases of epithelioma have been placed on record.[1] As stated in a previous page, the so-called villous cancer is usually of a non-malignant character, though cancerous tumors sometimes become covered by a kind of villous growth, exactly resembling the villi of the chorion. The anatomical difference between a malignant and non-malignant villous tumor consists, therefore, in the presence or absence of cancerous elements in the structures of which the tumor is primarily composed. A malignant growth infiltrates all the tissues, a villous growth affects the mucous membrane only. A cancer of any mucous membrane may sometimes be found covered on its surface with villous prolongations made up largely of blood-vessels.

Carcinoma may occur either in the form of a diffuse infiltration, or in isolated nodules, or as a local mass, polypoid in shape and sometimes covered, as above mentioned, by villous growths. In the last-mentioned form, its most common seat is between the openings of the ureters and the internal orifice of the urethra. The tumor is generally composed of several lobules, which are soft and spongy in texture and have an abundant epithelial investment. The remaining part of the mucous membrane of the bladder may be either normal, or studded over with nodules of schirrhus. Winckel gives an illustration (copied from Demme) of a carcinomatous tumor covered by villi having an epithelial investment. Out of seven cases of primary cancer of the bladder, described by Heilborn[2] three exhibited this appearance. In the St. George's Hospital Museum is a preparation consisting of a mass of malignant disease growing out of a cyst on one side of the bladder, and projecting to a certain extent into its cavity. Its outward appearance is exactly that of villous tumor, and it was only on microscopical examination that the difference

---

[1] Sir H. Thompson, Path. Soc. Trans., vol. xviii., p. 162.
[2] Heilborn, Ueber den Krebs der Harnblase. Inaug. Diss., Berlin, 1869.

could be perceived. The disease had existed for many years, and at times occasioned hemorrhage, the source of which could not be discovered. The same museum contains a specimen of villous tumor which had existed for at least twenty years.[1]

Malignant tumors usually grow very rapidly, and in this respect differ from the other new formations. The cavity of the bladder may be quite filled up, while the urethra and prostate remain unaffected. Adhesions, however, are frequently formed between the bladder and neighboring organs, as the uterus and the rectum. Peritonitis also is a frequent accompaniment. Tumors of all kinds, whether malignant or not, existing in the bladder for any length of time, are apt to become covered with an incrustation of phosphatic salts, and may, under such circumstances, be mistaken for calculi. It must not be forgotten that the bladder may become involved in cancer of neighboring organs, especially of the uterus. Under such circumstances, the cancerous growths between the two organs are usually continuous. Epithelioma of the bladder is characterized by the fact that it spreads locally and does not affect remote parts, though the glands in the neighborhood participate in the disease.

There are several points of interest in connection with epithelioma of the bladder. In Sir H. Thompson's case,[2] the disease had lasted for at least nine years, intermittent hæmaturia having been the first symptom. Frequency of micturition and some amount of pain gradually supervened, and the patient ultimately died from cystitis and exhaustion. On post-mortem examination, a large, soft mass was found in the bladder, loosely attached to a growth of wide base, circular in form, about three inches in diameter, and occupying the floor and part of the right side of the bladder. There was no enlargement of any of the neighboring glands. The microscopical appearances were those of epithelioma. In another case, reported by Dr. Hilton Fagge,[3] the disease was secondary to a perineal fistula of thirty years' standing, and was probably due to irritation from catheterism. On post-mortem examination, the posterior wall of the bladder was found to be the seat of an open ulcer, about the size of a five-shilling piece, having an irregular, sloughy surface, with thick, raised, everted edges, and with its floor and margins alike infiltrated with an opaque, soft, white growth. The microscope showed that this was a typical squamous epithelioma, with numerous and large "birds-nest aggregations." It is worthy of notice that the seat of the ulcer corresponded with the spot at which the point of the cathether would impinge upon the vesical wall, and the case is an example of the "development of epithelioma in an organ already suffering from the effects of chronic disease"—probably the direct result of irritation from the introduction of catheters. There were no secondary growths in this case, but in another, reported at the same time, the disease extended to tissues outside the bladder, and to glands near the bifurcation of the aorta.

From the above account of the various morbid growths to which the bladder is liable, it will be seen that the symptoms to which they give rise are mainly—pain, hæmorrhage, and disturbance in various ways of the functions of the organ. Other affections, however, are characterized

---

[1] Holmes, Principles and Practice of Surgery, 1st edit., p. 736.
[2] Path. Soc. Trans., vol. xxviii., p. 162.
[3] Ibid., vol. xxviii., p. 167.

by similar symptoms, which, when occurring singly, may depend upon a variety of causes. In all cases of suspected vesical tumor, it is necessary by examination to demonstrate the absence of other affections which give rise to the above-mentioned symptoms. Thus, for example, the patient should be carefully examined with the view of detecting calculus, stricture, prostatic hypertrophy, and disease of the kidneys; and not until these are proved to be absent should the presence of a morbid growth be suspected. It must be remembered that vesical tumors are far less common than the other affections with which they are liable to be confounded.

With regard to the manner in which the symptoms appear, they may either set in suddenly, or may become gradually developed. We can readily understand how the tumor may remain indolent for a certain time, until from its size, accidental displacement, rupture of its blood-vessels, etc., either hæmorrhage or impediment to the escape of urine suddenly ensues. Observation confirms this, and accordingly we find that in a few cases the disease manifests itself suddenly. Sometimes the earliest symptom is a sudden and perhaps copious discharge of blood from the urethra; this is soon followed by pain in the bladder and difficulty in making water. In other cases, the disease first shows itself by a sudden attack of severe pain, either in the region of the bladder or in the back and loins; in this case, also, the first symptom is soon followed by hæmaturia and painful micturition.

More frequently, however, in the proportion of about two to one, the disease is developed in a gradual manner. Pain is the first symptom; it is felt in or about the region of the bladder, and continues for an uncertain period, being sometimes intermittent, sometimes dull, but in the majority of cases severe and constantly progressing. After some time, the pain is attended by irritability of the bladder, and the urine soon contains more or less blood, sometimes merely streaking the urine, but often collected in clots or passed in considerable quantity. The irritability of the bladder gradually increases, and often gives rise to complete spasm of the organ. The desire to pass the urine becomes more and more frequent, until the patient is tormented every half hour, in some cases every ten minutes, enjoying no rest either day or night. The discharge of the urine is always painful, and the difficulty of micturating increases, until at last only a few drops may be passed at a time, the patient undergoing the most excruciating agony from violent straining, and spasmodic contractions of the bladder.

During the course of the disease, the constant discharge of urine is sometimes replaced by complete retention. On examination, this may be found to be due to the presence of a clot or fragment of the tumor at the neck of the bladder, or in the urethra; or to impediment from the tumor, which blocks up the internal orifice of the urethra, or to the pressure of the malignant growths on the orifices of the ureters.

The condition of the urine varies with the nature of the tumor and the progress of the disease. In almost all the cases it contains more or less blood. Hæmaturia, however, is not a characteristic symptom of fibrous growths, but it may occur if these become ulcerated on the surface. On the other hand, hæmaturia is the characteristic, and often the only symptom of villous growths. The urine, in these latter cases, is seldom free from blood, and is often deep red in color, owing to the quantity of blood present. In malignant disease, as the case progresses, the urine contains, in addition to blood, débris of various kinds, and is also highly purulent and offensive.

Having thus given a general sketch of the essential symptoms of vesical tumors, it may be useful to notice each in detail, and mention some points which may throw light on the nature of the complaint.

The pain which is first felt either sets in suddenly or is gradually developed. It presents several varieties with respect to its seat, intensity and character. The painful sensations experienced by the patient are, moreover, distinguishable into two kinds; the one is more immediately connected with the tumor itself, or its direct effects on the adjacent tissues; the other kind of pain evidently depends on the violent efforts of the bladder to expel the urine and foreign body, which act as constant irritants of the mucous membrane.

The former kind of pain alone can be said to be proper to the disease; the latter is a symptom common to stone, retention of urine, and many lesions of the urinary passages.

The essential pain now spoken of, when it commences suddenly, is confined to the bladder or to the lower part of the back; in this latter case, it soon extends to the bladder likewise; and, in both, it is of a severe nature. When the disease is less suddenly or is gradually developed, we find the patient complaining of some uneasiness about the region of the bladder, or of irregular or intermittent pains in that part. In other cases, the pain commences about the loins or hips. It may thus continue for several months, with alternations of severity and relief; or it may be very severe from the first, in the region of the pelvis and the loins. The character of the pain presents nothing of a special kind. In the early stages of the disease the patients usually refer merely to varieties in its intensity, and rarely, if ever, speak of it as being of the lancinating, darting kind.

As the disease advances, however, and as the irritability of the bladder becomes increased, lancinating pains are occasionally experienced, or there may be a cutting, darting pain, which radiates in several directions toward the rectum, into the perineum, or along the urethra.

It is obviously difficult to determine how far the pain, in this more advanced stage, depends on the tumor itself, or on the irritability of the bladder and secondary lesions, which the presence of the tumor may have determined.

In very few cases has the peculiar pain of the glans penis, so characteristic of stone been observed.

The seat of the pain is not generally well defined; but in some cases it has been referred to the neck of the bladder; in others, to the lower part of the back and vicinity of the rectum; while in others, again, the pain has been chiefly experienced above the pubes, and in the lower part of the abdomen. The different points from which the tumor springs will explain these varieties; and it should be observed that the bladder is sometimes painful to the touch, when examined with the finger through the rectum, or when pressure is made above the pubes. In almost every case the pain goes on gradually augmenting; but it is difficult to estimate this increase, the essential pain being, sooner or later, masked by severe spasm of the bladder.

The indications to be drawn from the presence of blood in the urine, and especially from the discharge of pure blood from the bladder, are sometimes of considerable value. Persistent vesical hæmaturia, in the absence of a calculus or other cause, is the most characteristic sign of the presence of a tumor. In some cases, the urine at first is merely tinged with blood, the hæmaturia being accompanied by only slight symptoms of vesical lesion; these, however, quickly become more severe, present few or no remissions, and are soon attended by more copious loss of blood. The *early* occurrence of hæmaturia in these cases is not to be overlooked; but the most significant character is its constant occurrence during the whole course of the disease. The quantity of blood discharged may greatly diminish, the urine may even become clear for a time, but the symptom recurs again and again until death supervenes.

Although the blood is usually discharged with the urine, because during micturition the tumor is especially exposed to pressure or irritation, yet hæmorrhage may occur while the patient is at rest and the bladder empty. Frequent and persistent oozings are characteristic rather of villous growth than of malignant disease. Another sign which must be taken into account is the great increase of hæmorrhage sometimes occasioned by the use of instruments during an examination of the bladder. Under certain circumstances the use of the sound may be followed by some discharge of blood, but the hæmorrhage is rarely abundant unless the bladder be the seat of a malignant or vascular tumor, or be in a state of ulceration. It has been justly observed that blood discharged with the urine "may come direct from the urethra, prostate, or kidneys, as well as the bladder, and that an error of diagnosis, as to the seat of the hæmorrhage, may be readily committed." The experienced practitioner will not often have much difficulty in determining whether the blood comes from the bladder or from the canal of the urethra. It is less easy to say whether the hæmorrhage may not be renal; we must look to the concomitant signs of renal disease; when the blood is derived from the kidney, the fluid discharged often contains the tubular casts indicative of disease in that organ.

Dr. Ultzmann[1] has drawn attention to the occasional occurrence of fibrinuria as a symptom of villous disease of the bladder. A full description of this condition will be found in the chapter on "Hæmaturia." Fragments of tissue are sometimes passed with the urine, and by examining these under the microscope, evidence may occasionally be obtained as to the nature of the case.

The sudden retention of urine, coming on after painful micturition and frequent hæmaturia, is a sign of great value; but it can serve as a diagnostic sign of tumors only in those cases where an examination has shown that the obstacle to the evacuation of the bladder does not depend on calculus, enlargement of the prostate, stricture, or any of the ordinary causes of retention.

It is unnecessary to discuss the points of similarity and difference between calculus and tumors of the bladder, because the rational symptoms of the former are just as deceptive as those of the latter, perhaps even more so. No surgeon would, at the present day, think of affirming the existence of calculus in the bladder without the presence of its positive signs, derived from the use of the sound. The absence, however, of these signs, does not entitle us to infer positively the absence of stone. The calculus may be in the bladder, yet out of the reach of the sound; the obstacles to complete examination are, however, now well understood, and though doubts may arise, errors of practice are seldom committed.

---

[1] Ueber Hæmaturia, Wiener Klinik, Hefte iv. u. v., p. 138.

As the rational symptoms of malignant or other morbid growths in the bladder tend only to suspicions, the surgeon will naturally turn to physical signs derived from an examination of the urinary organs, and of the tumor itself, or rather of any portions of it which may be passed with the urine. By careful examination we determine the absence of those lesions the symptoms of which are analogous to those of vesical tumor; by exploring the bladder in a methodical manner we can generally ascertain that it contains a soft tumor; and should any portions, or even débris of this tumor have come away with the urine, the microscope may enable us to ascertain whether it be of a benign or malignant nature.

It is a remarkable circumstance that the urethra and prostate gland seldom present any serious organic lesions in cases of malignant disease of the bladder. The mucous membrane of the bladder is also remarkably free from disease, the most common change of the organ observed being some degree of hypertrophy and contraction. On examination, then, the surgeon will frequently discover that the urethra, prostate and neck of the bladder are free from organic disease, and this discovery will considerably restrict the field of his inquiries, showing that the cause of the symptoms is confined to the bladder.

This organ should now be carefully explored. In thin subjects, a tumor situated at the upper part of the bladder may sometimes, as in Professor Billroth's case, be detected through the abdominal walls. Morbid growths, however, most commonly involve the base of the bladder, and are therefore situated deeply in the pelvis, and within reach of the finger in the rectum. For examining the interior of the bladder, a short-beaked sound should be introduced. The extremity of the instrument may be arrested a little beyond the internal orifice of the urethra, or it may pass freely into the bladder.

In the latter case, by slowly revolving the point and making the end of the instrument sweep, as it were, the fundus and sides, it will rarely happen that the existence of a tumor is not discovered. Unless the morbid growth be very large or occupy a particular position, the presence or absence of stone may be, at the same time, determined. The examination may be aided by introducing a finger into the rectum. By feeling for the sound, information may be derived as to the thickness of the coats at the base of the organ. By movements of the sound itself in various directions, any existing unevenness or roughness may be detected, and likewise any obstacle connected with the vesical walls. If the tumor is situated at the back of the bladder, the beak of the instrument will be directed forwards directly it enters the organ. An extremely soft tumor, such as a villous growth, will not offer any appreciable resistance to the movements of the sound. If the case be one of malignant growth, the tumor will usually be within reach of the finger introduced into the rectum, and the examination may be further aided by making pressure with the other hand above the symphysis pubis, so as to force the bladder as deeply as possible into the pelvis. A hard, nodulated swelling, extending far back so that its boundary cannot be reached by the finger, is strongly suggestive of malignant disease. Even if the tumor be situated in the upper part of the bladder, it may sometimes, as in Professor Volkmann's case, be detected on bimanual exploration. The existence of carcinomatous cachexia and of glandular enlargement will serve to confirm the diagnosis of malignancy, while the absence of these conditions, in a case presenting the symptoms of vesical tumor, will render it probable that this latter is of a non-malignant kind.

With regard to the evidence which may be obtained from an examination of the urine in cases of suspected vesical tumor, the reader is referred to what has been already stated in the foregoing account of the symptoms of these growths, and also to the chapter on "Hæmaturia." Small masses of tissue may be passed with the urine, and from an examination of these, valuable evidence may sometimes be obtained. This is particularly the case with regard to villous growths, the characteristic structures of which may be clearly identified under the microscope. With regard to malignant formations, it is generally impossible to distinguish any specific cell-growths in the multitude of variously formed cells which are discovered in the urine. If, however, soft, small, translucent masses are occasionally passed, and these, when examined under the microscope, are found to be composed of a multitude of large cells, each containing several nuclei, the diagnosis of malignant tumor is strongly confirmed.

A careful examination of the urine may enable us to arrive at a correct diagnosis in cases where many of the ordinary symptoms of malignant growths are absent. An instance of this kind is recorded by Dr. Dickinson.[1] Hæmaturia was, for some time, the only symptom. There was no unnatural sensation in the loins, no frequency of micturition, or irritability of the bladder, nor any other pain or discomfort. The urine, however, contained small, soft, buff-colored lumps, resembling fibrinous coagula, and these, on being placed under the microscope, were found to consist of an aggregation of large spheroidal cells, among which were many round bodies which had a radiating structure as though they were partially crystallized. These cells and masses evidently belonged to an organized cellular growth, apparently of an encephaloid character, and situated in the bladder. The blood was not diffused through the urine, but was passed at the end of micturition, almost pure. Other symptoms rapidly supervened and the patient died, and although, unfortunately, a post-mortem examination could not be obtained, there was no doubt as to the nature of the case. It seldom happens that such convincing proof of the nature of the affection is to be obtained from a microscopical examination of the urine as was the case in this instance. As Dr. Dickinson remarked, tangible masses of vascular cell-growth are rarely expelled with the urine. In the case of simple villous growths, loops of blood-vessels sometimes become detached, and are easily recognized under the microscope.

The following cases will serve to illustrate the symptoms and progress of malignant disease of the bladder. In the spring of 1848, Dr. Lankester was consulted by a gentleman, æt. 62, of small stature, active and temperate habits; he had, nevertheless, suffered much from dyspepsia, for which he had consulted medical men, and latterly had been compelled to attend to his diet with more than ordinary care. About twelve months before Dr. Lankester saw him, he was suddenly attacked, while working in his garden, with pain in his back, followed by a desire to make water, which, when passed, he found to be discolored with blood. He immediately sought for medical assistance, but the pain in the back continued, and there was a fresh appearance of blood in the urine. He lost flesh as well as appetite, and his nights were sleepless; the frequency of the desire to pass water increased, and the pain in the back continued, extending thence down the thighs. Under these circumstances, it occurred to his

---

[1] Path. Soc. Trans., vol. xx., p. 233.

medical attendant that stone in the bladder was the cause of his symptoms, and he was sounded more than once, but without any being found. When seen by Dr. Lankester he was very thin, with an anxious countenance, and peculiar stoop in his gait. He suffered much from dyspeptic symptoms and severe gastrodynia; the bowels were habitually constipated, and never relieved without medicine. He was restless at night, and obliged very frequently to get up and make water; there was severe pain in the back and in the region of the bladder above the pubes, where pressure caused severe suffering. The urine, passed every hour or hour and a half, was sometimes more colored than at others. It was acid; and after standing a considerable time, threw down a deposit, which, under the microscope, presented lithate of ammonia and organic shreds, apparently of mucus and fibrine. Blood-corpuscles and pus-globules could also be detected. On applying heat and nitric acid, albumen was precipitated. After a variety of treatment, he consulted an eminent surgeon, who passed a catheter, and endeavored to inject the bladder. The operation was attended with great pain, and followed by an increased amount of blood in the urine. It was found impossible to get the fluid to pass into the bladder, although the catheter was fairly beyond all possibility of prostatic obstruction. After a few months, he again came under the care of Dr. Lankester. He was greatly emaciated; the pain in the back and region of the bladder was constant and intense; the urine was passed more frequently than ever, being sometimes clear, sometimes mixed with blood. On standing for a short time, it deposited a thick sediment, which consisted of lithates, mixed with amorphous organic matter. The use of morphia afforded some relief; but the patient became gradually weaker, and sank in the month of January in the following year. Upon examination, the lungs and liver were found healthy; the heart of natural size, with slight deposits upon the mitral and semilunar valves. Both kidneys were enlarged, the left much more so than the right; the capsule was easily separated from the mass of the kidney; the tissue was soft and easily broken down, but there was no conversion of tissue. The pelvis contained some puriform fluid, which was traceable down the ureter. The right kidney was the smaller; its pelvis was highly injected, and parts of its surface presented granular degeneration. The bladder was empty, and when grasped from the outside, the coats were firm; while, on opening it through the urethra, a loose organized mass of yellowish color was found free within it. Attached to a point near the neck of the bladder, was a granulated mass highly injected, presenting portions of a red color passing into white. It was easily broken down with the knife, and yielded to slight pressure. Higher up the bladder, and on one side of the fundus, another tumor presented itself of larger size, but having the same character as that below. The mass that was loose had evidently been separated from one of the portions still in connection with the bladder. The coats of the bladder were very much thickened; the prostate was considerably enlarged, and on cutting into it, several points presented themselves, in which was found matter of precisely similar character to that of which the growth in the bladder was composed. The spleen, stomach, pancreas, and absorbent glands of the abdomen, both in the mesentery and along the course of great blood-vessels, were examined, but they presented no trace of malignant disease.

About the same time the following case came under Mr. Coulson's care. T. S. B., æt. 50, stout, and of a gouty habit, in May, 1848, passed bloody urine for the first time, having previously had no other symptom than severe pain in the back. He continued to pass blood in the urine for five or six months, at first without pain, but afterwards with intense suffering. The blood at the commencement was fluid and diffused in the urine, but after a short time coagula were passed, and the passing of these occasioned great pain. At the end of six months the appearance of the blood almost entirely ceased, but symptoms of chronic inflammation of the bladder came on. There was frequent desire to pass urine; great pain or scalding at the time of passing it, and afterwards the urine was loaded with stringy mucus. Some of the symptoms being those common to stone, he was sounded three or four times, but no calculus could ever be detected. The sound could scarcely be moved at all in the bladder, its motion being evidently impeded by some morbid growth, and the sounding was always attended by severe suffering. On one occasion, while the sound was being moved about, a portion of the morbid growth was detached and subsequently came away. On examination structures resembling cancer-elements were detected, and the diagnosis of malignant disease of the bladder was accordingly formed. His local symptoms increased in intensity; his strength and appetite began to fail him, and in the following May he died. A constant dribbling of the urine was the only symptom in the case for the last two or three weeks prior to his death. On examination of the body, the bladder was found occupying the greater part of the pelvis, and fixed in its position by morbid adhesions, which extended on all sides, and were very indurated. On passing the hand between the bladder and rectum, the glands along the course of the right ureter were found diseased. The right kidney itself was more than double its normal size, the other was only slightly enlarged. Both on section presented a clear surface, of an ashy gray appearance, the natural structure was obscured, and more or less destroyed by infiltration of disease. Towards that portion where the bladder is usually uncovered by the peritoneum, the part seemed much softer and readily gave way. The interior of the bladder was filled to a great extent by a morbid growth of a cancerous character, which seemed to take its origin from the mucous coat, and was much larger towards the neck. The mass bore a strong resemblance to the cauliflower excrescence of the uterus. In some parts, the submucous tissue and muscular coats were completely absorbed, in other parts they were not much affected.

The following case was complicated with stone. A man, forty years of age, was admitted under Mr. Coulson's care into St. Mary's Hospital, January 11th, 1855. He stated that he had been afflicted with uneasiness while passing his urine as long as he could remember; but that it was during the last two months only that he suffered great distress, under which his general health had begun to give way. On admission he complained of great pain along the urethra and at the end of the penis, especially after making water, and also of great uneasiness in the lower part of the abdomen; the urine, which he was obliged to void frequently, was always bloody, and contained crystals of the triple phosphate, but no cancer-elements. The presence of a calculus was readily detected by the sound. The man's countenance was pale, and expressive of great suffering. The lateral operation was performed, and two calculi were extracted, but not without some difficulty, for one was on the right side of the bladder, in a sulcus formed by the tumor. There was a good deal of venous hæmorrhage during the operation, which the patient survived only three days. The bladder was found to be adherent to the surrounding structures, and its cavity was occupied by a medullary tumor, which yielded a soft brain-like substance on pressure, and exhibited, under the microscope, the elements of a cancerous growth.

J. C., 22 years of age, applied, on the 15th of November, 1853, at St. George's Hospital, on account of some difficulty in making water, attended by incontinence of urine and hæmaturia. He stated that about a month previously he had fallen off a ladder, and struck himself, while falling, on the right side of the chest. Some three days after the accident he noticed, for the first time, blood in his urine. He then began to suffer from difficulty in making water, and he was suffering much on admission from inability to empty his bladder. The urine was dribbling away from the urethra, mixed with blood. The complexion was sallow and unhealthy, and the expression was indicative of much distress.

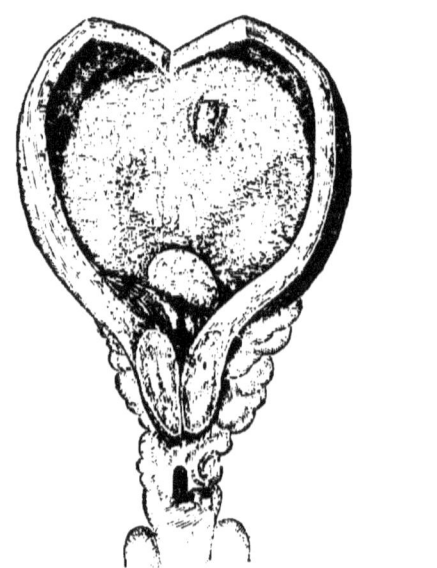

Fig. 3.—Showing Bladder with Medullary Tumor and two Calculi.

A small-sized silver catheter was passed by Mr. Pollock; there was no obstruction to its passage into the bladder, and only a small quantity of blood escaped through it. A small-sized gum catheter was then pushed far into the bladder. As it entered, a small quantity of blood and coagulum escaped; but on its further introduction clear urine came away; and after the bladder was emptied of water, blood again escaped through the instrument.

On examination per rectum, the bladder felt large and elastic; and it was inferred that this effect was produced by some diseased condition of its walls or by some growth in its cavity.

The discharge of bloody urine continued; the fluid evacuated became by degrees more offensive, alkaline, and purulent. Under the microscope, numerous blood-corpuscles, a copious deposit of crystals of the triple phosphate, and a large number of nucleated cells, were seen. These latter consisted of separate cells, with a delicate investing membrane, granular contents, and a large, highly granular nucleus. There were also small masses, which probably consisted of similar cells massed together, as their outline showed a similar granular arrangement. The patient sank gradually, and died eleven days after his admission.

The kidneys were enlarged, but not diseased. The pelvis and ureters on both sides were greatly dilated. The bladder was half filled by a fungoid growth, attached to the lower portion of its posterior wall, and contained some offensive urine, mixed with pus and blood. The mass of the tumor consisted of several independent growths attached to the walls of the bladder by separate pedicles; it had grown out into large flocculent tufts, shreddy at their surface, and very vascular in their structure. The surfaces of the masses were ulcerated. It was difficult to determine whether the original disease commenced in the mucous or sub-mucous tissues.[1]

In another case, under Mr. Coulson's care, the patient's symptoms were supposed to be due to the presence of a stone. He had suffered for a considerable time from pain in the region of the bladder, and frequent desire to pass the urine, which was frequently bloody and mixed with mucus. Various remedies were employed, and with some relief, for at the end of six months the pain in the bladder was greatly mitigated and the hemorrhage from the bladder much diminished. The patient, however, remained in a debilitated state from the frequent loss of blood which he had formerly experienced.

The symptoms latterly complained of seemed referable to the urethra rather than the bladder. The urine still contained mucus, and was occasionally tinged with blood. The discharge of urine was painful, but the patient referred his sufferings chiefly to the urethra, along the whole course of which he experienced intolerable pain while making water. It was most violent when the urine first flowed, and then gradually subsided, until the bladder was emptied, when nothing was felt except an aching sensation along the whole course of the urethra, which, however, soon passed off. The desire to pass water was experienced every hour; it was excited by standing or walking; the pain, also, was increased by exercise; but was much alleviated when the patient had remained for two or three hours in bed in a particular position on his back.

It was now thought that ulceration existed in the mucous membrane of the urethra. Balsam of copaiba produced irritation and hemorrhage; various kinds of injections were tried, some with relief, others with no result. The muriate of morphia, in half-grain doses, alleviated suffering, but produced no other effect. The patient sank in about a fortnight after coming under observation.

On examination of the body, the bladder was found to be somewhat contracted, and the muscular fibres irregularly thickened. Attached to the fundus was a tumor, about the size of a small orange. It was rather firm at the base, vascular, and of a cauliflower appearance in the body; the color of the tumor was whitish, interspersed with dark purple spots; its surface was covered with a soft, sandy deposit, consisting chiefly of the phosphate of lime and triple phosphate. The deposit was very friable, being easily crushed between the fingers; and it should be remarked that the urine occasionally contained a similar deposit during life.

At the right side of the bladder, and extending as far as its neck,

---

[1] Transactions of Path. Society, vol. vi., p. 258.

was another tumor, of similar structure but smaller in size: this was so situated that it must have occasionally blocked up the internal orifice of the urethra. The rest of the lining membrane of the bladder was dotted here and there with small, hard, reddish tubercles, between which the membrane itself appeared pale and flabby. The ureters were greatly dilated. The prostate was healthy, but a thin membranous band extended across the membranous portion of the urethra, and obstructed the passage. About three inches behind the glans penis the urethra presented a few points of ulceration; it was otherwise healthy.

The kidneys were much enlarged, and softer than natural. Their external surface was covered with elevated tubercles, about the size of hazel-nuts, which broke down under slight pressure. The internal structure of the kidneys was completely disorganized, and interspersed with small abscesses filled with pus and oil-globules. There was no calculus either in the kidneys or in the bladder.

With regard to non-malignant growths, the symptoms are nearly, if not entirely, the same as those of malignant tumor, but they are usually less severe. Copious loss of blood, however, except in the case of villous growths, more rarely attends the benign disease; irritation is more apt to extend along the urethra, and the progress of the disease is much slower in cases of fibrous or fibro-cellular polypi, etc., than when the bladder is the seat of encephaloid or epithelial cancer. Both the benign and malignant lesions are equally attended by pain in the region of the bladder, constant and painful micturition, and occasional retention of urine. But as the non-malignant tumors are much more frequently seated about the neck of the bladder than the cancerous, the pain and irritation which they excite are more often felt about that part, and retention of urine is a more frequent symptom. In many of these cases, also, hæmaturia is absent, or the urine is only slightly and from time to time tinged with blood.

An examination with the sound will sometimes enable the surgeon to detect a pedunculated tumor about the neck of the bladder, but a more complete diagnosis is generally impossible. The absence of hæmaturia, and of structures resembling cancer-cells, will assist in confirming an opinion that the tumor is of a non-malignant character. Examination through the rectum or vagina, as the case may be, may reveal the existence of a tumor, if this latter be of a large size and firm consistence. It has been mentioned in a previous page, that females are more liable than males to suffer from tumors of the bladder, and in the former sex, owing to the shortness and dilatability of the urethra, the diagnosis of these affections is often easy and their treatment simple and efficacious. We are mainly indebted to the late Prof. Simon, of Heidelberg, for showing how easily the internal surface of the female bladder may be explored with the finger, and how tumors may be removed therefrom. He has given a minute account of the various methods he employed for these purposes in one of Volkmann's series of clinical lectures.[1] We have already referred to this essay in the chapter on the various modes of examining the bladder. Examination with the finger will enable us to discover the position, size, consistence, mobility or the reverse of the tumor, and to come to a determination as to the possibility of successful operative interference. Simon also suggests that a fragment of the tumor should be scraped off with a sharp-edged spoon and submitted to microscopical examination.

[1] Volkmann's Sammlung Klinischer Vorträge, No. 88.

The symptoms of tubercular ulcer of the bladder are not well marked; they resemble those of chronic ulceration in general, and often simulate the effects of vesical calculus. In some cases on record the symptoms connected with the bladder were so slight as to have escaped notice during life. In these instances the ulcers were few, and of limited extent, while the kidneys, prostate, and several other organs were the seat of extensive tubercular disease.

The ordinary symptoms which attend the disease are pain about the neck of the bladder, with frequent and painful micturition; the urine at first contains mucus, then becomes albuminous, purulent, and is often tinged with blood as the disease advances; it is alkaline in some cases, acid in others. These symptoms, either masked by, or complicated with, symptoms of diseased kidney and of tubercular affections in other parts, gradually increase until the patient sinks in a state of hectic, or is cut off by some intercurrent malady.

There are no particular signs by which tubercular ulcer of the bladder can be made out with any degree of certainty. In the diagnosis of malignant affections we sometimes derive great assistance from the general aspect of the patient, but in suspected tuberculous disease deductions of a like kind cannot be drawn without a great danger of error. The disease occurs almost invariably in young subjects. In suspicious cases we should search for evidence of coincident tubercular disease of other parts, as of the kidneys and testicles, and of the prostate in adults. The appearance of elastic fibres in the urine is probably indicative of tuberculous ulceration.

With regard to the treatment of non-malignant tumors occurring in the male subject, such growths are for the most part beyond the reach of surgical efforts. We have to confine ourselves to alleviating the prominent symptoms, such as hemorrhage, painful and frequent micturition, and retention of urine.

Hæmorrhage is the predominant symptom of villous formations, and it may, as we have seen, lead to the death of the patient. It rarely occurs in connection with fibrous growths. When present, it may be combated by injections into the bladder, or by administration of remedies by the mouth. As a matter of course, all instruments should be used with the greatest gentleness, any roughness of manipulation will inevitably increase the mischief. As an injection, a weak solution of nitrate of silver (half a grain to the ounce) will probably be found the most efficacious for checking the hæmorrhage and allaying irritation. Continuous rest in bed and the avoidance of all excitement are also indicated. Sir H. Thompson recommends a weak solution of the tincture of the perchloride of iron (a drachm to four ounces) as an injection, with which he succeeded in arresting hæmorrhage when all other remedies failed. He also speaks favorably of an iced infusion of matico for the same purpose. Solutions of alum, acetate of lead, gallic acid may also be tried. Bladders of iced water applied to the hypogastrium and perineum, and the introduction of a plug of ice into the rectum, as recommended by Dr. Gross, will also aid in checking the hæmorrhage. As internal astringents we may give the tincture of the perchloride of iron, solutions of alum or infusion of matico. Dr. Gross speaks highly of a combination of turpentine and dilute sulphuric acid, ten drops of each, with five grains of gallic acid, administered every few hours. The various preparations of opium may be added to any of these remedies.

For the relief of painful and frequent micturition, anodynes are especially indicated. Opium, or some of its preparations, may be given internally or in the form of a suppository. Belladonna in the latter form often gives much relief. These and similar remedies should be administered freely. Should retention occur, recourse must be had to the catheter, which must be used with all gentleness.

The attempts which have from time to time been made to remove morbid growths from the male bladder deserve a passing notice. Mr. Crosse's case has been already alluded to. Mercier, Civiale, and others have devised instruments for removing pedunculated growths through the urethra, but the dangers of such an operation are such as to prohibit its performance. Uncontrollable and fatal hemorrhage, or laceration of the coats of the bladder, are results which might easily ensue, and although Civiale[1] has reported more than one successful case, few surgeons have followed his example. Such tumors occurring in the female bladder are, as we shall see presently, much more amenable to treatment of this nature.

Attempts have also been made to remove tumors from the bladder in another manner, namely, by cutting into the organ and exposing and removing the morbid growth. Billroth's[2] case has been already alluded to.

In connection with operations of this kind, Liston's[3] case may also be mentioned. He opened the bladder above the pubes, and removed a fibrinous cast which filled up the cavity of the organ, and caused complete retention. Nussbaum[4] also, in excising portions of a cancerous rectum, has repeatedly removed fragments the size of half-a-crown from the walls of the bladder which had become involved in the disease. The operations produced great relief to the patients, and in no instance were followed by fistulous communications.

Tumors of the female bladder are much easier of diagnosis and much more accessible to operative treatment. Several instances of successful removal have been placed on record. An operation is, of course, more likely to be successful when the growth is of the polypoid kind. In such a case, the nature of the tumor having been previously ascertained, the patient is to be placed under the influence of chloroform, and the index finger of the left hand introduced through the urethra into the bladder. A long slender pair of forceps is then passed along the finger, by which it is guided to the peduncle of the tumor. By gentle twisting or dragging movements the tumor may then be removed from its attachments, and withdrawn through the urethra. Should such an operation be likely to cause much hemorrhage, the wire of an écraseur may be passed round the peduncle,[5] or the galvanic cautery may be used, for the same purpose. If the tumor is very large, it may be broken up by the finger, or forceps introduced through a speculum in the urethra. A sharp-edged spoon may be used for the same purpose. Schatz (quoted by Winckel) treated a case by incising the urethra and dragging the tumor downwards by means of threads passed through its base. It was then divided and subsequently removed.

---

[1] Civiale, Traité Pratique sur les Maladies des Organes Génito-Urinaires. T. iii., p. 151 et seq.
[2] Langenbeck's Archiv, xviii., 411.
[3] Medical Times and Gazette, August, 1862.
[4] Quoted by Podraski, Von Pitha u. Billroth's Handbuch d. all. und spec. Chir., Band iii., Abth. ii., Lief. viii., s. 70.
[5] Braxton Hicks, Lancet, 1868, vol. i., May 30th.

Non-pedunculated growths can be only partially removed, but portions of them may be scraped away with the sharp-edged spoon as devised by Simon. The hemorrhage caused by such an operation is rarely considerable, as the vessels are torn and twisted. If, however, there be much bleeding, it may be arrested by injections of cold water or of infusion of matico, or by the application of the tincture of the perchloride of iron. In cases where the tumors or excrescences are situated in the upper part of the bladder, and are therefore only partially accessible to the finger introduced through the urethra, Simon recommended a T-shaped incision through the vesico-vaginal septum. The wall of the bladder is then to be drawn down and the tumor removed with the knife or scissors. The actual cautery may, if necessary, be used to check hemorrhage, and after the raw surface left by the removal of the tumor has healed, the opening in the septum may be treated as a fistula.

The treatment of malignant disease of the bladder is altogether palliative. The pain, irritability, and spasm are best relieved by full doses of morphia, pushed to narcotism if necessary. Hæmorrhage may be checked by various astringent injections, such as weak solutions of nitrate of silver, and of perchloride of iron, or by the local application of cold.

CHAPTER XII.

FISTULÆ OF THE BLADDER.

It has been shown, in a previous chapter, that laceration or rupture of the bladder is an exceedingly dangerous accident. The danger, however does not arise so much from the injury inflicted on the bladder itself as from the inflammation produced by the extravasated urine, and hence we find that, whenever the urine has a free outlet through any of the natural channels, the patient generally recovers, but a fistulous opening or passage often remains.

Affections of this latter kind do not appear to have been recognized by the earlier writers on medicine, and it is not until the commencement of the seventeenth century that we find any allusion made to them. A Spanish surgeon, Ludovicus Mercatus, in a book published in 1605, appears to have been the first to describe urinary fistulæ as occurring in women. He also indicated the various means by which they might be cured, and recommended that the callous edges should be removed, either by the application of cauterants or by a cutting operation; but the subsequent approximation of the margins by means of sutures did not form any part of his plans. This latter method was first proposed by a Dutch surgeon, Hendrik van Roonhuyzen, in 1663, but it does not appear that he himself ever adopted his own proposal. The affections in question, however, soon became fully recognized, and accordingly we find them alluded to by numerous writers on surgery, and diseases of women.[1] Coming nearer to our own times, Desault and Dupuytren may be mentioned as having published several cases successfully treated by cauterization.

According to Dr. Winckel, a new era in the operative treatment of these affections was inaugurated by Jobert de Lamballe in 1834. His method consisted in paring the edges of the fistula and bringing them together by means of suture. In the previous year, however, the same plan had been carried out by Mr. Gossett, surgeon to Newgate, for the cure of a fistula the result of vaginal lithotomy.[2] The success of the operation having been established, it was speedily adopted by surgeons in various countries, and numerous improvements and modifications were suggested and carried out by Dr. Simon of Heidelberg, who had been one of M. Jobert de Lamballe's assistants, Drs. Sims and Bozeman, of New York, and Mr. Baker Brown, of London. Quite recently Mr. Lawson Tait, of Birmingham, has shown that cases hitherto deemed incurable, on account of the extent of the lesion, may be satisfactorily dealt with by adopting certain modifications of the ordinary operation.

---

[1] For a résumé of the history of vesical fistulæ, and a copious list of authors who have written on the subject, see Winckel, Handbuch der Frauenkrankheiten, ix. Abs., s. 100.
[2] Lancet, November 29th, 1834.

Vesical fistulæ are of various kinds, according to the organ with which the abnormal communication is established. In the most common kind, vesico-vaginal fistula, the bladder communicates directly with the vagina through an aperture in the septum. The aperture may also be higher up, between the bladder and uterus, constituting utero-vesical fistula, while in a third form the urethra and neck of the bladder open directly into the lower part of the anterior wall of the vagina. It occasionally happens that the urethra alone is implicated, while in another variety, a communication is established between the ureter on one side and the vagina. Recto-vesical fistula has also been observed in female as well as in male patients, and it occasionally happens that the bladder communicates with some portion of the small intestine. That form of fistula caused by a pervious urachus has been described in a former chapter. Two or more of these fistulæ have been known to co-exist. Thus the bladder may communicate both with the vagina and the uterus, and a vesico-vaginal fistula may be complicated by an opening of the same kind established between the vagina and urethra.

With regard to the causation of these affections, the majority are due to ulceration occurring in the course of malignant disease of the genital organs, but such cases are rarely amenable to any form of treatment. Those cases, on the other hand, for the relief of which much may be accomplished by operative interference may, with reference to their causation, be divided into two classes, according as they are or are not connected with the puerperal state.[1] The larger of these two classes is composed of the puerperal fistulæ, and these again may be of spontaneous origin, or may be occasioned by various operative manipulations.

Puerperal fistulæ of spontaneous origin may be caused by laceration of the vesico-vaginal septum during the passage of the head of the child, especially if the bladder be distended; by injury to the same part during labor in cases of exostosis of the pelvic bones; or in cases where a calculus is present in the bladder. In these instances the laceration is immediately produced. In other cases, where the pelvis is contracted, the vitality of a portion of the septum may be destroyed by pressure; a slough is formed, the separation of which produces a communication between the bladder and vagina. When the pelvis is small, such injuries are more likely to occur in face presentations, as the vesico-vaginal septum is then compressed between the sharp edges of the child's inferior maxilla and the anterior bony walls of the pelvis. It was at one time believed that the majority of puerperal fistulæ were caused by operative interference on the part of the practitioner, the forceps being the instrument with which the mischief was alleged to be accomplished. On the other hand, it is asserted that in many instances the occurrence of the lesion might have been averted had an early recourse been had to instruments. Dr. Winckel, in commenting upon these discordant opinions, cites some statistics collected by Dr. Bouqué of Ghent. Out of 204 cases of urinary fistulæ, occurring in women, 118 were due to puerperal causes, and of these, 65 (55 per cent) had arisen after labors in which instruments had been used. It is worthy of notice that those fistulæ which occur spontaneously during labor are generally situated in the upper part of the vagina, and involve a portion of the uterus as well as the vaginal walls.

---

[1] This classification has been adopted by Winckel in his Handbuch der Frauenkrankheiten, a work from which much assistance has been derived in the revision of this chapter.

In a second class of puerperal fistulæ, the lesion has followed confinements in which instruments have been used for delivery. In Dr. Bouqué's cases, above alluded to, the forceps was used in 37 cases, the lever in 7, and the blunt hook in 3, while turning was had recourse to 5 times and craniotomy 12 times. The violent use of the forceps may produce direct laceration, or may cause such an amount of pressure that a slough forms, which on separating leaves a fistula. Injuries produced in this manner are usually situated in the lower part of the anterior wall of the vagina. The sharp fragments of bone after craniotomy may, unless due care be taken, cause considerable laceration of the vesico-vaginal septum, while in delivery by turning, the same parts may be so compressed as to lose vitality, and to be partially destroyed by sloughing and ulceration. Dieffenbach has reported a case in which the point of a catheter was thrust through the vesico-vaginal septum and produced a fistula. Of 14 instances of which the details are given by Dr. Winckel, only in 1 case instruments had not been used, in 2 cases the lesion was not attributable to the operative manipulations, in 3 it was probable that it had been thus caused, while in 8 such a causation seemed almost certain.

Fistulæ of non-puerperal origin may either be due to injury, or may arise spontaneously in the course of disease. The rough use of a sound or lithotrite, foreign bodies in the bladder, falls upon sharp-pointed substances, have been known to cause the lesions in question. A fistula sometimes remains after vaginal lithotomy, and the bladder has been also injured in operations for relieving an impervious vagina. The long-continued use of a pessary has been known to cause ulceration and fistula of the vesico-vaginal septum. Similar fistulæ of spontaneous origin are sometimes due to cancerous disease of the uterus and vagina. Tuberculous or syphilitic disease rarely, if ever, leads to destruction of the vesico-vaginal septum. Stone in the bladder, chronic ulcerative disease, as well as malignant affections of that organ, are occasional but rare causes of fistula.

It is not surprising that the anatomical condition of the abnormal opening should vary greatly in different cases, when we reflect on the various causes by which the accident may be produced. Thus, laceration from instruments, or sloughing from pressure, may destroy the soft parts at any point and to almost any extent, between the meatus urinarius and the upper extremity of the vagina. Urethro-vaginal fistulæ are usually mere openings, roundish in form and of variable size, and are generally situated about the middle of the lower wall of the canal. Vesico-vaginal fistulæ have usually a somewhat lateral situation with regard to the median line. When due to laceration from instruments they generally occur in the lower third of the vagina, but when of spontaneous origin, in women with narrow pelves, their most frequent seat is near the upper part of the vaginal canal. It is doubtful whether they are more frequent on one side than on the other. They vary very much in form and extent; they may be round, oval, crescentic or irregular. The destruction of the soft parts may be so extensive that the fistulous opening will admit several fingers, while in other cases it can be felt only as a slight depression. The form and size of the opening will depend upon the causes by which it has been produced, on the efforts which have been made to close it, and likewise upon the healing powers of nature. Nothing certain can be affirmed with regard to the direction which the fistulous communication may take; it is often transverse, or it may run longitudinally toward the neck of the uterus, which may participate in the laceration. In other cases it presents various degrees of obliquity.

The appearance of the fistulous opening will vary according as it has been produced by laceration or sloughing; and likewise according to the time at which it is examined, whether at a short or remote period after the accident. The practical point to notice, however, is the condition of the edges and of the tissues in the immediate vicinity. In old cases, and especially when the affection has been produced by sloughing, the edges are frequently callous or turned inwards towards the bladder, and the induration extends some way into the surrounding tissues.

When the fistula has been caused by direct injury, the edges generally exhibit a considerable amount of vitality, the lips of the wound are thick and tumid, the granulations vigorous, and the surrounding tissue appears succulent and reddish. When the fistula is adherent by a portion of its margin to the adjacent bone, the edges are apt to become much attenuated, a result which also occurs as a consequence of cicatricial contraction.

Fistulous openings between the vagina and ureters are of rare occurrence, and are generally situated laterally and at the upper part of the former passage. They are roundish in form and small in size. In a case recorded by Freund, the terminal end of both ureters had been destroyed. The opening, elliptical in shape and situated in the median line, was distant two centimeters from the meatus urinarius, and in its walls were seen two apertures into which a probe could be passed.

The uterus may also communicate with the bladder by a fistulous opening between the two organs, or a vesico-vaginal fistula may extend upwards and involve the neck of the womb. These fistulæ, as well as those between the bladder and various portions of the intestine, are extremely rare, and are of course generally beyond the range of surgical interference.

When the fistula has existed for any length of time, various changes occur in the bladder and adjacent parts. The former, no longer serving as a receptacle for the urine, diminishes in size, its walls remain in contact with each other, and the anterior wall often protrudes through the fistulous opening. Catarrh of the mucous membrane, thickening of the muscular coat, adhesions of the peritoneal covering to neighboring organs, are also frequent consequences. Obliteration of the urethra sometimes occurs, and even if the tube remains pervious, its calibre becomes much diminished in course of time. The external parts of the genital organs are constantly moist and sore, and may become eroded or even ulcerated. In consequence of cicatricial contraction the vagina may undergo various changes in shape; firm bands sometimes extend across it in various directions, and complete atresia may be produced by adhesions between the anterior and posterior wall above the fistula. Stony concretions sometimes form in the bladder, or in diverticuli from the vagina.

The single and characteristic symptom of a vesico-vaginal fistula is the escape of urine through the vagina. This occurs as soon as the communication is established—after laceration, immediately; after sloughing, in a few days, as soon as the slough has separated. In the former case, however, the occurrence of the injury may not have been suspected at the time of its infliction, and several days may elapse before the real nature of the case is ascertained. This may be the case, for instance, when the vesico-vaginal septum has been lacerated during an instrumental labor. The constant escape of fluid having a urinous odor, and the absence of any desire to pass urine will, however, sufficiently indicate the nature of the case. When the accident results from sloughing the symptoms are later in their appearance, and are preceded by pain and difficulty in passing water. After the slough has separated these latter symptoms will subside, and give place to a constant escape of urine. The possession of any capacity of retaining a portion of urine in the bladder, and of discharging it voluntarily, will depend upon the size and position of the fistula. The inconvenience resulting from incontinence of urine, and the troublesome excoriations produced in the neighboring parts are the chief effects of the accident; but they are sufficient to render life very miserable. The patients are usually sterile, but in some cases several children have been born after the establishment of the fistula, the size of which is usually increased by the act of delivery. As might be expected, the general health of the patient suffers in a variety of ways. Dr. Winckel[1] alludes to obstinate constipation as a common and prominent symptom. In most cases the urinary secretion is increased in quantity, though not to any great extent.

The diagnosis of vesico-vaginal fistula is, as a general rule, easily effected, but it is important that the situation, size, and other conditions of the fistulous opening should be carefully explored with the finger, catheter and speculum. In cases where the opening is the third of an inch or more in size, and is situated in the anterior wall of the vagina, its existence may usually be determined by passing a catheter into the bladder and a finger into the vagina. The point of the catheter is then cautiously passed through the opening and felt by the finger. When the fistula is lateral and small, there may be much more difficulty in discovering it. In order to examine the parts, the patient should be placed on her back, with the pelvis elevated. A Sims' speculum is then introduced, and the labia are kept apart by the fingers or spatulae. The orifice of the uterus is then plugged by means of a piece of soft wadding, and the walls of the vagina are carefully dried. Some colored fluid is injected into the bladder, and, should it escape into the vagina, the point where this occurs is carefully determined and examined. Should there be no escape of fluid, the supposition of the existence of a vesico-vaginal fistula will be negatived. The plug should now be withdrawn from the os uteri, and if fluid immediately escapes, a utero-vesical fistula may be inferred to exist. The cervix must then be gradually dilated, and the uterus further examined. If there be no doubt as to the passage of urine into the vagina, but if the colored fluid injected into the bladder escape neither into vagina nor from the os uteri, the probability will be that an opening exists between the ureter and the vagina or the ureter and uterus. This question must be determined by ascertaining whether the vagina remains dry when the os uteri is occluded by means of a plug. Should this be the case, an attempt should be made to catch the fluid which escapes from the os uteri. If the diagnosis still remain uncertain, the urethra may be dilated and the interior of the bladder examined with the forefinger. In all cases the examination should be as complete as possible; special attention should be paid to the condition of the edges of the fistula, and great care should also be taken to discover the position in which the opening can be best exposed. In some cases the examination will be facilitated by placing the patient on her elbows and knees, or by causing her to lie on her side.

---

[1] Loc. cit., p. 118.

The prognosis of vesico-vaginal fistula varies according to the extent, position, and other circumstances connected with the affection. The prospects of a cure have been much increased in recent times, owing to the various improvements which have been effected in the methods of operating. Rather more than twenty years ago it was stated on good authority, that, "out of the whole of the cases treated in the London hospitals during the last three years, complete closure of the fistula was obtained in two only; in all the rest, however, more or less benefit was obtained, and in not a few the patient was, to a very great extent, relieved from her distressing symptoms."[1] Much more favorable results, however, are now obtained, and during the last few years Mr. Lawson Tait has proved that even the most severe cases are not beyond the reach of surgical skill and ingenuity. A few instances of spontaneous closure of the fistula have been reported. Bouqué, quoted by Winckel, has collected sixty instances of this kind, and Dr. Winckel has himself met with several cases. The occurrence of a spontaneous cure is, however, very rare, and can be expected only when the fistula is very small and all the circumstances favorable for healing. With regard to the results of operations, Bouqué's statistics show that about 72 per cent of the cases are recorded as cured, while in rather more than 4 per cent the operation has terminated fatally, the causes of death being peritonitis, pyelitis, or pyæmia, the result of suppuration of the pelvic connective tissue. In cases where the operation succeeds the relief to the patient is immense. The general health speedily undergoes improvement, and the bladder gradually regains its function as a reservoir. The patient may become pregnant, and delivery be accomplished without injury to the cicatrix. According to the most recent statistics, operative interference is successful in from 85 to 90 per cent of all cases, while the mortality has been reduced and amounts to scarcely 3 per cent. In all cases, much will depend upon the position and size of the opening; when of moderate extent and seated near the orifice of the vagina, when of traumatic origin, recent date, and not accompanied by any considerable change in the adjacent tissues or surrounding parts, the fistulous opening presents the most favorable conditions for cure.

The treatment of vesico-vaginal fistula is either palliative or radical. The supposed impossibility of closing an opening through which urine constantly passes, long deterred surgeons from adopting any other than a palliative mode of treatment, and numerous appliances have been devised for this purpose. In the palliative treatment two objects are kept in view: the accumulation of urine in the bladder is prevented by the constant use of the catheter, and an attempt is made to close the fistulous opening mechanically, by the introduction of a solid plug into the vagina. For drawing off the urine, Desault employed a large elastic catheter, fixed to an apparatus resembling a truss, by means of a movable silver plate, provided with an aperture for the removal of the catheter; the vaginal plug was formed of a linen tent or a sort of glove-finger, stuffed with lint and smeared over with wax; the plug was applied rather with the object of pushing up the anterior edge of the opening against the posterior, than of merely blocking up the orifice; hence it was applicable only to transverse fistulæ.

The linen plug of Desault has been replaced by various contrivances intended to close the unnatural opening more or less completely. Thus, recourse has been had to pieces of sponge, elastic pessaries of various

---

[1] Medical Times, September 13th, 1856.

materials, hollow resinous cylinders, etc.; a good plug may be readily made with a piece of dry sponge inclosed in a cylindrical bladder; on moistening the sponge it expands and exercises a considerable degree of pressure. It has also been recommended that the patient should adopt the prone position for some weeks so as to facilitate the escape of urine. In some cases in which a vesical calculus had caused ulceration and perforation of the vesico-vaginal septum, Dupuytren succeeded in getting the fistula to close by keeping the parts as clean as possible, and numerous other cases have been reported in which a similar result was obtained by equally simple treatment. In cases, therefore, where the fistula is small and of recent origin, it is desirable to wait and see what amount of closure will naturally occur, and the process may be assisted by keeping a catheter in the bladder and plugging the vagina. For this latter purpose a plug of wadding, impregnated with salicylic acid and covered with linen, has been recommended by Esmarch. Should the catheter cause irritation, it must be withdrawn and introduced at short intervals. The plug of wadding should be renewed as required, and the treatment should be persevered in so long as any improvement is perceptible. Another form of apparatus may be applied for the purpose, not of closing the orifice, but of alleviating the effects of incontinence of urine. Mr. Barnes, of Exeter, employed in one case an elongated caoutchouc bottle, which, when placed in the vagina, presented an opening corresponding to the fistula. This opening was occupied by a piece of sponge, which conducted the urine into the bottle. The patient herself removed and emptied the bottle three or four times a day, and experienced much comfort from its use.[1]

A piece of sponge was employed for the same purpose in a case which had resisted all attempts to effect a cure;[2] and Dzondi describes various urinary receptacles which may be had recourse to.

When the fistula has become permanently established and the condition of the parts is such that no spontaneous improvement can be anticipated, the question of operative interference must next be considered, and the choice will rest between cauterization of the edges and union by suture after having pared the margins. The treatment by cauterization has fallen into desuetude, though in the hands of numerous surgeons excellent results have been attained by it. Substances of various kinds have been used for the purpose,—the actual cautery, the galvanic cautery, nitrate of silver, sulphuric acid, caustic potash have all been had recourse to, and these agents have been applied both to the vaginal and vesical surface and to the fistulous canal.

The caustic generally preferred is the nitrate of silver. In using it the patient is placed as in the operation for stone; a bi-valved speculum is then introduced, and the fistulous opening brought into view. The stick of nitrate of silver is fixed to a porte-caustique, and the whole of the free edges of the opening touched with it. It is unnecessary either to introduce a catheter or to plug the vagina.

Dupuytren, Delpech, and many other surgeons preferred the actual cautery to caustics. Dr. Kennedy, also, has related some interesting cases in which this method was employed with benefit.[3] Its application is simple, and attended with inconsiderable pain to the patient.

---

[1] Medico-Chir. Trans., vol. vi., p. 582.
[2] Trans. of the Prov. Med. and Surg. Assoc., vol. i., p. 542.
[3] Dublin Med. Journal, vol. ii.

The object is to stimulate the edges of the opening, and promote contraction of the wound, not to produce destruction of the parts. The number of applications will depend on the effect produced by each. Generally speaking, the slough will not be thrown off, and the contractile effects of a cauterization will not be over before two or three weeks. We must allow time for the contraction to take place. Hence if, on examining the wound after this interval, it be seen perceptibly contracted in all directions, a second application should be made, and renewed until the opening is entirely closed.

Contraction, after the use of the actual cautery, sometimes goes on for months. Delpech, who was a great advocate for the employment of the actual cautery, recommended the vaginal border of the opening only to be touched; and Velpeau, though giving a general preference to the nitrate of silver, still thought that in cases of long standing, with loss of substance and callous edges, the actual cautery would be best, as acting more rapidly and with greater energy. It is hardly necessary to mention that the method by caustic or cautery is suitable only for cases where the loss of substance is not considerable; even then a long time may be required, and the cure be imperfect. The caustic may also be applied to the canal of the fistula, as was suggested by Desault, and in another method the application is made at the same time to the mucous membrane of the bladder. A solution of chromic acid has been successfully used for this purpose by Bouqué.

The method of cauterization is particularly suitable for those cases in which the fistula is high up in the vagina, and where the peritoneum might be injured in the operation by suture. With regard to the time at which its performance is advisable, opinions differ, but on the whole it seems better to have recourse to it as soon as possible after the discovery of the lesion. In modern times Dr. Bouqué, of Ghent,[1] is the main advocate of the treatment by cauterization. Out of 35 cases thus treated by him 25 were cured, 4 relieved, and 3 died. The results, however, of the treatment by suture are more favorable, and it is probable that in the course of time this latter method will be universally adopted. There is this risk about cauterization, that if it fails, the edges of the fistula are in a worse condition for a subsequent operation than they would otherwise have been.

The most effectual method of cure in cases of vesico-vaginal fistula is that by suture, the edges of the opening having been previously pared off in order to insure their more speedy and complete union. This operation, proposed by Roonhuysen, about the middle of the seventeenth century, has been repeatedly performed with the most complete success even under unfavorable circumstances.

The operation is often difficult; it requires special instruments, great patience, and a practised hand; and hence innumerable efforts have been made either to simplify it or to overcome its difficulties. The operation consists of two parts, paring the edges of the opening, and keeping them together by various contrivances. Different kinds of instruments have been used to overcome the difficulties of the operation, which are in truth great. The surgeon has often but a narrow space to work in, and this is sometimes rendered still smaller by preternatural contractions or firmly cicatrized bands.

Instead of paring the edges of the wound with the knife or scissors,

---

[1] Quoted by Winckel, loc. cit., p. 129.

Lallemand destroyed them with nitrate of silver, and when the superficial sloughs came away, endeavored to keep the edges together by a peculiar kind of hook. The peculiarity of his method consisted in the circumstance that he operated on the vesical side of the fistula, not on the vaginal. His first operation was performed in 1825. The accident had been produced by a difficult labor; the fistula was about half an inch long, transverse, and seated fourteen lines beyond the meatus urinarius. Having removed the edges by a few applications of the nitrate of silver, Lallemand applied his instrument, which consisted of a silver catheter, four inches long, in which was concealed a stylet, bifurcated at the extremity, and there terminating in two short hooks. The other end of the stylet was connected with a screw, by turning which the hooks could be forced through the soft parts or drawn within the catheter. Round the catheter was a fine spiral spring, and the external spiral was attached to a silver plate which had its *point d'appui* on the pubes. In this way, when the hooks were forced by the screw through the posterior edges of the transverse fissure, they tended to bring them into contact with the anterior border, because the spiral spring, acting on the hooks, and having its *point d'appui* at the pubes, tended to bring together the movable and fixed points.

The action of the plate on the meatus was also supposed to push the anterior edge of the wound towards the posterior. In this case a complete cure was obtained, and likewise in another, operated on in 1833. Here a single application of the instrument for five days was sufficient to produce adhesion, as in cases of harelip.

Baron Dupuytren modified this method in the following manner. He employed a large female catheter, furnished at its sides with two wings or flaps, which could be closed by means of a spring stylet. These wings represented the hooks of Lallemand's instrument. When the catheter was introduced and the wings at its extremity opened, it was drawn outwards, bringing with it the posterior edge of the fistula; some lint was now placed on the meatus urinarius, the rim of the external end of the catheter was fixed on this lint, and the soft parts were thus pushed back with it. Dupuytren says that he succeeded, in one case, with this instrument.

For obvious reasons, however, it is preferable to act through the vagina, where so much more room is afforded, and the surgeon can see what he is about. When the fistulous orifice is seated low down, near the meatus urinarius, no great difficulties are experienced; but when it is far up in the vagina all the ingenuity and mechanical skill of the surgeon will be required to conduct the operation to a successful issue.

For removing the edges of the opening, knives or scissors may be employed. Naegele, Lallemand, and others employed peculiar shaped bistouries and scalpels; Roux used scissors, but the best and most convenient instrument is a narrow sharp-pointed knife, with a long handle, and cutting to a short distance only from the extremity. Much stress was formerly laid on the necessity of employing some contrivance whereby the knife might be made to act on some solid support placed beneath the mucous membrane; but this precaution is not necessary. In order to bring the edges of the fistulous opening fairly into view before proceeding to the operation, several methods have been employed. Malagodi dilated the urethra until it was capable of receiving a finger, which he passed into the bladder, and with it pushed down the edges of the fistula until they came within reach of the bistoury. Instead of the finger, a large-sized bougie, made of highly polished whalebone, as proposed by Dr. Hayward, of Philadelphia, may be used with advantage. In most cases, however, it will be quite sufficient to employ the speculum for dilating the vagina. When the vagina has been sufficiently dilated and a strong light admitted on the parts, the edges of the fistula may be drawn down by hooks or forceps, various kinds of which have been invented by Dieffenbach, Wutzer, and others.

The portion of soft parts removed by the knife sometimes comprehends the whole thickness of the edges, while by some operators it is confined to the vaginal membrane only. Thus Wutzer drew a line with the point of his knife around the fistulous opening and about three or four lines distant from the free edges. He then raised up the mucous membrane of the vagina with a hook, and removed it from the surface just designated, that is, between the line drawn by the knife and the free edge of the fistula; the surface everywhere exposed for union was hence about three or four lines broad. Dieffenbach's incisions comprehended the mucous surfaces of both the vagina and bladder. In a case operated on by Mr. Luke, of the London Hospital, the edges of the wound were so pared as not to enlarge the opening. This was effected by removing triangular slips, the angles of which, at the urethral surface, were only separated by the breadth of the fistula.

The edges of the refreshed wound must of course be maintained in contact for some time until union can take place between them. Sutures are the best means we can employ for this purpose; but as it is often most difficult to apply them effectually in so narrow a space as the vagina, a great variety of instruments and methods have been proposed to obviate the difficulties of the operation. The simple suture was first tried, but soon abandoned because the threads cut through the walls of the septum too quickly. The interrupted and twisted sutures were then had recourse to, and the former is the one now generally employed. Wutzer, however, preferred the twisted suture, whenever it could be conveniently applied, at it closes the wound more perfectly, and is better calculated to prevent infiltration of urine between its edges. Soft threads were formerly employed by surgeons; but metallic sutures of fine silver or platinum wire are preferable and are now always used. The flaps of mucous membrane having been removed from the vaginal surface of the edges of the fistula in the way described, the sutures are introduced three or four lines beyond the edge of each flap, and about two or three lines from each other. The needle should traverse, if possible, the muscular coat of the bladder as well as the septum, but not the mucous lining of that organ. To ascertain that the needle does not penetrate the cavity of the bladder, Mr. Brown recommended us to introduce the little finger of the disengaged hand into the bladder through the urethra. While passing the sutures it will often be necessary to fix the edges of the wound with a hook, as otherwise they may give way before the pressure.

As the insertion of the threads is a principal, and frequently the most difficult part of the operation, various kinds of needles have been proposed by different operators. For a description of these, reference must be made to special treatises on the subject. The surgeon, indeed, should be provided with a variety of needles, as the situation, size, and shape of the fistula will render it necessary to modify the kind of instruments employed. Generally speaking, the articulated needles figured by Mr. Brown,[1] will be found the most useful form.

---

[1] On some Diseases of Women admitting of Surgical Treatment, p. 99, f. 2.

Mr. Beaumont, also, has suggested an ingenious instrument for passing the sutures. It has the form of a forceps, one blade of which carries a needle curved towards its point, and close to the point is the eye of the needle. The other blade is broader on its opposing surface, less curved, and at its extremity has a hole, through which the needle-point and just the loop of the ligature are carried when the plates are closed. On the back of the broad blade is a spring, which, when pushed forwards, the blades being previously closed, catches the ligature on its point, and holds it at the extremity of the blade.

In using this instrument the operator has only to seize in its points, as he would with a pair of forceps, the border of the fistulous opening; the blades should then be closed, and the ligature will be carried through one lip of the aperture. The opposite border is then in like manner to be seized, and the blades are to be again closed and firmly held so. The spring on the back of the broad blade is now to be pushed forwards, by which the ligature will be caught and held at its point. The blades after this are to be opened and gently withdrawn, leaving a double ligature passed through opposite points of the fistulous aperture. Two or more stitches may be made in the same manner, leaving in each a double ligature, so that the quilled or other suture may afterwards be formed.

In many cases the fingers may be employed as needle-holders, and most surgeons will find it convenient to use them whenever they can. When the first suture has been introduced, the ends of the ligature should not be cut off immediately; several inches should be allowed to remain, as they will assist much in arranging the edges of the wound, while the remaining sutures are being applied.

As soon as the sutures have been made fast, the tension of the vaginal membrane and subjacent tissues should be relieved by free incisions, as recommended by Mr. Brown, made at four to six lines on each side of the closed wound. The principle on which they act requires no explanation.

The after-treatment will chiefly consist in obviating the effects of any inflammation which may arise. Mr. Brown advised that the patient should be placed on a water-cushion on her side, the hips being elevated, and the knees flexed on the abdomen. A catheter should be kept permanently in the bladder, the point being turned up behind the arch of the pubis, while the other extremity is fixed to an elastic bag, capable of holding four to six ounces. Wutzer and many other surgeons do not agree with the opinion that the catheter should remain permanently in the bladder; they prefer introducing it from time to time. To insure cleanliness, the vagina should be syringed once a day with cold water. Dieffenbach injected cold water every half-hour, and also applied cold to the lower part of the abdomen to prevent inflammation. On the third day after the operation the sutures should be examined, and in successful cases the wound will be found to have united perfectly about the fourth day; but as a general rule, the sutures should not be removed before the tenth or twelfth day: more frequently, however, only a portion of the wound unites, a fistulous orifice of small extent still remaining. Contraction may be excited by touching this orifice with caustics, or it may be necessary to repeat the operation of suture more than once.

With regard to the time at which the operation should be performed, there is no advantage in postponing it beyond the sixth or eighth week after delivery. Cases in which the operation has been performed thus early yield better results than others in which it has been deferred for several months.

But little preparation of any kind is necessary for the operation. The patient's condition should be improved as far as possible by nourishing diet, and the affected parts should be cleansed and soothed by means of warm baths. It is very desirable that the surgeon should thoroughly examine the fistula, and fully make up his mind beforehand as to the most suitable method of operating. With regard to the position of the patient, she may be placed either in the lithotomy position, as recommended by Mr. Baker Brown, with the pelvis raised by means of a pillow placed under it, on her side, as advocated by Marion Sims, or, as recommended by Bozeman, on her knees and elbows. Whatever may be the position selected it must be retained during the operation, the patient being firmly held by assistants.

In the preceding remarks some general points connected with the cure of vesico-vaginal fistula by suture, have been chiefly noticed, but it may be useful to describe more in detail the operation which was so frequently performed, and with such remarkable success, by the late Mr. Baker Brown, at St. Mary's Hospital ;. it is the same as the operation performed by Dr. Sims, of Boston, United States, and differs from preceding methods in many important particulars.

The patient is placed on her back as in the operation for stone, and the sides of the vagina are drawn firmly asunder by two assistants. The vagina having been thus dilated, and the parts brought well into view, the edges of the fistula are pared with a narrow-bladed straight knife : about a quarter of an inch around the edges is removed, but the incision never includes the mucous membrane of the bladder. Any slight discharge of blood which occurs is removed with a piece of sponge attached to the end of a whalebone rod. The next step of the operation consists in applying the sutures, and it is this part which is peculiar to Dr. Sims, who has named it the "clamp suture." The threads are composed of silver wire, not thicker than a horsehair, and they are fixed to two cylindrical cross bars, each about a line in diameter. These bars are made of silver, or highly-polished lead ; their length, as well as the number of the sutures will depend on the size of the fistula, and when drawn together by the metallic threads, they act on the principle of the clamp.

The needle employed for passing the sutures is tubular, and set in a handle. It is slightly curved near the point, close to which is the distal orifice for the passage of the wire. This needle, armed with the silver wire, is pushed into the septum about a third of an inch from the lower edge of the fistulous orifice, carried across the opening into the upper edge, without touching the mucous membrane of the bladder, and brought out into the vagina at a corresponding distance from the upper edge of the fistula. The wire being secured by means of forceps, the needle is withdrawn, to be used as before for the other sutures ; the operation should commence at the central part of the fissure. The intervals between the sutures should be about three-sixteenths of an inch. The next step of the operation is to secure the ligatures by means of the clamps. The distal ends are fixed to the cross bar by being passed through small holes, made in it to suit the distances between the several points of suture ; they are then either wound round the bar, or clamped to it by perforated shot. This done, the free ends of the sutures are drawn down until the cross bar is brought into contact with the vaginal wall, where it occupies a position just over the distal orifices of the ligatures. The second bar is now easily applied, by bringing the proximal ends of the ligatures through its openings, by pushing it against the vaginal wall, until the fistulous orifice is closed, and fixing the wires as before, on split shot, by twisting them with the forceps. The ends of the ligatures are cut off close to the shot, and the fistula is firmly bound together with a metallic clamp on either border. The object of employing the shot is to make them act as so many knots, by which the clamps are prevented from slipping. The sutures should not be removed before ten or twelve days, and it may be necessary to retain them longer.

This suture very rarely causes ulceration; there may be some trouble in removing it, for the clamps sometimes become imbedded in the mucous membrane of the vagina, which throws up granulations around them.

Another modification has, however, been introduced from America, the principal change being the substitution of a flat silver button for the cross bar clamps. The following is a short account of the operation:

Deborah P—, æt. 22, was admitted into St. Mary's Hospital, Sept. 22d, 1856. She stated that she was delivered of a still-born male child on July 15th, by instruments, after being forty-eight hours in labor. She was shortly afterwards seized with an attack of fever. Eight days after delivery she discovered that the urine dribbled away, after which it all seemed "to pass by the wrong passage." On examination per vaginam, a fistulous opening was discovered close to the os uteri, about the size of an ordinary director, which instrument could readily be passed within the bladder, through the opening into the vagina. It appeared that all the urine passed through this opening, and none through the urethra; in fact, that she was never able to retain any within the bladder, even for a short time. The health of the patient having been carefully attended to, Mr. Brown determined to operate after a method devised by Dr. Bozeman, of Alabama, in America. Accordingly, on October 15th, the patient being under chloroform, and placed in the position for lithotomy, a firm retractor was passed into the vagina, and pressed firmly backwards and downwards by an assistant. Each side of the vagina being held back by two other retractors, the bladder was seized, just at the juncture of its neck with the body, by a strong pair of vulsellum forceps, and held firmly upwards and forwards by the right hand of the assistant holding the left leg. The fistulous opening was then with great difficulty brought into view, and the mucous membrane divided, by a sharp knife, completely around the opening, about the eighth of an inch in depth. Three silver-wire sutures, eighteen inches long, were then passed by a needle held by the porte-aiguille, and which are shown in fig. 1, p. 151. The two ends of each wire were then brought together by an instrument, as represented in fig. 2, thus leaving the parts in apposition (fig. 3). A silver button, as shown in fig. 4, was then carefully passed over the end of each double suture, and a perforated shot passed over each wire, as shown in fig. 5, and pressed down upon the button, and then firmly pressed by a pair of long, strong forceps. The wires were then cut off close to the shot, leaving the parts as represented in fig. 6. A piece of lint dipped in sweet oil was then introduced within the vagina, the patient placed in bed on a water cushion, on her side, and a bent catheter with a bag attached to it was inserted within the bladder, and allowed to remain there. Two grains of opium were given directly, and one grain every four hours afterwards for the first twenty-four hours, and afterwards one grain every six hours, and a generous diet, with wine, allowed daily. The lint was removed on the second day, and the vagina washed out night and morning with tepid water. All the urine passed freely through the catheter. On the 24th, the button and the sutures were carefully removed, and the most perfect union was found to have taken place throughout the whole extent of the wound. On the 26th, the bowels were relieved for the first time, by castor oil and enema, and the catheter was removed for three hours, at the end of which time the patient passed urine comfortably, with no escape per vaginam. On the 27th, the catheter was removed entirely, and she was allowed to sit up and walk about a little. On the 28th, she was up and about all day; she was able to retain the urine for four hours, and to pass it well. On the 30th, a most careful examination was made, and the parts were found to be firmly and closely united so that no escape of urine took place.

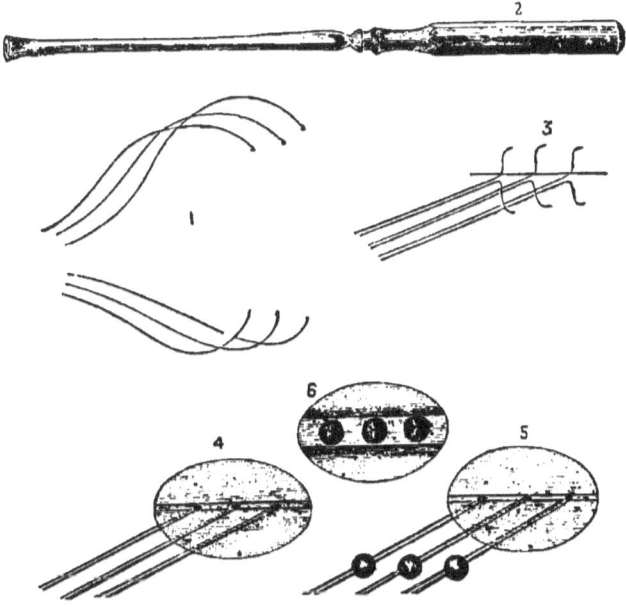

FIG. 4.—Method of applying sutures in the operation for vesico-vaginal fistula.

Mr. Brown observed that the result of this method had convinced him that cases hitherto deemed intractable would be found to be curable by this operation.[1]

Professor Simon, of Heidelberg, recommended that, after the introduction of the sutures, a catheter should be introduced and the bladder washed out with a weak solution of salicylic acid. By this means it can be easily ascertained whether the parts have been accurately brought together.

The catheter, however, is not allowed to remain, and is to be re-introduced only if the patient is unable to pass urine. He removes the sutures on the sixth day, and should any small openings appear they are to be touched with nitrate of silver. If the operation be only partially successful, it may be repeated at an interval of three or four weeks.

---

[1] Lancet, November 15th, 1856, p. 540.

When the loss of substance is considerable, the edges of the fistula cannot be brought together by any form of ordinary suture. For these unfortunate cases Dieffenbach and Jobert have employed special modes of operating. Dieffenbach had recourse to a metallic suture, which he applied in the following manner. The bladder was first returned though the fistulous orifice, and retained by a piece of sponge introduced into the fissure. One edge of the fistula was then drawn down, and a narrow slip was cut off from it; the edge of the bladder was also cut off at some lines' distance from the edge of the vagina. Two leaden threads where then carried through the edges of the vagina, the bladder being left untouched, and were drawn towards each other until the soft parts underwent a considerable degree of tension. To relieve this, and allow closer approximation of the edges of the wound, Dieffenbach now had recourse to the method employed in cases of cleft palate. With the knife he divided deeply the walls of the vagina on either side of the fistula, thus isolating a portion of the vagina, which was equal to about one-fourth of its whole width. These incisions liberated the edges of the fistula so that they could now be drawn closer than before by the suture; but to obtain further improvement the operator again divided the cellular tissue connecting the vagina to the pelvis, first on one side and next on the other. By steadily drawing the leaden threads together, the edges of the fistula were brought so far together that no further tearing of the parts was to be apprehended; the refreshed edges were, therefore, now united by the interrupted suture; the leaden threads were drawn closer together, and cut off, so that only two turns remained.

Mr. Lawson Tait[1] has recently shown that cases hitherto deemed incurable are capable of being dealt with by operation. Two such cases were detailed before the Obstetrical Society a short time ago. In the first case the loss of substance was enormous. When seen April 30th, 1877, two months after the confinement to which the affection was due, the anterior wall of the vagina was found to be entirely wanting; all that was left was three-eighths of an inch of the urethra. At the roof of the vagina a thick hard ridge ran across from side to side, and on the posterior surface of this the os uteri was discovered. In front of this ridge was a protrusion of mucous membrane, which was identified as the remains of the bladder by the fact that the two ureters were discovered upon it. The anterior edge of this protrusion was adherent to the rim of the pubis, and the whole of the mucous area was not much larger than a five-shilling piece. No trace of the anterior wall of the bladder could be found on careful search. Such being the state of things, Mr. Tait conceived that if he could make anything in the shape of a tube out of the cicatricial tissue in the vaginal wall he might, by releasing the ridge at each side, bring it and the uterus down, and folding the remains of the bladder upon itself and fastening it to the new tube, might at least make a receptacle for a small quantity of urine. Accordingly, on May 15th, he made two flaps, each about an inch long, out of the tissue behind the symphysis pubis and joined them in the middle by silver sutures. He did not attempt even to look at them again till July 14th, when he found that the operation had practically failed. He therefore, on July 18th, proceeded to make two similar, but larger flaps, consisting of everything he could raise from the bone, and again united them in the middle. On examining the parts two months later, he found that a canal, three-

---

[1] Trans. Obstetrical Society, April 3d, 1878.

quarters of an inch in length, and allowing of the passage of a No. 6 catheter, had been formed. A few days later he proceeded to make a raw surface on each side of this bridge at its upper end; he then made a deep incision at each end of the tense ridge at the upper part of the vagina, and, after arresting the hæmorrhage, pared the edge of the ridge and fastened it down to the raw surface with sutures, with the exception of one corner, where a free exit for the urine was left. On October 11th, he found that the whole of his proceedings had been successful, the artificial urethra leading into a bladder-cavity, and nothing remaining but to close the provisional orifice. This was done on November 17th, a canula being kept in the urethra for twenty-four hours. The patient left the hospital on November 28th, still unable to retain her water, but returned on January 1st, when it was found that the water escaped through one of the stitch-holes. A few days after this was healed she began to have a sense of desire to pass water, and was able to remain dry for about an hour. Since then the quantity retained had increased from half-an-ounce to four ounces, and she now described the feeling as being that almost of a new life. In commenting on the reappearance of sphincter-action after utter destruction of the sphincter muscle, Mr. Tait put forward two explanations—first, that some of the involuntary muscular fibres of the vaginal submucous tissue had taken on sphincter-action; and, secondly, that a valve-shaped opening had been made which yielded only on pressure being exerted upon it by the contraction of the muscular tissue in the remainder of the bladder when over-distended.

In the second case the lesion was of much older date, the confinement which had caused the mischief having taken place in 1862. In the interval the patient had been subjected to a variety of surgical proceedings, but without avail. In March, 1877, Mr. Tait found a large irregular opening, extending from within an inch of the cervix uteri to within a quarter of an inch of the meatus urinarius, admitting of the protrusion of the anterior wall of the bladder. This opening was narrow at its two extremities, but very wide at the point where the trigone and neck of the bladder should have been, and thus tissue was lost just where it was most wanted. On March 17th, Mr. Tait operated upon the narrow part of the opening above, and lifted two urethral flaps with the view of forming the basis of a new urethra. The second part of the operation failed; it was, however, repeated with perfect success on April 6th. On May 16th, he operated on the remainder of the aperture by making two large wedge-shaped flaps, the axis of which coincided with the circumference of the vagina. The free truncated ends coincided with the margins of the fistula, and the lateral margins of each flap were formed by somewhat divergent incisions, which travelled round quite one-fourth of the circumference of the vagina. The flaps were then carefully joined at their apices, and the lateral margins stitched down so as to ease the apices of as much strain as possible, a drainage-tube being fastened in at the outer end of the lower left-hand incision. Union took place except in a small portion, for which the same operation was repeated on a smaller scale. The stitches were removed a fortnight later, and in a few days the patient began to pass small quantities of urine with voluntary effort. By February 7th, 1878, she was passing 8 and 9 oz. of urine regularly, and was as well and as comfortable as she ever was in her life.

A few words must be added with reference to fistulous openings existing between the vagina, uterus, and bladder. When the abnormal communication is superficial, that is to say, when the upper border of the fistula involves the os uteri externum, the anterior lip of this orifice is pared and united by suture to the opposite margin of the fistula. When, on the other hand, the opening extends from the vagina into the uterus and involves the destruction of the anterior lip of the os uteri, the only method of remedying the defect consists in uniting the posterior lip of the os uteri with the margin of the fistula. The cavity of the uterus is thus connected with that of the bladder, through which the menstrual fluid must then escape. The patient, of course, is rendered sterile. Before deciding on this operation an attempt should be made to unite the remains of the anterior lip of the os uteri to the margin of the fistula. When the abnormal communication is between the bladder and uterus, an attempt may be made to produce a closure by applying caustics after a previous dilatation of the cervix. This latter part may also be incised, and an attempt made to expose the abnormal opening and to close it by suture. If neither operation be successful, the margins of the os uteri may be pared and united, a procedure suggested by Jobert. Sterility of course ensues, and menstruation takes place through the bladder. Fistulous openings between the ureters and vagina require a modification of the operation for vesico-vaginal fistula. Dr. Simon, of Heidelberg, advised that the bladder should be opened at the fistulous spot, and that a catheter introduced through the urethra should be passed into the ureter. The catheter serves as a director, and upon it the wall of the bladder and ureter is to be slit up on its internal aspect for about half an inch in an upward direction. The margins of the slit are to be kept separated by the daily use of the sound until cicatrization is complete. The vaginal opening, which will be at some distance from the new termination of the ureter, is subsequently to be pared and closed.

Another operation which has been proposed and performed for various kinds of vesical fistulæ, otherwise irremediable, is the closure of the vagina in the transverse direction. For this purpose the mucous membrane is pared away from the anterior and posterior walls of the canal, and the raw surfaces are brought together by several sutures. Care must be taken to produce complete and accurate co-aptation of the parts, especially at the angles. When the operation is completed the bladder should be injected in order to test the accuracy of the union. In a case in which this operation was performed by Dr. Winckel,[1] the result was only a partial closure, and the patient became pregnant and aborted at the fourth month. In several of these cases calculi have been known to form above the occlusion.

The accidents that may occur after any of these operations for vesical fistulæ deserve a brief notice. Hæmorrhage is sometimes very troublesome, and may take place into the vagina, bladder, or both organs. It is to be combated by injections of cold water and the application of a bladder of ice. If the bleeding comes from the vaginal surface and is obstinate, an attempt should be made to find the vessel. The bladder should be washed out so as to remove coagula. Pain is to be relieved by opiates internally or hypodermically. Peritonitis is an occasional consequence of these operations. It must be treated in the usual manner.

The only form of vesical fistulæ which has now to be considered is that which occasionally exists between the bladder and rectum. Wounds of various kinds, ulceration of the bladder, calculus, malignant disease are the ordinary causes of this affection; it may also arise from ulcers and

---

[1] Loc. cit., p. 153.

wounds which act from the side of the rectum towards the bladder. The recto-vesical operation for calculus was at one time a frequent cause of the accident, which has also followed simple puncture through the rectum for retention of urine. When the opening has become established, urine and feces indiscriminately pass through it. As the urine, however, finds its way into the rectum much more easily than the feces do into the bladder, the usual symptom of the accident is the discharge of urine through the anus. Flatus may pass through the urethra, and occasionally fluid feces. The unnatural discharge of the contents of the bladder and bowels does not, of course, prove that the lesion exists between the bladder and rectum, because another portion of the intestinal canal may open into the urinary reservoir; an examination of the parts must therefore be made; in most cases the opening will be detected by an instrument introduced into the bladder; and should any doubt arise, the speculum ani must also be employed.

The inconveniences which result from this accident are less than might be anticipated. The whole of the urine seldom passes away through the rectum, and the patient has the power of partially retaining it in the bladder. On the other hand, many years may elapse before the feces begin to make their way into the bladder. The dribbling of the urine through the anus, however, produces excoriation of the surrounding parts, and is very distressing, as it is difficult to apply any convenient reservoir for its reception. The passage of fecal matter into the bladder may give rise to irritation of that organ, retention of urine, the formation of calculous concretions, etc.; but it is astonishing how long the bladder will sometimes bear the presence of foreign bodies without apparently being irritated by them. Richerand and Cloquet attended a man who, thirty years previously, had received a gunshot wound in the groin, involving the bladder and intestinal canal; both external wounds healed in about a year, but after each evacuation of urine a quantity of flatus passed through the urethra; this continued for twenty years, when fecal matter likewise passed through the urethra whenever the bowels were loose. If the urine was retained for any considerable time in the bladder, a portion of it passed through the rectum on making water. The patient was thin and weak, but he had not suffered from his long infirmity to any dangerous extent.[1]

The mode of treatment to be adopted will be modified according to the extent and nature of the fistulous opening, its date, the age and general condition of the patient.

When the health is good, the accident of recent date, and produced by a cutting instrument, nature sometimes effects a cure without the aid of a surgical operation. The indications of treatment, in such cases, are manifest. The patient should be confined to the recumbent posture, and the bladder and rectum be kept as completely empty as is possible. These objects will be effected by the constant use of the catheter, the administration of opium, and the frequent washing out of the lower bowel by emollient lavements. In this way the fistulous opening may sometimes heal spontaneously; and the closure may be accelerated by the use of caustics, or the cautery if necessary. Dupuytren frequently employed caustics, if not with the effect of completely closing the fistula, at least with great relief to the patient, by diminishing the escape of urine. He repeated the cauterization after forty-eight hours, and five or six

---

[1] Jour. des Prog., t. x., p. 240.

applications were often sufficient to afford considerable relief. When the above means fail, or when circumstances render them inapplicable, we may have recourse to operative proceedings. The interrupted suture may be tried, as in cases of vesico-vaginal fistula. More commonly, surgeons are content with laying the parts open, either down to the anus, or through the sphincter muscles.

In a case of recto-vesical fistula after lithotomy, Mr. South divided the sphincter ani from the fistula into the perineum, with the object of producing a perineal fistula, which would be much more easily managed than the original one. After completely laying open the parts, the edges of the fistulous orifice were drawn down with Lisfranc's tenaculum, pared off, and united at their upper part by a single interrupted suture. The communication between the bladder and rectum was thus cut off, but the lower part of the wound was left open in order to establish a fistula in perineo. The after-treatment was continued for six months, the fistulous orifice between the bladder and rectum being freely touched with lunar caustic, until it had diminished to the size of a large goose-quill. Soon after the patient left the hospital, much relieved, but not cured by the operation.[1]

With regard to the occurrence of fistulous openings between the bladder and portions of the intestinal canal other than the rectum, in the majority of cases such lesions are due to malignant disease. The earliest sign of the existence of such a communication is the appearance in the urine of fragments of the constituents of the feces. Portions of vegetable and animal fibre may be discovered under the microscope. As the case proceeds the urine becomes distinctly fecal in odor and color. There is but little to be done for such cases in the way of treatment. The patient's condition, however, will be ameliorated by washing out the bladder, and by prescribing food that leaves only a small quantity of solid residue. All indigestible substances should of course be avoided. Amussat's operation has been performed with satisfactory results in some of these cases.

---

[1] Chelius's Surgery, by South, vol. i., p. 749.

CHAPTER XIII.

NEURALGIA OF THE BLADDER—IRRITABILITY OF THE BLADDER—SPASM OF THE BLADDER.

THE various so-called neuroses of the bladder are, in the great majority of cases, connected either with mechanical obstruction to the flow of urine, or with structural changes due to inflammatory action, and are therefore, as a general rule, only symptomatic of various pathological conditions of the urinary organs. In a smaller number of cases, there is no appreciable lesion, and the cause of the symptoms cannot be precisely determined. In such cases the symptoms constitute the disease, and though it would be inconsistent to regard the various neuroses of the bladder as distinct diseases, it is convenient to discuss them under separate headings, and to point out the causes upon which they may depend. They may be divided into two classes, according as the sensibility and contractility of the organ are increased or diminished. Neuralgia, irritability, and spasm of the bladder will belong to the former category, while the latter will include atony and paralysis.

Pain in the region of the bladder is a symptom common to many affections of the urinary organs, and in the majority of cases may be traced to some definite cause. In a few rare instances, however, the cause remains undiscovered, and the cases are regarded as examples of neuralgia. According to Dr. Gross, who has treated many cases of this nature,[1] the complaint presents itself in two varieties of form. In one, the suffering is more or less continuous, often remitting, but seldom intermitting; in the other, it is distinctly paroxysmal. The symptoms come on gradually, and when they are fully developed the pain is of a most agonizing nature. It is usually referred to the neck and base of the bladder, and it may extend to the rectum, urethra, glans penis, spermatic cord and thighs. The symptoms greatly resemble those produced by a calculus. Micturition is frequent and attended with difficulty, as the attacks are generally accompanied by spasm of the muscles at the neck of the bladder. The urine is clear and normal, unless catarrh or any other complication be present. The paroxysm may last from a few minutes to several hours, and then gradually subside, leaving behind it a feeling of soreness in the affected parts. Dr. Gross states that the attacks sometimes assume the quotidian type, and that under such circumstances they usually occur in the evening, during the night, or early in the morning. During the intervals there is occasionally a feeling of numbness in the perineum, scrotum, groins and thighs.

Very little is definitely known with regard to the causes of this affection. It has sometimes been attributed to miasmatic influences, and the attacks have been observed to occur in persons subject to other affections of a neuralgic character. Both sexes are liable to it, and the patients are usually middle-aged or advanced in life.

---

[1] Gross, Diseases of the Urinary Organs, p. 81.

If the attack be periodic, the treatment consists in the administration of those remedies which are known to exert a favorable action in similar cases. Quinine should be given in full doses, six or eight grains every four hours, during the interval. During the attack, in order to alleviate the pain, morphia should be injected subcutaneously, or administered in full doses by the mouth. The warm bath, followed by hot fomentations to the hypogastrium, will also be of service. If the patient be of a gouty or rheumatic habit, colchicum may be tried. In all cases the state of the general health should be carefully attended to, and every attempt should be made to discover and remove morbid conditions to which the symptoms may be due.

Irritability of the bladder is another combination of symptoms, indicating no special pathological condition, and often traceable to definite causes. The term merely expresses a fact, and there is no objection to its use so long as we clearly understand that a certain symptom, or group of symptoms, is all that is intended to be implied. A patient is said to have an irritable bladder when a frequent desire to pass water is the most prominent symptom, and in fact almost the only one of which he complains. There may be, and generally is, some amount of uneasiness about the neck and base of the bladder, but the absence of acute pain distinguishes the symptoms from those of neuralgia. Billroth[1] thinks that irritability of the bladder, occurring in the absence of any discoverable cause, may find its analogue in the spastic phenomena which accompany superficial affections of various other mucous membranes. He alludes to kerato-conjunctivitis, catarrh of the stomach and intestines, excoriations and fissures of the rectum, and various forms of catarrh of the air passages, all of which affections are accompanied by more or less spasm and irritability, and are characterized by marked hyperæmia of the mucous membrane with superficial epithelial detachment, and sometimes slight excoriation. He believes that similar pathological changes, of slight extent, occurring in the neck or base of the bladder, or in the ureters or pelvis of the kidneys, may, in the absence of other causes, sometimes account for the symptoms indicated by the term "irritability of the bladder."

In noticing the various affections of which this condition is a symptom, it will be convenient to divide them into three principal classes. 1. Diseases of the genito-urinary organs. It must be borne in mind that any source of irritation occurring in any part of the urinary passages, from the kidneys to the meatus urinarius, it apt to be referred to the bladder, or to show itself in the most marked degree in that viscus. 2. Diseases of neighboring organs. 3. Disorders of the general system, etc.

I. Irritability of the bladder is a symptom common to nearly all the diseases of the genito-urinary system, and is often the main cause of trouble to the patient. Thus it is a prominent symptom of stricture of the urethra, of the various forms of cystitis and their consequences, of diseases of the kidneys, of stone in the bladder and of diseases of the prostate gland. In any one of these affections the cause upon which the irritability depends is for the most part readily discoverable on careful examination. An analysis of the urine is, of course, indispensable, and strict inquiry should always be made as to the quantity of urine passed upon each occasion, and the aggregate amount voided in the twenty-four hours. When the urine is greatly increased in amount, as for example

---

[1] Chirurg. Erfahrungen, Zürich, 1860–1867. Arch. für klin. Chirurg., Bd. x.

in diabetes, a patient, overlooking the real nature of the case, may imagine, from the frequent calls to pass water, that he is suffering from irritability of the bladder.

The distinction between irritability of the bladder due to inflammatory affections of the organ, and the same symptom produced by other causes, is sufficiently marked by the absence of pain and of those general constitutional symptoms which characterize cystitis. Those cases in which the cause of irritability, supposed to be connected with the bladder, is in truth to be referred to the kidneys, though often somewhat obscure, may generally be elucidated by attention to the symptoms and careful analysis of the urine. Almost every form of organic renal disease is accompanied by irritability of the bladder, but it must not be concluded that the latter complaint invariably indicates the former. In a case related by Morgagni[1] the bladder was thus the seat of sympathetic pain; the disease being in the kidneys. The patient, he says, complained of very little pain in the region of the kidney; while he was tormented with pain in the bladder, so excruciating that five or six physicians who attended him entertained no doubt of the seat of the disease being in that organ. On dissection, however, no morbid appearance was discovered in the bladder; and there were large and ramifying calculi in the kidney. In certain renal affections, even where the urine is not very unnatural, the pain is confined chiefly to the neck of the bladder. In these cases, however, there is usually some amount of vesical catarrh.

Irritability of the bladder is a frequent accompaniment of gonorrhœa, and often remains after the latter affection has subsided. In such cases the permanence of the symptom is due to a chronic inflammatory condition of the neck of the bladder. The quantity of mucous deposit in the urine is inconsiderable, and the predominant symptom is the intolerance which the bladder exhibits towards the accumulation of its contents.

Irritability of the bladder is also a marked symptom of stone, and it sometimes remains after the calculus has been removed by lithotomy. It is also a prominent symptom of the various morbid growths to which the bladder is liable, and of enlargement of the prostate gland.

A habit of too frequently emptying the bladder may induce a permanent condition of irritability. Under such circumstances the sensitiveness of the mucous membrane becomes increased, and the presence of a small quantity of urine gives rise to an uncomfortable feeling. Sexual excesses are another cause of irritability. "Castus raro mingit" is an old proverb, frequently quoted by German writers on diseases of the urinary organs.

A temporary paralysis of the bladder may be followed by a condition of irritability, as exemplified by the following case. A gentleman during an attack of typhus was seized with inability to pass urine, and a catheter was used twice a day. As the patient gained strength the power of micturating returned, and the desire to pass urine became so frequent as to compel him to do so every half hour. He consulted Mr. Coulson for this symptom. There was no apparent disease of the urinary organs, and with the exception of occasional rheumatic attacks, his general health was good. The urine was very acid and scanty. He was ordered a grain of the acetous extract of colchicum at night, and a mixture containing bicarbonate of potash, sesquicarbonate of soda, and nitrate of potash twice or three times a day, soon after meals. By these means the urine

---

[1] On the Seats and Causes of Disease. Letter xlii.

became more abundant and less acid; but frequent micturition continued. The infusion of diosma, the decoction of the pareira brava, and various preparations of steel, were in turn tried for the relief of this annoying symptom, but without success.

Irritability of the bladder in children is frequently due to a contracted state of the prepuce, and to adhesions between the prepuce and glans penis. The following case is an example of this kind:—The boy, seven years of age, had for the previous eight months complained of a frequent desire to make water, attended with difficulty in passing it, and pains round the lower part of the abdomen. On examining the prepuce, it was found so contracted as scarcely to admit the point of a probe. The end of the prepuce was removed by circumcision. From that time the symptoms subsided, and the child recovered. Again, a boy, eleven years of age, had suffered during two or three years from pain in making water and incontinence during the night. The prepuce was found as contracted as in the previous case. Simple division was had recourse to, and all symptoms vanished. In cases of this kind, the division or removal of the extremity of the prepuce suffices for the cure of the complaint.[1]

Irritability of the bladder is not uncommon in children, especially at the period of dentition. We may infer in many of these cases that there exists something unnatural in the composition of the urine, although chemical analysis may not at all times succeed in detecting it. Spasmodic action of the bladder may also have some share in producing this form of the disease; for we know that the irritation of teething is often propagated to distant muscles. Considerable perseverance may be required to overcome this troublesome complaint, should it become permanent, especially if an involuntary discharge of urine occurs, as it commonly does, during sleep.

One of the most common causes of irritable bladder is a morbid condition of the urine. It rarely happens that this viscus, when free from organic disease or changes of structure dependent upon stricture of the urethra, is unable to retain healthy constituted urine, which acts as the natural stimulus to its mucous membrane. It has been well observed by Dr. Prout[2] that all deviations from the normal condition of the urine, whether in deficiency, in excess, or in kind, are recognized by the containing organs, and may prove a source of irritation in the kidneys and bladder. Hence, whenever the urine is very dilute or very concentrated, or is preternaturally acid or alkaline, or contains any unnatural ingredient, the urinary organs in general, and the bladder in particular, though perfectly healthy, are liable to become excited and irritable, and the individual has no peace till the unnatural secretion is discharged. In such cases the fault lies, not in the bladder, but remotely in the kidneys and assimilating organs.

Sometimes irritation is produced by taking alkaline remedies for too long a time; and, in such case, the urine becomes alkaline. Mr. Coulson was consulted for irritability of the bladder by a gentleman whose urine was alkaline, but whose appearance and state of constitution did not lead him to expect this state of the secretion. On inquiry, it was found that he had been in the constant habit of taking the sesquicarbonate of soda in large doses. He was ordered to discontinue its use, and he soon re-

---

[1] For an account of similar cases see Paget, Medical Times and Gazette, 1858, vol. i., p. 368; Bryant, Medical Times and Gazette, 1868, vol. i., 525; Forster, in same journal, 1872, vol. i., p. 593.
[2] On Stomach and Renal Diseases, p. 366.

covered. As the altered state of the urine is often the immediate cause of irritability, it should especially engage our attention. In other cases, the disease will be found to depend on the abuse of diuretic remedies, or the excessive stimulus of certain irritant substances, as cantharides, etc.

A transient form of irritability of the bladder, attended by slight catarrh, may be produced by drinking imperfectly fermented alcoholic liquors. Niemeyer suggests that, under such circumstances, the condition may be due to acute irritation of the urinary passages, caused by large sharp crystals of oxalate of lime, which sometimes appear after drinking unwholesome beers containing much carbonic acid.

II. In a second series of cases the complaint is due to some condition of neighboring organs. Thus the pressure of the womb during pregnancy, displacements and various affections of the uterus, diseases of the rectum (especially ulcers, fissures, and hæmorrhoids), ascarides and accumulation of feces in the bowels may give rise to symptoms of irritability of the bladder. Frequent micturition is also a symptom of various diseases of the brain and spinal cord.

III. Finally, there is a third series of cases in which the complaint forms part of some constitutional affection, such as gout or rheumatism, or depends upon causes operating upon the system generally. It often happens that a patient with irritable bladder applies to his medical attendant, who, on inquiry, finds him subject to occasional pains in the loins and limbs, to some scaly eruption on the skin, and various derangements of digestion; the urine is generally somewhat scanty, is very acid and contains a large quantity of urates.

The following case illustrates this common group of symptoms:—A gentleman, 43 years of age, and subject to rheumatism, complained of a very frequent desire to pass urine, from the annoyance of which he had for several years suffered. He was affected with a scaly eruption on several parts of the body, particularly about the elbows and knees; and he often felt severe pains in the hips and loins. The urine was very acid and scanty. He was ordered a mixture of diosma, henbane, carbonate of potash, and sarsaparilla, with blue pill and rhubarb at bed-time. After some time, the irritability of the bladder was much lessened, and the character of the eruption improved. The decoction of the pareira brava was then prescribed during the day, with a grain of the acetous extract of colchicum at bed-time. His complaint was much relieved but not cured.

Great irritability of the bladder and urethra, with increased secretion from the mucous membrane of these parts, is frequent with some persons shortly before an attack of gout. The disease assumes the character of gonorrhœa, and is attended by pain and scalding in making water, the calls to perform which act are frequent and urgent. These symptoms of irritability of the bladder are more likely to occur when the urethra is affected with stricture.

It sometimes also happens that, during a paroxysm of gout there is considerable irritation of the urinary organs, indicated by frequency of micturition, and a feeling of heat and discomfort about the bladder and urethra. On the other hand, Sir C. Scudamore[1] has remarked, that sometimes a patient who has for years been suffering from irritability of the bladder will have this symptom relieved or suspended during a paroxysm of gout. In such cases, the irritability is doubtless due to the ex-

---

[1] On Gout and Rheumatism, p. 18.

cessive acidity of the urine, and the relief may be accounted for by the fact that a diminished quantity of uric acid is excreted during the attack.

Hysterical persons sometimes suffer from irritable bladder, and great pain may also be experienced in passing urine. In these patients, the quantity of urine is often considerable, and the secretion possesses the aqueous character, or contains less than the usual proportion of solid matter, the nature and relative quantity of its constituents remaining, the same. This peculiarity in the quantity and quality of the urine is capable of explanation perhaps, to a certain degree, from the fact that the viscera apt to be deranged in hysteria all derive nerves from the same source, namely, the abdominal sympathetic system.

Various derangements of the digestive apparatus, by disturbing the functions of the kidney, may cause irritability of the bladder. In that condition of the system to which the term *oxaluria* has been applied, some amount of uneasiness or irritability of the urinary organs is often a prominent symptom. It also occurs with other forms of dyspepsia, as a result of some abnormal condition of the urine, which in such cases is often scanty and high-colored, and contains an excess of uric acid.

Exposure to extreme cold, or to great heat, sometimes produces irritability of the bladder. In the former case the cutaneous perspiration is arrested by the chilling of the surface, and a determination of blood takes place towards the internal organs. The kidneys, thus stimulated to increased activity, secrete the urine with greater rapidity than usual, and, as the bladder is generally intolerant of rapid distention, frequent micturition is the result. This condition sometimes occurs in persons who have returned to England after a prolonged residence in tropical countries. On the other hand, exposure to the rays of a tropical sun sometimes causes temporary irritation of the bladder. Increased frequency of micturition is a common symptom of sunstroke in its various forms. The suspended activity of the perspiratory glands, indicated by the abnormal dryness of the skin, causes an augmented secretion of urine. Dr. Gross,[1] in alluding to this cause of irritation of the bladder, states that he has met with several cases of the kind. The patients were all farm-laborers, who had been exposed to the rays of a hot sun while engaged in hard work, and the symptoms were a feeling of general prostration, an incessant desire to pass urine, and a sensation of heat and discomfort at the neck of the bladder.

In addition to neuralgia and irritability, the bladder is occasionally the seat of involuntary, uncontrollable and painful contractions, and these symptoms are designated spasm of the bladder. This affection arises from a great variety of causes, and differs from irritability inasmuch as the attacks occur at intervals between which the patient enjoys complete ease. In spasm of the bladder micturition may be affected in one of two ways, according as the fibres of the detrusor or those of the sphincter vesicæ are spasmodically contracted. In the former case the urine is expelled involuntarily, in the latter it is retained.

With regard to its causation, this affection, regarded as a symptom, is almost invariably a concomitant of stone in the bladder; it is frequently produced by malignant tumors; and it very often accompanies gonorrhœa, especially when injections have been employed either too strong in composition or too early in the treatment : in the latter instance the disease affects the sphincter vesicæ more especially, and is usually

---

[1] Gross, loc. cit., p. 75.

attended with more or less inflammation. Spasm of the bladder may also be caused by the presence of acid urine; it may be symptomatic of inflammatory or other disease of the kidneys, bladder or prostate gland, or it may be the result of irritation propagated from the intestinal canal, uterus or ovaries. As occurs also in the rectum, fissures in the mucous membrane of the neck of the bladder may be the cause of spasm. The affection is also an occasional symptom of disease of the spinal cord or brain. The motor nerves of the bladder have two centres of origin; a lower one in the lumbar part of the spinal cord, and an upper one in the crura cerebri, and Budge has demonstrated that these two centres are connected by strands of fibres running in the spinal cord.

A fit of stone, as it is familiarly called, is an example of a violent paroxysm of spasm of the bladder. The patient, who is suddenly attacked, experiences severe pain extending from the region of the bladder, along the urethra to the extremity of the penis. The urine may be expelled involuntarily, but more commonly there is retention with some amount of suppression from the orifice of the ureters being closed by the spasm. There is constant desire to void urine without the ability to do so, and the agony felt during the continuance of these attacks is excessive; while the pressure of the hard and contracted bladder on the rectum excites a feeling of desire to evacuate the contents of the bowel.

Cases are related in which the patient has expired with all the symptoms of suppression of urine, and upon examination after death, it has been found that the ureters, spasmodically closed at their vesical extremity, have been dilated by the accumulated urine; while the pelves of the kidneys were similarly enlarged, and the substance of the kidneys diseased. Indeed, after an attack of spasm from which the patient has apparently recovered, we may subsequently have to treat a new train of symptoms indicating the injury which the tubular and secreting substance of the kidney has received. Spasm may be confounded with acute inflammation of the bladder. In inflammation, however, the pain is constant, commencing with more of uneasiness than of positive pain, and becoming more intense by degrees; while in spasm, the seizure is as severe as it is sudden. In the former, the pain has the usual character of inflammation—it is lancinating and throbbing; whereas in the latter it is constrictive, resembling, in fact, labor-pains.

Spasm of the bladder, when of long standing, often gives rise to hypertrophy of the muscular tissue, and the paroxysms then may bear considerable resemblance to those of stone; but its frequent recurrence sometimes injures the tone of the bladder so much as eventually to induce an opposite state of the muscular fibres: in other words, the disease terminates in paralysis of the bladder.

With regard to the treatment of irritability and spasm of the bladder, the first step to be taken is to ascertain the cause or causes upon which the symptoms depend. Modern investigation has abundantly shown that cases of purely idiopathic irritability of the bladder are exceedingly rare, and that the frequent desire to micturate is only symptomatic of some primary affection. In dealing, therefore, with any given case, a general acquaintance with the different forms of renal and other urinary diseases is absolutely requisite, as well as the power of ascertaining, by the use of the microscope and by chemical analysis, the nature and relative proportions of the ingredients of the urine. It may, of course, happen that even after a careful examination the cause of the irritability or of the spasm remains undetected, and under such circumstances the treatment must be merely directed towards the alleviation of the symptoms.

The treatment of those cases in which a local cause is sufficiently obvious and at the same time removable is, for the most part, easily accomplished. Thus when the complaint occurs in children, an adherent or elongated and contracted prepuce should be carefully looked for, and relieved by breaking down the adhesions, or removing the redundant portion by the operation of circumcision. In like manner ascarides in the rectum or accumulated feces, if present, should be removed by suitable injections.

When the symptoms occur in gouty and rheumatic subjects and the urine is acid and scanty, the alkalies are indicated and they may be combined with saline purgatives, such as the sulphate of soda. A course of Carlsbad waters is especially indicated.

When the symptoms are due to a pre-existing gonorrhœa, small doses of balsam of copaiba should be assiduously persevered with. The same remedy will be found useful when the symptoms are connected with chronic catarrh of the bladder. A few grains of extract of henbane, or a quarter of a grain of extract of belladonna, may be taken night and morning, or used as a suppository, and will be found to diminish the irritability of the bladder. Alkalies are also serviceable when the urine is unduly acid, but care must be taken lest an opposite reaction be induced. The state of the prostate gland should be examined in cases of vesical irritability consequent upon an attack of gonorrhœa. Should prostatitis have been set up, a chronic inflammatory condition may remain for some time after all the active symptoms have subsided. Such a condition would be certain to give rise to the symptoms of irritability of the bladder. The most effectual treatment consists in applying counter-irritation locally, by painting the liquor epispasticus over a small surface of the perineum on each side of the raphé.

In cases of irritable bladder occurring in persons of a scrofulous habit, constitutional remedies are especially indicated. The combination of iodine with iron, which has been found very serviceable in many cases of nocturnal incontinence of urine, is especially indicated, and should be given in full doses. The tincture of the perchloride of iron may also be tried in combination with belladonna.

The treatment of spasm of the bladder must be conducted upon similar principles. The cause of the symptoms must be sought for, and if possible removed. The symptoms themselves may be relieved by the use of the warm bath and warm fomentations, and the administration of henbane or morphia, the subcutaneous injection of the latter drug being especially efficacious. It may be also used in the form of a suppository. If retention occurs and the symptoms become urgent, the inhalation of chloroform may be tried, and if this fails a gum-elastic catheter should be cautiously introduced. Bromide of potassium in full doses may be subsequently administered, and combined with bicarbonate of potash if the urine be acid.

CHAPTER XIV.

ATONY, PARESIS, AND PARALYSIS OF THE BLADDER.

THE muscular coat of the bladder may become paralyzed from any of those causes which induce loss of muscular power in other parts of the body, and the paralysis may affect either the detrusor urinæ, or sphincter vesicæ, or both these muscles simultaneously. There may be either simple diminution or complete loss of power; and the paralysis may be temporary or permanent. The muscles of the bladder concerned in expelling the urine, and in preventing its expulsion, are only in part under the influence of the will. Thus the contractions of the detrusor urinæ are altogether involuntary, and are due to the stimulus of the urine contained in the bladder. There is some difference of opinion with regard to the part played by the sphincter, and also as to whether a single or a double sphincter may be said to exist. The sphincter formed by the circular fibres of the bladder serves to close the cavity of the viscus at its neck, and while some physiologists regard the compressor urethræ as an external sphincter, others deny that it can act in this capacity, and regard it as a muscle the contraction of which aids only in the expulsion of the urine. It is one of those muscles which must relax before any urine can pass; it is also under the influence of the will, and there would appear to be much probability in the view which attributes to it the function of assisting to close the orifice of the bladder.

Atony and paresis of the bladder are terms applied to designate those conditions in which the contractions of the organ are slow, feeble and imperfect, while paralysis is said to exist when the action of the detrusor or sphincter is completely suspended. By some writers atony is regarded as something quite different not only in degree, but in kind, from paralysis. They refuse to admit that there can be any "paralysis" of the bladder without some change in a nervous centre,[1] and they attribute the condition termed "atony" to loss of contractile power by the muscular fibres, such as is liable to occur whenever the bladder is allowed to become over-distended. The coats of the bladder, however, are abundantly supplied with nerves, which must of necessity be stretched and injured when the organ is distended beyond its proper limits, and the "atony" which results in such a case is probably as much due to the local injury sustained by the nerves, and by which their reflex activity is interfered with, as to the loss of contractile power by the muscles. It is quite conceivable that a peripheral affection of the nerves may be caused by over-distention of the bladder. When the nervous irritability is merely diminished but not annihilated, the contractions are feeble, and a state of incomplete paralysis, which may be termed paresis or atony, is established.

Paralysis of the bladder, of various degrees, may be symptomatic of affections of other organs, or may be due to changes in the structure of the organ itself. In the course of diseases accompanied by loss or dim-

---

[1] See on this subject Sir H. Thompson, Diseases of the Urinary Organs, 310.

inution of sensibility, the presence of urine in the bladder no longer acts as a stimulus to provoke contractions of the detrusor, and more or less complete paralysis with retention of urine is the result. This condition is seen to occur in the course of fevers and in uræmia, and often tends to expedite the fatal issue.

Paralysis of the bladder may be a consequence of injuries to the head, or of apoplexy; but it comes most frequently under the notice of the surgeon as the effect of injuries or diseases of the spine; and these cases are attended by important changes in the whole urinary system as well as in the condition of the urine itself. The changes here referred to do not appear to be connected with the particular locality of the injury; they occur almost uniformly whether the injury affect the lumbar, dorsal, or cervical regions.

In some cases, the urine first secreted, though free from mucus, and with an acid reaction, has an offensive and disgusting odor. In other cases, the urine is highly acid, has an opaque yellow appearance, and deposits a yellow amorphous sediment. After slight accidents, such as a strain in the loins, this state of the secretion usually disappears under proper treatment in the course of a few days. But the most common change produced in the urine by injury of the spinal cord, is the following: when voided, it is turbid, and of ammoniacal odor; when allowed to remain at rest and to cool, it deposits much adhesive mucus; and, when tested with reddened litmus or turmeric paper, it is found to be highly alkaline, owing to the conversion of the urea into carbonate of ammonia. The urine thus rendered ammoniacal, excites inflammation of the mucous membrane of the bladder, which pours forth in consequence a large quantity of viscid ropy mucus. The phosphatic salt most frequently found mixed with this secretion is the neutral triple phosphate of magnesia and ammonia, the prismatic crystals of which, presenting different degrees of transparency, are easily recognized under the microscope.

If the patient live long enough, the inflammation thus excited by the decomposed urine in the paralyzed bladder spreads through the whole thickness of the vesical walls, which are rendered thicker than usual, lose all power of contraction, and sometimes form connections with the walls of the abdomen and adjacent viscera. The catheter under these circumstances fails in completely drawing off the contents of the bladder; a quantity of highly alkaline and fetid urine remains to add to the local disturbance, and to impregnate any fresh secretion as it enters the organ from the ureters.

Paralysis of the bladder may also occur from reflex irritation, as after operations, though perhaps in these cases the difficulty of emptying the bladder is due rather to spasmodic action of the compressor urethræ than to paralysis of the detrusor. It is not an unfrequent occurrence after an operation for hæmorrhoids, and it may continue for several days. Also after compound fractures, amputations, strangulated hernia, wounds of the abdomen and other serious injuries, the bladder occasionally loses for a time its power of contraction. Temporary paralysis of a complete character may also be caused by the use of certain narcotics, such as opium, belladonna, and hyoscyamus. The local application of these remedies, in the form of injections into the rectum or as suppositories, would be more likely than their internal exhibition to produce such an effect. The symptoms are usually transient, but Sir H. Thompson[1] has

---

[1] Holmes' System of Surgery, vol iv., p. 905.

met with a case in which permanent loss of power in the bladder was attributed to an overdose of belladonna.

In the cases above alluded to, the paralysis is symptomatic of some other affection, or is due to general causes, and is usually complete. In a second class of cases, the paralysis is due to local causes, and is more or less incomplete. The condition is that of paresis or atony, and it usually arises from over-distention of the coats of the bladder, ending in loss of tone of its muscular elements and diminished conductile power of its nerves. Such a condition may suddenly arise when the urine is retained voluntarily, but with great effort, for a considerable period; or it may gradually supervene, as in cases where some organic obstruction exists to the outflow of urine. Instances of the sudden occurrence of paralysis from over-distention are by no means rare. A person finding himself in some inconvenient situation neglects to empty his bladder, though the call to do so may be very urgent. The urine accumulates, the bladder becomes distended, and, after a while, on endeavoring to empty it, the patient finds that he has lost the power to do so. The use of the catheter affords immediate relief, and if the patient is duly careful in the future with regard to emptying his bladder, the symptoms may not recur. Such an attack, however, is likely to be followed by more or less intense catarrh of the bladder, which may prove very tedious,[1] and in addition to the immediate effects, a permanent condition of atony or paralysis may be induced by repeated acts of this kind.

Atony or partial paralysis from over-distention may, however, arise more gradually, and cases of this kind are of frequent occurrence. When organic obstruction has existed for any length of time at the neck of the bladder or in the urethra, the contractile power of the organ becomes gradually insufficient to expel the urine, and a portion remains behind after each act of micturition. Some amount of hypertrophy is the usual consequence, but as time goes on, and the obstruction increases, the coats of the bladder become gradually stretched by the retained urine, and loss of contractility is the result. A chronic catarrhal state also is produced by the residual urine, and this still further diminishes the tonicity of the organ. The distention may increase until the sphincter relaxes and the urine dribbles away. The involuntary flow of urine is nature's effort to relieve the over-distended bladder. This condition has been repeatedly mistaken for incontinence, from which, however, it differs in the most essential manner.

In illustration of this important point, the following remarks of Mr. Lawrence are worth quoting:—"I was sent for to see a gentleman laboring under an affection of the bladder; and the medical attendant who had lately seen him mentioned that the case was one of great irritability of the bladder, which would hold no water at all—the urine passing off as fast as it came into it. He said he had been doing all he could to get the bladder's natural power of retention restored; he had directed the patient to take diluent liquids; in short, he had done all he could to prevent it; but still the water ran off. It appeared to be a singular case. I put my hand under the clothes upon the abdomen; and I felt the fundus of the bladder forced up a good way above the umbilicus. I said that I had brought a catheter with me, and I might as well introduce it, to see if there was anything in the bladder. I introduced it,

---

[1] A case of this kind is mentioned by Podrazki. Billroth und Pitha. Handbuch der all. und spec. Chir., Band iii., Abth. ii., Lief. viii., p. 69.

and about 5 pints of urine immediately flowed off. The fact was, the bladder had been, in this way, allowed to distend for about five days before I saw the patient; and the consequence was, that he never afterwards recovered the natural power of emptying that viscus; but he acquired, after a certain time, the art of introducing the catheter, which he still employs: he can introduce it, and let the water off whenever he finds a desire to do so; but since that time he never has been able to empty the bladder by the natural powers."

Serious and even fatal symptoms have been known to follow the sudden abstraction of a large quantity of urine from an over-distended bladder, the patient being in the erect position. Sir H. Thompson[1] alludes to a case in which fatal syncope took place when 6 pints of urine had passed, just as might occur when a patient is tapped for ascites in the same position. It is, therefore, under such circumstances always advisable to withdraw the urine gradually, and while the patient is in a recumbent position.

In the majority of these cases of partial paralysis or atony, the affection is primarily due to the presence of some obstruction, such as is produced by an enlarged prostate, or by stricture of the urethra. In a smaller number of cases, the patients being advanced in years, no such obstruction can be shown to exist, and the symptoms may with probability be ascribed to fatty degeneration of the muscular coat of the bladder. In such cases the detrusor is the first to suffer, and incomplete evacuation of the contents of the bladder is the result. In the course of time the sphincter also becomes affected, and the patient experiences an increasing difficulty in retaining his urine.

Temporary paralysis of the bladder sometimes occurs in pregnant women owing to some displacement which the organ has undergone. It has also been known to occur after delivery, as a consequence of the pressure exerted by the head of the child upon the neck of the bladder. Sudden prolapsus of the uterus, after labor, may also cause displacement of the bladder and a bent condition of the urethra, with retention of urine and partial paralysis as the ultimate result. The functions of the detrusor may also be interfered with by pressure caused by interstitial œdema, such as may occur in ulceration of the cervix uteri, in parametritis and in peritonitis.[2]

The diagnosis of paralysis of the bladder is for the most part easy, especially if the affection be far advanced. When, however, the symptoms are due primarily to hypertrophy of the prostate, they may exist for a considerable time before being recognized. The patient experiences a difficulty in commencing to micturate, and the urine flows in a weak stream or falls almost perpendicularly from the orifice of the urethra. After some amount of urine has been thus passed, the patient feels relieved, but the dribbling continues for some little time in spite of his efforts to check it. As time goes on, the desire to micturate becomes more and more frequent, the quantity of urine passed on each occasion becoming less and less. Pressure upon the lower part of the abdomen causes more urine to flow. If, after the patient has exhausted his efforts at emptying the bladder, a catheter be introduced, a quantity of the residual urine will escape. If the bladder be very full, the urine may at first be ejected from the catheter with some force, but the stream will become weaker

---

[1] Diseases of the Urinary Organs, p. 106.
[2] Winckel, Handbuch der Frauenkrankheiten, p. 212.

and weaker until it finally ceases. Even then, however, more urine may often be obtained by making pressure over the lower part of the abdomen. The urine under such circumstances is generally cloudy, and often ammoniacal and fetid.

In the more advanced forms of the affection, complete retention of urine may occur, in spite of all the efforts of the patient. The distention increases until the bladder forms a tense, elastic, globular tumor which may rise as high as, or even extend above the umbilicus. When the distention has reached a certain degree, the relaxation of the sphincter generally permits some amount of urine to dribble away. The quantity of urine which may, under such circumstances, be withdrawn from the bladder, is sometimes enormous. The fatal case, alluded to by Sir H. Thompson, has already been mentioned. Other cases, occurring in women, have been reported, in which nine pints were withdrawn.[1] This enormous distensibility of the bladder has been explained by an experiment performed by Budge, who found that division of the spinal cord in the lower dorsal region was followed by increased reflex contraction of the sphincter and distention of the bladder, to a degree impossible to produce by artificial means after death.

The imperfect evacuation of the contents of the bladder in cases of atony, leads to catarrhal changes in the mucous membrane, and the inflammation, unless relieved, invariably extends up the ureters to the pelvis of the kidney. When the retention becomes more or less complete and relief is not afforded, uræmic symptoms may supervene and prove rapidly fatal.

The prognosis in all forms of paralysis of the bladder will depend upon the causes to which the affection is due. In cases of disease of the spine or brain, if the central lesion be irremediable, little or no improvement can be expected in the local symptoms. When the symptoms are primarily due to the presence of an obstruction to the escape of urine, e. g., a stricture of the urethra, if this latter can be cured, the bladder may be expected to recover its tone under appropriate treatment. In cases of enlarged prostate, the prognosis is not so favorable. If the sphincter as well as the detrusor be involved, the condition is usually too far advanced to admit of any expectation of relief from treatment. The age and constitution of the patient, the duration and degree of the paralysis, as well as the causes upon which it depends, are the main points for consideration in estimating the probabilities of recovery.

The morbid appearances presented after death, in cases of paralysis of the bladder, are dilatation of the organ, relaxation and attenuation of its coats, and a pale white appearance of the mucous membrane; these, at least, are the appearances found when the kidneys and the mucous membrane of the bladder have not been seriously involved in mischief, as, after injuries of the spine. In the latter case, there is great vascularity of the mucous membrane lining the bladder, ureters, the pelvis, and infundibula of the kidneys; the mucous surface of the bladder is thickened, of a slate color, and presents, here and there, dark red spots; and sometimes it is covered with phosphatic deposit, while the urine contained in the organ is fetid and ammoniacal.

In all cases of paralysis of the bladder, an early attention to the condition of the organ is of great importance. The longer the urine is allowed to remain in contact with the mucous membrane, the more likely

---

[1] Winckel, loc. cit., p. 213.

is it to excite disease ; while the frequent and continued distention of the muscular fibres gradually renders them more lax, elongates and makes them thinner, and diminishes the chance of our being able, by any treatment, to restore them to a healthy state.

The indications of treatment are sufficiently obvious. Our first care must be to relieve the distended bladder, in order to prevent distress and extension of mischief ; our next, to restore its contractile power, and remove all such causes as may appear to influence the paralytic condition of the organ.

In those cases in which the affection is due to some central lesion, such as disease or injury of the spinal cord, the regular and complete evacuation of the contents of the bladder by means of the catheter is the main point to be attended to. The catheter should be passed with the greatest care, so as, on the one hand, to avoid injury to the mucous membrane, and on the other, completely to empty the bladder. The sensitiveness of the parts being diminished or destroyed, considerable damage may be done by careless or rough manipulations without any pain being felt by the patient. A false passage may be easily made, and, once established, will render the subsequent passing of a catheter a matter of cónsiderable difficulty. Cases are also on record in which abscess of the bladder and ulcerative perforations have been the result of catheterization unskilfully performed. The condition of the coats of the bladder, in cases of spinal lesions, renders them especially liable to suffer from mechanical violence.

The same precautions are necessary when the functions of the detrusor become temporarily in abeyance, as in severe febrile affections. In chronic cases of paralysis of the bladder, due to cerebral or spinal lesion, we may, when all active symptoms have subsided, in addition to the regular use of the catheter, try small doses of strychnia, cold sponging or douches over the region of the bladder, counter-irritation to the spine, etc.

A gum-elastic catheter, or a French catheter *à boule*, is the best form of instrument which can be used for these cases. It is difficult to lay down any fixed rules as to the intervals which should elapse between each introduction of the catheter. The necessity of avoiding sudden evacuation of a distended bladder has been already alluded to. Independently of the danger which may arise from this proceeding, it is more expedient to withdraw the urine slowly, and to close the orifice of the catheter occasionally, for by this plan we excite the bladder to contract and awaken its dormant sensibility. The intervals between each introduction of the instrument will be mainly regulated by the capacity of the bladder, and the degree of the paralytic affection. Generally speaking, it will be sufficient to draw off the urine three or four times a day ; or, in cases of partial paralysis, when the patient begins to experience a desire to empty his bladder. When circumstances render it impossible for the patient to insure frequent surgical assistance, it will be prudent to leave a catheter permanently in the bladder, closing the external orifice so that the urine may be drawn off from time to time. For many reasons, however, the permanent use of the instrument should, if possible, be avoided. The vulcanized india-rubber catheter, which is eminently elastic and creates little irritation, is best adapted for those cases in which retention of the instrument is, on any account, necessary. Care should be taken that just so much as is sufficient of the extremity of the catheter remains within the bladder.

In cases of atony or partial paralysis, the relief of the distended bladder and the restoration of its contractile power are the main indications to be fulfilled. If the symptoms are connected with stricture of the urethra, steps should be taken to remedy the obstruction. In cases where the condition is due primarily to some enlargement of the prostate, nothing can be done towards removal of the cause; the symptoms, however, may be much ameliorated by treatment. The accumulated urine should be drawn off by means of an elastic catheter; and the operation should be repeated three or four times in the twenty-four hours (at intervals sufficiently short to prevent over-distention), until the bladder has recovered its contractile power.

In order still further to arouse the dormant contractility of the bladder, in cases of slight paralysis without prostatic enlargement, we may employ local stimuli, or make trial of various internal remedies. Injections of cold water into the bladder were recommended by Civiale. They may be commenced after the catheter has been employed for a few days, when its use will probably have ceased to excite any irritation about the neck of the bladder. Tepid water is employed for the first injection, care being taken not to attempt to introduce too large a quantity. Three or four ounces will be amply sufficient. The temperature of the water should be gradually lowered to about 60° Fah., and as soon as the patient is able to bear the contact of the cold water without pain, two or three injections are thrown into the bladder, one immediately after the other. These frequently have the effect of exciting the bladder to contract, and contraction once attained, an improvement may be expected. In mild cases, one or two injections daily for about a fortnight ordinarily suffice, and an improvement may appear on the second or third day. Generally speaking, the injections may be continued from fifteen to thirty days. Cold douches to the hypogastric region, to the perineum, and over the sacrum will also assist the treatment.

Electricity may also be employed with advantage. The induced current alone should be used, as the galvanic current might possibly cause severe irritation of the mucous membrane and electrolysis of the urine or mucus contained in the bladder.[1] It is useless to apply both poles externally; one of them must be introduced into the bladder. An ordinary electrode, covered with moistened sponge, is placed over the lumbar vertebræ or on the hypogastric region; the other electrode, which is introduced into the bladder, consists of an elastic bougie, with a metallic extremity, and traversed by a conducting wire. This insulated conductor should be gently moved about, with its point in contact with various portions of the wall of the bladder, and its application should not last for more than five minutes.

Another method suitable for cases in which the contractility of the bladder is only slightly impaired, has been recommended by Von Pitha.[2] This consists in the introduction of a wax bougie, which is to be passed down the urethra as far as the neck of the bladder, and permitted to remain *in situ* for a few minutes, until the patient experiences a desire to pass urine. It is then to be quickly removed, and its withdrawal is often followed by the passage of a considerable quantity of urine. The bougie should be introduced daily, the catheter also being had recourse to at intervals, in order to insure complete emptying of the

---

[1] Ziemssen, Die Electricität in der Medicin, p. 153.
[2] Handbuch der all. und spec. Chirurgie, Band iii., Abth. ii., Lief. viii., p. 72.

bladder. The patient should be directed to endeavor to pass his urine naturally, so that the use of the catheter may gradually be dispensed with.

With regard to internal remedies, the preparations of nux vomica may be employed, as in other forms of motor paralysis, and in a few cases the administration of this remedy has been attended with benefit. Lebert[1] asserts that the medicine in question exercises no specific influence over any form of motor paralysis, and he makes the same remark with reference to ergot of rye. On the other hand, Dr. Gross, of Philadelphia, strongly recommends strychnia, combined with cantharides and arnica, for chronic cases of vesical paralysis; and in that form of the disease which occurs after typhoid and other fevers he gives it as his experience that few remedies are so serviceable as arnica, administered in the form of the tincture, in doses of from forty to sixty drops three times a day. With regard also to ergot of rye, Dr. Gross has found it very efficacious in some cases. The dose is a drachm of the fluid extract three times a day.

The condition of the general system should of course be attended to in all cases of paralysis or atony of the bladder. A tonic regimen is usually indicated, and the preparations of quinine and iron will often be found to assist in improving the general health and the local affection. Constipation should be prevented by mild but efficient purgatives. The complaint is often attended by a sluggish or even paralytic condition of the rectum and lower portion of the intestinal canal. The use of purgative remedies is here indicated, not only because they stimulate the bladder to action, but from their effect in removing anything which may cause mechanical obstruction about the neck of the organ. The preparations of aloes are the most suitable for the purpose; these may be combined with extract of nux vomica, and their action may be assisted by enemata of warm or cold water.

In those cases of paralysis of the bladder, due to spinal disease or injury, in which the sphincter as well as the detrusor is affected, and the urine dribbles away involuntarily, there is but little to be expected either from general or local treatment. The parts should be kept as clean and dry as possible, in order to prevent excoriation and bed sores, and this object will be best accomplished by the regular use of the catheter, and the application of a well-fitting india-rubber receptacle.

---

[1] Ziemssen's Handbuch der spec. Path. u. Therapie, Krankheiten der Harnblase. Band ix., p. 368.

CHAPTER XV.

ENURESIS.—INCONTINENCE OF URINE.

INCONTINENCE of urine, occurring in the adult, is a symptom of various affections of the urinary organs. The term, however, has been often incorrectly applied to designate the escape of urine from an over-distended bladder. This, as has already been pointed out, is a common symptom of atony or partial paralysis of the organ, but is not an example of true incontinence. The urine dribbles away, not because the patient cannot retain it, but because he cannot pass it; the sphincter yields and allows it to escape, while the bladder still remains more or less distended. The condition, therefore, is one of retention of urine, with overflow of the surplus.

True incontinence of urine—meaning by that term, inability of the bladder to retain its contents—may depend upon cerebral or spinal disease inducing paralysis of the sphincter of the bladder. In such cases, the incontinence varies in degree according to the extent of the paralysis. Sometimes the urine dribbles away as fast as it reaches the bladder (*enuresis passiva*); in other cases, the bladder acts to a certain extent as a reservoir, but when a small quantity of urine has accumulated, there is no power of preventing its escape. This form of the affection has been termed *enuresis activa*. Some of these patients again are able to retain their urine fairly well during the day, but it flows from them involuntarily during sleep. In a few rare cases, incontinence of urine is produced by an enlargement of the middle lobe of the prostate gland, which projects forwards into the neck of the bladder and keeps the part constantly patulous. These forms of the affection are but little amenable to treatment, as the cause is usually irremediable. The application of a well-fitting urinal will, however, relieve the patient of much discomfort.

Females are more liable than males to suffer from incontinence of urine, and the complaint may be due to a variety of causes, principally of a mechanical kind. Thus it may be caused by injury to the urethra during difficult or instrumental labors, or it may follow artificial dilatation or division of the urethra for the purpose of removing a calculus. The incontinence may be partial or complete. These cases might properly be described as injuries to the neck of the bladder, the mechanism of which is no longer able to close the orifice. In some patients the urine flows involuntarily on sneezing, or coughing, or making any exertion—the action of the diaphragm and abdominal muscles overcoming that of the sphincter of the bladder.

Incontinence of urine sometimes results from a paralyzed condition of the neck of the bladder in women who have had very large families—ten or twelve children, for example.

In these cases, more especially if the child is large, or the pelvis small, and the labor has been severe, the bladder is apt to get so infirm about the neck, that it loses much of its contractile power, and, perhaps, from the moment of delivery, the woman is incapable of retaining her urine: or if at any time she chance to cough, laugh, rise suddenly, or in any other manner contract smartly the abdominal muscles, the urine gushes out involuntarily. The complaint frequently subsides spontaneously, but, if neglected, it may last for years, and even if the patient regains a fair amount of control over the bladder under ordinary circumstances, it often happens that gushes of urine take place involuntarily when any sudden effort is made. The treatment of these cases consists in the adoption of measures for the improvement of the general health; tonics and cold spouging, or cold sitz-baths if they can be borne, are especially indicated. Faradization may also be applied directly to the urethra and neck of the bladder.

The most common form of incontinence of urine is that which occurs almost exclusively in children, although instances of the affection are occasionally met with in adults. In these cases the urine is discharged at night during sleep, and without the knowledge of the patient. Such a condition, physiological among infants, is a source of great annoyance in after-life. Boys suffer more frequently than girls from this complaint, the causes of which are for the most part very obscure. It has been observed to be hereditary, and it is certainly more common among weakly, scrofulous, or rachitic children than among those who are well-nourished and healthy. It is often very obstinate, in some cases continuing up to or even beyond the period of puberty. Some of the patients have imperfect control over the bladder even during the daytime, but this condition is an exceptional one. The urine is generally natural, and it may be discharged once or more times during the night. In some cases there appears to be utter unconsciousness of the escape of urine, in others the distention of the bladder appears to give rise to a dream, and the patient imagines that he is passing his urine into some convenient receptacle.

With regard to the causes of this affection there is nothing definitely known. Lazy, careless habits, general want of tone in the system, small size of bladder, an irritable condition of the mucous membrane caused by acid urine, the presence of ascarides in the rectum, have all been considered as likely to play a part in the causation of the complaint. The condition may be regarded as one of local debility accompanied by increase of excitability in the affected structures. The irritation caused by a certain quantity of urine produces a contraction of the bladder, but the sensation is too feeble to arouse the patient. As before observed the complaint is most common in scrofulous children, and it is quite possible that the increased irritability may be due to superficial pathological changes in the mucous membrane of the neck of the bladder. Billroth's opinion with reference to the causation of some forms of irritability of the bladder has been already alluded to.[1] He suggests that they may be due to superficial epithelial detachment and slight excoriation of the mucous membrane of the bladder, and it is conceivable that a similar condition may produce, or at all events assist in producing, nocturnal incontinence of urine. The fact that scrofulous children are particularly prone to suffer from this complaint is in favor of this hypothesis, which is also supported by the remarkable effect, sometimes witnessed, of the treatment by iodide of iron. The complaint is usually aggravated by anything that disturbs the general health, particularly by unwholesome diet, and

---

[1] Vide supra, p. 158.

also when the patients are allowed to drink freely in the latter part of the day, and are not kept sufficiently warm at night. An elongated prepuce has been alluded to as a frequent cause of irritable bladder in children; and in all cases of nocturnal incontinence of urine occurring in boys the condition of the prepuce should be carefully examined. The diagnosis of the complaint is for the most part easy, the discharge of urine during sleep being the only tangible symptom. With regard to prognosis, some cases yield very readily to treatment, while others are more or less obstinate.

The remedies that have been suggested and employed for the treatment of nocturnal incontinence of urine are very numerous. In all cases the general health of the patient is the first point which requires attention. The complaint is generally associated with various other evidences of debility, and tonics are therefore indicated. The urine should also be examined, though it rarely happens that anything abnormal is discovered. Worms in the rectum may possibly cause or aggravate the complaint, and if present, must be got rid of by appropriate treatment. The diet of the patient should be generous, but easy of digestion, the quantity of fluid taken should be carefully regulated, and abstinence from liquids should be enjoined during the latter part of the day. The child should be made to pass urine at regular hours during the day, and particularly before going to bed. The bed-clothes should be sufficient for the purposes of warmth, but a soft feather-bed is undesirable. The child may be roused once or twice during the night and made to pass water. Trousseau and Bardeleben have recommended that during the day the child should be encouraged to retain his urine as long as possible.

With regard to local treatment, in the case of male patients the introduction of a catheter has been found to arrest the complaint, at all events for a time.[1] On the supposition that the bladder is deficient in capacity it has been proposed to dilate the organ by injections. When in the case of boys the prepuce is found elongated, it will be advisable to perform circumcision. By way of a mechanical contrivance, Sir D. Corrigan has suggested that the prepuce should be drawn forward over the glans, and then painted over with collodion, which when dry will form a cap. When a few drops of urine escape from the urethra, this cap will become distended, and the resulting sensation will probably suffice to arouse the patient. The late Professor Niemeyer[2] used to recommend adult male patients to wear at night an India-rubber ring to encircle the prepuce drawn forwards over the glans. When the urine begins to escape, the distention of the prepuce will be felt by the patient, he can then remove the ring and empty his bladder. The contrivance may then be laid aside, if the discharge of urine takes place only once during the night.

There are certain remedies classed as specifics which in some cases act admirably, but are by no means always to be relied on. Belladonna was highly recommended by Trousseau, in nightly doses of from $\frac{1}{4}$ to $\frac{1}{2}$ of a grain of the powdered leaves with an equal quantity of the extract. The effect of the drug is to diminish the sensitiveness of the mucous membrane of the bladder, and to cause partial paralysis of the detrusor urinæ. The remedy should in all cases have a thorough trial; if small doses do not succeed, the quantity should be increased until constitutional symptoms, such as dryness of the throat and dilatation of the pupils,

---

[1] Podrazki, loc. cit., p. 74.
[2] Lehrbuch der spec. Path. u. Therapie, Aufl. ix., Bd. ii., s. 92.

begin to appear. A great deal will have been accomplished if one or two nights pass without any involuntary discharge of urine; the old habit having been thus broken through may gradually wear off. Chloral hydrate is another remedy of this class, and its action is more prompt than that of belladonna. It is highly spoken of as a remedy for this complaint by Dr. Bradbury,[1] of Cambridge, who regards nocturnal enuresis as an epilepsy of the bladder, and who has noticed that persons who are incontinent in youth sometimes become epileptic after puberty. Another medicine which often produces very satisfactory effects is the syrup of the iodide of iron. In a series of cases reported by Dr. Barclay,[2] the majority were completely cured by the use of this medicine for several days. It has also been found very serviceable by Winckel,[3] Lebert,[4] and others. It may be combined with cod-liver oil. Dr. Gross recommends minute doses of strychnia (gr. $\frac{1}{16}$) with cantharides gr. $\frac{1}{20}$, and combined in atonic cases with tinct. ferri perchlorid. Tepid and cold baths, either local or general, are likely to be useful auxiliaries in effecting a cure. Where these cannot be obtained, recourse may be had to cold affusion or sponging with cold water over the lower part of the spine. In some cases the complaint is especially troublesome during cold weather. Care should be taken to keep these patients well clad and warm at night. The unfortunate sufferers from this affection are often severely punished by those in charge of them. Unless the habit is obviously the result of laziness, this treatment can only serve to make matters worse, and should therefore be deprecated.

---

[1] British Medical Journal, April 8th, 1871.
[2] Medical Times and Gazette, December 17th, 1870.
[3] Handbuch der Frauenkrankheiten, p. 216.
[4] Ziemssen's Handbuch der spec. Path. u. Therapie, Band ix., Hälfte ii., s. 373.

CHAPTER XVI.

RETENTION OF URINE.

RETENTION of urine is a symptom common to various affections of the urinary organs. The term expresses a condition in which the bladder contains a quantity of urine which, from some cause or other, cannot be expelled by the natural efforts of the patient, but continues to accumulate and to distend the organ. When the retention is complete, the symptom assumes a prominence far greater, at the moment, than that of the original disease of which it is the result.

Retention of urine may arise from a variety of causes, the majority of which may be classed under two main headings. Thus paralysis or atony may render the bladder incapable of expelling its contents; or, the expulsive power of the organ being normal, the outflow of urine may be prevented by some mechanical obstruction either in the bladder itself, in the urethra, or in the surrounding tissues. Civiale proposed the term "stagnation of urine" for those cases in which the symptoms are due to paralysis or atony of the bladder, reserving the term "retention" for the second class, which is characterized by the presence of some obstruction. It appears, however, unnecessary to make a verbal distinction of this kind between the two classes, and the more so because there is an intermediate class in which the symptoms are referable to a double cause, viz., to partial paralysis or atony, and to the existence of an obstruction to which the paralysis is due.

Retention of urine due to paralysis or atony of the bladder has been sufficiently discussed in a preceding chapter. It remains now to consider those cases in which the symptom is due to some form of obstruction, and also to point out various other causes of retention of urine which do not belong to the classification proposed.

The causes of obstruction to the passage of urine are tolerably numerous, but some of them are much more common than others. Thus, inflammation of the prostate, chronic enlargement of the same gland, and stricture of the urethra, are the causes most frequently met with, but besides these, cancer or polypus of the bladder, cancer of the prostate, a calculus or other foreign body in the bladder or urethra, the lodgment of coagula, laceration of the urethra, abscess of the prostate gland or in the perineum, extravasation of urine or blood in the perineum, may each of them produce an obstruction to the escape of urine, and thus cause retention.

A description of the majority of these affections, with the symptoms to which they give rise, will be found in various chapters of this book. It will here be sufficient briefly to glance at their symptoms, and to point out the means of distinguishing between the various circumstances under which retention of urine may occur.

When called to a case of retention of urine, the surgeon should inquire as minutely as possible into the patient's previous history. Did the retention come on suddenly, or had there been any previous difficulty in passing urine? Was there or had there been any discharge from the urethra? Was there any history of previous injury? These questions, together with others having reference to the nature and seat of pain, will usually elicit answers from which a preliminary diagnosis can be made as to the possible cause.

Inflammation of the prostate may occur during the course of gonorrhœa, and generally as a result of some indiscretion or dissipation on the part of the patient. In such a case the discharge usually ceases; pain, at first dull, but soon of a burning character, is felt at the neck of the bladder and in the perineum; the calls to pass urine are frequent and urgent, the pain being aggravated by each attempt as well as by movements of all kinds. Defecation also becomes very painful. These symptoms are accompanied by rigors and feverishness. The desire to pass urine becomes almost irrepressible, and complete retention may speedily supervene. On examination by the rectum the prostate is found to be swollen and exquisitely tender. The treatment of this form of retention consists in withdrawing the urine by means of an elastic catheter, and the administration of anodynes and saline aperients. Leeches, followed by hot fomentations, should be applied to the perineum. Perfect rest and low diet are of course to be enjoined. Abscess may form, and the pus usually finds its way into the urethra, less frequently into the rectum. It has been known to approach the surface of the perineum.

Chronic enlargement of the prostate gland is a frequent cause of retention of urine in elderly patients. After experiencing more or less difficulty for some time in passing water, the symptoms of retention suddenly supervene. On examination by the rectum some enlargement of the prostate may be detected, but inasmuch as the hypertrophy which causes the obstruction affects principally the middle lobe of the prostate, retention may occur from this cause without any great amount of enlargement being detectible by the finger. This affection and its treatment will be found fully described in a subsequent chapter. After an attack of retention the use of the catheter may be frequently required, as some time must elapse before the bladder can regain even its usual amount of contractile power. Malignant disease of the prostate may likewise cause retention of urine. The disease is, however, rare, and the history of the case will serve to distinguish it from other forms of enlargement. The symptoms are more rapid in their course than those of chronic hypertrophy; there is also frequent hæmorrhage and severe pain in addition to the constitutional symptoms of malignant disease. The induration is very considerable, and the inguinal glands are apt to be affected. Neighboring organs, the rectum especially, are apt to become involved. Stricture of the urethra is another frequent cause of retention of urine. The character of the obstruction may generally be determined by the history of the case, and the resistance experienced on instrumental examination will confirm the diagnosis. In cases of organic stricture, retention of urine may easily be caused by inflammatory swelling of the submucous tissue, a condition which may be produced by excesses of various kinds, or may supervene without any obvious cause. Spasm of the urethral muscles likewise aids in occluding the canal. The diagnosis of these cases is, as a rule, easily made, but the treatment may present great difficulties. The first step is to endeavor to introduce a catheter. A gum-elastic instrument should be first tried, and if this fails, a silver catheter should be employed, the greatest care being taken to avoid laceration of the urethra. If it be found impossible to pass an instrument, the effects of chloroform may be tried, or the patient may be placed in a warm bath, and a full dose of opium administered. After an hour or two, the irritation may be found to have subsided, and the spasm to have been relieved by the action of the opiate, and the patient may perhaps be able to pass urine without assistance, or if not, there may be little or no difficulty in introducing the catheter, which in that case may be allowed to remain.

In some cases the difficulty appears to be due rather to spasm than to inflammatory swelling, and for such Sir H. Thompson and others recommend repeated doses of tinct. ferri perchlorid.—15 or 20 minims every quarter of an hour, administered four or six times. The *modus operandi* of this empirical remedy, which was recommended long ago for so-called spasmodic stricture, cannot be explained, and as it is generally used in conjunction with other measures, such as the warm bath and opium, it is impossible to determine the amount of benefit which may be really due to it. Chloroform would appear to be especially indicated for such cases. Indeed, it may be tried in nearly all cases of retention of urine where no special circumstances forbid its administration.

Supposing, however, that all expedients fail, the patient's condition will soon become extremely urgent, and other steps must be taken in order to relieve him. The surgeon's choice usually rests between puncture of the bladder through the rectum, the same operation above the pubes, and cutting into the urethra through the perineum. The first method was strongly recommended by Mr. Cock, and it is the one that has been most generally adopted by English surgeons. In Ireland, however, according to Mr. Fleming,[1] the supra-pubic operation has been and is yet always selected. The necessity for the adoption of either method occurs but rarely. Mr. Fleming, during a very extended practice, has only twice had to puncture the bladder; in one case the retention was due to stricture, in the other, to malignant disease. Sir H. Thompson,[2] during a period of twenty years, has found it necessary to puncture the bladder twice for retention due to prostatic enlargement, and four times for retention from stricture. One of the former operations was a suprapubic puncture, all the others were by the rectum. The latter operation is simple and easy of performance. The instruments required are a curved trocar, seven or eight inches in length, and a canula. The rectum having been previously cleared out by means of an enema, the patient is made to sit on the edge of the bed, his back supported by pillows, and his legs kept apart by two assistants, with each foot resting on a low chair. One of the assistants makes firm pressure over the pubes, so as to press the bladder towards the rectum. The surgeon passes the forefinger of the left hand into the bowel, and feels for the prostate and the bulging of the bladder beyond it. The trocar, with its point retracted within the canula, is then passed along the upper surface of the finger, till the bladder is reached. The handle of the instrument is then depressed, and the point of the trocar is carried gently but firmly through the coats of the rectum and bladder, until it is felt to be free in the cavity of the latter. The trocar is then withdrawn and the canula is fixed in position with tapes. This operation is obviously unsuitable for cases in which the prostate is much enlarged, and when such a complication exists, the supra-pubic puncture may have to be resorted to.

---

[1] Injuries and Diseases of the Genito-Urinary Organs, p. 243.
[2] Diseases of the Urinary Organs, p. 122.

This operation was first performed by a French surgeon, Méry, in 1701. The patient having been placed on his back with his head raised, and his thighs slightly flexed, and the hair having been removed from the supra-pubic region, a vertical incision, about an inch and a half in length, is made through the parts in the median line down to the linea alba. Mercier[1] recommends that this preliminary incision should be omitted, on the ground that the division of the muscular and fibrous tissues only increases the risk of infiltration of urine. He directs that the surgeon, standing on the patient's right side, should place his hand over the hypogastric region, and make firm pressure with the thumb and forefinger at each side of the spot where the trocar is to be plunged in. This instrument, which should be long and curved, having been well oiled, is held in the right hand, its concavity directed towards the pubes. Its point is then plunged in at right angles to the vertical axis of the body, in the middle line, about an inch and a quarter above the pubes, and the point of the instrument, while passing through the wall of the abdomen, should be directed towards the posterior aspect of the bone. Penetration of the bladder will be indicated by absence of resistance. The trocar should then be withdrawn, the canula being kept in its place by the thumb and forefinger of the left hand. The patient should be placed on his side, in order to facilitate the escape of urine. The length of time during which the canula is allowed to remain in position must be determined by the circumstances of the case. A gum-elastic catheter may be passed down the canula, and this latter withdrawn, whenever it seems desirable to provide for the escape of urine by the wound for a lengthened period. In cases where the abdominal wall is thin and devoid of fat, a preliminary incision may not be requisite, but where there is any thickness of tissue between the integument and the bladder, the previous use of the scalpel will facilitate the passage of the trocar.

There is another operation which may be performed for the relief of retention of urine due to impermeable stricture, and it is one that has the advantage of dealing effectually and curatively with the cause of the symptoms. The operation consists in cutting down upon the urethra from the perineum, and it is termed "external urethrotomy," or "perineal section." It is distinguished from Syme's operation by the fact that for the latter the stricture must admit of the passage of a grooved staff which serves as a guide for the knife. When, on the other hand, the stricture is impassable, the extremity of a staff which has been passed as far as possible indicates the position of the urethra at the commencement of the stricture. The operation is thus performed. The patient is placed in the lithotomy position, and a large grooved staff is passed down to the obstruction in the urethra. If a false passage is known to exist, especial care is necessary to keep the point of the staff in the urethra itself. A free incision is then made in the raphé of the perineum, either from above downwards, towards the apex of the staff, or, the index finger of the left hand being placed in the rectum, the knife may be inserted with its edge upwards, and made to cut its way out towards the scrotum. The extremity of the staff having been thus exposed, the surgeon dissects through the thickened and indurated tissue, and endeavors to find the part of the urethra posterior to the stricture. In order to facilitate the dissection, the sides of the wound should be widely separated by means of tenacula, or loops of silk, one passed through each

---

[1] Anatomie et Physiologie de la Vessie, p. 17.

margin of the urethra, where that canal has been opened just in front of the obstruction. An attempt should also be made to pass a small filiform guide through the constriction, and if this can be done, the division of the tissues can be more readily and safely effected. If, however, no guide can be passed, the knife must be kept in the middle line, and the supposed direction of the urethral canal must be followed as closely as possible. The stricture having been divided, a full-sized catheter is to be passed down the urethra into the bladder. During the subsequent progress of the case the catheter must be retained until cicatrization has taken place, the instrument being changed every few days. For some time after the operation, undue contraction of the parts must be prevented by the frequent introduction of the instrument.

The bladder may also be tapped through the pubic symphysis. This operation was first proposed by Dr. Brander, of Jersey, in 1825, and it has been performed successfully in several instances. A hydrocele trocar and cannula, of medium size, is the instrument required. The patient should lie on his back and the trocar is introduced about the centre of the symphysis, and at right angles with the body. It should be passed somewhat obliquely downwards and backwards towards the sacrum. When the bladder is reached, the trocar is to be withdrawn, and a piece of flexible catheter is then to be introduced through the cannula and retained by a tape. The difficulty of the operation consists in the penetration of the inter-pubic cartilage, which in advanced life becomes firmly ossified. Failure arose from this cause in a case reported by Sir H. Thompson, who explains that "puncture through the symphysis, which is solid bone in an elderly man, blunts the trocar so much that when the point arrives at the soft tissues and bladder on the other side, it will not penetrate but pushes them away. At least this is what happened in three experiments made on the dead body, for the purpose of observing the result." This difficulty, however, might be obviated by drilling a hole through the symphysis by means of a suitable instrument.

A fourth method of puncturing the bladder has been suggested by a French surgeon, M. Voillemier. Meeting with a case in which all other means of relieving the bladder appeared peculiarly hazardous and unlikely to prove unsuccessful, he conceived the idea of passing a trocar through the suspensory ligament of the urethra. This, the so-called subpubic puncture, is a far more difficult and dangerous operation than any of the other methods. For its performance, the patient should be placed on his back with the pelvis raised and the thighs slightly flexed. An assistant draws the penis downwards and backwards so as to put the suspensory ligament on the stretch. The operator, having carefully examined the subpubic arch, plunges a curved trocar through the integuments and suspensory ligament, and passes it under the pubes onwards into the bladder. M. Voillemier's operation does not seem to have been repeated. It has no advantages over the other methods, and is obviously more difficult. There is often a considerable plexus of veins at the spot where the trocar would enter the bladder.[1]

There is yet another method which may be adopted for relieving a distended bladder. It may be punctured with a capillary trocar above the pubes, and its contents withdrawn by means of Dieulafoy's pneumatic aspirator. One advantage of this method is the avoidance of infiltration of urine, but it is liable, on the other hand, to certain objections. It

---

[1] Mercier, loc. cit., p. 13.

would have to be performed as often as the bladder became distended with the urine, and frequent repetitions of the operation might produce serious results, and would, moreover, be exceedingly troublesome both to patient and surgeon. In cases also in which the urine contained much tenacious mucus or epithelial débris, it would often be impossible to empty the bladder through a capillary tube.

Of the above-described methods of relieving the bladder, puncture through the rectum and external urethrotomy are the most suitable operations for the majority of cases. The supra-pubic operation may be had recourse to when the prostate is much enlarged. The indications for this procedure will be found described in the chapters on diseases of the prostate.

The other causes of retention of urine may be very briefly discussed. Malignant and polypoid tumors of the bladder occur but very rarely, and the history of the case will usually indicate the cause of the symptoms. Stone in the bladder sometimes causes retention, which, however, would at once be relieved by the passage of a catheter. The lodgment of a calculus in the urethra may also be a cause of retention. If in front of the prostate, its presence will be detected by carrying the finger along the urethra. Steps must then be taken for its removal, and various kinds of forceps are applicable for this purpose. If the calculus be impacted in the prostate, it must be removed by an incision similar to that for the lateral operation of lithotomy.

Abscess of the perineum, if of sufficient size to cause retention of urine, would be easy of diagnosis, and all the symptoms would be relieved by an immediate and free incision. Laceration of the urethra would also be easily discoverable if attention were paid to the symptoms and history of the case. The treatment consists in the immediate introduction of an elastic catheter, or, if this be impossible, an incision should be made into the urethra from the perineum.

The above are the most frequent causes of retention of urine in male patients. It is necessary to add a few remarks on the causes which may produce a similar condition in females. Retention of urine is an occasional manifestation of hysteria. Although diuresis is the most common symptom in such cases, hysterical patients not unfrequently declare that they have lost the power of voiding urine, and that none has been passed for several days. The main characteristic of these symptoms is that they are irregular and intermittent, and the patient experiences little real suffering in the region of the bladder, unless that organ has become greatly distended.

The treatment of such a case is obvious. The catheter must not be introduced too early. Indeed, these cases, if left to themselves, usually recover; but, if we once begin to pass the catheter, its use will be for a long time required, and the complaint will be protracted. Mr. Coulson met with a case of this kind, in which the catheter was introduced once or twice a day for some time, until at last a suspicion was excited that the malady was either feigned or nervous, and that it would be desirable not to introduce the instrument. After this, the young woman left the hospital, but it was ascertained that the affection of the bladder soon subsided. As a general rule, in these cases, the catheter ought not to be introduced. Care, however, must be taken, lest the bladder become truly paralyzed from over-distention. Morbid changes may then ensue, as in other instances of similar disease.

Other causes of retention of urine are disease or displacement of neighboring organs. Thus it may be due to retroversion of the gravid uterus. The accident is most likely to occur at or about the period of quickening, or shortly before delivery. The cervix uteri is turned upwards and forwards, and presses firmly against the urethra, the orifice of which is dragged upwards and backwards. The symptoms may come on suddenly, or slowly and gradually. Stricture of the urethra is of rare occurrence in the female, but it has been occasionally met with as a result of gonorrhœa or syphilis, or of injury to the urethro-vaginal septum, and it might cause retention of urine.

The same symptom has, in very rare instances, been caused by an imperforate hymen, which, by totally preventing the escape of the menses, had mechanically obstructed the urethra.[1] The various tumors of the bladder and of the urethra, and the pseudo-membranous formation which takes place in diphtheritic cystitis, may also obstruct the urethra and produce retention of urine.

---

[1] A case of this kind has been recorded by Mr. Coley, *Lancet*, 1833, p. 395.

CHAPTER XVII.

HÆMATURIA.

HÆMATURIA is a symptom common to many diseases of the urinary organs. The term merely signifies that blood, in greater or less quantity, is mixed with the urine and escapes during micturition. Hæmorrhage may arise from any portion of the urinary tract, and in all cases it is important to ascertain the source of the bleeding and the cause to which it is due.

Urine containing blood presents numerous diversities of color. These mainly depend upon the amount and character of the hæmorrhage. If only a small quantity of blood be present, the urine, to the naked eye, may appear unchanged; a larger quantity gives it a reddish hue, and if the hæmorrhage be profuse, the discharge may be almost pure blood. If a moderate quantity of blood has remained for some time mixed with the urine, the latter fluid often assumes a smoky tint, the depth of which varies according to the quantity of blood present. The urine may be uniformly tinged by the blood, or while the first portions that escape are normal in color, the last few drops may be considerably altered. The manner in which the blood is mixed with the urine is a point of importance for diagnostic purposes.

The microscope affords the surest means by which blood may be discovered in the urine. In true hæmaturia, blood-corpuscles are always to be found. These undergo variations in form and size according to the state of the urine and other circumstances. They very rarely form rouleaux, as is the case when the blood is obtained directly from the vessels. If the hæmorrhage be profuse, and the blood be discharged from the bladder soon after its escape from the vessels, the corpuscles will appear normal in size and form. If, on the other hand, they have been exposed for some time to the action of the urine, they will be found roundish or globular in shape, and brownish in color, and if the urine be very dilute, the corpuscles expand from imbibition, and appear " as pale circles with sharp, delicate outlines, and without any appearance of cell-contents. If the urine be more concentrated, they preserve more nearly their normal biconcave contour, and appear smaller and more deeply shaded. Sometimes they shrink and crumple and become mis-shapen in various ways."[1] When the urine is very concentrated, they sometimes become stellate in form. According to Ultzmann,[2] the blood-corpuscles are generally found to be much diminished in size when a small quantity of blood has remained for some time within the body mixed with a considerable quantity of urine. The action of the urine upon the blood-corpuscles is, first, to remove their red color, and secondly, to cause them to break up into smaller globular bodies of varying size.

---

[1] Roberts, Urinary and Renal Diseases, p. 129.
[2] Anleitung zur Untersuchung des Harnes, von K. B. Hofmann und R. Ultzmann.

Dr. Roberts states that "the marks by which blood-corpuscles are distinguished from other cells found in urine, are, the extreme tenuity of their outline, the absence of visible cell-contents, and especially of a nucleus, and their feeble refractive power. When the biconcave form is preserved, this, of course, is diagnostic." By these characteristics, they may be distinguished from confervoid sporules, the minute discoid forms of oxalate of lime and the nuclei of renal epithelium—bodies with which they are liable to be confounded.

As a matter of course, albumen is present in urine which contains blood-corpuscles, and may be detected by the ordinary tests. Other tests for blood in urine are the discovery of the coloring matter by means of the spectrum apparatus; very minute quantities of blood may be thus detected. Hæmoglobin in solution gives rise to two dark streaks in the yellow and green of the spectrum, between Fraucnhofer's lines D and E. This appearance is characteristic of the coloring matter of the blood. Another test is that devised by Heller, which consists in precipitating the phosphates by means of a solution of potash. These carry down with them the coloring matter of any blood that may be present. Crystals of hæmatin may also be discovered in the sediment by means of Teichmann's process.

It must not be forgotten that certain medicinal substances, taken internally, impart a red color to the urine. Rhubarb and senna produce this effect in patients whose urine has an alkaline reaction. The distinction between such a coloration and that due to blood is easily made. In the latter case, albumen and corpuscles will be detected in the urine, and the addition of nitric acid will cause no change of color; but when the redness is due to the coloring matter of rhubarb or senna, it will be removed by adding an acid, to re-appear on pouring in an excess of alkali.

In some cases of jaundice, the urine is brown-colored, owing to the presence of the coloring matter of the bile in an altered form. Such urine, however, contains no albumen, and if half its volume of sulphuric acid be added to it, it becomes opaque and deep-black in color.[1] Blackish urine is also occasionally found in cases where wounds have been treated antiseptically by means of carbolic acid. Such urine, when heated with half its volume of sulphuric acid, gives forth the characteristic odor of carbolic acid.

Hæmaturia being merely a symptom, it is to be considered in connection with those obvious divisions of the urinary tract from which the hæmorrhage may arise. Thus we shall consider the subject under the following heads: (1) Hæmorrhage from the urethra. (2) Hæmorrhage from the prostate and neck of the bladder. (3) Hæmorrhage from the bladder itself. (4) Hæmorrhage from the kidneys. (5) Hæmorrhage of a general character from the urinary tract.

Hæmorrhage from the urethra may be the result of violence or of disease. Thus it may be due to injury to the penis from blows or wounds; laceration by instruments; rupture of the corpus spongiosum during violent coitus or chordee; an impacted calculus, etc. It may also occur during severe gonorrhœa, stricture, ulceration, as in syphilis, and as a symptom of urethral tumors. Hæmorrhage from the urethra may be recognized by the fact that it occurs independently of micturition; the blood trickles more or less continuously from the urethra, and the quan-

---

[1] Ultzmann, Ueber Hæmaturie, p. 117.

tity can be momentarily increased by making pressure along the course of that canal. When the patient passes urine, the first portion comes away more or less tinged with blood, and often containing long, worm-like red coagula, while the last portions may be almost, if not quite normal in color. The same appearance will be observed if retention takes place, and a catheter has to be used. When the hæmorrhage occurs in connection with gonorrhœa, more or less pus will be found mixed with the blood.

Hæmorrhage from the prostatic portion of the urethra and neck of the bladder is characterized by the blood manifesting itself at the end of the act of micturition. Sometimes only the last few drops of urine are tinged with blood. There is no continuous trickling of blood from the urethra, and none can be made to appear by pressing with the finger along the canal. If a catheter be introduced into the bladder, the urine which first escapes is free from blood, which, however, shows itself as the catheter is being withdrawn. This form of hæmorrhage sometimes occurs as a symptom of chronic prostatitis, the result of gonorrhœa. The bleeding is very slight, and occurs at the end of micturition, or the last portion of the urine may be slightly tinged with blood. In addition, the urine is somewhat cloudy, owing to slight cystitis, micturition is frequent, and there is more or less pain at the end of the act. Hæmorrhage from the prostate and neck of the bladder may also be due to ulcerations, injuries, the lodgment of calculi or prostatic concretions, and may also be symptomatic of prostatic hypertropy or malignant disease.

Hæmorrhage from the bladder or kidney is distinguishable from hæmorrhage due to urethral or prostatic sources by the fact that the blood is uniformly mixed with the urine. In renal hæmorrhages especially, and in the majority of vesical hæmorrhages, this characteristic is extremely well marked. It is, however, often difficult to determine the exact source of the blood, but an examination of the urine will assist in the formation of a correct diagnosis. Much importance was formerly attributed to the reaction of the urine as affording a means of distinguishing between renal and vesical hæmorrhage. The reaction of the urine being alkaline, the bleeding was supposed to have its origin in the bladder; with an acid reaction, renal hæmorrhage was diagnosed. This distinction, however, is not true of all cases, for with profuse hæmorrhage either from the bladder or kidneys, the natural alkali of the blood masking any acidity of the urine, will communicate to the mixture an alkaline reaction. A copious secretion of pus, as in calculous pyelitis, mixed with a small quantity of blood and acid urine, may have the same effect, and, on the other hand, when the quantity of blood is slight, and no catarrhal state is present, vesical hæmorrhage may co-exist with an acid condition of urine. If, however, blood be found in alkaline urine, such a reaction being due to carbonate of ammonia, the hæmorrhage is, in all probability, of vesical origin.

The color of the urine is another point of importance. When the hæmorrhage is profuse, the blood will often be passed, with more or less urine, soon after its escape from the vessels, and before any change has taken place in its color. Under these circumstances, the bladder is the probable source of the hæmorrhage. On the other hand, urine containing blood may have a more or less brown, reddish brown, or almost black tint, and these alterations in color are due to the changes which contact for some time with urine has produced in the hæmoglobin of the corpuscles. These changes take place when the blood has remained for some time mixed with the urine, and either the kidneys or the bladder may be the source of the hæmorrhage. Thus a florid red color in hæmaturia is indicative of vesical mischief, while a brown or blackish tint shows only that the blood and urine have been for some time in contact, and it may occur both in vesical and renal hæmorrhage. A light brown, or smoky tint, however, is commonly indicative of renal affections.

The examination of any coagula that may be contained in the urine will often greatly assist the diagnosis. If these are the products of recently effused blood, they will be soft in consistence and dark-red in color; light-colored coagula, on the other hand, indicate that the hæmorrhage is not recent, and that the coloring matter has been removed by the urine. As regards the shape of the clots, long, worm-like coagula are usually of renal origin, whereas those resulting from vesical hæmorrhage are, for the most part, roundish or irregular in shape. The long, cylindrical, worm-like clots indicate that the source of the hemorrhage is the kidney, or the pelvis of the kidney; the coagula themselves are formed in the latter situation, or in the ureters. Blood, however, from the kidney, may form more or less irregularly-shaped coagula within the bladder, so that these are not characteristic of vesical hæmorrhage alone. Coagula do not always form when blood is effused into the urinary passages; they are not found when the blood is in relatively small proportion to the urine; on the other hand, they are generally found when the blood is in excess, a condition which is most likely to occur when the pelvis of the kidney or the ureter is the source of the hæmorrhage. In like manner, coagula are most prone to form in the bladder when a large quantity of blood is mixed with a small quantity of urine.

The microscopical examination of the sediment will often determine the source of the hæmorrhage. If the blood come from the substance of the kidney, tube-casts composed of blood-corpuscles or fibrin, with renal epithelium more or less deeply browned by the hæmatin, will be found in the deposit. In malignant disease of the kidney, the quantity of blood will usually mask all other structures. There is often nothing characteristic microscopically of hæmorrhage from the bladder. If catarrh, however, co-exist, there may be pus-cells in abundance, epithelium from the bladder, and crystals of the triple phosphate.

The sources of vesical hæmaturia are: wounds, new formations of various kinds, calculus in the bladder, ulceration, whether simple, tuberculous, croupous or diphtheritic, acute congestion, as in cystitis, parasites, a varicose condition of the neck of the bladder.

Vesical hæmorrhage, due to wounds, is generally easy of diagnosis. The history of the case, the presence of a wound of the abdominal parietes, perineum or rectum, or a fracture of the pelvis, together with escape of blood from the urethra, or bloody urine drawn off by the catheter will indicate the source of the hæmorrhage.

Hæmaturia is a marked symptom of some forms of tumor which occur in the bladder. These have been described in a previous chapter, and will therefore now be discussed only in connection with the symptom under consideration. The simple fibrous growths, which are very rare and occur chiefly in the form of polypi, do not necessarily give rise to hæmorrhage. If, however, their surface should become ulcerated, blood would appear in the urine.

Hæmaturia is a constant symptom of epithelioma, and the hæmorrhage is often profuse, and continues even when the patient is at rest. In addition to the blood-corpuscles, numerous small epithelial cells are sometimes to be found in the urine. These are very irregular in shape, or furnished with one or more prolongations. The nucleus is large and distinct, and each cell may contain several nuclei. According to Ultzmann,[1] the presence of these cells, co-existing with hæmaturia and vesical catarrh, is strongly suggestive of epithelioma.

In a third form of tumor to which the bladder is liable, hæmaturia is the most prominent symptom. The so-called "villous growth" is for the most part of a non-malignant nature, but cancerous tumors may be covered by a growth of a similar character. In either case blood appears frequently or constantly in the urine, the hæmorrhage occurring independently of bodily exercise or movement. The urine may appear like pure blood, or it may be brownish red or blackish in color. These variations depend upon the quantity of blood, and the length of time during which it has remained in contact with the urine. Pus-corpuscles may also be present.

The urine in these cases sometimes contains a larger amount of fibrine than corresponds with the quantity of blood apparently present. This condition has been called "fibrinuria." Dr. Ultzmann,[2] who has given a description of it, states that he has met with this symptom in three cases of villous tumor of the bladder. The urine, when passed, is quite fluid, and of a reddish yellow color, but after standing for a few minutes, it coagulates into a gelatinous mass, which adheres to the containing vessel. On being stirred, however, for some time, the fluid state is restored. Urine of this kind, as shown by its color, does not contain much blood, and it begins to coagulate almost immediately after it has been passed. In Dr. Ultzmann's cases the fibrinuria was accompanied by tenesmus and retention of urine, and the symptom occurred at an early period of the complaint. He accounts for its appearance by supposing that the vessels supplying the muscular coat are subjected to great pressure during the violent spasmodic contractions of the bladder. The veins being more affected than the arteries, congestion occurs in the vascular loops of the villous tissue. If the tension is very great, the walls of the vessels give way, and considerable hæmorrhage takes place. With a less degree of tension, only blood-plasma containing fibrine is effused, and this, mixing with the urine, coagulates on its escape from the bladder. For the same reason, in cases of villous tumor, the urine always contains more albumen than corresponds to the blood or pus in the sediment. This predominance of albumen might lead to the suspicion of an affection of the parenchyma of the kidney. In this latter case, however, the microscope would reveal the existence of numerous casts of tubes.

In these cases of villous tumor, besides blood and pus-corpuscles, the sediment often contains coagula and fragments of villous tissue. The former are darker in color than the latter, but the coagulated blood sometimes contains fragments of tissue, which can be recognized as "villous" under the microscope. Dr. Ultzmann, however, points out that the urine in these cases seldom or never contains such distinct fragments as are depicted in engravings. A characteristic and well-defined shred may come away in the eye of the catheter, but the shreds which appear in the urine, having been detached by a process of ulceration or

---

[1] Ueber Hæmaturie, Wiener Klinik, May, 1878.
[2] Ibid. The above account of fibrinuria has been taken from Dr. Ultzmann's exhaustive Essay.

mortification, are always more or less changed in appearance. The disintegration of the epithelium renders it difficult to distinguish the separate villi, and the meshes of the tissue often contain pus and blood-corpuscles and bacteria. Crystals of hæmatoidin may also be discovered in the shreds, as well as small colorless spheroidal crystals, which Dr. Ultzmann believes are composed of oxalate of lime. In an alkaline condition of the urine any shreds passed are likely to be coated with crystals of the ammoniaco-magnesian phosphate.

With regard to hæmaturia as a symptom of villous growths, the hæmorrhage is often very profuse, especially at the commencement of the disease, and the blood is intimately mixed with the urine. Later on pus-corpuscles also appear, but blood-corpuscles are never absent from the sediment even though the color of the urine may be almost normal.

Varix of the bladder is another affection of which hæmaturia is a symptom. This condition, which consists in venous enlargement of the vesico-prostatic plexus of veins, is somewhat rare, and is said by Dr. Gross to be generally associated with stone in the bladder, prostatic enlargement, stricture, or some other obstruction to the flow of urine. Dr. Ultzmann, on the other hand, states that he has met with several cases in which varix of the bladder co-existed with hæmorrhoids, the hæmaturia alternating with discharge of blood from the piles. His patients were in fair health, and the hæmaturia was the only symptom connected with the urinary organs. The bleeding comes on suddenly, and recurs at long intervals and is generally profuse. The diagnosis can be made only *per viam exclusionis*. There are none of the other symptoms of malignant disease or stone in the bladder, and nothing can be detected on sounding. During the intervals between the attacks the urine is normal.

Hæmaturia is a frequent symptom of stone in the bladder, but the quantity of blood lost is not so great as in cases of villous and other growths. Hæmaturia due to calculus is the result of injury to the mucous membrane of the bladder, and, as a symptom, it is especially characterized by appearing (when it occurs at all) after exercise, and subsiding with rest. The more violent the exercise, the greater the quantity of blood. When the hæmorrhage is only slight, the urine may be unchanged in color, but blood-corpuscles will always be discoverable in the sediment. Sometimes a few drops of almost pure blood are passed at the end of the act of micturition. This is especially likely to occur when the bladder is inflamed, and the stone is rough and hard. Other symptoms will, of course, be present, and their cause will be positively determined by sounding.

Hæmaturia, in certain countries, is caused by the presence of a parasite in the bladder. An affection of this nature is endemic in Egypt, the Mauritius, Madagascar, the Brazils, and some parts of Southern Africa. The parasite was discovered by Dr. Bilharz in 1851, and has been named the Bilharzia hæmatobia. Examples of this affection in persons who have returned from one or other of the above-mentioned countries sometimes come under the notice of practitioners in England. Dr. Harley has met with three cases, and Dr. Roberts[1] gives some details of the case of a groom, under his observation, who had been in the service of the Viceroy of Egypt. The parasite infests both the urinary and intestinal organs. It is found in the bladder, pelvis of the kidney and ureters, and

---

[1] Urinary and Renal Diseases, p. 586.

produces serious mischief in these localities. It is supposed to find its way into the body either through the intestinal mucous membrane or through the skin. Having passed into the veins, it appears to effect a permanent lodgment in those of the bladder, where it deposits its ova. These, by their presence, obstruct the vessels, and produce various morbid changes in the coats of the bladder. Portions of mucous membrane become detached, blood is effused on the internal surface and between the coats of the bladder, and chronic catarrh, with copious secretion of pus, is developed. The hæmaturia is the most prominent symptom. The sediment of the urine contains blood and pus-corpuscles, mixed with coagula and shreds of tissue in which latter may be found imbedded the ova and free embryos of the parasite. The eggs are oval bodies, about $\frac{1}{140}$th of an inch long, rounded at one end, and having a spiny projection at the other. Their contents are granular. The free embryos are furnished with cilia. In addition to causing hæmaturia, the ova and the fibrinous coagula often form the starting-points of urinary concretions, and the frequency of calculous disease in Egypt may be accounted for by the prevalence of these parasites.

The remaining vesical affections of which hæmaturia is a symptom are the tuberculous, croupous and diphtheritic forms of inflammation. In tuberculous disease, the diagnosis must be based upon other co-existing symptoms. The complaint is almost invariably associated with a similar affection of other organs, and the symptoms referable to the bladder are those of chronic inflammation and ulceration with copious discharge of pus and débris. Croupous and diphtheritic cystitis are of very rare occurrence; they sometimes accompany similar forms of inflammation of other mucous membranes. The hæmaturia follows the detachment of the sloughs, portions of which will be found in the sediment of the urine. It is this form of inflammation which is produced by turpentine and cantharides, and is occasionally seen after severe instrumental labors.

It now remains to say a few words upon hæmaturia as a symptom of various renal affections. Hæmorrhage from the kidney may be caused by external injury, or by the irritation due to calculous concretions. It is also a symptom of hyperæmia of the organ; of Bright's disease, acute and chronic; and of cancerous and tuberculous disease affecting the kidney. In hæmaturia from any of these causes the blood is usually intimately mixed with the urine, and the color of the mixture will depend mainly upon the quantity of the blood. A more or less deep smoky tint is the usual alteration when the blood is in moderate or small quantity; but when the hæmorrhage is profuse, as it may be after a severe injury, and in cases of malignant disease, the urine may resemble pure blood, or may be blackish-brown or chocolate color. The presence of casts of tubes in the deposit indicates that the blood comes from the substance of the kidney. These structures are found in the various forms of Bright's disease, and in cases where the hæmorrhage is due to injury. When the hæmaturia is due to a calculus in the pelvis of the kidney, it will be accompanied by the symptoms, more or less marked, of nephritic colic, and, in common with these, will be increased by exercise. The presence of long, cylindrical worm-like coagula indicates that the source of the hæmorrhage is the kidney or its pelvis. Malignant disease of the kidney does not always cause hæmaturia. This symptom was absent in 28 out of 59 cases alluded to by Dr. Roberts.[1]

---

[1] Urinary and Renal Diseases, p. 519.

When it occurs it is usually profuse, and the urine is blood-red, brownish-red, or dark red in color. The presence of a tumor in the loins, and the signs of cancerous cachexia are the other important symptoms upon which a diagnosis is to be based. In tuberculous disease of the kidney, the hæmorrhage is usually slight, the blood being present in quantity sufficient only to cause the urine to assume a reddish-yellow tinge. The hæmaturia is not increased by exercise, and any improvement in the patient's health is usually followed by a diminution in the quantity of blood in the urine. The other symptoms are those of chronic pyelitis and chronic cystitis, and the presence of tuberculous deposits in other organs can generally be demonstrated. In addition to the blood, the urine contains pus-corpuscles and débris of various kinds.

The last form of hæmaturia is that which occurs as a symptom of scurvy, hæmophilia, severe pyæmia, and putrid fevers, and after the inhalation of certain poisonous gases. The blood-corpuscles themselves do not appear in the urine, but undergo disintegration while in the vessels and yield up their coloring matter, which becomes changed by coming into contact with the urine. The symptom in question has been described as "hæmatinuria," or "false hæmaturia." The urine contains albumen, but no blood-corpuscles, and may be reddish-brown, or brownish-black in color. Any epithelial cells or débris contained in the sediment will be similarly tinged.

Hæmaturia being merely a symptom, the treatment, generally speaking, will be that of the disease or condition to which the hæmorrhage is due. It is, however, sometimes impossible to fulfil the causal indication, and when the hæmorrhage is very profuse it may have to be treated independently of the original disease. The discovery of the cause is, however, of the greatest importance in all cases, and this can generally be effected by a careful observation of the symptoms and an examination of the urine. When the primary object is to check the bleeding, the treatment mainly consists in the administration of certain astringent remedies, both by the mouth and by injection into the bladder, in the use of cold applications, and in keeping the patient absolutely at rest. In hæmorrhage, whether renal or vesical, due to calculus, rest in the horizontal position is the most effective method of treatment. Cold applications are especially indicated when the bladder is the source of the hæmorrhage. Bags of pounded ice over the pubes or to the perineum may be employed, or iced water may be injected into the rectum, or a piece of ice may be introduced into the bowel. Opium should be given internally in order to restrain the painful contractions of the bladder. It is undesirable to introduce a catheter so long as urine flows and there is no absolute retention. The coagulated blood helps to check further flow, and if left to itself will be dissolved by the urine. If retention occur, and especially if, owing to the nature of the original affection, the patient has been unable for some time past to void urine otherwise than by catheter, it may become necessary to evacuate the bladder. A full-sized catheter should first be tried, and if no urine issues, a Clover's exhausting syringe may be adapted to the catheter and an attempt made to withdraw portions of coagula. The utmost gentleness must of course be practised, as there is great danger of exciting fresh hæmorrhage.

The internal remedies which have been found most serviceable are alum, gallic acid, acetate of lead, ergot of rye, tincture of iron, turpentine, sulphuric acid and matico. These substances all possess styptic properties, and, generally speaking, they are more beneficial in renal than in vesical hæmorrhage. Dr. Ultzmann recommends the subcutaneous injection of extract of ergot of rye. His formula is three parts by weight of the extract to fifteen parts of a mixture of glycerine and water ; half a fluid drachm to be used at a time. Alum may at the same time be given internally in doses of ten grains with fifteen minims of dilute sulphuric acid. Sir H. Thompson recommends infusion of matico, two ounces every three or four hours if the bleeding is considerable, and for the local treatment of vesical hæmorrhage a weak solution of nitrate of silver, gr. i. to ℥ iv. water, or tincture of the perchloride of iron, ℨ i. to the same quantity. These latter remedies must be used with the greatest gentleness ; they are especially suitable for cases in which the hæmorrhage is due to the presence of a tumor in the bladder. The internal use of iron and quinine is indicated where there is a general predisposition to hemorrhage, and, in the majority of cases, opium may be added with advantage to other remedies.

CHAPTER XVIII.

THE CHEMISTRY AND STRUCTURE OF URINARY CONCRETIONS.

THE knowledge of urinary concretions in early times was very limited. Pliny, indeed, enumerates many remedies for stone and gravel; but these are commonly of a very ineffectual character. Paracelsus conceived that urinary concretions possessed an analogy with the tartar which urine deposits. He taught that a saline spirit unites with an earthy principle which is always present in the fluids of the body, and that by this union tartar or calculus is produced. Though Paracelsus in many respects was in advance of his age, especially on points of chemistry, he certainly added little, by these speculations, to the knowledge before possessed of urinary calculi. Van Helmont made the important observation, deduced from experiments, that urinary concretions are essentially different from stony bodies. He added, that such concretions are not derived from matter contained in the food and drink. He revived an experiment made some centuries before by Raymond Lully, viz., mixing what was then called spirit of urine, namely, solution of carbonate of ammonia, with rectified spirit of wine, by which a white precipitate was obtained. This white precipitate was termed the offa Helmontiana, and was plainly the carbonate of ammonia thrown down owing to its insolubility in spirit. After this period more correct views began to be entertained as to the nature of urinary concretions. Hales confirmed what Van Helmont had taught as to the great difference between common stones and urinary concretions, by showing how large a quantity of air the latter gave off under the action of heat. He directed his attention towards the discovery of a universal solvent of these concretions,—an attempt in which he necessarily failed. Subsequently, Boyle, Whytt, Alston, and Slare, engaged in the investigation of this subject. Whytt proposed lime-water as a remedy in calculous disorders. Alston showed that though lime-water is to a certain extent a remedy, yet that it is not a solvent of these concretions. The tendency of the urine to concrete, so as to encase foreign bodies accidentally introduced into the urinary passages, was fully investigated, and numerous histories of this character, in both sexes, were given during the last century. Early in the same century correct views of the nature of urinary concretions began to be entertained in the Leyden school. It was taught that the nuclei of such concretions are derived from the kidney, or from the bladder. Van Swieten held that the rudiments of calculi exist in the urine of the most healthy; that such concretions form by the coalescence of like elements; that if the urine be passed before the tendency to this coalescence of elements takes place, there is no concretion; that this tendency is exerted more or less slowly in different individuals, and that under this view the first concretion may take place either in the kidney or in the bladder; finally, that there is no ground for fear unless there is a too rapid tendency to the concretion of the elements of calculi within the urinary passages. Van Swieten pronounced those fortunate whose urine shows no such tendency to rapid deposit after being passed, and congratulated himself on being of the number.

In 1776, Scheele discovered uric acid. All the calculi that he examined contained this substance. He describes it as a peculiar substance, possessed of acid properties; soluble with effervescence in nitric acid, the solution affording a pink color when evaporated, without too much heat, to dryness. This substance was first termed lithic and afterwards uric acid. Scheele found all urine to contain uric acid, thus confirming what Van Swieten had remarked as to the presence of the elements of calculi in urine. Soon after, Bergman discovered lime in certain calculi. After this period little advance was made until Wollaston published his treatise in 1797. In his paper three calculi are described, besides that which Scheele discovered, namely, the fusible, the mulberry, and the bone-earth calculi. The mulberry calculus had been noticed by surgeons for some time, owing to its peculiarity of appearance. Smithson Tennant had remarked that the fusible calculus was not consumed by heat, like those examined by Scheele and Bergman; he found that this calculus passed into an opaque glass before the blowpipe, and conjectured that it contained phosphate of lime along with other salts. Wollaston showed that this calculus contained ammonio-phosphate of magnesia. He proved, besides, that oxalate of lime is contained in the mulberry calculus, also that the smaller calculi, termed hemp-seed calculi, contain the same substance. He also taught that the bone-earth calculus contains phosphate of lime, differing, however, from the phosphate of lime of the bones. A few years after, Fourcroy and Vauquelin made an important contribution to the chemistry of urinary calculi, though in their treatise no mention whatever is made of what Wollaston had discovered and taught. These chemists announced the presence of urate of ammonia and silica in urinary concretions. In 1810, Wollaston made known his discovery of cystic oxide as a urinary concretion. In 1817, Dr. Marcet published his valuable work on urinary concretions; in 1821, Dr. Prout, in his treatise, threw much light on this subject; while more recently, the works of Drs. Golding Bird, Bence Jones, Parkes, Beale, Thudichum, Roberts, Harley, M. Becquerel, Dr. Ultzmann and others have considerably added to our knowledge.

Urinary calculi, being more or less regularly deposited around a nucleus, usually assume the globular or ovoid form after remaining long in the bladder; but this shape is by no means universal, especially in the case of renal concretions, which are often very irregular, moulded to the pelvis of the kidney, and even branched like a coralline, while those found in the ureter are cylindrical. Even vesical calculi often affect shapes widely different from the globular or ovoid. In cases where several calculi exist at the same time in the bladder, some often exhibit flattened surfaces, produced by mutual attrition, which give them a polyhedral aspect. In other cases the globe or ovoid is compressed laterally, and changed to the amygdaloid shape, and in cases in which a portion of the calculus is embraced by a fold of the mucous membrane or encysted, the expost part continues to increase in bulk by successive deposits, so that ultimately the encysted portion forms a sort of pedicle to the entire calculus.

The character of the surface is as variable as the form; it may be either smooth, or even polished, covered with minute crystals, or more or less rough and tuberculated, as in the mulberry calculus.

The colors of calculi vary from white through pale yellow to brown, brownish-green, and even almost to black. Phosphatic calculi are often white; those of uric acid vary from yellow to brown; those of xanthic oxide having a cinnamon brown tint; while calculi of oxalate of lime vary from yellow to yellowish-brown, brownish-green, or even blackish-green.

The size and weight of calculi are even more uncertain. They ordinarily vary from a few grains to several ounces, the great majority being under 1 oz. in weight. The smallest which ever came under Mr. Coulson's observation, was removed by him in the presence of Mr. Buxton Shillitoe and others; it weighed only 5 grs. Dr. Gibson, of Richmond, United States, extracted one, the weight of which did not exceed 5 grs.; it was a phosphatic calculus, smaller than a middle-sized grain of coffee. Dr. Gross removed one of a similar weight, the patient being a boy, six years of age.[1] Phosphatic calculi are those which attain, as a general rule, the largest size; their increase is often very rapid.

The largest vesical calculus on record is said to have been in the possession of the French lithotomist, Morand. It weighed 6 lbs. 3 oz. An enormous stone weighing 51 oz. was long preserved in the hospital of La Charité, Paris. It was extracted after death, from the bladder of a poor curate, who died in that hospital, in the year 1690.

The largest calculus preserved in our collections is the one which Mr. Cline attempted to extract from the bladder of Sir William Ogilvie; it weighed 44 oz. and was 16 in. round one axis, by 14 round the other. Numerous cases are recorded where the stone weighed from 5 to 15 or 20 oz.; but how small a proportion they bear to the mass of cases operated on, may be gathered from the fact that, of the 703 calculi in the Norwich collection, weighed by Mr. Crosse, only two exceeded 6 ozs. in weight, and they were between 6 and 7 ozs.

The number of calculi existing at the same time in the bladder is subject to some variation. In the great majority of cases the calculus is solitary, but two or more may be found. Mr. Liston met with 7 out of 27 cases, in which more than 1 calculus existed. Klein met with 12 instances of 2 to 6 calculi in 79 operations. Mr. Crosse gives a list of 100 fatal cases of lithotomy, in 84 the calculi were single; in 7, 2 calculi were formed; in 6 cases, 3; in 2, 4 calculi; and in only 1 case, 5. Much greater numbers, however, have been extracted from the bladder; but it should be remembered that in such cases the calculi are small in comparison with their number. As many as 200 of these small calculi have been removed by Roux and Dupuytren. Sir A. Cooper extracted 243 at one time. Professor Eve, of the United States, removed 117 calculi by the lateral operation, weighing from one grain to three drachms. The case of a woman whose bladder contained 214 stones is recorded in the "Philosophical Transactions." It is stated that no less than 59 calculi were discovered in the bladder of the celebrated naturalist, Buffon; and Murat affirms, what one is forced to consider fabulous, that he found 678 calculi in the bladder, and nearly 10,000 in the kidneys of an aged patient. An authentic case, however, and the most remarkable on record, occurred to the late Dr. Physick, of the United States, who extracted more than 1,000 calculi from the bladder of Judge Marshall; they were of an oval shape, and varied from the size of a partridge shot to that of a bean.[2]

The internal structure discoverable on section is either uniform or

---

[1] Gross, loc. cit., p. 175.
[2] Gibson's Institutes, 5th edition, vol. ii., p. 220.

formed of concentric laminæ encircling the nucleus. In many calculi, in addition to the concentric laminæ, lines radiating from the centre are observed, as if the laminæ were composed of perpendicular crystalline fibres. In some calculi the concentric laminæ are easily separable, in others firmly adherent. In the latter case the laminæ are sometimes, as in the oxalate of lime calculus, indicated only by faint concentric lines.

The chemical composition of the nucleus and concentric coats may be either homogeneous, or the nucleus may differ from the superimposed strata; these again may be severally composed of different constituents, as in the alternating calculus, so that each concentric lamina must be subjected to a separate analysis.

As the chief constituent of healthy urine is urea, so does uric acid constitute the nucleus in the great majority of urinary calculi. If, as is now generally admitted, the phosphates be a secondary formation, and as Dr. Owen Rees asserts, oxalate of lime is to be regarded as uric acid, or urate altered after secretion, it would follow that calculous disease in the human race may be reduced to the deposit of a single element. "The other constituents," observes Dr. Rees, "are of such rare occurrence, that were it not for uric acid, calculus would be less frequently met with than tetanus."

However this may be, the predominance of uric acid, as a nucleus, is fully established. Of 212 specimens in the Museum of Guy's Hospital, examined by Dr. G. Bird, 128 contained a nucleus of uric acid or urates. Of 71 calculi in the Museum of the Transylvania University, 44 were formed round nuclei of the urates or uric acid.

Of Dr. Bird's cases, the oxalate of lime and phosphates form the nuclei in 69 specimens—in the Transylvania Museum they constituted 22; hence, out of the whole number of 283 specimens, we have only 19 in which the nuclei were formed of cystine and by foreign bodies, where the calculi were mixed. Round the uric acid nucleus is often deposited a layer of oxalate of lime, which we can readily explain if we admit, with Prout and Dr. Rees, the conversion of uric into oxalic acid; while, still more superficially, are frequently discovered layers of the mixed phosphates, deposited in consequence of the ammoniacal condition of the urine, an alteration due to the conversion of urea into carbonate of ammonia.

Recent investigations with reference to the composition of the nuclei of urinary calculi have fully confirmed Dr. Rees' observation as to the predominance of uric acid. Dr. Ultzmann, of Vienna, has paid considerable attention to this subject, and has published the results of an examination of many hundred calculi.[1] He divides urinary calculi into two groups. In the first group he places those in which the nucleus is formed of material deposited in acid urine. Calculi belonging to this group he attributes to a primary stone-formation. The second group contains calculi whose nuclei are composed either of the substances deposited in alkaline urine, or of a foreign body, and calculi of this class he describes as being due to a secondary stone-formation. Calculi of the first class originate almost invariably in the kidneys, and have as a nucleus, either uric acid, urate of soda, oxalate of lime, or cystine. Calculi of the second class originate for the most part in the bladder, and the nucleus is composed of the urate of ammonia, the ammoniaco-magnesian phos-

---

[1] Ueber Harnsteinbildung, Vienna, 1875.

phate, or the phosphate of lime. In examining 545 single vesical calculi, Dr. Ultzmann found

Nuclei of Uric acid	441 or 80.9 per cent.
" Oxalate of Lime	31 " 5.6 "
" Earthy Phosphates	47 " 8.6 "
" Cystine	8 " 1.4 "
" Foreign bodies	18 " 3.3 "

Of these calculi, 480 belonged to the primary formation, and of these

The nucleus was formed of Uric Acid in	441 instances, or 91.8 per cent.
" " " Oxalate of lime	31 " 6.4 "
" " " Cystine	8 " 1.6 "

From a more extended examination which included 73 multiple calculi, and 319 passed spontaneously. Dr. Ultzmann arrived at the conclusion that, in the *primary* stone-formation (a process to which 88 per cent. of all calculi are due) the nucleus is formed of uric acid in nearly 94 per cent. of the cases. The importance therefore of uric acid as a factor in the production of calculus cannot be over-rated. We shall return to this subject in the chapter on the "Causes of Stone."

It is interesting to compare Dr. Ultzmann's results with those obtained by Dr. Klien in Russia, and Dr. H. V. Carter in India. The former reports[1] that stone is extremely common in the central districts of Eastern Russia. At certain times patients suffering from calculous disease form one-fifth of the cases in the Moscow Hospital. In this hospital the yearly average of such cases is over 60. The patients are for the most part of the peasant class; children are three times as numerous as adults, and there were only four women out of 1,792 stone cases occurring 1822-60. Alternating calculi are the most common; the nucleus is generally uric acid or its salts, and is frequently covered by a layer of oxalate of lime. Phosphates sometimes form the entire stone, and generally the outer layer of the alternating calculi. Whereas calculi composed of uric acid are common in England, France, and Germany, in Russia the oxalate of lime and phosphatic calculi are more frequently met with. Of these, oxalates are found in children and young people, while phosphatic and uric acid calculi are more common in adults.

Dr. Vandyke Carter,[2] whose researches refer to Western India, has found that oxalate of lime is far more common, both as a nucleus and general constituent of calculi, in that country than in England. Calculous disease is frequent and peculiar in Western India, and the following table (from Dr. Carter's report) shows the percentage of calculi there and in England, having for their nucleus (A) uric acid; (B) urate of ammonia; (C) oxalate of lime; (E) earthy phosphates; and (D) cystine.

Series	Grant College, Bombay	Coll. of Surgeons, London	Guy's Hospital	Norwich
A	11.76	43.16	52.40	48.87
B	44.54	31.21	9.14	38.61
	56.30		71.79	
C	38.65	14.75	22.50	13.27
E	3.36	10.40	10.57	7.24
D	0.84	0.46	5.28	none

[1] Ueber die Steinkrankheit und ihre Behandlung in Russland, Archiv für klin. Chirurg., Bd. vi., s. 78.
[2] Calculous Disease in Bombay. St. George's Hospital Reports, 1871-72, p. 85.

Dr. Carter further observed the oxalate of lime occurs in 50 per cent. of calculi composed only of one ingredient, whereas in the College of Surgeons of London the proportion of such calculi is only 14 per cent. The proportion of calculi having uric acid or urates for their nucleus or substance is considerably less in India than in England, and in the former country urate of ammonia is more common than uric acid. The earthy phosphates are far more rare either as a nucleus or as forming the substance of the calculus. Of 170 Indian calculi, in 100 the nucleus was formed of oxalate of lime; in 38 of urate of ammonia; in 25 of uric acid, and in 7 of the mixed phosphates. From these figures we may safely conclude that over a large part of India as compared with home districts, oxalate of lime is the element most commonly met with in urinary calculi. We shall recur to these and to Dr. Klien's statistics when discussing the causes of calculous disease.

The disposition of the nucleus is not always the same. It does not necessarily occupy the centre of the calcareous mass, nor is it invariably continuous with the layers which may have been deposited on it. Sometimes the nucleus is inclosed in a hollow envelope of calcareous matter, like a kernel in a nut-shell; the deposit and subsequent desiccation of animal matter, blood, etc., on the surface of the nucleus may possibly explain this curious appearance. Mr. Crosse has given the figure of a stone, which resembled a small gherkin in shape; the nucleus, composed of oxalate of lime, occupied one end, while the remainder was composed of the phosphates. In this case the calculus had been encysted, and the peculiar formation of the calculus probably depended on the fact that the nucleus was deposited in the cyst, while the phosphates were superadded from the cavity of the bladder into which the elongated part projected. In several cases the calculus has been found to contain two or even more nuclei; but from the descriptions given, it seems reasonable to conclude that for the most part they were distinct calculi, formed, perhaps, at different periods, and united together by phosphatic cement.

Sometimes the calculus is formed round a nucleus of blood or mucus, and this may perhaps explain the curious circumstance, above alluded to, that calculi are occasionally hollow in the centre. The cavity, however, generally contains some pulverulent matter or small nuclei. A remarkable example of this kind was presented by Mr. Shaw to the Pathological Society some years ago. The calculus resembled in shape and size a small walnut, but was very light. On a section being made, the cause of this remarkable lightness became apparent. The calculus was nearly hollow, the thickness of the shell varying from one-sixth to one-half of an inch. Within the cavity was found lying loosely a small rugged calculus, about the size of a pea. The calculus, with its shell, was composed of uric acid.[1]

Several cases of the formation of calculi round mucus or clots of blood are recorded by Howship, Wilson, and others. The celebrated Frère Come made a fortunate guess in a case of this kind. Before proceeding to operate on the Archbishop of Paris, he announced that the stone inclosed a clot of blood, because the patient had been subject to hæmaturia for some time before the appearance of symptoms of stone. On examining the calculus after extraction, the predicted clot was found within it.

The nucleus may be formed by some foreign body accidentally introduced into the bladder. Almost every imaginable kind of substance has

---

[1] Trans. of Path. Soc., vol. vi., p. 251.

thus given origin to vesical calculus; and a detailed account of them would be more curious than useful. The foreign bodies are frequently introduced by accident, frequently also by design, especially in the case of females.

That the form of calculus is modified by its manner of growth, is apparent in many cases. A remarkable specimen was presented to the Pathological Society by Mr. James Salter, in 1854. The stone consisted of a central, oval mass, of uric acid, from which started three stalactite-like projections of the ammoniaco-magnesian phosphate. On examining the bladder, it was found that the uric acid nucleus had been lodged in an ulcer just behind the internal orifice of the urethra; and from this point the branches had passed into the three orifices of the bladder, the urethra and two ureters.[1]

The calcareous matter incrusting these foreign substances is usually composed of the phosphates, or the ammoniaco-magnesian phosphate; only one specimen in the college museum exhibits a coating with uric acid; the nucleus in this case was a piece of steel. M. Cloquet exhibited to the Academy of Medicine, Paris, an ivory pessary, one portion of which had penetrated into the bladder, and was incrusted with uric acid.[2]

Before giving an account of the chemical ingredients which enter into the composition of calculi, it may be well to notice the animal matter which is invariably present in them, and is supposed to bind their particles together, as by a cement. The nature of this animal matter has not been clearly determined, and it would appear to vary in different cases, being generally either mucus, fatty matter, or fibrine. It may, however, consist, in some part at least, of blood, pus, or epithelial scales. The mucous secretion of the bladder is the most obvious source, and to this Marcet referred the animal matter of calculi; Fourcroy and Vauquelin considered it as consisting sometimes of albumen, sometimes of gelatine, with an admixture of urea. Berzelius was unable to determine whether it was composed of fibrine, albumen, caseous matter, or mucus; Brande says it is composed of a mixture of gelatine with urea.

The rapidity with which calculi increase in size appears to be considerably influenced by the quantity of animal matter, whether mucus, pus, or epithelial débris, present in the urine. Professor Scharling, of Copenhagen, lays great stress on this point, and the following remarks, contained in a pamphlet quoted by Dr. Gross, are worthy of careful consideration.

"The degree of rapidity with which precipitation takes place depends on various causes. Among these may be enumerated the envelopment of the nucleus in albumen, blood, mucus, pus, or any other organic matter that chances to be present in sufficient quantity. These form a villous coating around the solid material, and their flocculi arrest, entangle, and ultimately determine the crystallization of the more insoluble ingredients of the urine. This explanation will go far to account for the animal matter contained in all calculi; the presence of which adds so greatly to the difficulty of distinguishing their constituents. It accounts also for the spongy interstices interposed between layers of a denser structure; and explains why certain calculi are full of small foramina.

---

[1] Ibid., vol. v., p. 203.
[2] Lancet, November 3d, 1855, p. 417. See also a case recorded by Mr. Furneaux Jordan, Trans. of Path. Soc., vol. xviii., p. 179.

"These organic substances, as they exist so constantly in calculi, may be regarded as the cement which binds calculous constituents together; and not only favors their increase, but in very many instances first lays the foundation for precipitation. If we attentively examine any of the fissured and perforated calculi so often met with, or those in which a central mass of crystals replaces the usual nucleus, we shall have evidence of the manner in which a clot of blood, or a flake of mucus or albumen, detains the solidifiable ingredients, the hydrate, as it were, and forms the elements of a nucleus, which consolidates, and in its turn constitutes a centre for future deposition."[1]

These observations are quite in accordance with those of Dr. Haskins who believes that no calculus can form without the aid of matter foreign to the urine in a chemical sense, and that this matter is uniformly of an animal character.

On placing under the microscope a small quantity of calculous matter, imperfectly pulverized, and partly dissolved, he found that the particles were enveloped by a pellicle of transparent animal matter, and when this was completely divested of salt, it bore so great a resemblance to epithelial scales, as to be easily mistaken for them. Dr. Haskins, also, frequently detected in the central parts of calculi, a large proportion of epithelial scales from the bladder and kidney, with fibrinous casts from the uriniferous tubes, and a peculiar fibriniform matter without any definite structure.[2]

The quantity of this animal matter has not been ascertained for the several species of calculi, and we can understand how it may vary according to several circumstances. Mr. Brande informs us that for uric acid calculi of the kidneys, the quantity of animal matter varies from a mere trace to two-sevenths of the whole weight. Uric acid calculi in the bladder contain a much larger proportion. The oxalate of lime calculus contains a greater proportion than any other. It would also appear that the nucleus contains less than the outer layers; at least in a phosphatic calculus examined by Morin, the nucleus contained only one-tenth, while the shell contained seven-tenths of animal matter.

Finally, we must not overlook the observations made by Dr. Bence Jones—that all calculi, while retained in the bladder, are permeated by moisture. This fact is of great importance in several points of view.

The chemical constituents of calculi may be conveniently divided into two classes, the first containing those which sometimes form entire calculi in a nearly pure state; the second, those which may be viewed as adventitious constituents.

A.—Forming entire Calculi or Laminæ.

1. Uric acid.
2. Uric or Xanthic Oxide.
3. Urate of Ammonia.
4. Cystic Oxide, or Cystine.
5. Ammoniaco-Magnesian Phosphate.
6. Oxalate of Lime.
7. Phosphate of Lime.
8. Carbonate of Lime.
9. Mixed Phosphate of Lime, and Phosphate of Magnesia and Ammonia.

To these must now be added uro-stealith and indigo.

---

[1] Gross, loc. cit., p. 186.   [2] Ibid., p. 186.

B.—Existing in small quantities associated with the preceding.
10. Urate of Potass.
11. "   " Soda.
12. "   " Lime.
13. "   " Magnesia.
14. Organic matter, fat, extractive albumen, vesical mucus, blood.
15. Carbonate of Magnesia.
16. Silica.
17. Oxide of Iron (?)
18. Phosphate of Iron (?)
19. Clay Mica.

The following method of examining calculi is that recommended by Loebisch;[1] it appears to fulfil every requirement.

With a fine saw the calculus is to be carefully divided into two parts, as nearly as possible through its centre; the layers, visible to the naked eye, are to be separated from each other, and each distinct layer is to be subjected to the following tests:

A small quantity is to be reduced to powder and exposed to a red heat on platinum foil; it will be either entirely or partially consumed.

I. The concretions which are entirely destroyed by heat contain only organic matters, and may be composed of:—uric acid, urate of ammonia; in rare cases, of cystine or xanthine; in still rarer cases, of protein substances and uro-stealith.

In order to distinguish these substances from each other, a small quantity of the powder is placed in a porcelain dish, dilute nitric acid is added, and heat carefully applied until evaporation has taken place. The result is:—

(*a*) A reddish yellow color. When cold a drop of liquor ammoniæ is to be added, a beautiful purple red color (murexide) is obtained. The stone contains uric acid.

A second portion of the powder is boiled with liquor potassæ; a strong odor of ammonia is developed, moistened turmeric paper exposed to the vapor becomes brown; a glass rod moistened with hydrochloric acid and similarly exposed, gives off white fumes. The stone contains ammonia, and consists, therefore, of urate of ammonia. If this test for ammonia yields negative results, the calculus consists of uric acid alone.

(*b*) Nitric acid and heat produce a lemon yellow color, which becomes reddish yellow when liquor potassæ is added, and yellowish red on heating. The stone contains xanthine.

(*c*) A reddish brown color. The powder is soluble in solution of ammonia and of carbonate of ammonia, and is precipitated from such solutions by the addition of acetic acid. Hexagonal plates are found to remain after evaporating the ammoniacal solution. The calculus contains cystine.

(*d*) The calculus shows no trace of crystallization, is insoluble in water, ether, and alcohol, but soluble in solution of caustic potash, and precipitable from this by acids. On the addition of acetic acid, it swells up; it is soluble in boiling nitric acid; when burnt, it gives off the odor of burnt horn; such a calculus is composed of protein substances.

(*e*) The calculus, when recent, is soft and elastic. On being dried, it diminishes in size and becomes hard. Warmth causes it again to become soft; on heating, it melts, swells, and gives off a very strong odor,

---

[1] Anleitung zur Harn-Analyse, von Dr. W. F. Loebisch, Wien, 1878, p. 221.

which resembles that of a mixture of shellac and benzoin. It is readily soluble in ether, the amorphous residue of which solution after evaporation becomes of a violet color when exposed to heat. The calculus is also readily soluble in warm caustic alkalies, forming a soapy solution. Nitric acid dissolves it with a slight evolution of gas, but without changing its color; the residuum becomes of a dark yellow color on the addition of alkalies. The calculus is composed of uro-stealith.

II. The calculi which, on being heated on platinum foil, leave behind them a more or less considerable residue, contain organic as well as inorganic substances. As an organic substance, uric acid is generally present, and is often accompanied by urates; oxalate and carbonate of lime may also be present. If the presence of uric acid has been demonstrated by the murexide test, it remains to discover the base with which it is connected.

(a) A quantity of the pulverized calculus is boiled with distilled water, and the solution is filtered while hot; the urates, soluble in hot water, are contained in the filtrate, from which they are precipitated on cooling.

To determine the base with which the uric acid is connected, the filtrate is evaporated and exposed to a red heat; the ashes contain the fixed alkalies.

A portion of the ash is exposed on platinum foil to the colorless flame of a Bunsen's burner; a yellow coloration of the flame indicates soda, while a violet color indicates potash. (Indigo solution or chloride of platinum may also be used as tests.)

If the uric acid is compounded with magnesia and lime, these remain behind in the state of carbonate if the residue be not too strongly heated.

To separate these bases, the heated powder is dissolved in dilute hydrochloric acid. The clear solution is neutralized with ammonia, and the precipitate is dissolved with a few drops of acetic acid.

Oxalate of ammonia being added, oxalate of lime is precipitated. The mixture is then filtered, and phosphate of soda and ammonia are added to the filtrate; the magnesia separates itself as a crystalline precipitate of phosphate of ammonia and magnesia.

(b) If, on applying the preliminary tests, nitric acid and ammonia, no uric acid reaction be manifested, the calculus may contain either oxalate or carbonate of lime.

A portion of the calculus is not affected by acetic acid, but dissolved by the mineral acids without the disengagement of gas, and precipitated by ammonia. On exposure to a red heat, the fragment blackens, owing to the combustion of the organic materials, but on further exposure to heat, the residue soon becomes white. This latter is alkaline, and effervesces with acids; a calculus thus acted upon is composed of oxalate of lime.

The fragment, on being heated, glows with an intense white light; previous to the application of heat, effervesces with acids, and is precipitated from the neutralized solution by oxalate of ammonia. Such a calculus is composed of carbonate of lime.

(c) The fragment, on being heated, gives off the odor of ammonia, and in a still more marked manner when warmed with a solution of potash; it is soluble in acetic acid without effervescence, and from this solution is precipitated in a crystalline form on the addition of ammonia. On being heated, the fragment fuses into a white enamel-like mass. The calculus is composed of phosphate of ammonia and magnesia.

(d) The fragment does not effervesce with acids either before or after exposure to a red heat; the residue is white, and precipitable from its solution in acids by ammonia. When dissolved in acetic acid and oxalate of ammonia is added, oxalate of lime is precipitated. The calculus is composed of basic phosphate of lime.

To distinguish between calculi which contain both the ammoniaco-magnesian phosphate and the basic phosphate of lime, we must test for the lime and magnesia in the way described in II. a.

(e) Calculi of neutral phosphate of lime occur but very rarely. According to J. Vogel, however, neutral phosphate of lime is often seen in the form of gravel, and, on a superficial examination, regarded as uric acid and treated with alkaline mineral waters which cannot fail to act injuriously under such circumstances.

We now proceed to a more detailed description of the several recognized species of urinary calculi. These have been arranged in the following order; and in treating of each the method of examining their chemical constituents will be explained :—

1. Uric or lithic acid.
2. Urate or lithate of ammonia.
3. Uric oxide, or xanthine.
4. Oxalate of lime.
5. Phosphate of lime.
6. Phosphate of magnesia and ammonia.
7. Mixed phosphates, or fusible calculus.
8. Carbonate of lime.
9. Alternating calculus.
10. Cystine or cystic oxide.
11. Fibrinous calculus.
12. Uro-stealith
13. Indigo.
14. Prostatic calculi and concretions.

Uric Acid Calculus.—Uric acid is by far the most frequent constituent of urinary calculi, either alone or in combination with bases, as urate of ammonia, lime, etc. The calculi consisting of uric acid alone, or merely in a state of mixture with a small quantity of coloring matter, are more common than those of any other single constituent. The relative number of pure uric acid calculi in the Museum of the Royal College of Surgeons is one-third of the whole collection ; in that of St. Bartholomew's Hospital, $1 : 11\frac{8}{17}$; in Guy's Hospital, $1 : 5$; in the Norwich Hospital, $1 : 4$; in Swabia, $1 : 11\frac{4}{5}$; and in Copenhagen, $1 : 5$. The general average deduced by Dr. Prout from these data is nearly $1 : 6\frac{1}{2}$.

The relative proportions of those composed chiefly of uric acid mixed with urate of ammonia, and of urate of ammonia with minute proportions of urate and oxalate of lime and phosphates is in St. Bartholomew's as $1 : 7$; in Guy's Hospital, $1 : 4$; in the Norwich Hospital, $1 : 1\frac{1}{5}$; in the Manchester Hospital, $1 : 2\frac{1}{2}$; in the Bristol Hospital, $1 : 3$; in Swabia, $1 : 10$; and in Copenhagen, $1 : 4\frac{1}{2}$; giving a general proportion of $1 : 3\frac{1}{4}$. If the calculi be included in which the nucleus is formed of lithic acid, we have the following as the proportions in these collections : in the Royal College of Surgeons, $1 : 1.4$; in St. Bartholomew's Hospital, $1 : 1\frac{3}{4}$; in Guy's Hospital, $1 : 4$; in the Norwich Hospital, $1 : 1\frac{1}{4}$; in the Manchester Hospital as $1 : 1\frac{3}{4}$; in Swabia, $1 : 1\frac{1}{4}$; in Copenhagen, $1 : 1\frac{1}{4}$; in the various collections examined by Dr. Ultzmann $1 : 1\frac{1}{4}$; the average proportion being $1 : 1\frac{1}{2}$. The proportion of calculi into which uric acid enters in larger or smaller proportions as one of the constituents in the collection of the College of Surgeons, is to the whole numbers as 1 : 1.36.

Of the 600 calculi examined by Fourcroy 500 were composed of uric acid. Of 374 calculi in the Museum of Guy's Hospital, examined by Dr. Golding Bird, 269 showed nuclei consisting of uric acid or urates. Of 545 calculi in several collections in Vienna, examined by Dr. Ultzmann 441 (or 80.9 per cent.) contained nuclei consisting of uric acid. In alluding to nuclei in another place the number of calculi examined by Dr. Bird was stated to be 212, but duplicate specimens were excluded from that estimate.

Two varieties of the uric acid calculus have been observed ; the one consisting of those in which the uric acid is deposited in more or less distinct concentric layers, giving the section of the calculus a laminated appearance; while in the other variety the acid is deposited in a confused mass of crystalline or amorphous grains; but these appearances are often mixed with or pass into each other.

The laminated uric acid calculus shows, on section, a series of concentric circles: a semi-crystalline and compact structure; its surface is generally smooth, but sometimes granular or even finely tuberculated, but if so the tubercles are smooth and polished. When broken, it separates into angular fragments whose surface is fibrous, as if it were composed of crystalline fibres radiating from the centre to the circumference. The fracture occurs both in the direction of the radiating fibres and of the concentric layers. The fractured surface of the more compact calculi has a vitreous lustre.

The calculi in which no lamellæ are seen, either consist of an aggregation of large crystalline grains firmly adherent to each other, and presenting a radiated appearance, or they have a porous and earthy structure as if formed of loosely-cohering fibres. These calculi are less regular, have a rough surface, a granular and unsymmetrical fracture, and are most frequent in the kidneys. The nucleus of the laminated variety frequently presents this character. This form of calculus is more liable than the compact laminated variety to spontaneous fracture in the bladder. Calculi are sometimes seen which present cracks in the direction of the radiating fibres, indicating that the manner in which this curious phenomenon occurs probably depends upon some unequal density in the deposits.

The color of the uric acid calculus varies from pure white to a deep brownish red. The exact nature of the coloring matter, which exists only in a small proportion in the deepest colored specimens, is uncertain, but it probably is of the same character as the coloring matter of the urine, so often seen in the crystals of uric acid, known as the red sand deposit. These calculi are usually of a flattened oval shape, and possess considerable hardness. They vary in weight, from a few grains to several ounces. One of the largest on record was removed by Sir James Paget.[1] It weighed 9 ozs. 1 drachm, and was 4 inches in length, $3\frac{1}{2}$ in width, and $2\frac{1}{4}$ inches thick. It was oval in shape, and its surface was smooth and hard. The symptoms had existed for twenty-three years. The stone was removed by the lateral operation.

There is another sub-variety of the uric acid calculus which is very common, especially in gouty persons. It is known as the pisiform calculus. This form is seldom solitary; many may exist at the same time in

---

[1] Lancet, vol. i., 1862, p. 198.

STRUCTURE OF URINARY CONCRETIONS.

the bladder, and some may be voided by the urethra. These calculi are rarely larger than a common bean or large pea; they have an irregular angular shape produced by attrition against one another, are crystalline at the centre, laminated near the surface, and often coated with a thin layer of urate of ammonia.

The specific gravity of the uric acid calculus is usually from 1.5 to 1.786; but specimens have been observed in which the density was so low as 1.276.

Uric acid, when quite pure, is white, very sparingly soluble in water, one part of the acid requiring about 15,000 parts of cold, and 1,932 parts of boiling water to bring it into solution, from which it is again deposited in the form of a granular powder composed of minute crystals of variable form.

It is absolutely insoluble in alcohol or ether, but dissolves very readily in solutions of caustic potash or soda, especially when heated; in phosphate and biborate of soda, and in weak solutions of the carbonates of potash and soda, and is precipitated from these solutions by acids. Nitric acid dissolves it with effervescence, the uric acid is decomposed, and equal volumes of carbonic acid and nitrogen are evolved. Sulphuric and hydrochloric acids do not affect it. When heated, it blackens and is gradually consumed, leaving no ash if perfectly pure. Uric acid is a feeble acid, but combines with bases and forms salts with them. The alkaline urates are sparingly soluble, but very much more so than the pure acid. Urate of potash dissolves in 140 parts of cold, and 85 of boiling water, according to Liebig, but Berzelius states that it requires no less than 480 of cold water for solution. Urate of soda, which forms the gouty concretions known as chalk-stones, is much less soluble than the corresponding salt of potash, requiring 372 parts of cold, and 124 parts of boiling water for solution. Urate of lime, small proportions of which exist in many calculi, forms shining white plates or needles, readily soluble in boiling water, but deposited from it as the solution cools.

Uric acid calculi are readily distinguished by a simple chemical analysis. When a fragment is heated to redness on a slip of metal, it blackens and gradually consumes, yielding hydrocyanate and carbonate of ammonia, and a small white ash remains, which is found to be pure lime produced by the presence of minute proportions of oxalate or urate of lime in the calculus. When reduced to powder, and heated with liquor potassæ, the calculus readily dissolves, yielding a clear solution from which a white precipitate is thrown down by the addition of hydrochloric, nitric, or acetic acid. The precipitate is at first pulverulent, and very bulky, but on being allowed to stand in the solution for a short time it contracts very much in bulk and becomes granular. In this state, it is composed of minute crystals of uric acid readily discoverable by the microscope. It has been stated that uric acid calculi, of moderate size, are dissolved in two days by an alkaline solution sufficiently weak to be borne in the mouth without inconvenience.

Another and striking test is the action of nitric acid on the calculus. A small portion of the powder is to be placed in a watch-glass or evaporating dish, and a few drops of nitric acid poured on it; the uric acid dissolves with effervescence; it is then gently heated over a spirit lamp until the greater part of the acid is driven off and the liquid dries into a yellowish-red residue. This latter, when cool, is touched with a rod dipped in liquor ammoniæ, when a bright violet hue (murexide) is immediately developed. This reaction is perfectly characteristic of uric acid.

The other substances found in uric acid calculi are animal matter, coloring matter, urate of lime, urate of ammonia, urate of soda and traces of the inorganic salts of the urine.

Urate of Ammonia Calculus.—Calculi composed of unmixed urate of ammonia are comparatively rare on account of its solubility, forming only 1 in 500 in the collection of the College of Surgeons. In Guy's Hospital Museum according to Mr. Bryant, there are only 7 such calculi out of 394 specimens. This calculus was discovered by Fourcroy and Vauquelin in 1798, but its nature was more satisfactorily demonstrated by Dr. Prout in 1823. Calculi composed entirely of urate of ammonia seldom exceed an inch in diameter, are flattened, ovoid, smooth, of a brownish-gray or clay-color, with frequently a tinge of green. These calculi are usually compact, earthy, and very brittle; they consist of thin consecutive layers so closely applied to each other as to appear homogeneous, but the laminæ are easily separated. They are almost peculiar to childhood, seldom if ever occurring after puberty, and their formation is accompanied by great constitutional disturbance and irritability. They have, however, been removed from adults. The Guy's Hospital Museum contains a remarkable collection of 142 calculi which Sir A. Cooper removed from the bladder of one patient; these calculi are all cubic in form, rounded at the edges and angles, and of the color of pipe-clay. A remarkable fact connected with this case is, that the patient had subsequently another calculus, of a different nature from these. Dr. Roberts believes that the calculi described by Prout as consisting of urate of ammonia were in all probability nothing more than fawn-colored uric acid.

Urate of ammonia is a common deposit from urine when that fluid has an acid reaction, and usually carries down with it a greater or less proportion of coloring matter, which gives it a yellow or brown tinge. In a pure state this salt is perfectly white; it is much more soluble in water than uric acid; it requires, according to Liebig, 1,727 parts of cold, and 243 of boiling water for solution, and when the latter solution cools the greater part of the salt is deposited in microscopic acicular crystals. The presence of chloride of sodium in the liquid, in the proportion of 2.5 to 1000, enables it to take up more than double the quantity of the salt, and it is no longer deposited by cooling or evaporation in the crystalline, but in the peculiar amorphous condition in which it is ordinarily deposited from urine.

The calculus decrepitates when exposed to the blow-pipe, and consumes much in the same manner as the uric acid calculus; the remaining ash is, however, more copious, and consists either of pure lime or lime mixed with phosphates. The presence of ammonia is shown by heating a small quantity of the powder with liquor potassæ, when ammonia is disengaged in abundance, which may be detected by its odor, and by holding a rod moistened with hydrochloric acid over the liquid; or the ammonia may be determined by heating the calculus with hydrochloric acid, which combines with the ammonia, forming hydrochlorate of ammonia; this may be afterwards precipitated by bichloride of platinum with the usual precautions. The uric acid is readily precipitated from the solution in caustic potash by either of the stronger acids.

On account of the more ready solubility of urate of ammonia, it may be separated from uric acid in cases where it exists in small proportions, by boiling the finely powdered calculus in water, pouring off the hot solution, and allowing it to cool, when the urate of ammonia is deposited either in small stellate crystals or in an amorphous condition.

Urate of ammonia is often associated with oxalate of lime in calculi; in fact, the oxalate of lime diathesis is frequently preceded by a condition of the urine which favors the deposit of urate of ammonia, so that the nucleus is perhaps formed of urate of ammonia, while the succeeding layers are a mixture of these salts. These calculi often contain a small quantity of urate of lime. The nucleus presents the character of urate of ammonia, decrepitating violently when heated, whereas the exterior layers are proved to contain both these salts by the tests already indicated for urate of ammonia, and those which will be given in the description of the oxalate of lime calculus. Urate of ammonia is also frequently associated with the earthy phosphates, either as a nucleus or in alternate layers or mixed in various proportions with the other constituents.

Some chemists regard the urate concretions as consisting mainly of urate of soda, and much confusion still exists with regard to their exact nature. The view most commonly adopted in Germany is that the lateritious deposit, so often seen in urine, and found to be amorphous under the microscope, is composed of uric acid combined with soda, while urate of ammonia is often found in ammoniacal urine, mixed with the earthy phosphates. This latter salt assumes a spherical or globular shape, with short or long projections. Dr. Roberts, on the other hand, considers that the lateritious deposit is composed of *mixed* urates, or uric acid combined with several bases—potash, soda, ammonia and lime—sometimes one base and sometimes another preponderating. The spiny crystals which are often deposited spontaneously from the urine of children, Dr. Roberts believes to be urate of soda. The distinguishing feature of this form of the salt is that it is precipitated within the urinary passages. "The spiny crystals irritate the mucous membrane of the bladder or urethra; and the latter canal may even be blocked up by impaction of masses of the deposit. It may also form a nucleus around which calculus matter may hereafter aggregate. The great comparative frequency of vesical calculi in children is not improbably owing to the occurrence of this deposit in the numerous fugitive febrile attacks to which children are subject."[1] We shall refer to these remarks in the chapter on the "Causes of Stone."

Xanthine, Xanthic or *Uric Oxide Calculus.*—This calculus is more rare than any other, only seven specimens having been discovered. It was first described by Marcet, who ascertained the properties of this curious substance, which in composition differs from uric acid only in containing one equivalent of oxygen less than the acid. The specimen examined by Marcet no longer exists, it weighed eight ounces. Some minute specimens were examined by Laugier; but one calculus weighing 338 grains, and about the size of a small hen's egg was extracted by Langenbeck, and examined by Liebig; another is mentioned by Dulk. The Museum of Guy's Hospital possesses a fragment of the calculus extracted by Langenbeck. A fifth specimen was discovered by Mr. Taylor, in 1866, in the Museum of the Royal College of Surgeons.[2] This calculus (of which the College possesses only half) when entire weighed 90 grains, and was given to Mr. Bransby Cooper by a surgeon in the East India Company's Service. It had been extracted from the bladder of a Mussulman child, four years of age, and was supposed to consist of uric acid, urate of ammonia, oxalate of lime, and earthy phosphates. Mr. Taylor, however, discovered that it was composed of nearly

---

[1] Roberts, loc. cit., p. 72.   [2] Path. Soc. Trans., xix. 275.

pure xanthine. It is made up of three concentric layers, closely aggregated, so that its cut surface presents a compact texture. It breaks with a conchoidal fracture. Its section possesses a peculiar flesh-colored or cinnamon tint; it has a yellow nucleus in which no trace of uric acid can be detected. Its external surface is slightly rough, and of a light brown or coffee color. Mr. Taylor has examined several uric acid calculi, which from their external appearance might be supposed to contain xanthine, but hitherto without success. Two other cases have been recently reported. One of these occurred to Mr. Fleming, of Dublin. The calculus was somewhat larger than a garden pea, and was passed spontaneously.[1] In the other and last recorded instance, Dr. Gaillard removed a xanthine calculus from the bladder of a boy aged 13. It was of a brick-red color; nearly 2 inches in length, and weighed about 350 grains.[2]

These calculi are of a light or cinnamon-brown color externally, and of a brownish flesh color in the interior; their surface is smooth and polished: they are compact, but consist of concentric layers easily separable, without fibrous radiations or crystalline structure, and they assume a waxy lustre when rubbed or scraped. In point of hardness they resemble calculi of uric acid.

Xanthine is soluble in liquor potassæ, from which it is precipitated by acids as a white powder, which becomes light yellow by drying, and acquires a resinous lustre by gentle friction. It differs from uric acid by dissolving slowly without effervescence in nitric acid, and the solution, when evaporated to dryness, yields a residue of a brilliant lemon-yellow color, partially soluble in water. This residue dissolves in liquid ammonia, forming a reddish-yellow solution. The same effects are produced when the xanthine is mixed with a considerable proportion of uric acid; indeed, its presence appears to prevent the formation of the violet color produced by the successive action of nitric acid and ammonia on uric acid, changing the color to a brick-red. Xanthine and uric acid are equally soluble in strong sulphuric acid, but the latter is precipitated by the addition of water, while the former is retained in solution, and on this difference the process for separating these substances is founded.

After dilution of the sulphuric acid solution, the precipitated uric acid is separated, and the xanthine may be removed by saturating with ammonia, evaporating to dryness, and dissolving the sulphate of ammonia with water. Xanthine is also soluble in warm hydrochloric acid, and the hydrochlorate thus produced presents characteristic crystalline forms. The solutions of xanthine in caustic alkalies are not precipitated by hydrochlorate of ammonia, which throws down uric acid. Caustic alkalies and alkaline carbonates dissolve xanthine, and it is more soluble in liquid ammonia than uric acid. Xanthine gives out a peculiar odor when burned, differing from that emitted by uric acid or cystine, and, unlike uric acid, it yields no urea by destructive distillation. By these characters this rare calculus may be distinguished from uric acid concretions, which it resembles very closely.

Oxalate of Lime Calculus.—Considerable difference occurs in the relative numbers of these calculi in different collections. The proportion in the Museum of the College of Surgeons, of calculi entirely composed

---

[1] Diseases and Injuries of the Genito-Urinary Organs, p. 332.
[2] New Syd. Soc. Biennial Retrospect, 1873-74, p. 222. The case is described in Gaz. Hebd., April 18th, 1873.

of this salt, is as 1 : 20 ; in St. Bartholomew's, as 1 : 15 ; and in Guy's Hospital, as 1 : 14½; but when those with a nucleus of uric acid or urates are included, the proportion is increased, in the collection of the College of Surgeons, to 1 : 3¼. If all the calculi could be taken in which oxalate of lime exists in any amount, then the proportion is, according to Dr. Prout, in the museum of St. Bartholomew's Hospital, 1 : 4¾ : in Guy's, 1 : 4 ;¹ in Norwich, 1 : 7½ ; Manchester, 1 : 4⅛ ; Bristol, 1 : 3⅓ ; Swalbia, 1 : 27 ; and Copenhagen, 1 : 2$\frac{1}{10}$. The average proportion in all these collections is 1 : 4⅓. Dr. Ultzmann's statistics yield the same results. Out of 545 calculi examined by him, 130 contained oxalate of lime. In India, as already observed, calculi composed of this substance are far more common.

It would seem, from the foregoing statements, that the oxalate of lime calculus is rarely pure, but is generally associated with urate of ammonia, uric acid, urate and carbonate of lime, coloring matter, and blood.

The oxalate of lime calculus has been long known under the name of mulberry calculus, so called from the tuberculated surface it usually presents, which gives it somewhat the appearance of that fruit. This calculus has usually a rounded shape: the color of its surface varies from gray to dark brown or almost black. The internal structure, when exposed by section, is imperfectly lamellated, the consecutive layers forming waving lines, and its color is such as to give the surface of the section a strong resemblance to knotted heart of oak, varying from white to bright yellow, yellowish-brown, or dark green. The structure is very compact and hard. The tubercles when divided appear to consist of stellate or acicular needles, arranged perpendicularly to the surface, and owing to the tendency of oxalate of lime to form nodular masses, it is by no means improbable that the whole calculus is made up of such nodules. Three other varieties of the oxalate of lime calculus exist; one of these is crystalline throughout, its surface studded with brilliant crystals of the oxalate in acute-angled octahedra, and these are nearly pure oxalate of lime.

Mr. Poland reports a case in which an oxalate of lime calculus crumbled to pieces immediately after extraction, a circumstance which was due to the absence of any binding material, there being no trace of animal matter.[2] He also mentions a second variety of this kind of calculus. "This," he says, "is of a milk-white color, possesses a highly polished surface, is of extreme rarity, and is generally, if not always, found in the kidney; its external surface presents no crystals, but is perfectly smooth, though it may be spinous." In the Museum of the Norfolk and Norwich Hospital are three specimens, one of which was found in a kidney after death, and the others were passed spontaneously. Mr. Williams, formerly House Surgeon to the Hospital in question, believes that the white color of this variety of calculus is due to the fact that it is met with only in the kidney, and does not lie continuously in urine. He points out that the tuberculated form of calculus, from its roughness, far more frequently than any other variety, causes the mucous membrane of the bladder to pour out blood, and that it is the coloring matter of the blood contained in the urine which gives to the ordinary oxalate of lime calculus its peculiar brown color.

---

[1] In Guy's Hospital Museum the number of calculi containing more or less of oxalate of lime is 73 in 374, or 1 in 5⅕. See Dr. G. Bird, Appendix. p. 447, etc.
[2] Holmes's System of Surgery, vol. iv., p. 1024.

The other form consists of small rounded masses, with a smooth polished surface, known as the hemp-seed calculi. These generally exist in considerable numbers. They are occasionally crystalline at the centre, and laminated towards the surface; but the laminæ are so fine as to give the section almost a compact appearance. They are composed of mixed oxalate and phosphate of lime in variable proportions.

Oxalate of lime calculi are, as a general rule, extremely hard, and when crushed by the lithotrite break into sharp angular fragments. They seldom attain any great size. The nucleus is often composed of uric acid; this was the case in 124 cases out of 130 oxalate of lime calculi examined by Dr. Ultzmann. Conversely (though this is much more rarely met with) a uric acid stone may have a nucleus of oxalate of lime. This occurred in 15 out of 208 such calculi mentioned by the same observer. Calculi, however, are often composed of alternate layers of oxalate of lime and uric acid. The urine is always acid during the formation of oxalate of lime calculi, but as a result of irritation of the bladder, the urine may become alkaline, and then phosphates are apt to be deposited on the pre-existing calculus.

The oxalate of lime calculus is distinguished from other calculi by swelling when heated on platinum foil, or in the blow-pipe flame, and leaving a copious white ash, which is either caustic lime, or a mixture of quicklime and carbonate of lime, according to the intensity and duration of the heat; but usually the latter. This ash exhibits an alkaline reaction with moistened turmeric or reddened litmus paper, turning the former brown, and restoring the blue color of the litmus; it is soluble with effervescence in hydrochloric acid, and the solution gives no precipitate when saturated with ammonia, but a copious white deposit of oxalate of lime on the addition of oxalate of ammonia. The calculus itself is insoluble in acetic, but soluble with the aid of heat and pulverization in moderately strong nitric and hydrochloric acids, without effervescence, and the latter solution yields a white precipitate with excess of ammonia. Caustic potash has little or no action on this calculus, but when it is reduced to a fine powder, and boiled for some time with a solution of carbonate of potash, mutual decomposition ensues, and oxalate of potash and carbonate of lime are produced. If the liquid be then filtered, the carbonate of lime remains on the filter, which must be washed, and being heated to dull redness indicates the quantity of lime. Acetate of lead is now added to the pellucid solution until it ceases to give any further precipitate: oxalate with carbonate of lead are thrown down, collected on a filter, well washed, suspended in water, and a current of sulphuretted hydrogen passed through the liquid until the whole of the lead is converted into sulphuret; the liquid, again filtered and evaporated, yields crystals of pure oxalic acid.

This calculus is distinguished from phosphate of lime or ammoniaco-magnesian phosphate by its insolubility in acetic and dilute hydrochloric acids, and the alkaline reaction of the ash on moistened turmeric paper. When oxalate of lime and phosphates exist in the same calculus they may be separated by dissolving the calculus in strong hydrochloric acid, precipitating by ammonia, and digesting the residue in acetic acid, which dissolves the phosphates, but leaves the oxalate untouched. Acetic acid will separate the urate from the oxalate by dissolving the former.

Phosphate of Lime Calculus.—Calculi of pure phosphate of lime are rare. They were first described by Dr. Wollaston in the "Philosophical

Transactions" for 1797. There are two varieties, the one evidently of renal, the other of vesical origin. Although entire calculi of pure phosphate of lime are uncommon, this substance is often found nearly pure in the laminæ of alternating calculi.

Calculi of phosphate of lime are stated by Dr. Prout to be in the ratio of 1 : 32¼ in the Museum of St. Bartholomew's Hospital; as 1 : 29 in Guy's; as 1 : 132⅔ in Norwich; and 1 : 155 in that of the Bristol Hospital. The general ratio of these calculi to others is as 1 : 117.

The calculi of renal origin are composed of the neutral phosphate of lime, those of vesical origin of phosphate of lime similar to that of bones, and are thence often called "bone-earth calculi." The latter are more common than the former.

Calculi of neutral phosphate of lime are usually pale brown with a smooth polished surface, regularly laminated, and the laminæ so slightly adherent as to be easily separable into concentric crusts. In some, radiating lines are seen in a direction perpendicular to the laminæ, as if the latter were composed of crystalline fibres. These calculi contain a considerable proportion of animal matter, which is precipitated from the alkaline urine with the phosphate of lime. When heated before the blow-pipe or on platinum foil in the flame of the spirit-lamp, they char, give out the odor of burnt feathers, and leave a white ash which fuses at a higher temperature into an opaque globule. They are soluble in hydrochloric and nitric acids, sparingly so in acetic acid, and the phosphate of lime is precipitated in a white gelatinous form by the addition of alkalies. Oxalate of ammonia throws down oxalate of lime when the solution is nearly or quite neutral. As the ammoniaco-magnesian phosphate is also readily fusible, it must be previously ascertained that none is present.

Bone-earth is said by Simon never to occur unmixed in the form of a calculus; but Dr. Taylor has described some specimens of vesical origin in the catalogue of the Museum of the College of Surgeons, as consisting entirely of this salt. These are in irregular masses resembling mortar, or a granular semi-crystalline powder, enveloped in a tenacious mucus. They are distinguished from the preceding by first charring, then leaving a bulky white ash which is infusible before the blow-pipe.

Ammoniaco-Magnesian Phosphate Calculus.—This double salt, commonly known as the "triple phosphate," rarely forms an entire calculus, but is a very common constituent of other calculi, either mixed with phosphate of lime, as in the fusible calculus, or forming layers in alternating calculi. The proportion of calculi consisting of pure phosphate of magnesia and ammonia, in the collection of St. Bartholomew's, is as 1 : 129; in Guy's as 1 : 43½; in the Bristol Hospital, 1 : 218; in Copenhagen, 1 : 19¼. Three calculi of this salt exist in the Museum of the College of Surgeons, which gives the proportion in that collection as 1 : 200. According to Mr. Bryant, Guy's Museum contains two calculi of this kind. One of these "has no nucleus, but a central cavity lined with delicate crystals of triple phosphate resembling the crystals of quartz in the cavities of flints." Another preparation shows a "section of a large calculus of the kind on a nucleus of a tobacco-pipe." Mr. Bryant also mentions another instance, which occurred in the practice of Dr. Kitchener; a piece of broken catheter served as the nucleus for the calculus.[1] The general proportions in all the collections excluding that of

---

[1] Bryant, Practice of Surgery, vol. ii., p. 85.

the College, is 1 : 126⅔. Ammoniaco-magnesian phosphate is a frequent deposit from alkalescent urine, or healthy urine which has been kept until rendered ammoniacal by decomposition of its urea. In such urine it is frequently deposited as a white brilliant crystalline powder, commonly designated *white sand;* but it is usually mixed with more or less phosphate of lime.

Calculi of ammoniaco-magnesian phosphate are generally white, uneven, and roughened by the projecting summits of the crystals, which are transparent and brilliant in the fresh state, but become opaque and lose their lustre after keeping for some time. They are either not laminated or imperfectly so, are friable, and easily reduce to powder. But occasionally they are hard, compact, and laminated; exhibiting a semi-transparent crystalline fracture, which gives them the aspect of alabaster. They sometimes attain a great size. Dr. Thompson mentions one which weighed nearly 2 lbs.

Two distinct salts are found in these calculi; one of them, the neutral phosphate of magnesia and ammonia, crystallizing, according to Wollaston, in rectangular prisms, the usual form being the triangular prism, terminated by three or six-sided pyramids, which forms a film on the surface, or is deposited in urine that has become slightly alkalescent, or is thrown down by the addition of a small quantity of ammonia. The other, the bibasic phosphate, is deposited in stellated or foliated crystals, from strongly alkaline urine, or by adding a large excess of ammonia to healthy urine, and forms an iridescent pellicle on its surface, usually mixed with phosphate of lime. These salts do not exist naturally in the urine; they are formed by the combination of ammonia, produced by the decomposition of urea, with the phosphate of magnesia, present in greater or smaller proportion in all urine.

These calculi are recognized by giving off ammonia and watery vapor when heated in the blow-pipe flame, leaving a large residue of phosphate of magnesia which fuses with difficulty into a white porous enamel. If a small quantity of phosphate of lime be added, the fusion is more easy and complete. They are readily soluble in dilute acids, including acetic acid, and the original salt is precipitated by ammonia as a white crystalline powder. When treated with caustic potash ammonia is evolved.

Fusible Calculus.—The calculi commonly known by this name consist of a mixture of phosphate of magnesia and ammonia, with phosphate of lime in very variable proportions.

These calculi bear a considerable proportion in all collections, being, in St. Bartholomew's, as 1 : 12; in Guy's, 1 : 3½; in Norwich, 1 : 19; in Manchester, 1 : 8½; in Bristol, 1 : 12; in Swabia, 1 : 11½; and in Copenhagen, as 1 : 19½; according to Prout. The relative number in the Museum of the College of Surgeons is 1 : 13¼, while that in all the collections, excluding College Museum, is 1 : 12¼.

Fusible calculi are white, gray, or dull yellow; they are more friable than any others, being sometimes so soft as to whiten the fingers like chalk when handled; these are not laminated. Others have distinct lamellæ with sparkling crystals of triple phosphate between them, which are so slightly adherent as to be readily separated; others again are composed of crystals aggregated into a confused mass. The calculi themselves are usually globular or ovoid, but sometimes very irregular in shape, being occasionally moulded to the cavity in which they are formed; they often attain a large size, and sometimes fill the whole cavity of the bladder; in this case impressions of the folds of the mucous membrane are visible on the surface, and when two or more exist in the bladder, they take a cubic or tetrahedral form. Excrescences composed of triple phosphate are sometimes found resembling pearls in texture. The ammoniaco-magnesian phosphate is most abundant in those which have a shining crystalline texture, the phosphate of lime in the earthy and amorphous variety. The fusible calculus is found in all parts of the urinary organs, and in large cysts or cavities in the prostate gland.

The earthy phosphates are, according to Mr. Taylor, rarely succeeded by any other deposit; the only exceptions, that in the Museum of the College of Surgeons, being one in which layers of mixed phosphate exist in an oxalate of lime calculus, and one in St. Bartholomew's Hospital, in which a fusible calculus is surrounded by a layer of uric acid.

The nucleus is often situated at a distance from the centre of these calculi. This was the case in a large specimen removed by the late Mr. Southam, of Manchester.[1] The calculus weighed nearly 5 ozs., and its greatest circumference measured 8 ins. It contained at one extremity a small, round, alternating calculus (the nucleus), composed of uric acid and oxalate of lime. The recto-vesical operation was the one selected, and the patient made an excellent recovery.

The deposition of the mixed phosphates is due to the conversion of the urea into carbonate of ammonia, and the consequent alkalinity of the urine. The transformation of the urea is caused by the presence of the ropy mucus which is abundantly secreted by the mucous membrane of the bladder in a state of catarrh. This latter condition, as a direct consequence of the presence of the foreign body, often occurs in cases of calculus, and hence it happens that concretions, especially if long retained in the bladder, become coated over with phosphatic incrustations. So long as the urine remains acid, no deposit of phosphates takes place, but the change to an alkaline reaction is speedily followed by the deposit in question. Incrustations of the same character are often found covering foreign bodies of various kinds that have been accidentally introduced into the bladder. The incrustation of phosphates commonly contains a large proportion of animal matter.

The fusible calculus is characterized, as its nature implies, by giving off ammonia, and fusing readily before the blow-pipe into a grayish-white transparent, or semi-opaque globe; and this when moistened does not yield an alkaline reaction with the turmeric paper. It is thus distinguished from the oxalate of lime calculus. The two chief constituents may be roughly separated, and their relative proportions estimated, by digesting a portion of the calculus, previously reduced to fine powder, with very dilute acetic or sulphuric acid, which dissolves the ammoniaco-magnesian phosphate, leaving the phosphate of lime untouched.

In a complete analysis of this calculus, fat and extractive matter must be first removed from the powdered calculus by ether and alcohol. The urates must be separated by repeated boiling in distilled water, in which they are soluble. The residue is next dissolved in hydrochloric acid, and if anything remain it is either uric acid or oxalate of lime. The uric acid may be separated from the oxalate by liquor potassæ, and the latter recognized by the tests already described under the head of oxalate of lime calculus.

If carbonate of lime be present, effervescence occurs during the solution of the calculus in hydrochloric acid. In this case the mixed phos-

---

[1] Med. Chirurg. Trans., vol. xlii., p. 427.

phates are precipitated by ammonia, and to the filtered liquid oxalate of ammonia is added to throw down the lime. The precipitated phosphates are next digested in strong acetic acid, which dissolves the ammoniaco-magnesian phosphate with a small proportion of the phosphate of lime. The lime is first precipitated by oxalate of ammonia, and the phosphate of magnesia, as ammoniaco-magnesian phosphate, by ammonia. The animal matter, which resembles vesical mucus, is seen floating in the solution of the calculus in hydrochloric acid, and may be separated by filtration.

These salts may be readily detected by dissolving a small fragment of the calculus in hydrochloric acid, precipitating by ammonia, and examining the precipitate under the microscope, when the triple phosphate is seen in the form of stellate crystals interspersed with the amorphous granular phosphate of lime.

Carbonate of Lime Calculus.—Calculi consisting of this salt, associated with only a small proportion of animal matter, are extremely rare in the human subject, but common in herbivorous animals, in whose concretions the carbonate of lime is usually mixed with carbonate of magnesia; but carbonate of lime is often found in variable proportion in the oxalate of lime and phosphatic calculi. Brugnatelli was the first to recognize this calculus; the specimen analyzed by him was about the size of a pea, and contained some oxide of iron. Dr. Prout analyzed some small calculi, consisting entirely of carbonate of lime, which were white and friable. Mr. Smith, of Bristol, describes 18 removed from the bladder of a young man, and Brugnatelli 48 from a similar source, and 16, the size of a nut, from a woman.

Dr. Roberts[1] has reported the case of a gentleman from whose bladder myriads of minute calculi of carbonate of lime were voided with the urine, which was ammoniacal and contained a good deal of pus. The largest of the calculi was about the size of a mustard seed, but the majority were much smaller; they were mostly spherical in shape, many rudely cubical or pyramidal, of a full amber color and finely translucent. In some, the nucleus was an object resembling a glandular cell; in others, a prismatic crystal; in others, amorphous earthy-looking matter. Dr. Roberts also refers to a case of Dr. Haldane's in which calculi, identical in every respect with those just described, were found in the kidney after death.

Mr. Wagstaffe[2] has reported a similar case in which numerous calculi, composed of carbonate of lime, were found in the right kidney. Some of these concretions were spherular in shape and laminated, and closely resembled those found in the prostate. On the addition of dilute hydrochloric acid a copious evolution of gas occurred, and an animal matrix, the shape of the original spherule, was left behind.

These calculi may be described as small—the largest not exceeding the size of a large almond, white or gray, but sometimes yellow, brown, or red, and the surface dusted over with a white powder. The concretion shows no concentric laminae, but irregular waved lines similar to those in the mulberry calculus. The largest of the calculi in Mr. Smith's possession was so hard as to require the assistance of the lapidary's wheel to divide it, and was then capable of receiving a high polish. Others, passed by the urethra, are of a rounded and flattened form, and compact, lamellar, and of a light brown color. Carbonate of lime calculi are dis-

---

[1] Roberts, loc. cit., p. 277.   [2] Path. Soc. Trans., xix. 270.

tinguished by dissolving with effervescence in dilute hydrochloric acid, the solution yielding no precipitate with pure ammonia, but a copious white precipitate of oxalate of lime when oxalate of ammonia is added, and also a white precipitate of carbonate of lime, by the addition of alkaline carbonates. When such a calculus is placed in acetic acid and left for a few days, the mineral matter is dissolved, and the animal matter is left as a soft, light ball, preserving the stratified appearance of the original calculus, but with diminished translucency (Roberts). If phosphate of lime be present in the calculus, it is precipitated from the solution in hydrochloric acid by ammonia.

Alternating Calculi.—A very large proportion of the calculi formed in the human urinary apparatus are of this character, owing to the successive changes in the condition of the system, and that of the urinary secretion. These calculi exhibit a series of alternating layers, differing in chemical composition. They may be composed of two layers : e.g. a calculus may have a nucleus of uric acid, covered by a deposit of urate of ammonia, oxalate of lime, phosphate of lime, or mixed phosphates; or the nucleus may consist of oxalate of lime, followed by uric acid, urate of ammonia, phosphate of lime or mixed phosphates. Other calculi have three layers; thus a nucleus of uric acid may be followed by deposit of oxalate of lime and mixed phosphates, or oxalate of lime and uric acid; or a nucleus of oxalate of lime may be covered by uric acid and urate of ammonia, or a nucleus of mixed phosphates followed by phosphate of lime and mixed phosphates. Others, again, consist of four or even more layers; thus a nucleus of uric acid may be incrusted by urate of ammonia, uric acid, and urate of ammonia; or oxalate of lime may be followed by uric acid, oxalate of lime, phosphate of lime, etc.

The proportion of alternating calculi of two layers to the whole collection, in St. Bartholomew's, is as $1 : 2\frac{1}{2}$; in Norwich, $1 : 2\frac{2}{3}$; in Manchester, $1 : 2\frac{2}{5}$; in Bristol, $1 : 3$; in Swabia, $1 : 1\frac{1}{3}$; and in Copenhagen, as $1 : 2\frac{1}{4}$, giving a general proportion of $1 : 2\frac{2}{3}$.

Those of three deposits are, in St. Bartholomew's, as $1 : 6$; in Norwich, $1 : 6$; in Manchester, $1 : 26\frac{2}{3}$; in Copenhagen, $1 : 4\frac{1}{2}$, yielding a general average of $1 : 8\frac{1}{3}$.

Those of four deposits are stated to be, in the Norwich collection, as $1 : 26\frac{1}{4}$.

The deduction to be drawn from all these data is, that alternating calculi of all kinds form somewhat more than half of the whole.

The order in which different deposits succeed each other in these calculi is of considerable interest, as elucidating the successive changes in the condition of the body, corresponding with the progress of the disease. Dr. Prout has calculated that oxalate of lime succeeds uric acid in the ratio of $1 : 15\frac{2}{3}$, whereas uric acid succeeds oxalate of lime in $1 : 13\frac{5}{13}$; so that the alternation of these two substances is nearly equal. He states that oxalate of lime follows urate of ammonia more frequently than uric acid, in the proportion of $1 : 9\frac{2}{3}$, whereas urate of ammonia follows oxalate of lime in $1 : 38$. Phosphates succeed uric acid in the ratio of $1 : 9\frac{1}{2}$; the urate of ammonia, $1 : 12\frac{1}{2}$; and oxalate of lime, $1 : 7\frac{1}{2}$; whereas only three instances occur in the museums already mentioned in which phosphates are succeeded by uric acid or urate of ammonia, and oxalate of lime incrusts the phosphates only in the proportion of $1 : 253\frac{1}{4}$. The general ratio in which phosphates succeed other deposits in all the collections is $1 : 4\frac{1}{23}$. From these results Dr. Prout educes the law, that in urinary calculi, a decided deposition of the mixed phosphates is not followed by other deposits. The cause of the predominance of the mixed phosphates as an external investment has been already explained.

Cystic Oxide or Cystine Calculus.—This calculus is comparatively rare. It was found only in four museums mentioned by Dr. Prout in his work on urinary diseases, while the large collection of the College of Surgeons contained, in 1849, only three specimens. The general proportion is said to be about 1 in 304. In Dr. Golding Bird's analysis of the 374 calculi contained in Guy's Hospital Museum, the number of cystine calculi is set down as 11. M. Civiale, in the course of his extensive practice, had up to 1851 met with only 8 examples. Dr. Ultzmann met with 6 cystine calculi out of 545 specimens, and 2 others in which it formed the nucleus, the outer portion being composed of secondary phosphates. But very few specimens exist in America. Dr. Gross has never met with a cystine calculus. The Lexington Collection contains but two specimens. In the Museum of Grant College, Bombay, according to Dr. Vandyke Carter, the proportion of cystine calculi is .84 per cent.

Some years since, Dr. J. Risdon Bennett exhibited two specimens of cystine calculi to the Pathological Society. They were taken from the kidneys of a female, one having been formed in each kidney, and were of considerable size. These calculi were surrounded by coagulated blood, and were of an irregular figure, being moulded to the shape of the distended infundibula of the kidney. The larger one measured 2 inches in length, and rather more than an inch in breadth at its larger extremity; its weight was 204 grains. Of a very irregular form, it closely resembled a piece of crystallized ginger, having a yellow, smooth, waxy appearance externally, and a crystalline character internally. The surface of one of its nodules was white, smooth, and polished apparently from friction. The smaller calculus was of a similar irregular form, but differed considerably in its general aspect. It measured 1½ inches in length, and ¾ inch in breadth, and weighed 116 grains. Its external surface, blackened by dried blood, was rough, channelled, and presented the general appearance of being worm-eaten; it was of much less density than the larger one. It had, however, the same crystalline structure. Both calculi, on analysis, were found to consist of pure cystine.

Dr. Hanfield Jones exhibited another specimen of this calculus to the Harveian Society, in 1854. The patient, like the subject of Dr. Bennett's case, had been insane. The four calculi exhibited had come away spontaneously, with very little pain, and some small, sand-like fragments were subsequently discharged. The calculi cut easily, and their surfaces had a waxy lustre; they presented under the microscope crystalline prisms of various sizes, generally much elongated and mixed with rather fatty epithelial scales. Ammonia dissolved the calculus completely, leaving, on evaporation, hexagonal plates and elongated prisms.

At a meeting of the Pathological Society, November 2d, 1875, Mr. Christopher Heath exhibited 19 specimens, all removed from one patient, a man aged 28. A large number of them were unquestionably imbedded in the prostate, and there were three larger ones in the bladder. These latter were peculiar in having a distinctly laminated structure, the outer layers being still pure cystine.

It is remarkable that the disposition to the formation of these calculi appears to be hereditary. Out of 22 cases, 10 occurred in four families, and in 3 cases the subjects of the complaint were brothers.

In a very marked case, however, recently reported[1] by Dr. Ultzmann, a most careful examination failed to obtain any evidence of hereditary predisposition. The patient was a boy about three years of age, and a cystine calculus weighing nearly 80 grains was removed by the lateral operation. Symptoms of pyelitis were present at the time. The patient was under observation for two years afterwards, during which time he continued to suffer from cystinuria and pyelitis, the bladder remaining free from concretions. The symptoms pointed to the presence of a renal calculus composed of cystine. Cystine calculi are originally formed in the kidneys, and not in the bladder. They therefore belong to the primary stone formation. They are not peculiar to the human body, a calculus of this kind having been found in the bladder of a dog.

Cystine calculi are usually small, rounded, of a yellowish color, with a smooth, semi-transparent, and glistening surface. Sometimes, however, vesical calculi of cystine attain a weight of three or four ounces. They resemble those composed of ammoniaco-magnesian phosphate in general appearance, but are more compact in structure. They have a soft consistence; the section is confusedly crystalline and not laminated. The fracture exhibits a crystalline structure. The largest known calculus of this substance is in the museum of University College, London; it weighs 850 grains, and is formed of concentric semi-crystalline layers. Another specimen, weighing 740 grains, is in the museum of St. Bartholomew's Hospital. They are said, by Dr. Golding Bird, to undergo a change of color from yellowish-brown to a fine bluish-green when long kept and exposed to daylight. Cystine calculi are much more friable than those of uric acid or oxalate of lime. Dr. Roberts mentions a case reported to him by Mr. Southam in which 90 grains of the fragments of a pure cystine calculus had been voided by a little girl after a single crushing.

Cystine is not unfrequently found as a deposit in the urine, which then has a greenish-yellow appearance, a disagreeable odor, a neutral or slightly acid reaction. It rapidly becomes alkaline, and deposits cystine, or an oily film forms on the surface, consisting of cystine mixed usually with ammoniaco-magnesian phosphate, and at the same time uric acid and urea are sometimes more or less deficient in quantity.[2] Dr. Golding Bird obtained .340 of a grain from 1,000 grains of urine of sp. gr. 1.014.

Cystine is remarkable for its solubility in a large number of menstrua. It is alike soluble in the mineral acids and in the caustic alkalies, and the solutions yield colorless granular crystals by spontaneous evaporation. It is sparingly soluble in pure water, and insoluble in alcohol, acetic, tartaric and citric acids, and in solution of bicarbonate of ammonia. During the evaporation of the solution in caustic ammonia, the ammonia escapes, the crystals of pure cystine are deposited either as hexahedral prisms or plates. From its alkaline solutions it is precipitated by acetic acid in the form of a white crystalline powder. Cystic oxide is remarkable for containing 26.58 per cent. of sulphur, and it was ascertained that the large calculus in University College yielded 19 per cent. of this substance. When heated cautiously on platinum foil, it is surrounded by a blue flame, which disappears on the application of a stronger heat,

---

[1] Wiener Medizinische Presse, July 21st, 1878.
[2] Loebisch (Wiener Med. Jahrb., 1877, Liebig's Ann., 182) has pointed out that, inasmuch as cystine contains only one equivalent of nitrogen, the diminution in the uric acid and urea cannot be due to the removal of the nitrogenous elements by the cystine.

and at the same time it gives out a peculiarly disgusting odor, resembling that of garlic.

These calculi are distinguished from all others, except those of xanthine, by their ready solubility in acids and alkalies, especially by dissolving readily in liquid ammonia. They are best distinguished by dissolving a fragment in liquor ammoniæ, and allowing the solution to evaporate spontaneously on a slip of glass. The residue, examined by the microscope, is found to consist of minute six-sided tables or prisms, characteristic of cystine. The peculiar nauseous odor when burnt is also diagnostic of this calculus. It is distinguished from xanthine by its solubility in carbonate of potash and dilute hydrochloric acid, and from uric acid by its solubility in dilute hydrochloric, phosphoric and oxalic acids.

Cystine calculi are usually free from admixture of other substances, but uric acid sometimes forms the nucleus, and a layer of uric acid has been found surrounding a nucleus of cystine. It is also occasionally associated with ammoniaco-magnesian phosphate and carbonate of magnesia. Dr. Roberts gives an engraving of a specimen in the Manchester Infirmary Museum, in which "the central nodule is uric acid; around this is a body of pure cystine; overlying this a layer of mixed uric acid and cystine; and enveloping the whole, a crust of secondary phosphates mixed with cystine."

Silicious Concretions.—Although, as Dr. Golding Bird informs us, silicic acid has been found in crystals, forming part of a calcareous concretion, it does not appear that any calculus has ever been discovered into the composition of which silicic acid enters to any considerable degree. The cases on record of silicious pebbles being extracted from the bladder admit of being explained in a simple manner. The foreign bodies were introduced into, not formed in the bladder. A curious example of this kind is related in Mr. Coulson's work on "Lithotrity and Lithotomy." A female was supposed to labor under calculus; lithotrity was about to be performed, when it was found that the foreign body could be removed by dilatation of the urethra. This was effected, and the supposed calculus turned out to be a silicious pebble which the woman had slipped into her bladder.

Fibrinous and Blood Concretions.—This rare calculus was first recognized and described by Dr. Marcet. It was about the size of a large pea, and was passed, after much suffering, by a gentleman about 50 years of age. He had previously passed three similar concretions, and had suffered from symptoms of calculus for several years. Dr. Roberts met with a similar instance; the concretion was "about the size of a small pea, and was passed by a man of 35, whose urine was not albuminous. Its texture was hard and brittle, its external surface rough, and its color dark reddish-brown." Dr. Roberts possesses a very fine blood concretion, as large as a small walnut, taken from the bladder of a sheep. It has the appearance of baked clay, and breaks with a dull fracture, like a piece of catechu. Dr. Prout states that he has seen several specimens, but none are enumerated in the collections to which reference has been repeatedly made. Their relative frequency is, therefore, unknown.

Fibrinous calculi are described by Dr. Prout as being usually of an amber color and waxy consistence, with more or less of a fibrous texture. They are, when unmixed, wholly combustible, giving out an odor of burnt horn, and leaving only a small quantity of ash; insoluble in water, alcohol, or ether, but soluble in caustic potash, from which a precipitate is thrown down by hydrochloric acid. They are soluble in acetic acid, and ferro-cyanide of potassium produces a white precipitate in the solution. Nitric acid slowly dissolves these calculi. Dr. Prout asserts that they really consist of fibrine, while others consider that they are formed of vesical mucus or albumen. The formation of these calculi appears to be connected with the occurrence of renal hæmaturia. Such concretions may easily serve as nuclei for the formation of true calculi. They have been found lying loose in the pelvis of the kidney in cases of rupture of that organ, and also associated with calculous pyelitis, as in a remarkable case described by Dr. Alison.[1]

Uro-stealith.—This name has been applied to certain fatty or saponaceous concretions, which have been found in the bladder. But few instances are on record. The Museum of the College of Surgeons contains two specimens of vesical calculi, composed of a central fatty or saponaceous mass, covered by a thick layer of phosphates. Both belonged to Hunter's collection, and are described in the catalogue, published in 1842, as consisting of the earthy phosphates deposited upon a mass of oleate and margarate of lime. They were supposed to originate from the soap contained in injections which had been used to wash out the bladder, the earthy bases of the urine having become precipitated in combination with the fatty acids of the soap, in the form of a semi-gelatinous, sparingly soluble compound. Around this the phosphates had become deposited. Heller[2] has described similar concretions. His patient was a man, 24 years of age, who complained of difficult micturition and pain in the region of the right kidney. He passed a number of concretions of a rounded form, soft and elastic, and varying in size from that of a nut to that of a hemp-seed, the majority being about the size of a pea. The concretions, as they were taken from the urine, were soft and elastic like india-rubber, but they dried up into hard wax-like masses, some greenish-yellow, others light brown or blackish in color. Some were covered with crystals of the triple phosphate. They dissolved readily in liquor potassæ, forming a soap. These were almost entirely soluble in ether, insoluble in boiling water, and nearly so in alcohol. They melted with heat, and burned with a bright yellow flame, giving off a peculiar odor resembling a mixture of shellac and benzoin. The residue, somewhat bulky, consisted of lime. The patient's urine contained no uric acid. Dr. Moore,[3] of Dublin, has also described similar specimens obtained from a single patient. Two of these were passed in the urine, and two others were removed after death from the patient's bladder. The two former were small concretions, dark-brown in color, and soft and waxy in consistence, and partially soluble in liquor potassæ. Dilute nitric acid had little effect upon them. Heated in the flame of a spirit-lamp, they fused and afterwards burnt with a bright flame, leaving a blackish ash which became incandescent and white before the blow-pipe. A considerable portion of the calculus was soluble in boiling alcohol and separated on cooling as a whitish deposit, exhibiting, under the microscope, numerous fat corpuscles, but no crystalline plates. Of the calculi removed after death, one was large, being the size and nearly the shape of a small hen's egg; in the centre was a cavity containing, but not filled by, a quantity of the dark-brown substance already described. The

[1] Quoted by Mr. Poland, Holmes's System of Surgery, vol. iv., p. 1030.
[2] Heller, Archiv, Bd. ii., p. 1, 1845.
[3] Observations on Urine and Urinary Calculi, by W. D. Moore, A.B., M.B., Dublin Quarterly Journal of Medical Science, vol. xvii., p. 473, 1854.

layers of calculus next adjoining this substance were composed of ammonaco-magnesian phosphate and phosphate of lime; the second calculus was composed entirely of this latter substance. The nucleus of the larger calculus was composed of uro-stealith, which, as Dr. Moore points out, is distinguished from cholestearine by the fact that it is deposited from its alcoholic solution, when cooled or partly evaporated, in the form of oily globules. In Dr. Moore's opinion, the lime and the fatty or waxy substance form a combination resembling a soap, and, as illustrating this combination, he refers to the existence in the feces of soaps of lime and magnesia, formed by the decomposition of a portion of the fat met with in the intestinal tube.

A Calculus containing Indigo.—A very remarkable calculus was exhibited by Dr. Ord at a recent meeting of the Pathological Society (March 5th, 1878). It was contained in the pelvis of the right kidney of a patient under the care of Dr. Bloxam. The left kidney was sarcomatous, and contained a calculus of carbonate and phosphate of lime. The calculus containing indigo was partly of dirty dark-brown color, and partly covered with a blackish-blue matter which made a blue mark when rubbed on paper. The stone had a circumference equal to that of a shilling, and weighed 40 grains. When heated, it gave off a sooty smell, like that of burning indigo; a great part of it disappearing as smoke. The blue portion gave a blue solution with hydrochloric acid; when heated in a test-tube to a degree below redness, it gave off a bulky vapor, resembling that of iodine, which deposited crystals on a cool surface; the crystals so deposited were characteristic, being elongated six-sided tablets with pointed ends. Powdered and rubbed up with strong sulphuric acid, the substance yielded a blue solution which, when diluted and filtered, furnished a beautiful blue fluid. Spectroscopically examined, the fluid stopped the yellow by a defined line, the centre of which corresponded to the sodium-line. This appearance is characteristic of solutions of indigo. Various theories have been advanced with regard to the origin of indigo in urine. With reference, however, to this calculus, the first of its kind placed on record, Dr. Ord suggested that probably certain products of the disease in the left kidney had been absorbed into the blood, and excreted by the right kidney as indigo-blue. These products had no natural exit, on account of the obstruction in the left ureter, which was blocked up by a soft round-celled sarcomatous tumor. Dr. Roberts observes that indigo-blue is frequently seen in putrescent urines, forming glistening blue shreds and fibres on the sides of the glass and the surface of the urine. In cases of cystitis it sometimes colors the precipitated urate of ammonia, and it has also been found, as a deposit in such cases, in the form of small rhombic crystals. Beneke observed a blue-colored urine in a patient suffering from Bright's disease and general dropsy.[1] According to Jaffé, normal urine contains indigo in the proportion of 4.5 to 19.5 milligrammes in 1,500 C.C. The urine of horses contains 23 times as much. No vesical calculus of this kind has as yet been reported.

Prostatic calculi and concretions will be described with other affections of the prostate gland.

---

[1] Loebisch, loc. cit., p. 92.

CHAPTER XIX.

THE CAUSES OF STONE.

THE importance of acquiring a knowledge of the causes of urinary calculi must be apparent to every one who considers that only by such aid are we likely to discover means for preventing the occurrence of this most painful disease, as well as its recurrence after the sufferings and hazards of an operation.

In former times, the causes and mode of formation of calculi in the bladder were very imperfectly understood; but much has been done of late to elucidate these subjects by careful examination of the chemical constituents of the urine, of the concretions which form in it, and the changes which take place during respiration and digestion. Much, however, remains to be done, before a complete account of the etiology of calculous complaints can be given.

The great determining cause of the formation of urinary calculi is the presence in the urine of some sparingly soluble substance, which is capable of concreting into a solid mass, either by itself or with the addition of an animal cement. The precipitation and agglomeration of the calcareous principles are physical phenomena; but vital acts likewise concur, by producing the precipitable matter in excess, and also by furnishing the animal substances which serve to bind its particles together.

In endeavoring to trace the formation of calculi to its causes, the following method appears the most logical, and in pursuing it we may inquire: What are the sparingly soluble substances which enter into the formation of urinary calculi? Whence is the excess of insoluble matter derived? What are the circumstances which promote or determine its precipitation from the urine? And lastly, By what means, or under what circumstances is it collected into a mass so as to form a stone? If we omit from consideration the few cases in which the stone is formed of uro-stealith, xanthine, or carbonate of lime, it will be found that all the remaining species of calculi are composed of uric acid or urates, the earthy phosphates, oxalate of lime or cystine. These are the four sparingly soluble substances, the agglomeration of which forms urinary calculi.

The great majority of urinary calculi are composed of uric acid either alone or combined with bases. Uric acid is a constituent of healthy urine in which it exists under the form of urate of ammonia, or, as some chemists hold, of urate of soda. From eight to ten grains of this substance are daily excreted with the urine during a state of health; urine likewise contains the well-known substance urea, which closely resembles uric acid in composition, but on account of its solubility never is precipitated to form a stone. About 500 grains of urea are separated daily from the blood by the kidneys.

The composition of urea is $C_2H_4N_2O_2$.
That of uric acid is $C_{10}H_4N_4O_6 + 4$ aq.
The principal difference is made up by carbon and oxygen, the ele-

ments of carbonic acid, and, by a theory in vogue some years ago, the excessive formation of uric acid, to which calculi were supposed to be due, was referred to some defect in that process by which carbonic acid is produced; viz., in the function of respiration. Uric acid was regarded as a stage of the retrograde metamorphosis through which the nitrogenous tissues of the body have to pass before their final transformation into urea, and an excess of uric acid was regarded as evidence of the incompleteness of the process.

There are, however, no sufficient reasons for regarding uric acid as a substance in an intermediate state of transformation, and physiology has not as yet informed us as to the precise seat of its formation, or explained the processes involved in its production. In normal urine it is probably combined with a variety of bases,—soda, ammonia, potash, lime and magnesia. Its union, however, with these substances is very feeble; the presence of almost any acid causes it to be precipitated. The readiness with which the precipitation takes place, and the insolubility of uric acid in neutral or acid solutions are, as we shall presently find, the main determining factors in the formation of gravel and calculus. The variations in the quantity eliminated are, according to Ranke, "in great measure independent of differences in age, sex, height, weight or temperature, but stand in close connection with the ingestion of food, diminishing to a minimum of 3.7 grains with abstinence, and rising to a maximum of 32.5 grains on a full meat diet."[1] The influence of muscular exercise is uncertain; sometimes the quantity of uric acid is increased, and sometimes it is diminished. Dr. Bence Jones has shown that "there is no relation whatever between the acidity of the urine and the absolute amount of uric acid which it may contain; for in the urine which is most acid and which deposits the largest uric acid sediment, very little uric acid may really exist, whilst that which contains most uric acid may hold it in perfect solution, and may have but a feeble acid reaction." A concentrated state of the urine favors the precipitation of uric acid and its salts, and the same effect is sometimes produced by a reduction of temperature. Within the body, however, this latter cause can scarcely be supposed to operate, and the precipitation of uric acid in any part of the urinary organs may be regarded as dependent upon one or both of the following conditions: (1) increased acidity of the urine, owing to the presence of some other acid; (2) increased proportion of uric acid to the water of the urine, either positively or relatively. Uric acid, thus precipitated, becomes "gravel," and the agglomeration of particles of gravel into a compact mass produces a calculus or stone. Dr. Golding Bird attributed the formation of uric acid deposits to the following causes: "(1) The waste of the tissues being more rapid than the supply, as in fever, rheumatism, etc. (2) The supply of nitrogen in the food being greater than is required for the reparation of the tissues, as in over-indulgence, especially in the use of animal food. (3) The process of digestion being insufficient to assimilate an ordinary and normal supply of food, as in dyspepsia. (4) Obstruction to the cutaneous outlet for nitrogenous secretions, as met with in diseases of the skin, or due to variability of climate, etc. (5) Congestion of the kidney from injury or disease." With regard to these causes it may be observed, that the increase of uric acid in the urine may be due to any of the first four, and that disease of the kidney might prevent the excretion of uric acid, par-

---

[1] Carpenter's Human Physiology, p. 531.

ticles of which might then be deposited in the renal tubules. The desquamation also of renal epithelium, which occurs in some diseases of the kidney, would probably favor the formation of calculus by entangling the lithic acid or lithates. It has been already explained, however, that an increase in the proportion of uric acid is not an essential condition for its precipitation, a phenomenon which is more often due to the presence of some other acid.

Uric acid is by far the most frequent, and therefore the most important constituent of urinary calculi. About 80 per cent. of all calculi are formed of it or its salts, whilst a still greater proportion of the nuclei are composed of this substance. We may refer to Dr. Ultzmann's researches on this subject described in a previous page (197). The primary stone-formations, which include uric acid and its salts, oxalate of lime and cystine, are the most difficult to account for; the secondary formations, those which occur in alkaline urine, can for the most part be easily traced to definite causes.

The occurrence of calculi in early life has been attributed to those infarctions of uric acid which are often found in the kidneys even of new-born children. The urine of infants, during the first week of life, contains a large quantity of uric acid, crystals of which often form spontaneously when the fluid is left at rest. During the second week the quantity of uric acid becomes diminished, and in the third or fourth week is reduced to a normal amount. The uric acid infarctions have been very carefully studied by Virchow, who believed that they were to be found in the kidneys only of those infants in whom respiration had taken place. He thought, therefore, that such formations had a medico-legal significance, inasmuch as they were to be regarded as a sign of live birth. Other observers, however, have discovered these infarctions in the kidneys of infants whose lungs had certainly never been inflated with air, and their occurrence during intra-uterine life has been placed beyond dispute. It is, however, admitted that the infarctions are comparatively rarely found in the kidneys of children who have died immediately after their birth; and, on the other hand, that they are more common when death has occurred after a few hours, and that they are most frequently met with when life has been prolonged for several days.[1] From that time onwards they occur with diminished frequency, but they have been found by Virchow in children from the third to the fifth month. According to Ebstein, they have been discovered in about 47 per cent. of all the cases that have been examined. Under normal conditions, they are gradually dissolved by the urine and removed from the kidneys, and the persistence of deposits of uric acid, either free, or combined with bases, in the urine of infants must be regarded as abnormal. Such deposits, however, are very prone to occur in the urine of weakly infants, and especially of those who suffer from the various forms of intestinal catarrh. The urine becomes altered in composition, and does not contain the requisite proportion of solvent material. Ultzmann has drawn attention to the fact, often noticed in foundling establishments, that on the napkins used for weakly infants a pulverulent substance, resembling brick-dust, may frequently be found.[2] He has very carefully examined this powder,

---

[1] Ebstein, v. Ziemssen's Handbuch der Spec. Path. u. Therapie, Bd. ix. Hälfte ii., s. 216.
[2] Ultzmann, Ueber Harnsteinbildung, p. 152. In the preparation of this chapter, much assistance has been obtained from Dr. Ultzmann's admirable essay, which forms No. V. of the Wiener Klinik.

and the results at which he has arrived are as follows:—Uric acid is one of its constituents, as evidenced by the muroxide test. Placed under the microscope, and with a power magnifying 300 diameters, the deposit is found to be composed of brown, spherical or dumb-bell-shaped masses. These being still further examined with an immersion lens, are found to present two forms, in one of which the globules are larger and somewhat lighter in color, distinctly spherical, and without any evidences of structure; in the other, the masses are smaller, disk-shaped, flat, and with a central darker portion from which radiating streaks proceed. This latter form is found to occur in recent preparations, but when the substance of the kidney has undergone putrefactive change the larger spherical globules are found to prevail. Both forms dissolve gradually in acetic or hydrochloric acid, and in these solutions crystals of uric acid soon appear; in regular rhombic prisms when the former acid is added, while in the hydrochloric acid solution the uric acid crystallizes in larger spiny clumps.

From these reactions Dr. Ultzmann is led to believe that the kidney-infarction in question consists in part of urate of ammonia, and in part of urate of soda. The disk-shaped masses he regards as being composed of the latter salt, while the larger globules consist of the former. He thinks that urate of soda is the primitive ingredient of these infarctions, and that the urate of ammonia is the result of decomposition. Other observers, however, have held different views, and the question is a very difficult one to decide. The important fact is that these infarctions contain a large amount of uric acid, and that they often block up the pyramidal tubules to such an extent as to make them appear as if injected.

Under normal conditions, as before alluded to, these infarctions are dissolved by the urine and removed from the kidney. If, however, the quantity present is too great to be dissolved, or if the urine does not contain the proper solvents, or if some of the uriniferous tubes are so blocked up by deposit that the urine cannot make its way through them, masses may remain and form starting-points for the deposition of crystals, and thus a nucleus may arise around which, by further deposition, a renal calculus may at last form.

Such a calculus, at first very minute, gradually increases in size, and remains in the pelvis of the kidney until its own weight and the flow of urine cause it to descend into the ureter, and thence into the bladder. It then takes one of two courses: either it remains in the bladder and forms the nucleus of a vesical calculus, or it is discharged, with the urine, by the urethra.

Uric acid infarctions are in all probability the principal cause of calculi in early life, and to their occurrence may be attributed the frequency of calculous disease among children. Civiale's tables show that more than half of the entire number of cases of stone occur in persons under twenty years of age, and that in 36 per cent. of all cases the patients are less than ten years old. Dr. Ultzmann draws attention to the fact that infarctions in the kidneys of infants are never found to consist of oxalate of lime, but always of uric acid or its salts.

The occurrence of calculi in adult life cannot, however, be accounted for by supposing that they originate from these infarctions; a primary stone-formation must undoubtedly occur quite independently of such a causation. On this subject numerous theories have been in vogue from time to time. According to one of these the formation of gravel was supposed to be due to certain diatheses, that is, there was assumed to be either an excessive secretion of some of the normal constituents of the urine, or the development of a morbid product which was excreted by the kidneys. These products, whether normal or abnormal, were deposited in some part of the urinary passages, and formed gravel or calculi. A urate, oxalate, and phosphatic diathesis was supposed to exist, and the production of gravel was always attributed to constitutional causes. It is well known, however, that calculi often form where no such "diathesis" is demonstrable, and without there being any excess in the urine of those substances which form the concretion. Phosphatic deposits also occur from purely local causes. A chronic cystitis, for example, leads to production of mucus. This acts as a ferment and causes the urea to be converted into carbonate of ammonia, the presence of which in the urine causes the precipitation of the phosphates, which may subsequently form a calculus. In such a case there is no diathesis, no increase in the production of the constituents of the stone. Again, with regard to uric acid, its presence in the urine as a deposit is no certain evidence of increased production; such a deposit more frequently indicates that the urine contains materials which interfere with the ordinary solvent action of the fluid. The deposition of uric acid is commonly due to the presence of the acid phosphate of soda, which decomposes the urates of ammonia and soda and sets the acid free. The presence also of lactic acid, which may result from the acid fermentation of urine, is sufficient to produce a similar result. Increased production of uric acid is, of course, another factor which determines the formation of gravel and calculus, and it probably often occurs contemporaneously with increased acidity of the urine.

Deposits of uric acid are not unfrequently found in the uriniferous tubules of adult kidneys. Frerichs (quoted by Ebstein) found in one case the pyramidal tubules occupied by little columns of deposit consisting of urate of soda. He found also, in a case of Bright's disease, in the tubuli both of the cortex and medullary substance, which were blocked up with amorphous fibrinous coagula, large brown crystals of uric acid, some separate, others united into nodules the size of a pin's head. The cortical substance was studded over with sand-like granules. The pelvis of the kidney contained gelatinous coagula of fibrine which entangled numerous crystals of uric acid. Uric acid and the urates of soda and ammonia are often found in dilated tubules, and in cysts of the kidney. These deposits are of frequent occurrence in the kidneys of gouty patients. They occupy the interior of the tubules and the intervals between them, and may attain the size of a hemp-seed or pea. Ebstein has also noticed crystals of oxalate of lime in the casts of the tubules, and the same deposit has also been found in the substance of the kidney; its origin is by no means evident. It is doubtful whether it is a constituent of normal urine, though Schultzen,[1] Furbringer and others assert that under ordinary circumstances 20 milligrammes are excreted daily. It is one product of the decomposition of uric acid, and, as Dr. Roberts observes, it constitutes probably one of the penultimate stages in the series of decompositions through which the effete tissues pass preparatory to their final exit from the body. It possesses a great affinity for lime, its combination with which is extremely insoluble. According to Neubauer the acid phosphate of soda is the solvent which the urine contains. The greater the proportion of this phosphate in the urine, the less will be the

---

[1] Quoted by Ebstein, loc. cit., p. 208.

quantity of oxalate of lime precipitated, and this quantity is therefore no indication of the amount actually present.

In accordance with the theory which connects calculous disorders with certain diatheses, a close alliance has been supposed to exist between these affections and gout and rheumatism. Gouty patients doubtless excrete an increased amount of uric acid, but it does not appear that they are particularly liable to suffer from calculus. The same may be said with regard to rheumatism, and it is evident that something more than an augmented excretion of uric acid is required for the formation of a calculus.

Another theory is that supported by Meckel.[1] He believed that the origin of calculi was due to the operation of a peculiar catarrh. He thought that all primary calculi originated from acid fermentation of the urine, and were composed of oxalate of lime and mucus; that all urates were the result of metamorphosis, and that phosphates were a later development. Oxalate of lime he regarded as a constant and normal constituent of the urine, and his theory was that in catarrh of the urinary passages from any cause, this salt united with the mucus and formed a new compound which was the starting point of a calculus. There are numerous objections to this theory. Calculi are often passed by persons who have never shown the least symptom of catarrh of the urinary organs, and it is far more probable that catarrh, when it does occur, is the consequence of the presence of the gravel or calculus, and not the cause of their formation. The catarrh, however, doubtless assists to increase the size of the calculus by furnishing materials for its growth.

Heller[2] divided calculi into two classes, a primary and a secondary. Those formations which were connected with the decomposition of the urine he placed in the latter class, while the first contained those calculi which had a different origin. Heller believed that calculi of the first class might originate in the bladder as well as in the kidneys.

Ultzmann's classification has been already briefly referred to, but we will now explain his views at greater length. He regards a primary stone-formation as that which is due to the presence of those substances which form deposits in acid urine. These are uric acid, urate of soda, oxalate of lime, and cystine. Formations of these substances occur always in the kidney. Secondary stone-formation takes place in the bladder, and is due to the presence of those substances which form deposits in alkaline urine. These are the urate of ammonia, carbonate of lime, the amorphous phosphate of lime, phosphate of ammonia and magnesia, and crystalline phosphate of lime. The most important part of every calculus is its nucleus, around which all the other layers are deposited. If the nucleus consists of any of the substances deposited in acid urine, the stone is due to a primary formation, no matter what may be the composition of the outer layers; if, on the other hand, the nucleus is composed of any of the substances which form sediments in alkaline urine, the formation of the stone is to be referred to a secondary process, even though the outer layers consist of uric acid.

From Dr. Ultzmann's researches (of which an account has been given at p. 197) it appears that in the primary stone-formation, the nucleus is formed of uric acid in nearly 94 per cent. of all specimens. What are the special peculiarities which characterize this substance as a constituent of

---

[1] Meckel, Microgeologie, Berlin, 1856.
[2] Heller, Die Harnconcretionen, Vienna, 1860.

calculi, and why is the primary stone-formation induced by it in so large a majority of instances?

Dr. Ultzmann refers to the common belief that a predisposition to the formation of calculus exists when an excessive deposit of uric acid or urates, or of oxalate of lime is frequently found in the urine. He does not, however, think that any special predisposition is thus indicated. Many patients constantly pass large quantities of uric acid or oxalate of lime without the development of any form of calculus, and, on the other hand, patients may suffer from all the symptoms of renal calculus without any evidences of abnormal deposit in the urine. From these facts he infers that the formation of calculus does not depend upon the quantity of uric acid actually present in the urine. He has instituted a series of experiments which render it probable that the formation of a calculus is mainly conditional upon the form which the uric acid crystals assume. When hydrochloric, acetic, or any other acid is added to normal urine, the uric acid is deposited, and the form of its crystals depends upon the extent to which the urine has been acidified. The addition of a few drops of acid causes the uric acid to appear in normal rhombic prisms or lozenges, but with hydrochloric acid in the proportion of 1 to 20 or 1 to 10 the crystals are deposited in the form of elongated pointed rods, attached to each other, and radiating like the ears of corn in a sheaf. Sometimes two such fan-like masses are joined in a common centre. Between these and the lozenge-shaped crystals there are many intermediate variations, and the important point is that the spinous character of the crystals becomes increased in proportion to the quantity of acid. All these forms occur as natural deposits in urine, the shape being influenced, as in the experiments, by the quantity of acid present. The spinous sheaf-like crystals, or masses of crystals, may easily be supposed to form the starting-point of a calculus. Owing to their shape they would be much more prone, than crystals of the rhombic form, to adhere to the tubuli uriniferi or the pelvis of the kidney. It is these spiny crystals, Dr. Ultzmann observes, which are often found in the urine in cases of calculous pyelitis and albuminuria.[1] He attributes primary stone-formation, as a general rule, in adults, to the presence of these crystals, and he refers to the fact that out of many hundreds of spontaneously voided calculi, a very small minority are composed of oxalate of lime, and that some even of these latter have a nucleus of uric acid.

With regard to secondary stone-formation, Dr. Ultzmann's view is that this takes place always in the bladder, and is due to a process of incrustation. Foreign bodies, mucus, pus, and coagulated blood may serve as nuclei around which various layers are deposited. Calculi which have descended from the kidneys may also become thus incrusted with secondary deposit. This incrustation usually consists of those substances which form deposits in alkaline urine, but sometimes the deposits of acid urine. If, for example, a smooth uric acid calculus reaches the bladder and fails to produce inflammation, the calculus continues to increase in the acid urine, just as the growth of crystals goes on in saline solutions, and the secondary layers of the calculus are formed from the deposit of acid urine. If, however, catarrh be induced by the presence

---

[1] On this subject see a paper by Dr. W. M. Ord, St. Thomas's Hospital Reports, 1870, p. 335. Dr. Ord's experiments show that the forms of uric acid, as deposited from urine, are often determined by the nature of other constituents. Various conditions of albumen appear to exercise a remarkable influence in determining the kind of crystal deposited.

of the calculus, the urine speedily becomes alkaline, and the secondary layers are formed of those substances which are deposited in alkaline urine, viz., urate of ammonia, ammoniaco-magnesian phosphate, and phosphate of lime. The secondary formations are especially prone to occur where there is any disease of the bladder, prostate, or urethra.

With regard to the metamorphosis of urinary concretions, we find that when calculi of primary formation remain in contact with alkaline urine, a marked change takes place. According to Meckel, there are two kinds of metamorphosis: a central or centrifugal, and a peripheral or centripetal. Meckel believed that all primary calculi had their origin from oxalate of lime and mucus, and it was therefore necessary to assume that the oxalate of lime became converted into uric acid, afterwards into urate of ammonia, and finally into the earthy phosphates. There is, however, no evidence in favor of any such metamorphosis; but, on the other hand, if all calculi consisted primarily of oxalate of lime, we should expect to find that all renal concretions passed spontaneously would be composed of this substance. Moreover, if the metamorphosis proceeded centrifugally, we should find nuclei of phosphates covered by layers of uric acid, and such calculi are of very rare occurrence.

On the other hand, as Dr. Ultzmann points out, a peripheral or centripetal metamorphosis is by no means unfrequent. This change commences on the surface of the calculus, and gradually involves the deeper layers. It occurs when the urine becomes alkaline. The carbonate of ammonia acts upon the uric acid, urate of ammonia is formed; this becomes dissolved in the excess of alkali, and the phosphates take the place of the uric acid. This process much more rarely occurs when the calculus is composed of oxalate of lime, which is far less readily acted upon by alkaline solvents. The fact that phosphatic calculi often contain a nucleus of uric acid is explained by the occurrence of this metamorphosis.

Evidences of this kind of metamorphosis are often to be found in large renal calculi, which are composed of phosphates, with a nucleus of uric acid, and it is a noticeable fact that persons with such a calculus in their kidneys (as shown on examination after death) have been known to pass spontaneously small concretions of uric acid. Such an occurrence is sufficient to prove that the primary formation was due originally to the same constituent. It is the production of this centripetal metamorphosis which is the object sought to be attained when alkalies are administered for the relief of renal calculus.

Dr. Ultzmann's general conclusions with regard to the formation of calculi are as follows:

1. The formation of a calculus may be either primary or secondary.
2. Primary stone-formation takes place always in the kidneys, whereas the secondary stone-formation is confined almost exclusively to the bladder.
3. The primary stone-formation originates, with a few exceptions, from uric acid; the secondary stone-formation, from the earthy phosphates.
4. The spiny clumps of uric acid are those forms which are most prone to be the starting-points of calculi.
5. The oxalate of lime is adapted for the formation of layers rather than nuclei.
6. Uric acid infarctions in the kidneys of new-born infants are the cause of the frequent occurrence of urinary calculi among children.

7. The metamorphosis of primary calculi is always centripetal.

Such are Dr.. Ultzmann's views on several points connected with the formation of calculus. It has been thought desirable to give a somewhat full account of them, as they are the result of long-continued and minute investigation, and up to the present time they have not been referred to by English writers on the subject.

Having discussed the various theories which have been advanced, to account for the frequency with which uric acid is found both as a nucleus and as a constituent of calculi, it is necessary to add a few remarks with regard to oxalate of lime and cystine. During the formation of calculi composed of these substances, the urine is always acid. Oxalate of lime is often found entering into the composition of the layers of a calculus, but less frequently as a nucleus. Dr. Ultzmann found that, out of 224 urate concretions, 15 had a nucleus of oxalate of lime, whereas, out of 136 oxalate of lime calculi, the nucleus was composed of uric acid in 124 instances, and of oxalate of lime only in 5. As before mentioned, oxalic acid is said to exist in normal urine, though only in very minute proportions. It can be formed by a series of processes from uric acid, the intermediate stages being alloxan, parabanic acid and oxaluric acid. Oxalic acid can be produced in the laboratory from a large number of substances which are constantly taken as food, *e. g.*, fats, starch, sugar, etc., and a similar change may be supposed to take place within the organism. The consumption of certain vegetable substances, such as rhubarb and sorrel, which contain large quantities of oxalate of lime, might also account for its presence in the urine. Crystals of this salt, accompanying infarctions of uric acid, have been found in the kidneys, and Dr. Beale has often observed a little mass of such crystals occupying the centre of a nucleus of uric acid.

With regard to the origin of oxalic acid, Dr. Roberts points out that "it may be partly derived from the blood and appear in the urine at the moment of secretion, and partly be produced after the urine is secreted, by conversion from uric acid." Dr. Owen Rees,[1] however, some years ago expressed his disbelief in the origin of oxalate of lime from the blood, and contended that oxalic acid and the oxalate of lime are produced after the urine has been secreted by the kidneys, and are artificially derived from uric acid or the urates.

The close relation between uric and oxalic acids has been long known; the urate of ammonia contained in guano is frequently converted into oxalate of ammonia during a voyage, and it can readily be imagined that a similar conversion may take place in the urine. Dr. Rees adduced the following reasons for his opinion: Oxalate of lime, on account of its insolubility, could not exist dissolved in the blood; its presence in the urine is due to the action of heat, which converts the uric acid or urate, more or less completely, into an oxalate.

In many cases, the presence of oxalate of lime can be demonstrated by applying a gentle heat to urine containing a full deposit of the urates, and then subjecting the deposit obtained to microscopic examination; but, according to Dr. Rees, there is no evidence to show that the oxalate of lime was originally present in these cases, while the employment of heat would sufficiently account for its appearance.

In other cases, however, the oxalate of lime has been found unmixed with the urates. Here the same explanation may be given: the uric was

---

[1] On Calculous Disease, Croonian Lectures, 1856.

converted into oxalic acid by the mode of analysis employed. In confirmation of these views, Lehmann observed that morning urine, after standing for some hours, often contains oxalate of lime in quantity, although the fresh urine did not afford any trace of it, while Wöhler and Frerichs found that the urates, when injected into the blood, produced oxalate of lime in the urine. Finally, Dr. Rees holds that there is no peculiar pathological condition connected with the oxalic diathesis, but that the latter is an accidental and unimportant modification of the uric acid diathesis. From the preceding remarks, it will be observed that, on Dr. Rees' theory, the oxalate of lime calculus is formed by decomposition of uric acid, or an urate in the kidneys or bladder; these are converted into oxalic acid, which unites with the lime, but we are not informed on what the conversion depends; we are allowed to infer that it may result from heat. If this be the case, it would be important to determine what degree of heat is required for the conversion of uric into oxalic acid.

Dr. Roberts,[1] however, refers to several experiments, the results of which sufficiently controvert Dr. Rees' theoretical objections, inasmuch as they prove "that oxalic acid and its compounds, even the insoluble oxalate of lime, pass through the blood into the urine when introduced into the stomach. Wöhler found that oxalic acid given to dogs caused oxalate of lime to appear in the urine. Piotrowsky confirmed these results by experiments on himself." He found that from 8 to 14 per cent of oxalic acid taken internally appeared in the urine as oxalate of lime, and that when oxalate of lime itself was taken, about 1½ per cent. could be found in the same fluid. It may, therefore, be regarded as absolutely certain that oxalate of lime, in spite of its great insolubility, can find its way from the blood into the urine.

The origin of cystine is enveloped in obscurity. No trace of this substance is found in normal urine, and there are no special symptoms connected with the occurrence of cystinuria. Ebstein[2] refers to a case, described by Marowsky, in which cystinuria co-existed with chronic atrophy of the liver. Marowsky suggested that the kidneys acted vicariously, and that the cystine took the place of taurine. In the case in question, however, the excretion of cystine was not constant, while the acholia and the disturbance of the functions of the liver continued to increase. As regards its composition, cystine is closely analogous to taurine.

The origin of calculi belonging to the second class is far more easily traceable. These are formed from constituents which exist as deposits in alkaline urine. If the alkalinity be due to fixed alkali, the carbonate of lime and the amorphous phosphate of lime are liable to be precipitated, but if the same reaction be due to carbonate of ammonia (produced from the decomposition of urea) the secondary phosphates are precipitated. These latter substances often form the external layers of calculi which have remained for some time in the pelvis of the kidney or in the bladder. They are also frequently seen to be deposited around foreign bodies which have gained access to the bladder. Their causation, under those circumstances, is the same as when they occur as the outer investment of urinary calculi in general. The amount of phosphates present in the urine is not increased, but their precipitation is determined by the change of composition which that fluid has undergone. The vesical mucus has an important share in the formation of these calculi; it acts as a cement and binds the earthy particles together.

---

[1] Loc. cit., p. 77.   [2] Loc. cit., p. 269.

The chief causes of the precipitation of the insoluble elements of the urine having been thus pointed out, and the various theories with reference to the production of these substances having been discussed, it now remains to consider other conditions which influence the formation of calculi. The conditions are connected with age, sex, habits of life, diet, climate, locality, occupation, hereditary influence, etc.

According to the views commonly taken, calculus is much more incident to early years than to any other period: and to old age more than the prime of life. Hippocrates noticed the fact, that infants at the breast are not exempt from the disease; and our countryman, Philip Barrough, observed, "Stones in the bladder do engender oftener in children than in older folk."[1] Some authors have of late denied that children are so subject to calculus as is commonly represented. It is certain that uric acid gravel is very frequent in children under some circumstances, and it is known that calculi have been found in the bladder even at the time of birth, and, in some rare instances, as an intra-uterine affection.

Professor Langenbeck, of Berlin, informed Mr. Coulson that he once met with a small uric acid calculus in the bladder of a six months male fœtus. Geyer[2] has recorded the case of a boy who suffered from calculus from birth: he was cut in his twelfth year, when the stone had acquired so large a bulk, that it had to be broken before it could be extracted. The whole mass weighed ten ounces. Stahl found a calculus the size of a peach-kernel in an infant of three weeks, who had from birth suffered great distress in passing water. Henoch[3] has recorded the case of a child, five months old, who at frequent intervals, during a period of four weeks, passed uric acid concretions. The child had suffered, from its birth, from symptoms of difficult micturition, and the escape of the concretions was accompanied by convulsive attacks with contraction of the extremities and rigidity of the muscles of the neck. Several similar cases are on record. This early occurrence of vesical calculi is to be accounted for by the predisposition which so many infants exhibit to the formation of uric acid infarctions in the kidneys.

Mr. Coulson, upon many occasions, extracted calculi from children under three years of age. His youngest case was a boy of eighteen months. The oldest patient upon whom he operated was eighty-six years old, a patient of Mr. Wilkinson's, of Sydenham. The case did perfectly well.

The following tables have been drawn up for the purpose of determining, in a more precise manner than has hitherto been done, the relative frequency of calculus at different periods of life. The first table includes only cases which had been submitted to operation.

Age.	No. of Cases.	Proportion.
1 to 20	2327	71.20 per cent.
21 " 40	395	12.10 "
41 " 60	355	10.87 "
61 " 80	187	5.72 "
Total	3264	

From this table it would appear that 71 per cent. of persons submitted to lithotomy are under 21 years of age.

---

[1] Methode of Physick, lib. iii., c. 41, editio quinta, 1617.
[2] Miscel. Nat. Curios., Dec. 2, An. V. p. 450.
[3] Quoted by Ebstein, loc. cit., p. 211.

To ascertain, however, the real proportion of calculous affections at different ages, we must include persons not operated on as well as those submitted to lithotomy, and by this method we obtain a somewhat different result.

From another table, arranged from Civiale's work (see below), it would appear that 55½ per cent. of calculous patients are under 21 years of age. Dr. Prout estimated the proportion at 62 per cent.

More recently an analysis by Mr. Bryant[1] of 230 cases of lithotomy, shows that nearly one-third of this number occurred in children under five; one-fourth in children between the ages of five and ten, or more than one-half (56 per cent.) in children ten years old and under. Mr. Bryant states that, in every succeeding period of five or ten years after the first decade, stone becomes rare.

Frequency of Calculus at different ages, arranged from M. Civale's work.

Age	Cases	Total	Proportion
Up to 5 years	896		
From 6 " 10 "	1150		
" 11 " 15 "	555	2989	55.56 per cent.
" 16 " 20 "	388		
" 21 " 25 "	242		
" 26 " 30 "	218		
" 31 " 35 "	162	790	14.69 "
" 36 " 40 "	168		
" 41 " 45 "	167		
" 46 " 50 "	224		
" 51 " 55 "	159	804	14.95 "
" 56 " 60 "	254		
" 61 " 65 "	316		
" 66 " 70 "	261		
" 71 " 75 "	138	776	14.43 "
" 76 " 80 "	61		
" 81 " 89 "	17	17	
	5376	5376	

The above tables exhibit the numbers of calculous persons at different ages of life, but they do not show the liability of individuals to be attacked at these different ages. Here, as in so many other statistical tables, an error has kept in from not taking care to distinguish between absolute and relative numbers. To determine liability, the absolute numbers should be corrected by the numbers of persons living at the several periods of life enumerated. Thus, if all persons under twenty were affected with a certain disease, and all persons over seventy were affected with the like complaint, it is evident that the liability would be the same, though the absolute number of persons attacked would be very different.

Corrected in this manner, the tables would show that children and young persons are less liable to calculous disorders than has been commonly supposed: and that from twenty years upwards the tendency goes on increasing in a very remarkable manner to the end of life. Thus, out of every 100 persons living, the numbers are—

---

[1] Med.-Chirurg. Trans., vol. xlv., 327.

Years.	Persons Living.
Up to 20	46
From 20 " 40	30
" 40 " 60	16
60 and upwards	7

On the other hand, the proportion of calculous patients at these different periods is, according to M. Civiale's table, up to 20 years, 55 per cent.; from 20 to 40, 14 per cent.; 40 to 60, 14 per cent.; 60 and upwards, 14 per cent. or thereabouts. Hence it follows that the liability to attack amongst young persons is below 25 per cent.; while, during the bidecennial periods that follow youth, the liability increases in the relation of 7, 16, and 30. From 40 to 60, persons are twice as liable to stone as those between 20 and 40, while beyond 60 they are four times more liable to the complaint. In fact, the liability will be obtained by dividing the numbers attacked by the numbers living at each period.

If this be done, the liability to stone at different epochs of life will be expressed by the following numbers: In the young (up to 20 years) 65; in adults (20 to 40) 26; in middle age (40 to 60) 50; and in the aged (over 60) 111.

Sir Henry Thompson has given the statistics of 1827 cases. Of these there were, under 20 years of age, 60.42 per cent.; between 21 and 40, 10.18 per cent.; between 41 and 60, 17.56 per cent., and between 61 and 81, 11.83 per cent. The cases came under treatment during given periods of time, in nine English hospitals.

With reference to all these tables, it must be remembered that patients usually suffer from the symptoms of calculus for a considerable time before they place themselves under the care of a surgeon, and therefore that the period of formation is usually long anterior to that of extraction. Dr. Gross has also pointed out that locality appears to influence the period of life at which calculi occur. In some countries the proportion of young patients is much greater than in others.

All authors who have written on this subject concur in stating that females, notwithstanding their sedentary habits, are much less subject to calculus than males; the reason for this is apparent, besides their more temperate and regular mode of living—namely, the shortness and less complexity in the form and construction of the urethra, its more dilatable nature, and the absence of the prostate gland. When a calculus descends from the kidney into the bladder, or when a calculus by any accident is formed in that viscus, it is frequently expelled during micturition by the mere efforts of the bladder to empty itself, and it is surprising to what an extent the female urethra allows of being distended. Tulpius[1] relates the case of a lady, aged 89, who spontaneously passed a calculus, *per urethram*, weighing three ounces and two drachms; but it appears to have paralyzed the sphincter vesicæ, as she was ever afterwards troubled with incontinence of urine.

In another work[2] it has been shown that out of 2834 patients, 123 were females, making a proportion of one female to twenty-three males. According to Dr. Prout, the proportion is the same. In some countries the proportion is even much less than this. Dr. Klien's statistics, already alluded to, show that among 1792 cases of calculus, treated 1822–60, in the Moscow City Hospital, only four were females.

Another reason why vesical calculus occurs far less frequently in

---

[1] L. iii. Obs. 7.   [2] Coulson on Lithotrity and Lithotomy, p. 247.

females than in males is to be found in the fact that women are far less liable to certain conditions and affections, *e. g.*, stricture of the urethra, atony of the bladder, etc., which predispose to the formation of calculi and are obstacles to their escape. The difference between the sexes as regards liability to renal calculus is by no means so decided.

The conditions and habits of life doubtless act, in various ways, as predisposing causes of the development of calculus. That the affection is very much more common among the children of the poor than among those of the richer classes of society is a fact that admits of no dispute, and on the other hand, it is equally true that the well-to-do classes furnish the large majority of elderly patients. In the case of children, the decreased liability of those belonging to the upper classes is in all probability due to the greater care taken with regard to diet, clothing, and cleanliness. Owing to unsuitable food or irregularities in the mother's diet, the children of the poor are especially liable to suffer from derangements of the digestive organs, while insufficient clothing, exposure to cold, and uncleanliness interfere very seriously with the action of the skin. From the fact that uric acid infarctions are frequently found in the kidneys of new-born children (see p. 223), we may infer that the foundation of a calculus is occasionally laid during intra-uterine life. Mr. Cadge is strongly inclined to the belief that the tendency to calculous disorders among the children of the poor is still further fostered by an insufficient supply of milk as an article of diet, and several facts might be adduced in support of this theory. With regard to elderly patients the increased liability of the upper classes may be traced to the influence of sedentary habits, luxurious and excessive eating and drinking, inducing derangement of the liver, indigestion, and various affections of the alimentary canal.

It is difficult to estimate the amount of influence which diet, *per se*, exercises in the formation of calculus, or to what extent the constituents of urinary concretions are derived from or depend upon the food habitually taken. That, generally speaking, food influences the nature of calculous formations, is a fact which admits of little doubt. Thus in countries where the diet consists largely of nitrogenous materials, uric acid is the prevailing ingredient; where vegetables form the staple food of the community, oxalate of lime calculi are more common. With regard to the relative frequency of calculous disease among flesh-eaters and vegetarians nothing positive can be stated, and even if reliable statistics could be obtained, it would be impossible to exclude the influence of other factors. We know that the disease is very prevalent in several European countries where animal substances enter largely into the composition of diet, and we know also that the same may be said of many parts of India where the great majority of the natives adopt an exclusively vegetable diet. There must therefore be other factors, common to these countries, which predispose to the formation of calculus. It is, moreover, indisputable that calculous affections, in all countries, are more common in certain localities than in others, and it is impossible, at present, to show to what extent, if any, this difference is due to differences of food. Take the English counties for example; calculus is very common in Norfolk, very rare in Cheshire. That this difference is in part to be attributed to difference of diet (*e. g.*, the abundance of milk in the latter county and scarcity in the former) is quite supposable, but there must be other factors at work. The nature of these, in the case of Norfolk, will be subsequently discussed. It remains to notice briefly the various articles of diet, the presence or absence of which in a given locality, has been considered to favor the development of calculus. It is a well ascertained fact that an excessive supply of nitrogenous food increases the quantity of uric acid in the urine; a diet corresponding with that description would therefore aid in the formation and growth of calculi, by furnishing materials of which the large majority of them are composed.

In the second place, the precipitation of uric acid is promoted by the use of articles of food which are liable to undergo acid fermentation. Even before the chemistry of calculous concretions was understood, it had long been believed that acids were among the most active causes of gravel, and pathologists have been able to trace a new cause of urinary concretions in acidity generated in the alimentary canal, as a result of some morbid conditions of the digestion. Wherever fermented liquors are a common beverage, stone and gravel have been observed to be more than usually frequent, and to this cause, namely to the use of acescent wines in France and of malt liquor in Britain, the prevalence of calculi in the two countries has often been ascribed. Majendie, in particular, while acknowledging the influence of a rich animal diet, insisted upon the effect of acids in the production of urinary concretions. The late Mr. Crosse, of Norwich, also considered the prevailing source of urinary calculi to be dyspepsia, leading to the generation of acid in the stomach, and to excess of uric acid in the urine.

It is a popular belief that water strongly impregnated with calcareous matter is apt to produce stone. The salts of lime found in water are the carbonate and sulphate, the former of which doubtless undergoes decomposition in passing through the system. The constant use of such waters would tend to make the urine alkaline, and hence might be supposed to favor the deposition of phosphates, and, in the second place, the salts of lime may possibly irritate the urinary organs, and increase the formation of products, such as mucus, the presence of which favors the precipitation of the salts of the urine, and promotes the formation of calculus. Calculous diseases, however, prevail in many localities where the drinking water is comparatively free from such impregnation, and are uncommon in some mountainous districts where the water is highly charged with salts of lime. It is therefore difficult to trace in a general way any causative connection between the use of calcareous waters and calculous disease.

With regard to the prevalence of stone in Norfolk, Mr. Cadge[1] is inclined to believe that it is *in part* due to the hardness of the water. He points out, however, that the presence of lime-salts in excess does not tend to the formation of oxalate of lime or phosphatic calculi. Out of nearly 200 cases of stone in the bladder, operated upon by him, only in three instances was the calculus composed of oxalate of lime, and in each of these the patient had for some time resided out of Norwich. In Norfolk, as elsewhere, the great majority of calculi are composed of uric acid, and it is not easy to explain how the presence of lime in drinking water can promote the precipitation of this substance. There are, however, certain facts which support the theory, that with an improved water-supply calculous diseases would diminish in frequency. Dr. Roberts[2] states that the suburban district of Hulme has supplied considerably

---

[1] Address in Surgery, British Medical Association, August, 1874.
[2] Loc. cit., p. 265.

fewer cases of stone to the Manchester Infirmary, since the pipe-water has replaced the old pump-water supply, and Professor Gamgee reports that sheep are particularly liable to be affected with calculus in the limestone districts. On the other hand, in Dr. Richardson's[1] opinion, the excess of cases of calculous disease in Norfolk and Norwich is not attributable to the water. His inquiries induced him to believe that exposure to cold, and rheumatic disease, not necessarily acute, but chronic and diathetical, were the main cause; the predisposition, thus engendered, being sustained by peculiarities of diet. He assumed the endemic prevalence of a febrile condition characterized by the formation of lithates and phosphates in excess.

With regard to the use of alcoholic liquors, it seems probable, as before mentioned, that the development of calculi is influenced by the use of beer and the stronger kinds of wine. Some wines, however, appear to possess an antilithic property. The exemption, pointed out by Soemmering, of the inhabitants of the Rhenish Provinces from calculus and gout, was said by Liebig to be due to the character of the wines of those districts. The bitartrate of potash, which is abundant in the Rhine wines, is converted into the carbonate as it passes through the system, and thus, as an alkali, promotes the solution of uric acid. Calculous diseases, according to Van Swieten, are common in Holland, and the circumstance is the more curious considering that gin, a powerful diuretic, is the spirit most in use in that country. Nevertheless, among the Dutch who come from Europe, and inhabit Batavia (Java), stone is a rare disease, although their manner of living does not differ from that pursued at home. Denys, who resided in that island for some years, tells us that he could find only two persons who were obliged to submit to the operation of lithotomy. He further observes that the water they drink, flowing from the neighboring mountains, is much impregnated with earthy matters. Mr. Charles Hawkins has drawn attention to the infrequency of stone in cider-producing countries.[2]

It has been usually stated that calculus is more common in temperate than in warm or cold climates, but a further examination of this statement is required. Humid countries, in particular, of moderate temperature, have commonly been supposed to be the chief seat of calculous complaints; Holland, France, England, and Germany are enumerated as exhibiting the prevalence of calculous disorders in the most striking degree. During the last few years, however, numerous reports have been published which prove that calculous disorders are far more wide-spread than was commonly believed, and that they are very common in some countries which were supposed to enjoy total or comparative immunity. It is certain that they are much more common in England and Scotland than in Ireland, and are more prevalent in some parts of England than in others. Thus during the five years ending 1871, the mortality from stone in England in proportion to the population was more than double that in Ireland, while in Scotland the mortality from the same cause was twice that of England. With regard to Ireland, Mr. Fleming[3] states that for the last hundred years the average annual number of cases of stone operated upon, by lithotomy or lithotrity, in the various hospitals or infirmaries throughout the country, has not exceeded twelve. In

---

[1] Medical History of England, Medical Times and Gazette, 1864, p. 100.
[2] Lancet, 1874, vol. ii., p. 289.
[3] Injuries and Diseases of the Genito-Urinary Organs, p. 208.

England the number of deaths from stone varies very greatly in different districts and counties. According to Mr. Cadge's tables it appears that the greatest mortality occurs in the following six counties, in the order named:—Norfolk, Huntingdonshire, Kent, Sussex, Buckinghamshire, and the West Riding of Yorkshire; that the following six counties show the smallest mortality:—Cheshire, Cornwall, Cumberland, Hampshire, Durham, and Devonshire, while the other counties hold a varying position between these extremes. In Norfolk there is one death from stone yearly to 42,744 of the population, while in Cheshire the proportion is one to 425,520. The museum of the Norfolk and Norwich Hospital contained in 1874 about 1,050 calculi, the total number removed, either during life or after death, since the foundation of the institution, rather more than a hundred years ago.

This remarkable prevalence of calculous diseases in Norfolk has attracted considerable attention. The late Mr. Crosse, of Norwich, was of opinion that neither the food, the soil, nor the beverage used by the inhabitants of Norfolk, had great effect in producing the tendency to calculous complaints in that county; and he was inclined to ascribe some part at least of that tendency to the variableness of the climate of Norfolk, and to the prevalence of north-east winds. He illustrated this opinion by stating that he had known persons belonging to another part of the country, and quite free from any calculous tendency, contract a disposition to the disease during a temporary residence for a few weeks at Norwich. Mr. Cadge, however, believes that climate exerts but the smallest influence in the causation of calculus. He points out that the disease is uncommon in many localities, the climate of which is even much more inclement than that of Norfolk. From a consideration of all the circumstances connected with the locality, he has arrived at the conclusion that the local favoring influences consist, to a very trivial degree, in the coldness of the climate: in the universal consumption of malt liquor; possibly, or even probably, in the constant daily use of exceedingly hard drinking water, and lastly in the accumulated effect of hereditary predisposition. Mr. Cadge also thinks that the want of milk largely influences the development of lithuria, and to this cause he attributes the great frequency of calculous disease among the children of the poor in districts where milk is scarce. He points out that in Ireland the poor can obtain an abundance of milk, whereas in Norfolk, Warwickshire, Suffolk, Kent, and other counties it is difficult to obtain a proper supply. The rarity of stone in Cheshire, where milk is abundant, tends to support this theory. It is certainly a remarkable fact that as many cases of stone occur annually in Norfolk among a population of 438,656 as in the whole of Ireland (pop. 5,412,377).

When we turn to other countries, we find that the degrees in which calculous disorders prevail are extremely various. In the United States, according to Dr. Gross, cases of stone are far more common in Kentucky, Virginia, Tennessee, and Ohio, than in other parts of the Union. They are comparatively infrequent in New York, Georgia, the two Carolinas, Florida, Texas, Mexico, and California. "In New Jersey, Delaware, and the New England States generally, stone in the bladder is proverbially rare. The malady is also uncommon in Canada and the other British Provinces of North America."[1] The negroes in America are far less subject to these affections than the white population, the proportion of operations among the two classes being about one to six.

---
[1] Gross, loc. cit., p. 167.

The remarkable prevalence of stone in some parts of Russia has been already alluded to (p. 197). Dr. Klien[1] is unable to trace the peculiarity to any special cause. He thinks that the vegetable diet and the wide-spread use of acid drinks may account for the frequency with which oxalate of lime calculi occur. The nucleus, however, in the majority of cases, consists of uric acid or the urates; and it is worthy of notice as supporting the theory that uric acid infarctions form the starting-point of calculi in young subjects, that in the districts referred to, children at the breast often suffer from urinary troubles. Dr. Klien states that the patients generally do not present themselves for treatment until the symptoms are very severe; large stones are often removed from children, urethral concretions are also common in young subjects. Cases of stone sometimes form one-fifth of the total number of patients under treatment in the Moscow Hospital. The disease is also common throughout the central parts of European Russia, in the districts watered by the upper tributaries of the Volga, whereas in the northern, southern, and western provinces it is comparatively rare. The experience of Dr. Beketow,[2] the senior surgeon to the Government Hospital in the Kazan district, fully confirms the account given by Dr. Klien. He has had 275 calculous cases under his care during a period of thirty years. The calculi were almost invariably composed of oxalate of lime with a uric acid nucleus. The prevalence of the former constituents is attributed by Dr. Beketow to an acidulated drink named *kwass*, in common use among the peasants, and also to the diet, which is largely of a vegetable nature. On the other hand, we learn from Dr. Estlander,[3] Professor of Surgery in the University of Finland, that calculous affections of a primary character are almost unknown in that province. The case-books of the University Hospital show that during the last forty-four years only one case of stone has occurred in a Finlander, though several operations have been performed upon Russian subjects coming from another province. That a similar immunity exists throughout the country generally may be inferred from the fact that the surgeons attached to eight other hospitals, in various districts, reported to Dr. Estlander that not a single case of stone had occurred during the last ten years. The hospital-books contain no record of renal calculus. Cases also of secondary stone-formation are extremely rare in Finland; they occur almost exclusively among the upper classes, and are generally due to affections of the bladder and prostate. Dr. Estlander has met with only twelve such cases during a practice of seventeen years.

With regard to the causes of their immunity, some importance may be attached to the purity of the water, which is derived from granite mountains, and contains a very minute proportion (only 7 parts in 100,000) of saline matters. The food of the people consists chiefly of fish, potatoes, cereals, and milk (which last is almost always taken sour); very little meat is consumed. The Finlanders, even the poorest, as a very general rule, pay considerable attention to cleanliness. Their custom is to take frequent baths in rooms filled with hot steam. It seems probable that their immunity from calculous diseases is in great part due to

---

[1] Ueber die Steinkrankheit und ihre Behandlung in Russland. Archiv für klin. Chirurg., Bd. vi., p. 78.
[2] Medical Examiner, vol. iii., p. 150.
[3] In a paper read at the Philadelphia Medical Congress, and published in the Boston Med. and Surg. Journal, Nov. 2d, 1876.

the simplicity of their diet, and to the attention they pay to the condition of the skin.

When we turn to Eastern countries we find similar differences with regard to the prevalence of calculous diseases. In certain parts of India, for example in the Punjab, North-western, and Central Provinces, the disease is very common; we learn from Dr. Fayrer[1] that in these districts, during six months of 1863, there were no less than 554 cases submitted to lithotomy. On the other hand, in Lower Bengal, according to the same authority, vesical calculus is not of frequent occurrence. Dr. Harris[2] attributes the great frequency of calculous disease in the North-western Provinces of India, in part at least, to the fact that these districts are exposed to the influence of the cold winds from the Himalayas, and also in part to the diet, which consists of heavy unfermented bread, made from an admixture of various kinds of cheap flours.

Reference has been already made (p. 197) to Dr. Vandyke Carter's report on the prevalence of calculous disease in Bombay. In that Presidency the disease is of frequent occurrence, and the cases show a decided predominance of oxalate of lime as an ingredient of the calculi. This predominance, which is very striking when compared with the experience of England, France, and Germany, illustrates the influence which diet exerts on the nature of calculi. The majority of the natives of Bombay live almost exclusively on vegetable food, and abstain for the most part from alcoholic liquors. The climatic conditions, however, under which the Hindoos live, constitute another factor of importance. Dr. Carter suggests that probably in the tropics there is a constant suboxidation of products within the body, the hydrocarbons are imperfectly consumed, and oxalic, rather than carbonic, acid is eliminated. Oxalic acid has a great affinity for lime, with which it forms a very intractable salt. The lime may be in excess in the water or the food, and its presence may perhaps favor the increase of the acid. Besides this, the copious perspiration, induced by the excessive heat, enables the system to throw off by the skin much organic matter which would otherwise pass through the kidney in the form of urea or the urates. In this way, Dr. Carter thinks, the predominance of oxalate of lime, as an ingredient of calculi, among the natives, may at least to some extent be explained.

The frequency and endemic prevalence of calculous disorders in Egypt are in all probability due to the mischief produced by a parasite which is common in that country. This creature (which has been named the Distoma hæmatobium) gains entrance into the body through the skin or intestines, and inhabits chiefly the minuter veins of the urinary organs. It commits great ravages in the coats of the bladder, producing destruction and detachment of portions of mucous membrane, and causing persistent hæmaturia. The ova of the creature and the fragments of mucous membrane and coagulated blood act as foreign bodies and become coated with uric acid and other urinary deposits, and in this way large concretions are often formed. Large quantities of ova are sometimes found contained in the substance of the calculi.

A report has recently been published, giving some information with regard to the prevalence of calculous disease in China.[3] We learn that, at the Mission Hospital at Canton, the total number of operations for

---

[1] Clinical and Pathological Observations, p. 385, etc.
[2] Lithotomy and Extraction of Stone, etc., pp. 4 and 5.
[3] Calculus in China, by J. Dudgeon, M.D., of Pekin. Med. Times and Gazette, Sept. 2d, 1876.

stone from 1854 to the end of 1875 was 432. Of this number 345 patients were subjected to lithotomy, and the remaining 87 to lithotrity. The total weight of stone removed was 408 ounces, or 25½ lbs. Uric acid would appear to be the prevailing ingredient. The mortality is not given, but is stated to have been very small, owing to the freedom from excitability characteristic of the Chinese, and the absence of inflammation and shock. With this exception of Canton and of Takow in Formosa, the experience of the Mission Hospitals would indicate that stone is an infrequent affection in China. No explanation has been given of its almost exclusive prevalence at Canton. It is stated that the common water at Pekin contains an abundance of lime, and is, in consequence, very brackish. At Canton, on the other hand, river water is almost exclusively used, and these facts, as Dr. Dudgeon observes, militate against the frequently repeated statement that calculous diseases are most common in places where the water contains much lime. The Chinese are said to be very subject to dyspepsia arising from acid fermentation. They might therefore be expected to suffer from gout and lithuria, and in the north of China the disposition to these affections ought to be even greater than in the south, as the diet in the former districts is much more of an animal nature than it is in the latter. Opium-smoking, a habit which prevails throughout the country, is not adduced as a factor in the estimate of causes. In the Pekin Hospital from 1864–76 three cases of urethral calculus occurred, but there was no instance of stone in the bladder. Dr. Dudgeon, however, performed three median operations for the extraction of foreign bodies.

From the above account of the prevalence of calculus in various parts of the world, it is evident that very little can be positively asserted with regard to any direct influence of climate in the causation of calculous disorders. It seems, however, probable that this influence, *per se*, is very slight. The mode of life, comprehending under this head the diet, clothing, and the general hygienic conditions of a population, would appear to exercise a far greater influence than mere climatic peculiarities, though obviously these latter cannot be altogether excluded from a general consideration of the causes of stone.

When we inquire into the influence of occupation upon the prevalence of calculous diseases, we find that very few facts exist which would support any theories as to causation. The difference in liability between the rich and poor has been already alluded to, but this difference is not connected with occupation, but rather with variations in diet and habits of life. Certain classes of society have, however, been supposed to enjoy almost complete immunity from calculous disease. In an essay published more than half a century ago, by Mr. Copland Hutchinson,[1] an attempt was made to prove that calculous diseases are extremely rare among sailors and seafaring people. He estimated the annual average number of seamen and marines in the British Navy, from 1800–1815, to amount to 162,000, nine-tenths of these being men who had served at sea from very early life. Out of these there were, in the 16 years, only eight cases of calculus admitted into hospital. In the three great naval hospitals of Haslar, Plymouth, and Deal, the proportion of calculous cases to patients in general was only one to 10,750. Of the eight cases, two occurred in boys under 14. This supposed immunity Mr. Hutchinson

---

[1] Practical Observations on Surgery, more particularly as regards the Naval and Military Service. by Alex. Copland Hutchinson, London, 1826, chapter vii., p. 308.

was inclined to ascribe to several causes connected with the food, drink, and general habits of seamen. He laid stress upon the fact that animal food is supplied to the men only in the form of salt beef and pork, that they have few opportunities of drinking wine or malt liquor, that diluted spirits constitute their daily drink, that they pass their lives freely exposed to the atmosphere, and therefore exercise the functions of respiration with an unusual vigor, and above all, that they seldom allow the bladder to become distended, but pass urine frequently during the day. Mr. Hutchinson also thought that the moist state of the atmosphere of the lower deck, in which the men slept, served to promote the functions of the skin, and hazarded the conjecture that there may be some essential connection between the state of the cutaneous functions and the greater or less prevalence of calculous disorders. So far as his facts go, Mr. Hutchinson certainly succeeded in showing that calculus is not a disease to which seamen are particularly subject. Doubts, however, have been entertained as to whether the data on which he proceeds are sufficient to show that sailors possess a special exemption from this complaint. Sir George Ballingall has made it appear that soldiers enjoy a similar exemption. On this point, Mr. Samuel Cooper remarks: "Mr. R. Smith, of Bristol, has published an interesting statistical inquiry into the frequency of stone in the bladder in Great Britain and Ireland, though strictly it is a comparative estimate of the number of operations for stone in different parts of the kingdom in given spaces of time, and in all the number of calculous patients. So far as I can judge from the facts stated in Mr. Smith's paper, and from what I know about the average number of operations for stone in London, not more than 180 can be fairly reckoned as the annual total of Great Britain and Ireland, which is about one for each 100,000 of the population taken at 18,000,000." Now, if this fact be recollected in reasoning about the rarity of stone-operations in the navy, and if we also bear in mind the fact that the great majority of sailors are men in the prime of life, the cause of the marked infrequency of stone in the navy will be tolerably clear. Notwithstanding these objections, Mr. Hutchinson's paper must be regarded as a valuable contribution to the medical history of calculous complaints; for, though seamen may not be more exempt than some other classes of the community from this disorder, it is at least satisfactory to know that so large a body of men, living in a special manner, and on a special kind of diet, should enjoy an almost complete exemption from this malady, the more so because the ordinary diet of seamen cannot be regarded as particularly favorable to the healthy exercise of the organs of assimilation. It is interesting to notice that Mr. Hutchinson's opinion with regard to the prophylactic action of vapor-baths (for such he considers the lower decks of a ship to be at night), is supported by that of Dr. Estlander, already referred to, who attributes the immunity of the Finlanders, in part at least, to their habit of taking steam-baths. The diet of sailors, consisting as it does largely of salted food, may afford some protection against the development of calculus, inasmuch as chloride of sodium taken in full doses diminishes the quantity of uric acid in the urine.

A few facts have been placed on record which indicate that hereditary tendency sometimes plays an important part in the development of calculous affections. A remarkable instance has lately been reported by Mr. Clubbe,[1] of Lowestoft. A fisherman's three children, aged two, three,

---

[1] Lancet, Feb. 10th, 1872.

and eight years respectively, were all operated on for stone. The father and mother suffered frequently from lithuria, and a grandfather and grandmother were known to have passed several concretions. There was also a similar history connected with other members of the family.

The manner in which concretions become formed around foreign bodies of various kinds, introduced into the bladder, will be described in a subsequent chapter. A few remarks will now be added with reference to the occurrence of calculus among the lower animals. A few facts have been collected by veterinary surgeons illustrative of the prevalence and causation of urinary calculi among domesticated and other animals.

Man is not the only animal subject to urinary calculi: in the higher mammalia we often notice that their formation, passage into the bladder, and gradual increase in size are attended by the same train of symptoms as those with which we are familiar in our own species. The museum of the Royal College of Surgeons contains specimens from the monkey, dog, rat, rabbit, hog, horse, ass, ox, sheep, elephant, whale, eagle, ostrich, fowl, tortoise, iguana, and sturgeon. The collection shows that graminivorous are more liable to calculous disorders than carnivorous animals; also that calculi from the lower vertebrata consist in greater part of carbonate of lime. According to Mr. W. Williams,[2] urinary deposits are not unfrequent in the horse, ox, and sheep; calculi being most common in stallions and geldings, sabulous deposit in mares, and calcareous incrustations on the mucous membrane of the bladder in oxen and sheep. In the horse and ox, the carbonate of lime calculus is that generally found; while in the sheep and pig, the ammoniaco-magnesian phosphate, and in the dog, uric acid concretions, are most commonly met with. Mr. Williams thinks that the composition of the calculus is mainly influenced by the character of the food. He states that turnips, on which sheep are often fed for long periods, contain much phosphoric acid with ammonia and magnesia. With regard to horses, he points out that those which, from the nature of their work, are compelled to retain their urine for many hours at a time are most subject to calculous deposits; and that for this reason hunters are more commonly affected than other horses. Foreign substances, introduced into the bladders of mares, are not unfrequent causes of calculus. With regard to food, it is said that some kinds of clover cause the formation and deposit of large quantities of urine-salts, and the secretion of much mucus, the presence of which predisposes to the formation of a calculus. Mr. Gamgee states that sheep are especially liable to calculus in limestone districts.

---

[2] Principles and Practice of Veterinary Surgery, p. 587; see also Gamgee, Our Domesticated Animals, vol. iii., p. 55.

CHAPTER XX.

SYMPTOMS AND DIAGNOSIS OF STONE IN THE BLADDER.

In the great majority of cases, urinary calculi are first formed in the kidney; while small, they descend along the ureter into the bladder, where they increase in size by a succession of layers deposited from the urine. If the calculus be smooth and small, its passage down the ureter excites little or no inconvenience; it may even remain for a long time in the bladder without the patient being aware of its presence. If, however, the renal calculus be rough, and of some size, its descent into the bladder is attended with most excruciating suffering—the pain extending from the kidney toward the bladder, down the thighs and along the course of the spermatic cord, being sufficient to "double the patient up," and make him roll in agony on the ground. The descent generally occupies from twelve to twenty-four hours. The usual symptoms in the milder cases are, nausea, rigors, restlessness, and some alarm; pain in the back, coming on in severe paroxysms, and extending down the thighs; frequent attempts to pass urine; retraction of the testicles; coldness of the extremities, and extreme prostration.

The calculus may become impacted in the vesical orifice of the ureter: the flow of urine from the corresponding kidney will then be prevented; dilatation of the ureter and distention of the tubular structure of the kidney will ensue, terminating in absorption of its substance; but most of the usual symptoms of stone in the bladder will be absent.

When a calculus is loose in the vesical cavity it falls to the most depending part behind the prostate; but it shifts about according to the posture of the patient, and gives rise to symptoms which are more or less marked in different cases.

The early symptoms of stone in the bladder are by no means uniform, and a great difference exists in their intensity at the two extremes of life. In childhood, the symptoms are ordinarily very severe and distinctly marked; in advanced age the patient's troubles are generally far less acute. The following symptoms are those commonly met with in children, even in cases where the calculus is very minute:—intense irritability of the bladder, evinced by frequent micturition and perhaps incontinence of urine; during and after micturition, severe pain which they endeavor to alleviate by pulling the prepuce; severe straining during the same act, often causing prolapse of the rectum and discharge of fæces; occasional sudden stoppage of the stream of urine, and occasional hæmaturia.

In the case of adults, during the earlier stages of stone in the bladder, and while the organ is healthy, the patient experiences a dull sense of weight about the neck of the bladder, with pain or uneasiness extending to the hypogastric region, perineum, groin, or thighs. The bladder becomes irritable and cannot retain its contents; the calls to make water are frequent, and the expulsion of the last drops is attended with pain, which shoots along the perineum and penis, and centres in the glans. The frequency of micturition is more marked during the daytime, when the patient is moving about; it is less troublesome at night, under opposite conditions. Sometimes when the urine is flowing in a full stream, it is suddenly stopped, evidently in consequence of the foreign body falling across the orifice of the urethra, and mechanically obstructing it. The patient finds he cannot take any violent exercise without suffering pain; the jolting of a carriage increases his symptoms, and is frequently followed by a discharge of bloody urine, while in some constitutions, hæmorrhage to a considerable extent takes place. As the stone increases in size, micturition becomes more frequent and distressing, and the pain or uneasiness at the end of the penis becomes more constant and severe. The last drops of urine are expelled with a spasmodic effort of the bladder, lasting for some time, and the contents of the rectum and vesiculæ seminales are often involuntarily passed, adding materially to the distress and exhaustion of the patient. There is occasionally priapism, with or without sexual desire. Some patients experience the sensation of a foreign body rolling about in the bladder.

These symptoms come on slowly; indeed, if the stone be smooth, and the bladder healthy, there may be nothing for a considerable time to direct the patient's attention to his case. Sooner or later, however, indications of disturbance of the urinary organs ensue, and the features of the affection become well marked. In elderly patients, the symptoms are usually less severe, owing probably to diminished sensibility of the bladder. When, in these patients, the symptoms are acute, we may expect to find a large or rough stone, with more or less inflammation of the mucous membrane.

In the early diagnosis of stone, in adult patients, very reliable and valuable information may generally be derived by noticing closely the effect of exercise, which invariably aggravates the most prominent symptoms, the hæmorrhage, pain and irritation. These symptoms are due to the mechanical action of the foreign body which the bladder contains.

With regard to hæmorrhage, the presence of blood in the urine is more common than is generally supposed, as an early symptom of stone, and the variations in the amount at different times are most significant. It may be little or much; but, if it occurs at all, it will appear after exercise, and will be proportionate in quantity to the duration and character of that exercise. It is frequently the first symptom that causes alarm and induces patients to seek medical aid. Sometimes a discharge of florid blood occurs with the last few drops of urine. The hæmorrhage then arises from the stone having been brought into closer contact than before with the walls of the bladder, and it is especially likely to occur when the mucous membrane is inflamed, or when the stone has a rough, tuberculated surface. The smooth concretions of uric acid seldom give rise to much hæmorrhage, which, on the other hand, is common when the stone is composed of oxalate of lime, or studded on the surface with points of triple phosphate crystals. If the hæmorrhage is only slight, the quantity of blood present may be insufficient to discolor the urine. In such a case, however, the blood-corpuscles will be easily detected, by means of the microscope, in the sediment of the urine, after this has been allowed to stand. These will usually be found normal in shape, but they are sometimes globular, especially in acid urine. They will always be discoverable in greater abundance after exercise. The jolting of a carriage, for example, is likely to cause somewhat profuse hæmaturia, and a patient obliged to ride much on horseback would notice the existence of this symptom at a very early stage of the complaint. The discoloration of the urine may last several days; but it generally disappears with rest and the recumbent posture. When there is ulceration of the bladder the case becomes complicated, and the hæmorrhage is more profuse.

The next important and prominent symptom that declares itself early is pain, either in the region of the bladder, rectum, perineum, or penis. The pain varies in character, being sometimes sharp and acute, sometimes dull and heavy. It is commonly influenced by various conditions, such as the presence or absence of urine in the bladder, the state of contraction of that viscus, exercise, posture, etc. It is described by some authors as sometimes periodical; and Deschamps relates the case of a man, in whom for upwards of eight months pain of a sharp cutting character, produced by stone in the bladder, came on at every return of the new moon. Although, as a general rule, the larger and rougher the stone the more acute the pain, exceptions to this are occasionally met with. Mr. Coulson removed from a thin man, of advanced age, an oval stone, of this character, which had never excited severe suffering; and there had been great difficulty experienced in detecting its presence by the sound. It is probable that in all these cases there is no congested or inflammatory condition of the bladder. When such a state ensues, the sufferings of the patient become greatly aggravated, and, except by operation, there are no means of affording effectual relief. Dr. Gross[1] thinks that a spiculated stone, or one studded with long spines, produces often slight suffering, because the spines admit of the more ready passage of the urine, in the same manner that a rough body lodged in the bronchial tube will occasionally cause less distress than a smooth body, as it produces less obstruction to the entrance of the air.

When pain is present, a careful consideration of the circumstances which tend to relieve or aggravate it, will add much to the diagnostic value of the symptom. It is, however, by no means always present as an early symptom, and it is sometimes absent even when the calculus has attained a considerable size. This absence of pain is generally due to one or more of the following conditions:—An enlarged prostate, with a pouch behind it; a feeble, atonic state of the bladder accompanied by imperfect emptying of that viscus; a sacculated bladder; and lastly, a bladder rendered less sensitive by age.

The first condition, hypertrophy of the prostate, has the effect of raising the neck of the bladder above the level of the trigone, and thus occasions a pouch in this situation. A small stone in this position rarely occasions marked pain, unless during violent exercise, inasmuch as the prostatic tumor prevents the stone from falling down on the neck of the bladder, and interfering with or obstructing the internal meatus during micturition. This condition also prevents the bladder from completely contracting, and the patient is thus saved the pain he would otherwise experience, from the compression of the stone between the coats of the bladder, after the escape of urine.

The second condition, a feeble atonic state of the bladder, produces a similar effect; the coats of the organ are never brought into close contact with the calculus, and unless the bladder becomes over-distended,

---

[1] Gross, loc. cit., p. 193.

there is, as a rule, an absence of pain. Under these circumstances acute pain may be produced by drawing off the urine with a catheter.

A sacculated bladder may permit a stone to become encysted in a pouch formed by the protrusion of a portion of the mucous coat between the muscular fasciculi; a stone thus lodged may occasion but little uneasiness.

In elderly patients, the bladder often loses to a great extent its normal sensibility, and, in such cases, the presence of a calculus may give rise to little, if any pain.

In the absence, however, of these four conditions, pain is often one of the earliest symptoms of calculus, and, when occurring under certain circumstances, and accompanied by other evidences of the presence of a foreign body, it is almost certainly pathognomonic. Thus it often happens that, during micturition, the stone falls across the internal orifice of the urethra: it then gives rise to great pain and suddenly arrests the flow of urine, and at the same time induces violent straining. The arrest in the flow of urine may not last more than a few moments; it disappears upon some change of posture, or of its own accord. This, as an early symptom, is far more commonly met with in young than in elderly patients. If the calculus be so small, or of such a shape, as to admit of becoming impacted in the urethra, it may produce complete retention. Patients have been known to lie on the back during micturition, and to elevate the pelvis; the object being to allow the calculus to fall from the urethral orifice into a less sensitive part of the bladder.

In the case of children, the pain during and after micturition is usually very severe, causing the little patients to cry or scream in agony, and to run about the room in the hope of obtaining relief. They also constantly apply the hand to the end of the penis, and elongate the prepuce in their attempt to deaden the pain. This habit, when acquired, of pulling at the prepuce, is strongly indicative of the presence of a stone in the bladder.

The diagnostic value, as a symptom, of pain occurring after micturition, cannot be too strongly insisted upon, depending as it does on the mechanical effect of the foreign body. The stone being movable, it is a symptom almost invariably present with all patients who possess the power of emptying their bladder. The pain may be felt at the neck or over the region of the bladder, but more commonly it is referred to the end of the penis and lasts for some minutes after micturition. That irritation of the bladder, arising either from the presence of a foreign body, or from disease, should so commonly occasion pain at this spot is to be accounted for by supposing that the impression made upon the branches of the sacral plexus distributed to the mucous membrane of the bladder is referred to the ultimate ramifications of the filaments from the same plexus distributed to the extremity of the penis. The mechanical effect of sudden jolts, of horse exercise, of driving, etc., in inducing pain, is generally a striking symptom and often an early one; the pain is sometimes felt in the neck of the bladder and in the region of the rectum, but more commonly toward the end of the penis. In cases of phosphatic calculi there is often more suffering than when the stone is composed of uric acid or oxalate of lime, because more or less catarrh of the mucous membrane of the bladder generally co-exists with the former variety. If the patient be of a nervous and excitable temperament, or of a gouty or rheumatic habit, the pain experienced is often of an anomalous character, and is described by the sufferer in vehement or exaggerated terms.

The third symptom that has been referred to, as indicating mechanical injury, is irritation of the bladder. This is shown by increased frequency of micturition, and occasional incontinence of urine. These symptoms, by themselves, are of little import; they are certainly not to be regarded as in any sense characteristic, though, for diagnostic purposes, they become very valuable when taken in conjunction with other symptoms. To make them of value, it is necessary, by a process of exclusion, first to ascertain on what they do not depend, as one or both of these symptoms may be due simply to an acid condition of the urine, and they are, moreover, commonly met with in nearly all diseases of the bladder and prostate, and in most cases of long-standing stricture. At the same time, it must be borne in mind that vesical calculus often occasions inflammation of the bladder, and it may even produce urethritis with all the symptoms of acute gonorrhœa. The irritability of the bladder, evidenced by the frequent calls to micturate, is a source of great annoyance to the patient. Sometimes these occur every hour or even every half hour. In adults, erections of the penis and involuntary emissions of semen often add to the existing discomfort. Prolapsus of the rectum, especially in young children and old men, is by no means an uncommon complication. It varies considerably in degree; but is at all times a source of great inconvenience, because so frequently attended with the discharge of flatus or fæces.

Should the calculus have caused ulceration of the mucous membrane of the bladder, the desire to make water becomes proportionately more severe, so as to be almost if not quite uncontrollable.

Besides the hæmaturia, the presence of a calculus induces other changes in the condition of the urine. These mainly depend upon the duration and amount of irritation to which the bladder has been subjected, and the character of the stone. If the calculus is small and smooth, an increased secretion of mucus may be for some time the only change.

The urine retains its acid reaction, and there is neither albumen nor pus present. If, however, the stone is large and rough, in the majority of cases it speedily induces more or less vesical catarrh. The urine is cloudy, owing to the presence of pus and blood; its reaction becomes alkaline from the conversion of the urea into carbonate of ammonia, and it contains albumen, in a quantity corresponding to the pus and blood. The sediment, at first loose and flocculent, becomes stringy or ropy, more or less fetid, and forms a dense, tough, yellow layer at the bottom of the vessel. Examined under the microscope, the sediment will be found to consist of blood- and pus-corpuscles, epithelium from the urinary passages, and the various amorphous and crystalline deposits. The character of these latter affords an indication of the nature of the stone, or rather of the substances which compose its crust. A uric acid calculus, for example, often becomes coated with a layer of the triple phosphate; in such a case, crystals of the latter substance will be found in the urine, though the bulk of the stone has a different composition. The same substance may invest a concretion of oxalate of lime, and this again may form the coating of a nucleus of uric acid. As a general rule, therefore, the deposits occurring in the urine indicate the nature only of the external layer of the calculus. If the symptoms be of recent date, if the urine have an acid reaction and the deposit consist of uric acid, we may safely infer that the calculus is composed of this last-named substance. If, with a similar reaction of the urine, oxalate of lime be deposited, the calculus will probably be of a similar nature, or at least the external layers will have this composition. If the urine when passed be alkaline from carbonate of ammonia, the crust or surface of the calculus is certain to be composed of the mixed phosphates. If the same reaction be due to fixed alkali, a condition extremely rare, the calculus will be composed either of phosphate or carbonate of lime. The discovery of crystals of cystine in the sediment places the nature of the calculus beyond a doubt.

A few cases have been placed on record which show that a calculus may undergo spontaneous disruption within the bladder. The late Mr. Southam, of Manchester, has reported several instances of this kind. In two cases the calculus was composed almost entirely of uric acid and oxalate of lime, materials not likely to be affected by any degree of force which could be applied to them in the movements of the body, or from the action of the muscles of the bladder. The cause of the disruption must therefore be sought in the calculi themselves. Mr. Southam believed that the fracture was due to the generation of some gaseous agent, owing to chemical changes in the earthy constituents of the calculi, or to decomposition of the mucus and animal matter. His view with regard to the cause of the disruption is supported by a case of Dr. Ord.[1] This patient, aged 83, passed spontaneously several calculi and numerous fragments, all of which had the appearance of being segments of spheres having a small central cavity. Dr. Ord believed that the calculi were broken some time before they were passed, and that the fracture occurred when the urine underwent a change in reaction from acid to alkaline. He also suggests that the disruption was due to the expansion of the nucleus at the time when this alteration in the reaction took place.

With regard to the rate at which the growth of a calculus takes place, nothing positive can be stated. According to Mr. Crosse, primary calculi increase in weight by one or two or even four drachms yearly; and the larger the stone, the more rapid the increase. Meckel thought that primary stones generally increase in diameter from two to six lines every year. Dr. Ultzmann[2] has instituted numerous experiments from which it would appear that by keeping uric acid stones in urine and renewing the latter daily, the increase in weight in a year's time was equal to nine times the original weight of the calculus. He thinks that small stones may, within the body, increase at this rate, but that for large concretions the increase is much less rapid. The rate of increase depends, of course, mainly upon the quality of the urine; if the latter is rich in the constituents of the calculus, the growth will be more rapid than under opposite conditions. Phosphatic concretions, which are secondary formations, are the most rapid in their growth. Phosphates are excreted in urine in the proportion of 45 grains in the 24 hours, while uric acid exists only in the proportion of 8 or 10 grains. When a stone is present, if from any circumstances the urine contains less of the constituent, the growth of the stone will as a matter of course become less rapid than before.

When a vesical calculus has existed for some years, and has induced severe symptoms, more especially if complicated with stricture of the urethra or enlarged prostate, the kidneys though previously healthy are apt to become involved, and the symptoms of chronic pyelitis begin to be developed. The lining membrane of the pelves and infundibula becomes

---

[1] Path. Soc. Trans., vol. xxviii., 170.
[2] Ueber Harnsteinbildung, Wiener Klinik, 5 Heft, s. 162.

thickened from repeated attacks of inflammation, and these parts become enlarged and altered in shape. Owing to the pressure, the parenchyma of the kidney undergoes more or less absorption ; the papillæ then become flattened or obliterated ; cysts filled with pus and formed of dilated uriniferous tubules, occupy the cortical part ; and all the most serious evils that belong to renal affections follow in the train, and carry off the sufferer.

There is a look of great distress in a patient suffering constitutionally from the presence of a stone in the bladder combined with disease of the kidneys. He is anxious; the brows are partially contracted and expressive of suffering, both bodily and mental. He becomes thin and wan ; the skin is dry ; the extremities are cold ; the pulse small ; the stomach does not readily receive food ; the secretions are vitiated, and the bowels become irregular and irritable. Pains may manifest themselves in parts of the body quite removed from the seat of disease, such as the knee, heel, foot, or arm.[1] Many circumstances may cause temporary remission of the symptoms, and there is always the possibility of the stone becoming encysted. Several cases, too, are recorded, where calculi had existed for years in the bladder without giving rise to any symptoms, and were discovered for the first time upon examination after death. In the great majority of cases, however, local and general symptoms make their appearance in a more or less regular order of sequence, and increase in intensity as time goes on.

Persons affected with stone, and at the same time subject to gout (a coincidence by no means unfrequent), require the greatest care as regards an operation proposed for their relief ; for any increase of irritation may bring on an attack of gout, in which the bladder sometimes participates. It is in such cases as these that operations must be promptly performed ; for useless and long-continued manipulations are apt to be followed by serious results.

With regard to the diagnostic value of the subjective symptoms of calculus, it must be borne in mind that many of them are common to other affections of the urinary organs. Irritability of the bladder, evinced by frequent micturition, accompanies various affections of the prostate gland. It is a very common symptom of hypertrophy of that organ. In that case, however, it is especially troublesome during the night, whereas in calculous disease the irritability is more marked during the day-time (if the patient is moving about), and diminishes at night during rest. Various tumors of the bladder, stricture of the urethra, pyelitis, and other affections of the kidneys are all characterized by irritability of the bladder. In children, the same symptom may be caused by worms in the rectum, or an adherent prepuce. Sudden stoppage in the flow of urine may be caused by a pedunculated tumor, and is therefore not pathognomonic of calculus. The pain in calculus is most marked at the end of the act of micturition, because the mucous membrane is then brought into closer contact with the foreign body. In cystitis, on the other hand, the pain is usually worse before micturition, and diminishes after a little urine has been passed. In acute inflamma-

---

[1] Symptoms of this kind occasionally present themselves without there being any serious affection of the kidneys or other organs. Professor von Pitha relates that Dr. Reisich, of Prague, when the subject of a vesical calculus, used to complain of a severe burning pain in the sole of the left foot. This pain finally disappeared only after the last fragment had been removed. Med. Times and Gazette, 1875, vol. ii. 358.

tion of the prostate, the pain resembles that which is caused by stone, but is less severe; in hypertrophy, any pain that may be present depends upon the distention of the bladder or a chronic inflammatory condition of that organ, and is relieved by the escape of urine. The pain due to a calculus is, moreover, exaggerated by movement, whether active or passive, and this characteristic is, as already mentioned, very important for diagnostic purposes.

The pain that accompanies tumors of the bladder varies very considerably as to its seat and character. It usually depends upon the amount of obstruction occasioned by the growth, and its character, generally speaking, is uninfluenced by movement. With regard to the other leading symptom of calculus, hæmaturia, this, as before mentioned, occurs in the large majority of cases at an early stage of the complaint. When, after exercise, a small quantity of nearly pure blood is passed, the symptom is strongly indicative of the presence of a calculus. There is, of course, nothing pathognomonic in the other appearances presented by the urine. Considerable alterations often occur, but these are due, as already stated, to the acute or chronic cystitis which accompanies the calculus. As regards diagnosis, it may be briefly said that the most important symptoms of calculus in the bladder are, the hæmaturia, the severe pain at the end of the act of micturition, the aggravation of the symptoms after exercise, and the occasional sudden stoppage of the stream of urine. It must, moreover, be laid down as a rule, that we cannot positively pronounce upon the presence of a calculus in the bladder unless it be ascertained by the sound; and there are few operations which differ more in the suffering which they cause to the patient according to the manner in which they are performed, than this apparently simple proceeding.

The best kind of sound is that introduced some years ago by the late Sir William Fergusson. It has a thin shaft, about No. 5 size, and a short bulbous extremity. This instrument is freely movable in the urethra, and its short lithotrite-like bulb can be readily brought into contact with the interior of the bladder. It is advisable to have other sounds of this shape, but perforated like catheters, so that urine can be drawn off, or the bladder injected, if required.

Previous to sounding, the bowels should be opened by an enema. If the rectum be distended, it occupies too much of the pelvic cavity, and prevents a satisfactory examination being made *per anum*, should the case require it.

The patient having been placed in the semi-recumbent position, the instrument, previously well oiled, should be passed carefully and gently along the urethra into the bladder, the cavity of which is then to be explored. First, the handle should be slightly raised and the convexity of the sound passed in a sweeping direction from before backwards, along the inferior surface of the bladder from the neck to the fundus. Next, the bladder should be explored laterally by revolving the handle from side to side between the fingers and thumb. Lastly, the instrument is to be withdrawn to the neck of the bladder, and the point is to be directed first downwards to the space behind the prostate, and secondly upwards to the surface behind the pubes. If the bladder be suddenly emptied during the operation, we may withdraw the sound, and slowly inject through a catheter a moderate quantity of warm water by means of a properly adapted syringe. The patient should be supported on a table or sofa so as to enable us to depress the shoulders and bring the fundus of the bladder to a depending position, into which the stone may gravitate. Sounding a patient is an operation which cannot be satisfactorily performed in a hurry; and it may be necessary to repeat the proceeding several times before we can form a positive opinion as to the presence or absence of stone. It may excite in some patients such severe suffering that it becomes advisable to exhibit chloroform, which, by producing insensibility, greatly facilitates the examination. The introduction of the finger into the rectum will often bring the calculus into contact with the sound; in children especially this additional manipulation will be found of great service. In adult patients the lithotrite will be found to be a very suitable instrument for the detection of calculus. By its means the surgeon is enabled to discover the exact condition of the bladder, its capacity and sensibility, and the size of the stone, when present. He may also ascertain whether the symptoms are occasioned by a foreign body of another description, such as a polypoid growth, and again whether the bladder is sacculated. Also, if the circumstances are favorable, the stone may be crushed as soon as detected, for the examination and crushing, when properly conducted, need occasion no more suffering than a patient would experience from the ordinary examination with a sound. If, however, it be thought advisable to postpone the crushing, valuable information may be gained as to the position of the stone, its size, the character of its surface, etc., and also as to whether the bladder contains more than one calculus.

The use of the lithotrite enables the surgeon, as above stated, to measure the stone. Other instruments have been devised for the same purpose. Sir H. Thompson[1] uses a sound furnished with a slide and scale near its handle. This is introduced into the bladder, and its end is to be passed beyond the distant extremity of the calculus. The collar is then to be slid down the shaft to the end of the penis, so that it touches the external meatus. The sound is then to be partially withdrawn until its bulb touches the near extremity of the calculus. "The distance of the collar from the end of the penis is the diameter of the stone in the direction passed over." Mr. Fleming[2] has devised an instrument on the principle of Mr. L'Estrange's gauge-sound. This he calls a lithometer. It resembles a small lithotrite, with a scale near the handle, by which the separation of the blades can be measured. This instrument he uses for children only. In all cases, a calculus readily found is probably a large one.

A hard stone, as the oxalate of lime or the lithic acid calculus, rings when struck, and communicates through the instrument a peculiar sensation. A soft calculus, composed of the phosphates, conveys a more equivocal feeling to the surgeon; the instrument strikes against a hard unyielding substance, but the sound is less sharp, sometimes almost inaudible. The larger and harder the stone, the more audible in general the sound communicated by the instrument. It is quite different from the grating sensation indicative of a fasciculated condition of the bladder, and may often be heard at the distance of several yards. The introduction of the finger into the rectum for the purpose of lifting the surface of the bladder behind the prostate, though very often practised, is requisite chiefly in old persons and in children.

Sounding is an operation not wholly free from danger. It has been

---

[1] Diseases of the Urinary Organs, p. 149.
[2] Injuries and Diseases of the Genito-Urinary Organs, p. 327.

known to cause fatal inflammation of the bladder. Mr. Fletcher[1] relates the following case:—"A healthy boy, six years old, was brought to the hospital with symptoms of stone, for which he was twice sounded by the surgeon, who satisfied himself, on both occasions, of the existence of the complaint. On the day, however, appointed for the operation, he could not detect it, and several of his friends tried successively with no better success. The surgeon at length put a stop to these proceedings; but his interference was too late. The boy was put to bed, complaining that his belly ached. Active peritoneal inflammation followed; and, notwithstanding the most energetic means to control it, death took place on the fourth day after sounding. The inner membrane of the bladder was spotted deep red everywhere, and its peritoneal coat was glued to that of the intestines, which were on all sides inflamed and covered with lymph." Dr. Prout informed Mr. Coulson that a patient of his in fair health died forty-eight hours after being sounded by one of the best surgeons of the day.

Dr. Gross sounded a young man, in 1844, who had suffered for twenty-four years from stone in the bladder, attended with chronic cystitis, and great disorder of the general health. The operation was performed with all possible gentleness, and yet three weeks elapsed before he was sufficiently recovered to justify the performance of lithotomy. Severe cystitis ensued, accompanied with violent spasm of the bladder, and the bowels became tympanitic and exceedingly tender on pressure, peritonitis having evidently supervened. These symptoms gradually yielded to the ordinary antiphlogistic remedies, but not without inducing the belief at one period that the patient would die. Sanson, Civiale, Crosse, and others, have recorded similar and even fatal cases. "In consequence," says Mr. Crosse, "of persevering and unsuccessful attempts to discover a stone with the sound, in a little boy, inflammation of the bladder came on attended by vomiting, and extending to the peritoneum; the most antiphlogistic treatment failed to arrest it, and death ensued in four days."

The practice of auscultation, by applying a stethoscope to the pubic region, the sacrum, or the perineum, as recommended by Laennec, is not adopted by modern surgeons, for the unassisted ear can detect, if possessed of its proper power, the contact of the steel instrument with the stone.

Before the operation of lithotomy be commenced, the presence of the stone should be established upon the most unequivocal evidence, and that can be only obtained by the process of sounding, as commonly performed.

Considerable difficulty is often experienced in determining on the presence of stone in the bladder. Numerous instances are on record of large stones being found after death, the existence of which could never be verified during life by the most careful examination.

Some years ago Mr. Coulson saw a gentleman who had been sounded by an experienced surgeon, who was unable to say whether there was a stone in the bladder; the examination caused a good deal of pain, and was never repeated. Sixteen months after, the patient died, and a stone was found in the bladder weighing 4½ oz., too large, as we may fairly suppose, to have formed since the sounding; the bladder was rather small and the prostate not enlarged. The museum of St. Bartholomew's Hospital contains a large calculus which was taken from a patient who was often sounded, but in whom the presence of stone could never, during

---

[1] Medico-Chirurgical Notes and Illustrations.

life, be ascertained, although there was little or no doubt in the mind of the surgeon of its existence. In the museum of the Royal College of Surgeons there is a bladder taken from the body of a man, who committed suicide after long and unalleviated suffering from stone, the presence of which had never been detected, though he was repeatedly sounded by the late Mr. Abernethy and others. The bladder contains a large, rough calculus, of an oval form, which measures in its several diameters about 1¾, 1½, and 1 inch. The mucous membrane is thickened, indurated, and superficially ulcerated. The muscular and cellular coats also are thickened, and the peritoneum and other tissues round the bladder appear condensed and unnaturally adherent. The ureters are dilated and thickened; both the bladder and prostate are of ordinary size.

Repeated examinations of the bladder with the catheter, observes M. Civiale,[1] have failed, even in the hands of the most experienced surgeons, such as Cheselden, Pelletan, and Dupuytren, to determine the existence of stone. Verzascha,[2] Benevoli,[3] Duretus,[4] Riverius,[5] Marcellus Donatus,[6] Chesneau,[7] Valentin,[8] Riolanus,[9] Morgagni,[10] Covillard,[11] Tolet,[12] Colot,[13] Morand,[14] Deschamps,[15] and Chopart,[16] state facts which prove that large calculi, of the size even of a hen's or duck's egg, nearly filling the bladder, have escaped the most minute and careful examination. Even where there are numerous calculi, they cannot always be detected. Colot says he sounded a man at sixty, and though unable to ascertain the existence of stone, operated on him, and extracted twenty-two calculi, which were hard, and of the size of a hazel-nut. Lapeyronie died of stone; his bladder contained a calculus weighing more than three ounces, which was not discovered during life, although he had been sounded several times. Portal and Distal also died of this affection, but in neither could the stone be detected with the sound until it was too late to operate. M. Kruger Hansen[17] relates the case of a man who suffered for a long time from dysury, and in whom a stone could never be felt. After death a very large calculus, covered with small-pointed elevations, was found in the bladder. M. Leroy[18] also mentions similar instances. With regard to all these cases, it must be borne in mind that up to comparatively recent times the instruments in general use for sounding were not adapted for the discovery of small stones, and large concretions doubtless often escaped detection owing to a similar reason.

---

[1] Traité de l'Affection Calculeuse, p. 477. Paris, 1838.
[2] Obs. lii., p. 109.
[3] Masotti, Lettera sopra gl'instrumenti, etc., p. 31.
[4] Schol. in J. Holler. de Morb. Int., lib. i. c. xlvi., p. 143. Two calculi, each weighing two ounces, were found after death.
[5] Praxis Med., i. 14, c. ii., p. 179.
[6] De Medic. Hist. Mirab., 1, 4, c. xxx., p. 453. A large stone was discovered after death.
[7] Obs. 1, 3, c. x.; Obs. 2, p. 354. A calculus as large as a goose-egg was found after death.
[8] Chirurg. Med. Sec., iii. c. vii., sect. ix., p. 352.
[9] Anthrop. 1, 2, c. xxviii., p. 150. Riolanus quotes the case of Casaubon in reference to this.
[10] De Sedibus, Ep. 42, art. 10.
[11] Obs. Iatro-chirurg., p. 42.
[12] Traité de la Lithotomie, p. 77.
[13] Traité de l'Op. de la Taille, pp. 167, 170, et 172.
[14] Traité de la Taille, p. 276.
[15] Ibid., t. i., p. 254.
[16] Traité des Maladies des Voies Urin., t. i., p. 55.
[17] Beiträge Mecklenb. Aerzte, t. i., p. 115 et 123.
[18] De la Lithotripsie, p. 56.

Cases not unfrequently occur in which a stone is felt distinctly at one time, and not at another, although the examination is made by the same surgeon, and, as nearly as possible, under the same circumstances. Mr. Crosse, in his work (fig. 9, plate i.), refers to "two calculi found after death in a gentleman who was about to submit to lithotomy; but the stone, previously ascertained to be in the bladder, could not then be felt."

A boy, aged twelve years, was admitted into one of the metropolitan hospitals with symptoms of stone; on the first examination, a stone was felt, but when the boy was put on the table to be operated upon, the stone could not be detected. He was brought a second time for the same purpose, but no stone could be felt. The parents shortly after removed him from the hospital. Not long after, the boy was brought to the General Dispensary, and the circumstances of the case related by the parent. It was determined to prepare the boy for the operation for stone, and then sound him, so that in the event of a calculus being discovered the operation might at once be performed. On introducing a sound, the calculus was distinctly felt, and, when removed, was found to weigh a little more than a drachm. Two cases occurred to Mr. Coulson in which considerable difficulty was felt by himself and others in detecting a stone in the bladder. One, a child, three years old, had been taken to a hospital and examined three times, but no stone was detected. He saw the child afterwards and sounded it three times, allowing an interval of a month between each examination, and it was only on the third occasion that the calculus was detected. Just before operating, considerable difficulty was again experienced in feeling the stone; and it was not until the finger was introduced into the rectum, and the neck of the bladder raised, that the stone could be felt. The operation was then performed, and a stone composed of lithate of ammonia, and weighing one drachm and six grains, was removed. It was situated in a pouch at the neck of the bladder, so that the sound, on entering the bladder, passed over the calculus towards the fundus. It is probable that if the finger had been introduced into the rectum when the child had been previously sounded, the stone would have been felt.

The other case was that of a boy, five years old; a stone could be distinctly felt on the first examination; but on repeating it a month afterwards Mr. Coulson was unable to detect the calculus. The child was sounded by one of the most experienced lithotomists in the metropolis; but, after a careful examination, he could not detect a stone. After waiting three months, the child was sounded again, and at the extremity of the sound, towards the fundus of the bladder, the stone was felt. The operation was then performed, and a mulberry calculus, weighing one drachm and forty-three grains, was removed. The operation revealed the cause of the difficulty: the bladder was very capacious for a child of his age, and the stone was situated at its fundus. The operator could not touch the stone, as in ordinary cases, with his finger, nor seize it with the forceps, until pressure had been made over the bladder, and the child's body raised.

In a case in which Sir B. Brodie operated, he could discern no stone when he first introduced his finger into the bladder; at last he felt it on the anterior part of the bladder, behind the pubes. It was not lying loose in the cavity of the bladder, but evidently contained in a cyst, communicating with the bladder by a round opening. By means of a probe-pointed bistoury, he carefully dilated the orifice of the cyst, and then, introducing his finger, separated the membrane from the calculus until he was enabled to take hold of the stone with the forceps. It was both encysted and adherent, for it was brought away with a portion of the membranous lining of the cyst closely attached to it. The boy recovered, and the calculus is preserved in the museum of St. George's Hospital.

A calculus, at one time loose in the bladder and easily felt by means of the sound, sometimes eludes detection by becoming lodged in a pouch or cyst formed by the protrusion of the mucous membrane between the muscular fasiculi; or, when such a pouch exists, it may happen that the calculus is sometimes contained in it, and at other times makes its escape into the bladder, a peculiarity which will readily account for a calculus being sometimes easily felt, and at other times not felt at all. Sir B. Brodie relates the following case illustrative of this condition of the bladder. "The case is, in many respects, remarkable. I discovered a stone in a gentleman's bladder; but he was advanced in years, and as, for the most part, he suffered very little inconvenience from the disease, he did not wish to go through any dangerous operation for the sake of obtaining relief ; nor did I think it right, considering all the circumstances, to urge him to submit to it. He went on, in general suffering little or nothing. He was a convivial man, dining a great deal in society, as if he had no ailment. Every now and then, however, he was suddenly seized with the usual symptoms of stone in the bladder, and very severe ones too; he then sent to me. I kept him in the horizontal posture, prescribed him an opiate clyster, and in the course of a few days, sometimes sooner, sometimes later, the attack subsided; he was again at his ease, and enabled to return to his usual habits. I had been occasionally in attendance on him for three or four years, when he was seized with a severe cold, which ended in a pleurisy, of which he died. On examining the body, I found the stone imbedded in a cyst near the fundus of the bladder. The cyst was formed, in this case, not by the protrusion of the mucous membrane between the muscular fibres, but by a dilatation of both tunics of the bladder, the muscular as well as the mucous. The stone was not so closely embraced by the cyst as to prevent it occasionally slipping out of it; and I suspect that this actually happened, and that it was when the stone lay in the cyst that the patient was free from the usual symptoms of calculus, and that his sufferings took place when the stone escaped from it into the general cavity of the bladder." Again, the bladder may ulcerate, and the stone either wholly or partially escape.

The causes of the difficulty in detecting stone in the bladder may be arranged under the following heads:

1. Various morbid conditions of the bladder, as fungous growths, sacculation of its coats, irregular contraction of the muscular fibres and displacements of the organ.

2. Enlargement of the prostate gland, accompanied by the formation of a pouch or cul-de-sac in the fundus of the bladder. Other abnormal conditions of the prostate.

3. Certain peculiarities of the calculus itself.

When a fungous growth exists at the neck of the bladder, it may interfere with, or entirely prevent the detection of a calculus, and when the stone becomes encysted, the difficulties of detection are often very great. In a case exhibiting some of the symptoms of calculus, an encysted condition may be inferred to exist when the stone can be sometimes, but not always found; when it appears to occupy always the same place in the bladder, and when only a small portion of it can be touched by the sound; when little or no increase in the patient's symptoms is produced by exercise, and when these and the effects on the patient's health are comparatively trivial. An encysted calculus rarely occasions much pain, but it may, as in Sir Benjamin Brodie's case, sometimes leave its pouch and become free in the cavity of the bladder, and subsequently return to its former position. Several encysted calculi may exist at the same time. In the forty-third volume of the "London Philosophical Transactions," Mr. Nourse relates the case of a patient in whom the calculi, nine in number, and contained in six separate cysts, were detected on the first sounding, but never afterwards. Ellerus relates a case in which a stone was contained between the coats of the bladder.

In children and grown-up persons with irritable bladder, the contraction of the organ renders it difficult to detect the calculus. In this state, it is desirable not to examine the bladder in an empty state, and if this organ be irritable, the irritation should be subdued by appropriate means, to enable the bladder to contain a fair quantity of urine. In one of the cases just detailed, there was little doubt that the fear of the operation caused the escape of the urine, and that the stone was consequently embraced by the folds of the bladder, and could not be felt. When the operation was performed, the boy had no previous knowledge of what was to take place, for the operator was in a state of uncertainty, from what had occurred, whether he should be able to feel the calculus.

Owing, as may be supposed, to irregular contraction of the muscular fibres, the stone is sometimes found at the upper part of the bladder, in close contact with the anterior wall, and above the symphysis pubis. Von Pitha has met with several instances in which the calculus was constantly found in that position.[1] He supposes that the walls of the bladder, in a state of spasm, closely embrace the stone, and that, owing to the roughened surface of the latter, and the possibly fasciculated condition of the muscular coat, the stone remains adherent to the bladder. Podrazki, however, states that in a case of this kind which came under his notice, the stone (phosphatic) had a perfectly smooth surface, and presented no points which might become entangled in a fasciculated bladder. In this case, by gentle friction above the symphysis, the stone could be dislodged from its position, and the patient asserted that he was conscious of the change, and that he could feel the calculus at the base of the bladder. Podrazki infers that irregular contraction of the muscular fibres is the cause of these anomalous symptoms. In the third case, just alluded to, of Mr. Coulson's, the situation of the stone was the cause of the difficulty in detecting it. In suspected cases of this kind, the patient should be examined in various positions, and the bladder should contain a moderate quantity of fluid. In cases of cystocele, a calculus has occasionally been found in the protruded portion, and, under such circumstances, it would be impossible to discover the stone by sounding.

In examining a patient with enlargement of the prostate, who is suspected to be the subject also of stone, the advantage of using a sound of the shape and curve of the lithotrite is sufficiently obvious. The short beak of the instrument can be turned round until it points towards the trigone, the probable situation of the stone in these cases, and this por-

---

[1] V. Pitha u. Billroth, Chirurgie, Bd. iii., 2 Abth., 8 Lief., s. 87.

tion of the bladder can then be thoroughly explored. This form of instrument was recommended many years ago. Mr. Crosse,[1] mentions that Mr. Pearson occasionally used a catheter of this shape, and found it enter the bladder when an instrument of a different shape would not. No examination of the bladder, when the prostate is enlarged, should be considered complete in which an instrument of this shape has not been employed. Again, the introduction of the finger into the rectum is an auxiliary, which, especially in these cases, and indeed, in most other difficult cases of stone, should not be neglected. It is singular that M. Civiale, whose experience in calculous affections is so great, should have spoken so lightly of this form of examination.[2] In one of the cases just related, the stone could not be felt with the sound until the finger was introduced into the rectum, and the operation was almost on the point of being abandoned. Mr. Crosse says, "I should deem myself as little justified in omitting this method of exploration as in operating without having sounded at all."

There are other abnormal conditions of the prostate which may render it difficult to detect a calculus. Thus Müller mentions the case of a boy in whom the following morbid condition of the urinary organs existed. The prostate was converted into an immense pouch, in which the catheter was arrested; the true vesical cavity, in which a stone was lodged, was not reached by the instrument. The patient was sounded twice without any stone being discovered. The third time, however, it was detected, and the operation was performed; a large quantity of pus escaped, but no calculus was found. The patient died, and, on dissection, it was perceived that the bladder had contracted tightly round a concretion the size of a small lemon. The prostate was partly destroyed by suppuration, and presented an enormous cavity, which had been mistaken for the bladder, and into which the instrument had wandered during sounding.

With regard to the stone itself, its small size may render it difficult of detection, or it may have a coating of lymph, mucus, or coagulated blood, and emit scarcely any sound when struck. Under such circumstances, frequent and very careful sounding may be necessary.

From the preceding remarks, it is clear that, though calculi are contained in the bladder, it may be difficult, if not impossible, to ascertain their existence during life, that the bladder may be sacculated, and that an excavation may exist immediately behind the prostate, or on one side of it, in which the stone may easily escape detection. In persons far advanced in years, a secondary cavity, occupying the fundus of the bladder, bounded above by the prostate, behind by the orifices of the ureters and the intervening portion of the parietes of the bladder, may contain the stone firmly fixed in it, this state being generally accompanied with

---

[1] Treatise on the Formation, Constituents, and Extraction of Urinary Calculus, p. 55, 4to. Lond., 1835.

[2] En effet, veut-on explorer la partie prostatique de l'urètre et le col vésical? le doigt se trouve séparé de la partie malade par un plancher épais, dur et résistant, à travers lequel le tact le plus délicat ne saurait rien discerner. S'agit-il d'atteindre la pierre située dans le bas-fond de la vessie? pour peu que la prostate soit engorgée, et même alors qu'elle ne l'est pas, le doigt ne saurait arriver jusque là ; sa pulpe appuie sur la face inférieure de la prostate, et rarement elle s'étend au bord postérieur, entre les vésicules séminales et les conduits spermatiques ; les cas dans lesquels il serait le plus facile de palper cette région de la vessie sont précisément ceux où l'on n'a aucun intérêt d'introduire le doigt dans le rectum."—*Traité de l'Affection Calculeuse*, p. 476.

ulceration of the coats of the bladder. In other cases, the stone may be, as it were, pinched between two folds of the bladder, or be fixed in a cul-de-sac, formed either by a hernia of the mucous membrane in consequence of separation of the muscular fibres, as so often happens in columnar bladders, or by a real cyst round the stone.[1]

A variety of anomalous symptoms may combine and simulate those of stone. Numerous cases have occurred in which so many of the symptoms of calculus were present that the patients were submitted to operation, but without any stone being discovered. The only unequivocal evidence of the presence of a calculus is that furnished by the sound, but even after the most careful sounding, numerous errors have arisen.

The symptoms of calculus are closely imitated by several diseases to which children are especially liable. In the case of a child with tubercular disease of the bladder, there is likely to be frequent micturition, irritation of the glans penis, pain in the hypogastrium and perineum, pus and occasionally blood in the urine. The bladder is also liable to become coated with phosphatic deposit, which, on sounding, may communicate the sensation of a calculus. It will not, however, give the clear ring of a stone. In dealing with a case presenting these symptoms, if a calculus cannot be detected, it is advisable to examine for tubercular disease of other parts, as the kidneys or testicles, or of the prostate gland, if the patient be an adult. It very rarely happens that the bladder alone is affected with tuberculous disease.[2] In some cases the disease commences in the testicle and spreads to the spermatic cord, vesiculæ seminales, prostate, and bladder, and under these circumstances the nature of the complaint would be sufficiently obvious. The symptoms of other morbid growths more or less closely resemble those of calculus,[3] but in the majority of cases careful examination with the sound will clear up the difficulty. Among other affections which simulate calculus may be mentioned phymosis, with contracted preputial orifice, occurring in children: irritability of the bladder, dependent upon morbid states of the urine, ascarides in the rectum, or other sources of irritation. In some of these cases the imitation is so close, that only after most careful sounding can the absence of a stone be demonstrated. It is worthy of notice that all these affections are more common before eight years of age than they are afterwards, and that even those not connected with the urethra are more common in boys than in girls. Stone in the kidney may also give rise to many of the symptoms of calculus in the bladder, but in the former affection the hæmorrhage is usually more profuse, the pain in the loins more decided, and the irritation of the glans penis less troublesome.

A very remarkable case, in which many of the symptoms of stone were present, and for the relief of which cystotomy was performed without finding a calculus, has been reported in detail by Mr. Paget,[4] of Leicester. The patient was a boy, three years and eight months old, who had suffered for two years from pain in micturition, stop-

---

[1] Velpeau, Médicine Opérative, 2d edit., 1839, tom. iv., p. 484.

[2] One instance in which the bladder alone was affected with tuberculous disease has been recorded by Mr. Prescott Hewett, Clin. Soc., November 27th, 1874. The patient was a girl aged nine, and the symptoms closely resembled those of stone.

[3] For a marked case of this kind reported by Mr. Savory, see Medical Times, vol. ii., 1852, p. 106. The tumor was a pedunculated growth so situated as to obstruct the internal orifice of the urethra, and the symptoms closely resembled those of calculus.

[4] British Medical Journal, December 14th, 1861.

page of stream of urine, and prolapsus of the bowel as the consequence of prolonged straining. The child was healthy looking, and there had never been hæmaturia. On examination with a sound, an audible click was produced, but it was not sufficiently clear to encourage an operation being performed. After repeated attempts the sound was found to be producible at will, but it did not impress all who heard it as being the click of an uncovered stone. In this dilemma, influenced by the character and intensity of the symptoms, and other considerations, it was decided to open the bladder. The median operation was performed, but there was no stone discoverable, and there was no click heard on introducing the forceps. The child died forty-eight hours after the operation, and on examination of the body, neither stone nor calculous deposit was found in any part of the urinary organs. Both ureters were tortuous and greatly dilated; the left formed a complete pouch, nearly as large as a pigeon's egg, at the lower end. There were evidences of slight peritonitis, and the mucous membrane of the bladder was injected posteriorly. The click heard in sounding was found to arise from the point of the sound impinging upon the iliac portion of the brim of the pelvis, the edge of which was unusually thin and sharp. There was extensive disease of the brain. The cause of the symptoms could not be satisfactorily explained, but it appeared to be connected with the dilated and pouchy state of the ureters. The lesson, as Mr. Paget observed, to be drawn from this case is not to operate unless the clearest and sharpest click is given to the sound. In another very obscure case,[1] in which the symptoms of calculus had existed from infancy, they were found to be caused by a large coagulum beneath the membranous portion of the urethra. The origin of the fibrinous concretion was attributed to the impaction of a calculus, and the formation by its side, of a cyst, which gradually enlarged, and became filled with mucous and fibrinous exudation and deposits from urine. The retention and difficult micturition were apparently due to the obstruction of the urethra caused by the solid mass.

In Mr. Paget's case and in the one just alluded to, examination with a lithotrite would doubtless have prevented the mistake. With regard to the sensation communicated when the sound impinges against a bony prominence of the pelvis, it is only necessary to remember that a calculus, unless encysted or adherent, rolls about in the bladder and changes its place, whereas a bony projection always maintains its relative position to the parts around.

A curious case, recorded by Mr. Fenwyck,[2] may be mentioned as an instance of the rarer conditions, the symptoms of which may simulate those of stone. In this case an aneurism of the abdominal aorta gave rise to symptoms which misled the surgeons attending the patient. There was violent pain at the extremity of the penis and in the pubic region, during and after micturition; there was pain in the kidneys and epigastrium; the urine was red, and deposited a thick sediment; the tongue was white, and the appetite bad. The symptoms appearing to indicate the presence of stone in the bladder, a sound was introduced, but no calculus could be detected. There was nothing which created a suspicion that the disease was aneurism. The patient, soon after he was admitted into the infirmary at Newcastle-upon-Tyne, died under symptoms of great exhaustion, preceded by sudden fainting. The abdomen

---

[1] Medical Times, 1852, vol. ii., p. 565.
[2] Lancet, 1846, January 24th.

was found filled with blood, which had proceeded from the rupture of an aneurism of the aorta, about the size of an orange; it had produced caries of the bodies of the vertebræ on which it rested, and was accompanied by hypertrophy of the left ventricle of the heart and thickening of the tricuspid valves. There was no calculus in the bladder, and no disease in any portion of the urinary organs.

CHAPTER XXI.

FOREIGN BODIES IN THE BLADDER.

The number of foreign bodies, besides proper urinary calculi, which have been found in the bladder, is immense. In males and females, hair, beans and peas, fruit-stones, ears of corn, portions of bougies, nails, bullets, small bones, pins, needles, string, stalks of flowers, etc., etc., have all been extracted from the bladder, and many of these foreign bodies have frequently formed the nucleus of a stone.[1]

The manner in which they may sometimes find their way into the bladder is curious. In many cases the accident has arisen from an attempt to satisfy morbid desires. Numerous cases of this kind are on record, especially in females; at other times the foreign body is introduced by accident or design. Some years ago Mr. Hayes Walton, exhibited, at the Medical Society of London, five calculi, which had been removed by Dr. Mackenzie, of Edinburgh. The calculi were of prismatic shape and uniform size. On inquiry, it was found that a number of beans had been forced into the patient's urethra and rectum by some companions in a drunken quarrel. Each calculus was formed round one of these beans. In other cases the foreign body has slipped into the bladder by accident. This is very likely to occur during the use of bougies, catheters, etc., and in a great many instances it has been found that the foreign body was a tobacco pipe or some cylindrical body, employed by the patient as a substitute for a bougie and for the relief of stricture. Sir A. Cooper once cut a woman for stone which had formed round a female catheter; the calculus was nearly an ounce in weight. Some months previously a surgeon had allowed the catheter to slip into the bladder, but said nothing about the accident for fear of getting into trouble.

Mr. Toogood, of Bridgewater, has published, in the *Medical Gazette* two interesting cases where a female catheter slipped into the bladder. In both cases the instruments were extracted by dilating the urethra by the sponge-tent, so as to enable the fore-finger to be introduced into the bladder.

Mr. Coulson was once summoned to a case in which a female catheter had slipped into the bladder, and after an hour's trial with various instruments, he at last succeeded in laying hold of it with a long, straight, narrow forceps, and withdrawing it. He also, in consultation with the late Mr. Key, saw a case where three inches of a thin gum-elastic catheter broke in the urethra, near to the bladder, and although called to the gentleman immediately after the occurrence of the accident, they were unable to lay hold of the broken portion. Three weeks afterwards the fragments (for there were two, one an inch in length and the other two

---

[1] I would refer the curious on this subject to the Dictionnaire des Sciences Médicales, t. vii., p. 38, and also to M. Civiale's Traité de l'Affection Calculeuse, p. 78.

inches) were voided by the urethra. In another case, seen by Mr. Coulson with the late Mr. N. Ward, of the London Hospital, a male silver catheter had become broken during introduction, and a part of the instrument, five inches in length, had made its way into the bladder. With the lithotrite, the broken part was extracted without difficulty, and the patient suffered but little from the accident and operation.

The use of gutta percha bougies has given rise to several accidents of the kind now alluded to; the substance very possibly be converted into a hydrate, which alters its qualities, and renders it brittle. A remarkable case occurred some years ago at Guy's Hospital, in the practice of Mr. Cock. The patient, a man forty-six years of age, had been under the care of Mr. Lawrence, at St. Bartholomew's Hospital, in 1854, with stricture, and had been in the habit of wearing a gutta percha bougie; but on one occasion the instrument became tightly embraced by the stricture, and gave way on an attempt being made to withdraw it. Mr. Lawrence made several fruitless efforts, with forceps, etc., to extract the foreign body, a portion of which projected from the bladder into the back part of the urethra. The canal was then incised through the perineum, but the fragment of bougie again broke, a portion remaining entirely in the bladder. The man was put to bed, and in a few days the fragment escaped spontaneously through the wound, which afterwards healed.

Eighteen months afterwards this man entered Guy's Hospital from a similar accident. He again had had recourse to a bougie of his own manufacture, and broke it a second time. He supposed that about five inches of the instrument was left in the urinary passages. The urethra was dilated by means of bougies, but no foreign body could be felt; and as the bladder was extremely irritable, Mr. Cock did not deem it prudent to employ the forceps or lithotrite. In a fortnight after the accident, the symptoms had become very severe, the urine contained blood, and ropy mucus mixed with phosphates; the pain was very great. The operation for lithotomy was therefore performed, but the division of the prostate was limited, being just sufficient to allow the introduction of the little finger. Three separate portions of bougie coated with phosphatic deposit were extracted with a small forceps; they made up about five inches, and evidently constituted the whole of the retained fragment. The wound progressed favorably, and was completely healed before the expiration of a month.

About the same time a similar case occurred in the practice of Mr. Callaway. The patient had been addicted to onanism, and introduced a cylindrical piece of French chalk, about an inch and a half long, into the urethra. It got into the bladder and excited great irritability of that organ, the patient making water every quarter of an hour with pain. A vertical incision was made through the raphé, and the membranous portion of the urethra opened—but still no body could be felt; the wound was therefore enlarged until the finger could pass into the neck of the bladder and dilate it; a small pair of lithotomy forceps was now introduced, and the foreign body removed. The patient did well.[1]

In a third case, observed by Mr. Hutchinson, the fragments of a gutta percha bougie came away spontaneously. The circumstances under which this occurred are worthy of attention; all attempts to discover the foreign body failed,—the urethra was therefore dilated, and the bladder

---

[1] Medical Times, Aug. 30th, 1856, p. 213.

afterwards injected with warm water to distention, until it began to rise above the pubes. About one-third of the quantity was at once rejected by micturition, in a full stream, but nothing else came away; in six hours, however, the patient again made water, and felt something in the perineum; on kneading this body forwards, he succeeded in removing a fragment about an inch long. The same fortunate accident occurred five hours afterwards, and again the patient was enabled to push forward a fragment and remove it as before. The two fragments were nearly three inches in length, and were thinly coated with phosphates.[1] It is proper to observe that, in these three cases the bougies had been imperfectly manufactured by the patients themselves, and not by an instrument-maker.

Various other foreign bodies have been known to gain entrance into the bladder by perforating its coats. In this way bullets and other projectiles, as well as fragments of bone, whether detached by violence or the effects of disease, have found their way into the bladder. Such bodies speedily become coated with phosphatic matter, and thus lead to the formation of calculi. Dr. Gross[2] states that during the last war in America, there were twenty-one examples of lithotomy for the removal of concretions consequent upon wounds of the bladder. In these cases, leaden bullets, a fragment of a grenade, an arrow-head, splinters of bone, pieces of cloth, inspissated mucus, etc., formed the nuclei of the concretions. Sir H. Thompson[3] has given an account of two cases in which fragments of exfoliated bone had passed into the bladder. In one case there was a history of injury to the pelvis, and in the other of abscesses in the neighborhood of the hip-joint.

The symptoms occasioned by the presence of a foreign body in the bladder more or less resemble those of stone. These, taken together with the history of the case and supplemented by careful examination with the sound, will usually establish the diagnosis. When, however, as is sometimes the case, no satisfactory history can be procured, information as to the nature of the foreign body may sometimes be obtained by introducing a lithotrite into the bladder, or passing the finger into the rectum or vagina.

The course to be pursued when foreign bodies, and especially fragments of bougies, etc., are retained in the bladder, is indicated in the preceding cases. The most favorable termination is, of course, that in which the foreign body is expelled spontaneously. Hence, whenever circumstances will permit, when the symptoms are not urgent, and the foreign body may be presumed to be of moderate size, Mr. Hutchinson's practice is worthy of imitation. The canal of the urethra should be dilated, and the bladder distended by injections. The change in the form of the bladder, produced by forced distention, is a circumstance which permits the elongated body to present itself at the neck of the organ, in a position best calculated to promote its passage into the urethra.

In other cases, when the end of the foreign body projects through the neck of the bladder into the urethra, an attempt should be made to extract it by means of Weiss's instrument for the removal of urethral calculi.

The surgeon familiar with lithotrity is evidently the most likely to

---

[1] Loc. cit., p. 213. [2] Loc. cit., p. 308.
[3] Diseases of the Urinary Organs, 4th edit., p. 288.

succeed in attempts of this kind; and hence we are indebted to M. Civiale for the most detailed account of the manner in which these foreign bodies may be removed from the bladder. The following remarks are condensed from his work on "Lithotrity."[1] M. Civiale had recourse to lithotritic instruments for the extraction of such foreign bodies as admit of being removed in this manner, and with them he has at various times extracted a haricot bean, a pea, corn beards, stems of plants, bougies, pieces of wood, and a barometer tube. In most of these cases, the operation presented much less difficulty than might have been anticipated, while the pain was not excessive.

Generally speaking, the patient can sufficiently explain the nature of the body which has been introduced into the bladder; but its shape, size, and consistence may be modified by its sojourn in that organ, phosphatic concretions almost always forming round it.

The bodies may be distinguished into three kinds. Some are small and oval, more or less soft, and capable of being flattened or extracted whole. The second kind is composed of flexible bodies, as bougies, etc. In the third species we have long inflexible substances which can be extracted only in the direction of their long axis.

With the first class the great difficulty is to lay hold of the foreign body; and for this purpose M. Civiale preferred his three-branched instrument. In employing it, care must be taken to close the branches on the foreign body by pushing forward the sheath, not by drawing back the branches. The reason of this precaution is obvious. In certain cases, a curved lithotrite with broad flat blades may be employed; from one to two ounces of fluid having been previously injected into the bladder.

Flexible bodies of the second class may often be seized and extracted with a small lithotrite; as the flexible body bends, great caution is required, the operator being necessarily ignorant of the manner in which the foreign body may be placed between the blades of the instrument. For long and inflexible bodies, various instruments have been constructed. In one of these, after introduction into the bladder, a hook is made to protrude by moving a portion of the handle. When the foreign body is felt, by withdrawing the hook, it is brought into a position which coincides with the axis of the instrument, which is then withdrawn.

Should these measures fail, or not be deemed applicable, an operation somewhat similar to that of lithotomy should be performed without any further delay. The phosphates are deposited on the foreign body with surprising rapidity in some cases, and it is always dangerous to wait too long. Delay is attended with two bad consequences: it gives time for the formation of calculous deposits, and it increases the risk of disease of the bladder or kidney, whereby the danger of any operation subsequently necessary must be greatly augmented.

In the case of females, the extraction of a foreign body from the bladder rarely occasions much difficulty. The urethra admits of considerable dilatation, which may be accomplished either by the forefinger or by the aid of Weiss's dilator, or Simon's specula. A pair of forceps can then be introduced and the foreign body seized and drawn out, the finger being used to assist the extraction. If the foreign body is of considerable size, it may be broken up with a lithotrite before extraction. Tedious and very careful manipulation may be necessary for the removal of a sharp-pointed foreign body. Mr. Bryant[2] reports a case in which a

[1] Traité pratique et historique de la Lithrotritie. Baillière, Paris, 1847, p. 233.
[2] Practice of Surgery, vol. ii., 122.

stiletto had been passed into the bladder of a young woman, and the point of the instrument was found to be projecting forwards and upwards, and fixed in the symphysis. With the left forefinger in the urethra, Mr. Bryant pressed the stiletto back, and after some little manipulation through the rectum with the right index finger, he managed to fix its point against the tip of the finger in the bladder, and then with a pair of forceps he drew the stiletto out. The extraction of a hair-pin may present some difficulties. When passed into the bladder with the rounded end foremost, its points separate from each other and are liable to fix themselves in the mucous membrane. In such a case a blunt hook should be passed along the finger introduced into the bladder, and an attempt made to catch the loop of the hair-pin. If, owing to its size or other causes, the foreign body cannot be withdrawn through the urethra, an operation similar to that for the removal of a calculus may become necessary.

An ovarian cyst has been known to empty itself into the female bladder; from which in such a case, pus, serum, and colloid matter have been discharged for a considerable period. In other cases, fragments of various kinds, such as hairs, teeth, and bones, have been discharged in a similar manner from dermoid cysts of the ovary. In cases too of extrauterine fœtation, a communication is sometimes established between the bladder and the sac containing the fœtus, and fragments of the latter may pass into the bladder. Six cases of this kind have been collected by Giessler.[1] In a case recorded by Sir Henry Thompson[2] numerous fragments of a fœtus were removed from the bladder of a patient.

Hydatids have also been discharged from the urethra, and after death, numbers of them have been found floating loose in the bladder. In the majority of such cases, it is most probable that they have descended from the kidney, along the ureter, but they have been known to find their way into the bladder from a cyst which had opened into it by ulceration. In a case referred to by Winckel, the posterior wall of the uterus was the original seat of the cyst, the contents of which found their way eventually into the bladder and rectum. Other parasites which invade the bladder are the Bilharzia hæmatobia, the ova of which have been repeatedly found in the urine of patients suffering from the endemic hæmaturia of the Cape of Good Hope and Egypt. This parasite has been already alluded to in the chapter on "Hæmaturia." Other worms have been reported as occurring in various parts of the urinary organs, but, as pointed out by Dr. Roberts,[3] many of the cases have been clearly proved to be examples of imposition—consciously or unconsciously—on the part of the patients. In this category may be classed a case reported by Mr. Curling,[4] and mentioned in former editions of this book. Intestinal worms, however, occasionally find their way into the bladder; in women through the urethra, and in both sexes through fistulous openings.

Several instances of hair voided with the urine are mentioned by Sir Hans Sloane; one particularly of a brewer, who suffered from the passage of long hairs, matted or woven together, passed with great pain, but with little or no calculous matter attached to them. Mr. Powell relates a case of a middle-aged lady, who after being teased with disordered

---

[1] Inaug. Diss., Marburg, 1856. Quoted by Winckel, Handbuch der Frauenkrankheiten, p. 158.
[2] Lancet, 1863, Nov. 22d, p. 621.   [3] Urinary and Renal Diseases, p. 591.
[4] Med.-Chir. Trans., vol. xxii., p. 274.

stomach and bowels, and evacuations of whey-colored and fetid urine, passed little masses of hair mingled with a peculiar viscid mucous substance, partially crusted with calculous matter. The excretion of these masses was attended with aggravation of the distress and pain in the bladder, from the urine bringing them into contact with the orifice of the urethra. The complaint, which in this case continued long, induced great weakness and much wasting of the body. Dr. Wallace also met with an instance in which hair was several times voided with the urine; and on the body being examined after death, a stone was found in the bladder, as large as a goose's egg, from parts of which hairs had grown out. It was thought that the hairs voided during life, which were very numerous, and some of an extraordinary length, grew out of that stone; because when the hairs hung out of the urethra, as they frequently did to the great torment of the patient, they were obliged to be pulled out, which was always done with a resistance, as if plucked by the root.[1]

---

[1] Howship, op. cit., p. 107.

CHAPTER XXII.

LITHOTRITY.[1]

VARIOUS efforts have been made, from time to time, to supersede the serious, and, before the introduction of chloroform, the painful operation of lithotomy.

All these attempts, however, failed in their object, and the removal of stone by cutting into the bladder was the only effectual mode of getting rid of the complaint, until the discovery by M. Civiale of the operation about to be described, viz., that of breaking up the stone in the bladder so as to allow of its discharge by fragments through the urethra.

Lithotrity is not suited to all cases of stone; its successful application will mainly depend on its being employed only in such cases as are suitable for this mode of operating. The indications or contra-indications of the operation will be pointed out at the close of this chapter; for the present it will be assumed that the subject about to be operated on does not present any conditions which would render lithotrity improper. These conditions are ascertained by a careful examination of the general health, and of the state of the urinary organs in particular. Having decided as to the propriety of the operation, the next point to be considered is, whether it be necessary that the patient should undergo what is called preparatory treatment.

When the calculus is small, the bladder healthy, and the urinary passages show little signs of irritability, preparatory treatment is not to be regarded as necessary. In most cases, however, it will be prudent to correct any unfavorable conditions which may exist, before proceeding to operate. Thus, the digestive organs must be attended to, and any derangement of the intestinal secretions improved as far as possible. Irritability of the bladder must be relieved by opiates, etc.; in a word urgent symptoms must be alleviated or got rid of. As a general rule, however, even where there are no unfavorable conditions, before submitting a patient to examination, and also before operating, he should be kept in bed for several hours, and directed to retain his urine for two hours, if possible. This precaution will enable him to bear a lengthened and thorough examination without distress. At the same time care must be taken not to overtax the bladder; the length of time during which the urine can be retained without causing distress must be ascertained, for if, at the time of the operation, there exists a strong desire to pass urine, with an irritable bladder, involuntary escape will almost certainly take place the moment the instrument comes into contact with the coats of the organ. Usually a sufficient quantity of urine, perhaps about four ounces, will accumulate in the bladder in an hour, or

---

[1] Several of the additions made to this chapter have been taken from the editor's work on Stone in the Bladder, London, 1868. The account here given o lithotrity refers to the operation as practised before the introduction of Dr. Biglow's method, which forms the subject of the next chapter.

an hour and a half, and most patients will be able to retain this quantity, if kept perfectly quiet. It is the more necessary to attend to these points, because before proceeding to operate the surgeon should, by a careful examination of the urinary organs, obtain some notion of the capacity and contractile power of the bladder; the size of the stone; the condition of the prostate; in a word, of various circumstances connected with the substance on which he is about to operate, and of the organs that inclose it or give passage to his instruments. Should the examination have given rise to any degree of irritation, it will be prudent to defer the operation until such iritation has subsided.

Previous to examination it is important to ascertain that no renal complication exists. This will be determined by a microscopical examination of the urine, which is always advisable before an operation. When there is evidence of kidney affection, this does not necessarily contra-indicate lithotrity, but on the other hand is a strong reason for preferring it to lithotomy. At the same time, however, it is a warning that very little should be done at each sitting, and that all unnecessary instrumentation should be avoided.

It is also necessary to ascertain whether any constriction exists in any part of the urethra. When there is a stricture it must of course be treated before any attempt is made to introduce the lithotrite, and it will be for the surgeon to decide whether to treat it by ordinary dilatation, rapid dilatation, or by internal urethrotomy. In the choice of the most suitable method, he will be mainly guided by the urgency of the bladder symptoms, by the circumstances of his patient, and by the seat and character of the stricture. When the stricture is confined to the orifice of the urethra, it may be divided by Civiale's urethrotome, and the operation of lithotrity may safely be proceeded with at once.

Before operating, however, an important question presents itself. Should the patient be submitted or not to the influence of chloroform? Both M. Civiale and Sir Benjamin Brodie pronounced against the use of chloroform in lithotritic operations. In the majority of cases, the operation, properly performed, is almost painless, and the occurrence of pain is generally an indication that the proceedings are being continued beyond what the bladder and constitution of the patient are capable of enduring. Under such circumstances the feelings of the patient become a most useful guide to the surgeon, who, however, must not require such assistance in order to determine whether or not his manipulations are injuring the coats of the bladder. As a general rule, therefore, chloroform is unnecessary, but in certain cases the use of an anæsthetic is attended by many palpable advantages. In some patients the urethra is abnormally sensitive—a condition not unfrequently co-existing with calculous affections and other diseases of the bladder. In well-marked instances of this kind, the passage of an instrument causes intolerable pain, and a thorough examination becomes impossible. In such cases some have recommended the preliminary introduction of bougies, in order to diminish the hyperæsthesia. The practice, however, is open to the objections that much time is lost, and the patient is subjected to much unnecessary suffering, not only from the repeated introduction of instruments and the acute symptoms occasioned by the stone, but also from the mental anxiety arising from the dread of impending operations.

To avoid these evils of delay and unnecessary suffering, it is better to have recourse to chloroform, and this state of the urethra is one of the few conditions in which anæsthetics are indicated during the operation of lithotrity. The hyperæsthesia in no way contra-indicates lithotrity; on the contrary, it often progressively diminishes with each operation. In fact, this abnormal sensitiveness, in many cases, seems to depend upon the presence of a foreign body in the bladder, and consequently subsides with the other symptoms, *pari passu* with the diminution in the size of the calculus. In cases, also, where the stone has been partly crushed without the aid of chloroform, and great irritability of the bladder supervenes, chloroform may become indispensable in the subquent operations. A remarkable instance of this kind occurred in the late Mr. Coulson's practice. A gentleman, æt. 52, in whom he had crushed a lithic acid calculus several times without the aid of chloroform, suffered after a time so severely from irritation of the bladder that malignant disease of the organ was supposed to exist. On examination of the bladder without the aid of chloroform, no stone could be detected. The symptoms becoming more severe, it was determined to examine the bladder under the influence of chloroform. This was done, when a large calculus, with lithic acid nucleus, covered with the phosphates, was readily seized and crushed. The operation was repeated four times under chloroform, with an ultimately successful result.

Another condition that renders the administration of chloroform desirable in lithotrity is cystitis, where this has arisen from the presence of the stone and does not subside with rest. In such cases, the stone is usually a rough one, and nothing but its removal will cure the cystitis, and the more rapidly and safely this is effected, the greater the chances of recovery. In such a case, the chloroform is of great assistance. The examination of the bladder, at the close of the treatment, for the purpose of ascertaining whether any fragments remain, is also sometimes greatly aided by the use of anæsthetics.

In performing lithotrity on children, chloroform is absolutely necessary, and, in operating upon women, also, the use of an anæsthetic is attended by many advantages.

Various instruments have been invented from time to time for the operation of lithotrity; as with most other inventions, suggestion followed suggestion, until, after many years of patient labor, M. Civiale succeeded in devising the instrument, which, having undergone certain small modifications, is now generally in use. It consists of a round steel shaft, with a curved and flattened extremity projecting about one inch from the line of the shaft at an angle of 120° or thereabouts. The shaft is adapted in diameter to the calibre of the urethra, and its length exceeds that of the latter canal by three or four inches. Throughout its whole length runs an open channel, in which fits a sliding steel rod, freely movable. The extremity of this sliding rod is curved and flattened, like the extremity of the shaft, only narrower; these curved extremities are called the blades of the lithotrite, the end of the shaft being called the female blade, and the end of the sliding rod the male blade. At the opposite end of the sliding rod is a screw turned by a wheel. This screw can be connected with or disconnected from the shaft at pleasure, by means of a disk fitted at the upper extremity of the shaft, containing a kind of double bolt, which, by a quarter turn of the disk, is made to press and rest against the groove of the screw. When the screw is connected with the shaft, the male blade, by turning the wheel, can be closed upon the female blade, so as to crush any stone grasped between the two, and again, by a quarter turn of the disk, the screw can be instantly disconnected from the shaft, and the male blade released from its position without the interruption that would be caused by unscrewing, thus enabling the operator to proceed at once to renew his search for the stone. The inner surface of the male blade, namely, that side which fits on the female blade, is roughened, to facilitate the grasping and crushing of the stone, and there is a small hole at the angle of the shaft, to prevent accumulation of débris. In one class of instruments used when the stone is very hard or large, and it is required to crush without pulverizing it, the female blade is open or fenestrated, as it is termed, to admit of the escape of fragments through it on crushing. For all ordinary purposes, however, the lithotrite with plain blades is sufficient. An instrument of this kind, owing to the flatness of its blades, requires less space in the bladder to enable it to grasp the stone, which, when the screw-power is applied, it reduces to fine fragments or débris. The fenestrated lithotrite, on the other hand, having wider blades, requires more room in the bladder, and, by its action, the stone is reduced to larger and more angular fragments. This form of instrument is suitable for large and hard stones, but an efficient substitute for it, and one without its disadvantages, is a non-fenestrated instrument, which has the male blade somewhat wedge-shaped. In dealing with large and hard stones, this latter form of lithotrite may be used for the first sitting, to be replaced subsequently by the ordinary flat-bladed lithotrite. With regard to other modifications of the instrument, Mr. Weiss has invented a method of changing the sliding into screw action, and *vice versa*, by touching a button in the handle. Sir H. Thompson has adopted this contrivance, and has introduced a cylindrical handle, which he regards as a very important improvement, claiming for it that it enables the operator to execute more rapid and delicate movements than are possible with handles of other forms.

Preliminary injections are, as a rule, unnecessary. The introduction of a catheter renders a patient less able subsequently to tolerate the presence of an instrument for any time in the bladder. The opinion formerly prevalent was, that unless the bladder contained several ounces of fluid, there was considerable risk of injuring the coats of the organ. With modern instruments, however, this risk is reduced to a minimum; the bevelling off of the male blade, so that the edge does not coincide with that of the female blade, renders it almost impossible to injure the coats of the bladder. Besides the lithotrite, the instruments sometimes required are, a pair of urethral forceps, and a large-sized catheter and injecting bottle for patients unable completely to empty the bladder.

The position of the operator and of the patient is of much importance. The surgeon stands on the right hand of the patient, so that it may not be necessary for him to change his position after the instrument has been introduced and when he proceeds to search for the stone. The patient should be on his back, his body being slightly inclined towards the operator. The operation may be performed on the ordinary hospital bed, but as this involves stooping on the part of the surgeon, it would be better if the patient's body were about thirty inches from the ground.

Many operators, following the example of M. Civiale, are in the habit of placing a firm cushion under the sacrum, so that any foreign body may roll back from the neck of the bladder. This plan, however, is really necessary only when there is considerable hypertrophy of the prostate, occasioning a regular pouch behind it, in which the stone lodges. The patient's head and shoulders should be supported by pillows, so that there may be no strain on any part of the body, and the patient may feel himself quite at ease. The legs are to be separated, and the thighs slightly flexed upon the abdomen.

The lithotrite, having been warmed and well oiled, is now to be passed into the bladder. To effect this part of the operation without injury to the urethra is only second in importance to crushing the calculus without injury to the bladder, for there is no accident in lithotrity that is attended with more distress to the patient, both at the time and after the operation, than that occasioned by laceration of the urethra. A minute description of the best method of introducing the instrument will therefore be given, as by careful and cautious practice of the various manœuvres required, injuries can in all cases be avoided.

On comparing the lithotrite with the catheter, it will be at once perceived that the difference in the size, shape, and weight of the two instruments will necessitate a corresponding difference in the method of introducing them. And it is from a disregard of these differences, and from an attempt to pass the lithotrite in the same manner as the catheter, that failure sometimes occurs, and that the patient is often subjected to much unnecessary suffering and occasional injury. The lithotrite, owing to its more abrupt and irregular curve, and the shortness of its bent portion, cannot be carried, like the catheter, with one continuous sweep into the bladder. And these differences, together with its greater weight and greater volume, render it essential that, in the use of this instrument, we should be much slower and more deliberate in all our manipulations.

The operator stands with his back towards the face of the patient; the lithotrite being held horizontally in the right hand, and this hand being placed immediately over the right anterior superior spinous process. The penis is taken between the middle and ring fingers of the left hand, and the lips of the meatus are opened with the thumb and index finger; the instrument is then introduced into the meatus, and the penis gently drawn over the short curve of the lithotrite, the instrument being at the same time allowed to decline slowly towards the perineum. To accomplish this, it is necessary to raise the lithotrite and at the same time to draw the penis towards the hand holding the instrument. When the extremity of the instrument reaches the bulbous portion of the urethra, the handle will be still inclined towards the right side; though it will have approached nearer the median line of the body. The lithotrite is now swept slowly round till it is brought to a perpendicular direction, the shaft forming a right angle with the body of the patient, and the result of this movement will be that the curved part of the instrument is brought into the direction of the canal under the pubic arch. Keeping it in this position, the penis is drawn upwards, and the lithotrite allowed to penetrate slowly by its own weight the membranous portion. When this part of the urethra has been reached, the left hand is withdrawn from the penis, and the instrument is lowered carefully and without force towards the patient's thighs, at the same time that it is pressed forward into the bladder. It is this last part of the introduction which requires the greatest care, for it is here the greatest difficulty arises, especially when the prostate is enlarged. It is only by constant practice on the dead subject that the natural amount of resistance to be expected can be learnt. The instrument has to be lowered in proportion to the hypertrophy of the prostate; cases have occurred to the editor in which this organ was so large that it was found impossible to lower the instrument sufficiently until the pelvis had been raised by a bolster; the handle of the lithotrite coming into contact with the couch before the instrument was sufficiently depressed to enter the bladder. When the lithotrite enters the membranous portion of the urethra, the left hand of the operator is free and should gently press on the pubic region. This proceeding has the effect of relaxing the triangular ligament, and of diminishing spasm of the abdominal muscles; and it is always found useful in proportion to the amount of resistance offered by muscular spasm and hypertrophy of the prostate.

The instrument having safely entered the bladder, the next business of the surgeon is to find and crush the stone. The best method of performing this part of the operation is that introduced by M. Civiale; it presents many advantages over the method originally practised in England, and even now adopted by some. In this latter method, the lithotrite is introduced fairly into the bladder, against the floor of which the convex portion of the instrument is pressed, so as to cause a hollow at that part. The blades are then widely opened, and the stone, naturally gravitating to the lowest point, comes within their grasp, when it is secured and crushed. It is sometimes necessary to give the instrument a smart shake in order to dislodge the stone. This proceeding, however, always occasions great pain and severe shock to the patient, and, after the first crushing, a sharp fragment may at any time get behind the instrument, and by pressure be made to lacerate the mucous membrane of the bladder.

The essential difference in the French operation is that, instead of the lithotrite being pressed against the trigone so as to produce a hollow into which the stone may roll, the instrument avoids contact with the vesical walls, and its blades are directed in search of the calculus. In one method, it is attempted to make the stone come to the instrument; in the other, the instrument seeks for the stone. In the French method, the patient being placed in the lithotrity posture, the instrument is passed into the centre of the bladder, in the manner already described, and its convexity is directed towards the trigone. In this position the blades are opened, and are allowed to fall half over to the right or left side and then closed. If the stone is not included, the instrument is brought back to its first position, viz., that which it occupied when first introduced; the blades are opened as before, turned to the other side and again closed, care being taken that the axis of the instrument is still maintained in the median line. In this way the central part of the floor of the bladder is explored. To examine further than this from the middle line, the instrument should be again opened in the first position, turned laterally, and gently swept over the floor of the bladder, until the convexity of the instrument is brought close to the concavity of the right side of the organ, so that the sides of the blades shall all but rest against the extreme right of the floor of the bladder, the points, of course, being directed inwards: the blades are now closed. The movements are repeated, if necessary, on the opposite side, with the same precautions. These manipulations will certainly find a stone when present, unless the prostate is so large that a pouch is formed behind it. When this is the case, the instrument is opened and turned round until the open blades point towards the trigone, the handle of the instrument is then raised until the points of the blades just touch the floor. When a stone is seized during any of these manipulations, the instrument should be brought back to the first position, and the stone crushed. The fragments will then fall immediately under the lithotrite, and a half turn of the instrument, without any lateral movement, will be sure to include some of them. Cases sometimes occur in which the operator experiences considerable difficulty from the insufficient length of the lithotrite as ordinarily manufactured; and many English instruments made on the French principle have been found to be more than an inch shorter than those of French manufacture. Where, however, difficulty arises from this cause, it generally happens that the penile portion of the urethra is unusually long and the prostate gland hypertrophied. The editor of this work has recorded a case of this kind, in which had the calculus been at all larger than it was, it would have been impossible to open the instrument wide enough to seize it.[1] The patient also suffered more than he otherwise would have done, as, in order to crush the stone, it was necessary to press back the curved portion of the female blade deeply into the bladder.

Another instance occurred to Mr. Coulson, where, from the unusual length of the urethra, he was obliged to have a lithotrite expressly manufactured some inches longer than his largest instrument.

All the manipulations should of course be conducted slowly and in a cautious manner. In searching, the instrument should be held as lightly

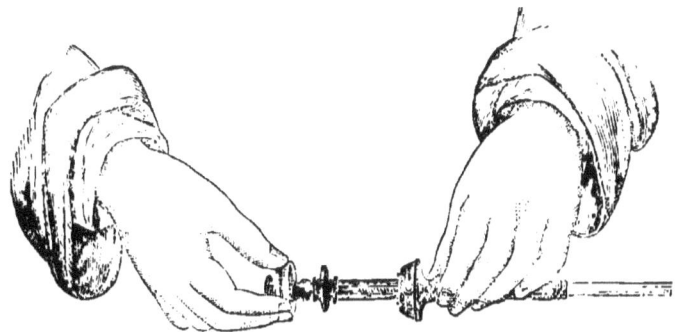

FIG. 5.—Figure showing the lithotrite held in the manner just described.

as possible, the right hand being applied to handle the extremity of the male blade; the left hand holding the handle of the female blade. When the stone is to be crushed, the left arm is pressed to the operator's side, while the left hand grasps the handle or wide part of the female blade as firmly as possible, so as not to allow of any giving or lateral movement, the right hand screwing the male blade home. It is of course desirable that the calculus should be seized, as nearly as possible, in the direction of its short axis.

When the calculus is large, the surgeon has little difficulty in finding the stone, but it may not be so easy to seize it. In such a case, the two blades of the instrument should be applied, slightly separated, to the surface of the stone; they are then more widely separated, by drawing one branch forwards and pushing the other backwards, until they reach the edges of the foreign body. The instrument, thus opened, is pressed laterally on the stone, until its branches become placed between the foreign body and the walls of the bladder.

Before withdrawing the instrument from the bladder, it is absolutely necessary to ascertain that the branches are perfectly closed, and that no

---

[1] Stone in the Bladder, by Walter J. Coulson, p. 38.

fragments of the stone or detritus are retained between the blades. On looking at the scale marked on the handle of the instrument, we shall see at once whether the blades are closed, or to what extent they are separated. It is scarcely necessary to say that any increase of the size of the instrument, or the projection of any angular fragments between its blades, might produce serious injury to the neck of the bladder and narrow parts of the urethra, while the lithotrite is being withdrawn. It requires several movements of the screw to get rid of the detritus from the scoop, but the detritus must be got rid of. A few turns of the screw backwards and forwards will usually suffice to clear the instrument, or the action of the screw may be suspended, and percussion be exercised slightly on the end of the male branch.

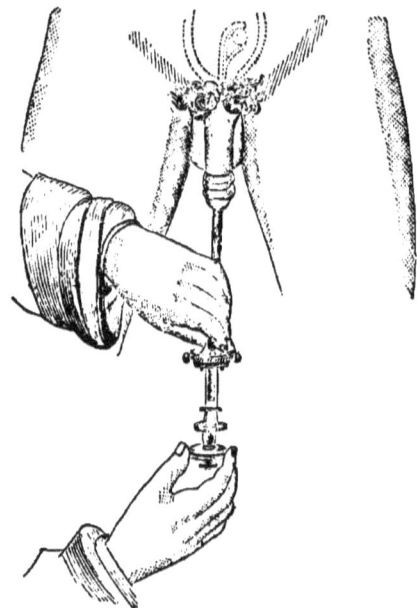

Fig. 6.—This figure shows the stone between the blades of the lithotrite, the screw of the instrument being in action.

The first operation in lithotrity should be regarded as a kind of test, and should not occupy more than a couple of minutes; in this time it is usually easy to crush the stone two or three times, and this is quite sufficient for the first sitting. By watching the results, an insight will be obtained into the constitution of the patient, and valuable knowledge will be gained as to the amount of tolerance the bladder exhibits towards operative interference. Much suffering and great loss of time are sometimes occasioned by attempting too much at the commencement. After a cautious experiment in the first instance, much more can be afterwards done with confidence. As a rule, at subsequent operations, the instrument may be kept in the bladder for five or six minutes without occasioning marked distress, and in this time many fragments may be crushed. With respect to the interval to be allowed between the different sittings, four days may be usually suffered to elapse between the first and second, and afterwards the surgeon should be guided mainly by the coming away of the detritus. Where, however, the early symptoms of calculus have not been disregarded, two operations should suffice. After an efficient crushing, so long as the débris continue to escape in considerable quantity, it would be useless to do more, but as soon as this ceases to be the case, the operation should be repeated. The interval of a few days allows ample time for all symptoms of irritation and soreness of the urethra to subside, and for the patient to recover from the slight shock of the operation.

It has frequently been observed that a first operation has afforded marked relief to patients, who, before the crushing suffered acutely from the more distressing symptoms of calculus; but that others, who had previously suffered comparatively little from the presence of the stone, had all their symptoms materially aggravated after the first crushing. Increased irritation is to be anticipated with hard calculi, uric acid, and oxalates, where the fractured portions are sharp and angular, and in this case it is particularly important to pulverize the fragments as much and as soon as possible; and not only is there a strong reason for speedy operation, but it is useless, and even hazardous, to wait for the subsidence of the irritation, when the latter clearly depends on such a cause. The following is the practice which the editor's experience has dictated in this matter. If the pain and other symptoms have begun distinctly to diminish before the usual time for the next sitting, it is well to wait until the fourth day for their further subsidence. If, however, they continue, and especially if there are rigors, and much pain of a cutting character, the cause is almost certainly irritation by angular fragments, and it is desirable rather to anticipate by a day or two the usual period. This course has been pursued by the editor on several occasions with marked benefit, and at other times its non-adoption gave cause for regret. When operating under the above-mentioned circumstances, the use of chloroform is to be recommended.

The removal of débris by injecting and washing out the bladder was formerly the rule; at the present time such an attempt is rarely made. At one time, the method was generally practised immediately after each operation, and it occasionally happened that patients who bore the lithotrity with perfect heroism, completely broke down under the unnecessary infliction of syringing. The word "unnecessary" is used advisedly, since it is only the smaller fragments that will escape through a catheter, while these, and much larger fragments than these, will pass readily and without pain through a healthy urethra during micturition. The method therefore is to be regarded as a mischievous one, since it involves an unnecessary introduction of instruments—a practice which should always be avoided.[1]

The first indication in the after-treatment is absolute rest after each operation. So long as the surgeon has reason to believe that a fragment

---

[1] It is necessary to state that the above remarks do not apply to the method of operating introduced by Dr. Bigelow, in which the stone is crushed, and the fragments removed by means of a powerful aspirator through a large evacuating tube, at a single sitting. An account of this operation is given in the succeeding chapter, and the remarks in the present one with regard to the removal of fragments by syringing and washing out the bladder, refer to attempts made with this object prior to the introduction of Dr. Bigelow's highly-successful method; which differs, both in its essential principles and in its scope, from those hitherto practised.

of any size remains in the bladder, the patient should be kept perfectly quiet.

The fact has been already alluded to that marked relief to all the painful symptoms of stone sometimes occurs after the first crushing: occasionally to such an extent that before any fragments come away, the patients imagine they are cured of their disease. This disappearance of painful symptoms before there is any material diminution in the bulk of the stone, depends most probably on the irregular shape of the broken fragments which permits the urine to filter through them, and so offers little impediment to micturition; so far, however, from the pain subsiding, the symptoms would all be very much aggravated if the patient attempted any exercise.

The operator will always know from the size of the original stone, from the quantity of débris passed, from the state of the urine, and the facility with which fragments are seized at the subsequent sittings, how much remains to be done. Towards the termination of the case, when all irritation has subsided, and when the urine passed shows very little trace of vesical mucus, the patient should be directed to try himself with gentle exercise—walking, driving, or riding. Should any such exercise induce irritation, or give rise to the appearance of mucus or blood in the urine, it is essential that another examination should be made with the lithotrite.

The necessity of a very careful trial, before dismissing a patient as perfectly cured, is shown by the well-known fact that recurrence of all the symptoms of stone is much more common after lithotrity than after lithotomy. In many cases, the obvious reason for this is, that some fragment has remained after the last crushing. The speedy recurrence of stone appears most commonly in the old, and is often the fault of the surgeon for trusting too implicitly the sensations of the patient, and not bearing in mind that, with advancing years, the normal sensibility of the bladder is, to say the least, impaired. The absence of irritation sometimes induces the surgeon to neglect what is of so great importance, viz., the microscopical examination of the urine, and this as affected by exercise or repose. When either the sensations of the patient, or the appearance of the urine under these two conditions, indicate that a fragment is left behind, a further exploration becomes necessary, and it is at this stage of lithotrity that the most difficult and careful manipulations are required. In the old, when the prostate is very large, or when from partial atony of the bladder, a portion of urine is always left behind, some difficulty may be experienced; and, in rare cases, when the bladder is sacculated, this difficulty is materially increased. Under either of these conditions, a careful examination with the lithotrite will enable the surgeon to deal with the offending fragment. In conducting this final examination, it is important not to have too much urine in the bladder; three ounces is in most cases quite sufficient to admit of all the necessary manipulations.

When the prostate is hypertrophied, it is advisable for the final examination to have the pelvis raised by means of a bolster. The patient having been thus placed in the most favorable position, the examination then becomes simply a matter of manipulative dexterity; at the same time, the previous experience of the operator in this particular case will prove of the greatest assistance in enabling him to select the most suitable instrument, and to judge the position in which the fragment is most likely to be lodged. The instrument generally most convenient and handy for the final examination, is a lithotrite with a rather short and broad curved extremity. With such an instrument, the search is far more likely to be successful than with the trilabe which the late M. Civiale was in the habit of using. The latter instrument is most untrustworthy, and possesses all the disadvantages of the sound, as compared with the lithotrite, for the purposes of exploration. It has also another disadvantage in being perfectly straight, and consequently in those cases where the prostate is large and there is a pouch in the bladder behind, the very cases in which trouble is to be anticipated from the retention of a last fragment, it is impossible, by its means thoroughly to explore the most important area of the bladder, whereas no difficulty is encountered in efficiently exploring it with a lithotrite. Even if unsuccessful once with the lithotrite, it is better to make a second attempt with it before resorting to any other instrument.

Having described the operation of lithotrity under ordinary and favorable circumstances, certain obstacles to its performance must next be considered. These obstacles often occur, and unless they be overcome with judgment and dexterity, various unpleasant or even fatal accidents may result.

The obstacles to lithotrity may be divided into four principal classes, for a difficulty will be created by anything which impedes the introduction of the instrument, the seizure of the stone, the manipulations in the bladder, and the subsequent discharge of the detritus. The effects of strictures, etc., in contracting the canal and impeding the introduction of the lithotrite are obvious. In other cases, the same obstacle arises from a morbid degree of sensibility in the urethra and neck of the bladder.

These parts are excessvlely sensitive in many calculous patients, while the bladder is at the same time endowed with increased contractility. Should these conditions depend on organic disease of the prostate or bladder, it will be prudent to abstain from operating. When they are merely nervous, appropriate remedies will often relieve the nervous sensibility, and the operation may be had recourse to, but with great caution. The administration of chloroform is to be recommended for these cases. The fact that the symptoms are often considerably relieved by the first operation has been already alluded to. Should severe rigors, followed by fever, or nervous symptoms supervene, the operation must not be persisted in.

Excessive irritability of the bladder is another obstacle of a serious nature, and which it is often difficult to overcome. Unless the bladder can be made to retain two or three ounces of fluid, we ought to give up all thoughts of crushing the stone rather than run the risk of exciting fatal symptoms, by operating in an empty and a disordered organ. The irritability, however, may often be much lessened by keeping the patient in bed, and paying attention to the state of the secretions, for some days before the operation.

Enlargement of the prostate impedes not only the easy introduction of the lithotrite, but its working in the bladder; still, we are not to conclude that every degree of prostatic enlargement presents an insuperable obstacle. Each case must be judged according to its own merits. Slight enlargement of the prostate is certainly no great impediment; on the other hand when the prostate is considerably enlarged, all the circumstances of the case must be carefully considered; the patient will have at least one point in his favor, viz., that his urethra has probably become so much accustomed to the passage of instruments, that, if the fragments are not passed by the natural efforts, the introduction of a full-sized catheter, and the use of injections to wash out the bladder, will not be productive of much irritation.

The condition of the bladder itself may present certain obstacles which must be taken into account. In many old cases of stone the bladder is hypertrophied, the fleshy columns project from its surface, while the organ is at the same time much contracted, and very irritable. There is no room to work in a bladder of this kind, and this state is extremely unfavorable. Lithotomy is a preferable operation in such cases. The sacculated bladder presents another serious obstacle to the performance of lithotrity. When the calculus is firmly impacted in, and more or less covered by the sac, it is difficult to understand how injury to the mucous membrane of the bladder can be avoided; yet a few cases are recorded in which it is presumed that encysted stones were successfully crushed, from the circumstance that the detritus had always to be sought for at a particular spot in the bladder.

The size and consistence of the stone are the last points to be noticed. When the calculus is beyond a certain size, it cannot evidently be seized by any instrument; but beyond this it is difficult to lay down any absolute rule as to the exact size of the calculus which renders lithotrity inadvisable. The consistence of the foreign body must here be taken into account, as well as its size. Phosphatic calculi, of any size, are easily broken up, provided they are not too large to be seized. Even when the stone measures an inch and a half or two inches and a half, it may be crushed, provided it be not very hard, and the urinary organs be free from disease.

Mr. Massey, of Nottingham, crushed two uric acid calculi, the detritus of which weighed about two and a half ounces; Mr. Cæsar Hawkins also operated on a patient 75 years of age, the detritus amounting to the same weight of two and a half ounces.

Mr. Key says, "I know of no limit to the size of a calculus removable by lithotrity but the power of the lithotrite. If a powerful instrument can be brought to embrace it, and the bladder be healthy, the operation may, as far as my experience goes, be attempted with propriety."

Excessive hardness of the stone is no absolute impediment to the performance of the operation, although it may be a counter-indication. There is no stone, however hard, which could resist the power of modern instruments, provided that its size was not too great to prevent it from being fairly grasped.

The accidents which sometimes supervene in the operation of lithotrity are, if we believe some authors, numerous and severe. M. Velpeau, who was always hostile to lithotrity, enumerated no less than 29 accidents occurring after this operation. Many of these belong to the past history of lithotrity, and occurred before the operation had attained its present state of perfection. At the present day we seldom hear of laceration of the urethra, preforation of the walls of the bladder, fracture of the instruments, etc. With regard to accidents connected with the instruments, the editor has met with a case in which the working of the screw caused the female blade to yield, and much difficulty was experienced in releasing the fragments and withdrawing the lithotrite. He also witnessed, on another occasion, the giving way of the worm of the screw, an accident which caused much difficulty in withdrawing the instrument. Other accidents, however, evidently depend on the operation, and may be fairly attributed to it. Thus, lithotrity sometimes produces more or less pain and uneasy sensations. These are generally slight; in more difficult cases it is impossible, unless we employ chloroform, to avoid inflicting a certain amount of pain. When the operation has been restricted within its proper limits, and due precautions are used, the pain is almost always of a nature to be readily borne by the patient. The contrary may occasionally happen. The bladder and urinary organs may be apparently healthy, the calculus small, every precaution has been used, yet the neck of the bladder is excessively irritable; the passage of an instrument will, in such cases, excite great pain, and if the operation be persevered in, dangerous constitutional disturbance may ensue. These are cases well suited for chloroform.

Many other cases are on record, in which the pain is described as having been of the most violent nature, and its effects of a disastrous kind; but on examining them, we shall generally find reason to believe either that one or more conditions were present, which rendered the adoption of lithotrity improper, or else that the operation was unskilfully performed.

The same causes which produce pain, may likewise give rise to more or less discharge of blood from the urethra, or the bladder.

A patient operated on at the Hotel-Dieu, in 1832, died of hæmorrhage from the bladder, and Mr. Key relates a case where the bleeding was so copious as to render necessary an immediate recourse to lithotomy. In one or two cases, operated on by Mr. Coulson, the amount of blood lost was somewhat considerable, but he never met with a case in which dangerous hæmorrhage occurred. The editor's experience has taught him to regard the occurrence of hæmorrhage after lithotrity as a very rare exception. Sir B. Brodie also states that in the 115 cases operated on by him, he did not meet with a single example of serious loss of blood from the urinary organs. Sir H. Thompson [1] has recorded one case in which fatal hæmorrhage occurred from the bladder, after a short sitting, conducted with great care. All treatment proved unavailing. No sign of injury could be discovered at the autopsy; the hæmorrhage was attributable to constitutional causes, and had been a prominent symptom in the history of the case. It may be doubted whether the patient was a fitting subject for any operation, but the details as given by Sir H. Thompson are too scanty to admit of any conclusion being drawn from them.

In a few cases, the operation has given rise to suppression of urine, and the consequences attendant on that serious condition. It is probable that, in some of these cases, the operation merely called to light some latent disease of the kidney which previously existed. Such a causation, however, cannot be inferred in all cases in which the passage of instruments has been followed by suppression of urine, proving rapidly fatal. Dr. Roberts [2] has recorded a case in which death in 20 hours—with total suppression—was the result of catheterism in an old case of stricture, where instruments had been repeatedly used before without any ill effects. Intense congestion of the kidneys was the only *post mortem* appearance of consequence. Another very similar case is recorded by Sir H. Thompson. Such an accident, therefore, inasmuch as it may occur after the simple passage of a catheter, is not to be regarded as connected in any special manner with the operation of lith-

---

[1] Practical Lithotomy and Lithotrity, 3rd Ed., p. 174.
[2] Urinary and Renal Diseases, p. 25.

otrity. There appears to be no clew as to the pathology of these cases. Intense congestion of the kidneys would seem to be the cause of the suppression. Dr. Roberts is inclined to attribute the suspension of the secretion to disturbance of the innervation of the organs, and points out that death, in these cases, is too rapid to be due to the non-elimination of the urinary excreta.

Acute cystitis is of rare occurrence, but in hospital practice we hear of many cases of subacute or chronic cystitis, resulting from, or at all events following lithotrity. The patient is then left in a very distressing condition, and the results of lithotrity, if judged of from these cases, would appear anything but favorable. It is by no means easy to determine the circumstances under which the subacute inflammation now alluded to has been produced. In private practice we do not frequently meet with cases of the kind, and it is certain that catarrh of the bladder is more often relieved than aggravated by a properly conducted operation of lithotrity. It would appear probable that in many of the unfortunate cases alluded to, the severe chronic inflammation and other accidents have arisen from retention of imperfectly pulverized fragments in a diseased bladder.

Acute symptoms of cystitis may occur after crushing a hard calculus, being due to the irritation caused by the presence of sharp fragments. Under such circumstances, nothing but the removal of the fragments will cure the cystitis, which, so far from indicating any cessation of operative procedures, urgently demands their repetition, as the only effectual method of treatment. Especial care and gentleness are, of course, indispensably necessary, and the operations should be conducted under chloroform.

The constitutional accidents of lithotrity may be comprised under rigors, febrile attacks, and nervous disturbance. Many patients experience a smart attack of rigor immediately after the operation, especially after the first sitting. It is not known on what the rigors depend; they have been attributed to sympathy and other indefinite states. They do not bear any proportion to the amount of pain, but experience teaches us that they are apt to be severe when the operations have been prolonged. Where disease of the kidney co-exists, rigors are very apt to occur, and under such circumstances they indicate an aggravation of the original mischief. In the majority of cases, however, they are not of serious import. Warm drinks, such as tea, or a little warm wine and water, are the best remedies; their action will be assisted by keeping the patient in bed, and covering him with a warm blanket, and applying hot bottles to the feet and abdomen.

The rigors may be followed by a temporary attack of fever which resembles the irritative fever of some authors; or a more severe and permanent attack may ensue, rapidly assuming a typhoid character, and terminating in death. In these latter cases, the fatal result is due to the aggravation of a previously existing affection of the kidneys, the symptoms of which have been altogether latent, or else masked by those of the calculus. Chronic pyelitis may exist to a considerable extent without producing any marked symptoms. There are not necessarily any tube-casts in the urine in this affection, and the quantity of albumen may correspond only to the pus from the bladder. In other fatal cases, in which the symptoms assume this type, the condition is due to præmia or septicæmia. However slight the first rigor, the patient should be closely watched, and, and if it be followed by frequent desire to pass water, and pain over the pubic region, hot flannels should be kept constantly applied to the hypogastrium and opium administered in the form of a suppository. No danger is to be apprehended unless there is reason to fear the co-existence of some renal affection. Should the symptoms, however, indicate that the inflammation is extending along the ureters to the kidneys, counter-irritation should be applied to the loins by means of hot flannel sprinkled with turpentine. Opium should be discontinued, and the patient placed on a strict regimen. If the surgeon is on the watch, and the symptoms are early observed, they will generally yield to appropriate treatment. When there is reason to believe that a renal complication exists, though it may not contra-indicate lithotrity, it should warn the operator to do very little at each sitting, and should prove an additional reason for using the greatest gentleness and deliberation in all the manipulations.

The surgeon who undertakes lithotrity, should watch very carefully the condition of the bladder after the first operation of crushing, for retention of urine may at any time supervene. The use of the instruments almost invariably determines an urgent desire to empty the bladder; but no sooner is the operation over, than the desire ceases in a great many cases, and may be replaced by retention of urine. The expulsive power of the bladder should therefore be carefully watched, and the causes of retention determined. It may depend on coagula, impaction of a fragment in the urethra, etc., but more frequently it arises from some loss of power in the bladder, combined with a spasmodic state of the urinary passages. This latter condition often exists in elderly persons affected with enlargement of the prostate. In these patients also, a slight temporary swelling of the neck of the bladder and prostate, superadded to the obstruction caused by the enlargement, may render micturition all but impossible, and interfere considerably with the passage of fragments.[1] This condition is better treated at first by retaining an elastic instrument in the bladder, than by the frequent introduction of the catheter. An instrument thus retained is very apt to become clogged by the accumulation of pulverulent débris, and it is necessary every three or four hours to inject a little warm water for the purpose of keeping the catheter perfectly open. It is also desirable, when this complication exists, to wash out the bladder after each sitting, by means of Clover's apparatus. Other cases sometimes occur where the retention is only partial; a condition that has probably existed prior to the operation.

Another accident to which patients who have undergone lithotrity are liable is the lodgment of calculous fragments in the canal of the urethra. This accident, rarely, if ever, occurs, unless the urethra has been torn or lacerated, a complication which can always be avoided by proper care on the part of the surgeon in the introduction of the instrument, and by taking time and pains to empty the lithotrite before withdrawing it. Stricture of the urethra or contraction at the orifice may also give rise to impaction of fragments. In the editor's work on "Stone in the Bladder,"[2] will be found an account of one case in which the complication occurred. The patient was the subject of a stricture, which had been ruptured previously to lithotrity, and at this ruptured part a fragment had become impacted.

The fragment may be stopped at any part of the urethra, but the

---

[1] For a report of a case of this kind see the editor's Lectures on Stone in the Bladder, p. 54.   [2] P. 52.

most common seats of impaction are the neck of the bladder and the membranous portion of the urethra. The symptoms are usually severe and well marked. A rough sharp fragment of any size soon excites pain about that part of the canal where it is arrested. The patient suddenly discovers that he is unable to pass water freely, or he may have complete retention; the desire to urinate soon becomes frequent and urgent. These symptoms should always arouse the surgeon's attention: if relief be not given, and that speedily, dangerous constitutional disturbance may ensue. The patient becomes irritable; rigors often set in; these are followed by irritative fever of a severe kind ; the irritation extends to the bladder, perhaps to the kidneys, and the condition of the patient soon becomes serious.

An accident of this kind requires prompt and effectual means for its relief. As soon as the surgeon has any reason to suspect its existence he should pass a sound into the urethra to ascertain the seat and nature of the fragment. A large metallic sound may be used in the first instance; and if this fails to detect the foreign body, a large-sized soft bougie may be introduced, for it is possible that the fragment may be lodged in one of the lacunæ. If this be the case, the edge of the fragment, as M. Civiale observes, seldom fails to leave a mark on the bougie.

For the removal of the fragment various means have been proposed ; thus it may be pushed back into the bladder, extracted whole, crushed in the urethra, or removed by external incision. The choice of a method will be governed by the circumstances of the case. Where the fragment is impacted in the neck of the bladder, or behind the triangular ligament, great difficulty will be experienced in extracting it, and the best practice will be to push back the foreign body into the cavity of the bladder. When the fragment is small, and seated close to the neck of the bladder, this can easily be done with the sound or bougie ; a fragment more firmly impacted will, of course, present more resistance ; in such a case, a large silver catheter, with the end cut off, and the opening closed by means of a knob attached to a stylet, is passed down to the fragment; the stylet is then withdrawn, and warm water forcibly injected into the urethra.

If the fragment is firmly impacted in the prostatic portion of the urethra, and cannot be dislodged, an accident which is especially likely to happen where the prostate is enlarged, it is better to make an incision through the perineum and remove the fragment, which otherwise may cause inflammation and abscess.

Impaction often takes place in the membranous portion of the urethra, and in this situation the fragment may be extracted whole or crushed. The expediency of crushing fragments seated so far back in the urethra seems, however, very doubtful. For extraction, various kinds of forceps are employed ; or a blunt hook may, perhaps, be passed behind the fragment, and the same means may be had recourse to whenever the fragment is seated at any point between the bulb and the glans penis. A convenient instrument for the extraction of fragments is a small urethral lithotrite, made expressly for the purpose, the female blade of which can be depressed and brought in a line with the shaft of the instrument, by a turn of the disk to the left. By this contrivance the female blade can be passed beyond the impacted portion, and with a somewhat similar instrument, but without a male blade, impacted fragments may be removed from the penile portion of the urethra. If it is found absolutely necessary to cut down on this portion of the canal, the edges of the wound should be brought together by means of a wire suture, and a catheter should be retained until cicatrization takes place, otherwise a urethral fistula is inevitable. The extraction of fragments by incision should be resorted to only when every other method has failed, for apart from the danger of a permanent fistula, an opening in the urethra materially interferes with the subsequent stages of lithotrity.

Great caution is required in all cases where the fragment is seated in that part of the spongy portion of the canal which corresponds to the scrotum. Division of that part of the urethra which is covered by the scrotum is liable to be followed by infiltration of urine and urinary abscess. The danger of this occurrence was long since pointed out by Deschamps, and in later times has been dwelt on by Sir B. Brodie and Mr. Liston.

The only other complication, which remains to be alluded to, is orchitis. This does not often occur, and when it does, it can usually be traced to some laceration at or near the prostatic portion. It occurred, for example, in the patient already referred to, whose stricture was ruptured previous to performing lithotrity, and who had a fragment impacted at the seat of the stricture. It happened also to another patient under the editor's care, who, after leaving the hospital, injured himself in attempting to pass a cathether. The only treatment necessary is perfect rest, with hot local fomentations. Sometimes, though rarely, suppuration occurs. Inflammation of the testicle, apart from the inconvenience and pain, would be of no great moment, were it not that it delays the convalescence of the patient, as it renders further operative interference for the time inexpedient.

There are some conditions of the urinary organs which require certain modifications of treatment. When, for instance the bladder is in a state of atony, and unable to empty itself completely, the urine should be drawn off at least six times in the twenty-four hours succeeding an operation, and the bladder injected night and morning with lukewarm water. The object of these injections is not so much to bring away fragments as to prevent irritation from decomposition of the urine. Where there is complete paralysis of the bladder, or rather where the bladder has no power to expel its contents, it is often advisable to retain the catheter in the bladder for twelve or fourteen hours after each operation. The instruments which are most convenient for this purpose, and which patients bear with the least annoyance, are the black elastic French catheters; they are very readily retained, and permit of perfect freedom of movement on the part of the patient, from their lightness and extreme suppleness. They are also the most suitable where frequent introductions of the catheter are necessary, as the patient can readily learn to pass them himself without pain or difficulty.

In connection with these cases another point may arise, and it is one of the greatest importance. It is whether, under any circumstances, an attempt should be made to bring away fragments in the lithotrite or scoop. In the opinion of the editor this practice is to be unreservedly condemned. Fragments cannot thus be removed without great risk of lacerating the mucous membrane of the urethra, or of the neck of the bladder. This does not merely cause pain and constitutional irritation, but often more serious mischief, by entangling fragments which would otherwise pass freely; it is a direct cause of impaction, and sometimes of extravasation of urine. Sir B. Brodie, in a paper on Lithotrity,[1] men-

---

[1] Med.-Chir. Trans., vol. xxxviii.

tions four cases in which this practice was followed by extravasation and urinary abscess, and two of the patients died, notwithstanding that the abscesses were freely opened. The editor of this work met with a case in which a small calculus was removed entire in the lithotrite without being crushed, but it caused laceration of the prostatic portion of the urethra; this occasioned great pain at the time, and gave rise to a persistent tenderness, which even after a considerable interval rendered the introduction of an instrument an exceedingly painful procedure.

The only cases which offer any justification for this method of treatment are those of so-called paralysis of the bladder; but, even in these, this method is still objectionable, as from the inevitably attendant laceration of the urethra, the subsequent and necessary introduction of the instruments is in all cases rendered painful and sometimes difficult. It is better, in these cases, to introduce the non-fenestrated lithotrite twice or three times, and to pulverize the fragments as much as possible. Each time the instrument is withdrawn, some quantity of débris will be brought away, but care should be taken that it is not too loaded, and successive half-turns should be made with the screw in order to empty the lithotrite as much as possible, and to prevent any fragments from protruding beyond the sides of the blades. Dr. Bigelow's method, a description of which is given in the succeeding chapter, is admirably adapted for dealing with these cases.

In cases where, from the character of the detritus, there is reason to suppose that the stone is composed for the most part of uric acid, great advantage will be derived from the administration of the citrate of potash, taken fasting three or four times a day, in doses sufficiently large to keep the urine constantly alkaline. The following prescription, suggested by Dr. Roberts, yields a solution containing a drachm of the citrate in each fluid ounce:—

℞ Potass. Bicarb. ʒ xij
Acidi Citrici ʒ viij+gr. xxiv
Aquæ ad ʒ xij. Solve.

"A fluid ounce of this solution, mixed with three or four ounces of water, makes a draught which has scarcely any taste, and which even children take without any difficulty.

In a patient treated in this manner, under the care of the editor, the use of the citrate had the effect of bringing away a large quantity of uric acid in a state of solution, and it not only lessened the number of operations, but materially relieved the bladder irritation by rounding off the edges of the angular fragments.

When the calculus is phosphatic, and the urine ammoniacal, it is well, after the second or third sitting, to try the effect of injections of the bladder with one drachm of dilute nitric acid to the pint of lukewarm water. But it is only in such cases, and under the other exceptional conditions already referred to, that injections are either useful or necessary.

The various points connected with the indications and contra-indications of lithotrity require to be considered with the utmost care; for the success of lithotrity, more perhaps than of any other operation, mainly depends on its being applied in a proper manner; that is to say, on a strict observance of the indications established by experience.

When lithotrity is confined to those cases which are suited to it, and when the cases are of a simple nature, no other important operation is performed with less difficulty, and no other is attended with such brilliant results. This conclusion we are now entitled to draw from facts on which no suspicion rests. Mr. Charles Hawkins assisted at sixty operations performed by Sir B. Brodie, all of which terminated successfully. On the other hand, we know that lithotrity becomes extremely difficult in its performance, and is often attended with fatal results if it be applied to cases which are not suitable for the operation. Of the first twelve cases operated on by M. Velpeau, three died, five were cured; and in four cases the operation was abandoned, the surgeon being unable to terminate it. There must be some reason for such extraordinary differences between results, and these differences are probably due to the care taken by some operators not to violate the rules of science by employing an operation in cases where it should not be had recourse to, and also to the varying degrees of skill with which the operations were performed.

In order to discover the indications and contra-indications of lithotrity, we must bear in mind the various steps which, taken together, constitute the operation. Reflecting on these we shall find that certain conditions are requisite, before this operation can be performed in a safe and proper manner, or we can obtain from it the results which it is capable of yielding.

Thus, we must be able to introduce our instruments into the bladder; having introduced them, we must be able to seize and fix the calculus; having fixed the stone, we must be able to crush it into small fragments; these latter must have a ready exit from the bladder; and, finally, the various steps of the operation must be performed without exciting any serious disorder in the urinary organs, without aggravating in a dangerous way any disorder which may already exist, and without inflicting any serious injury on the general health of the patient. To perform lithotrity in a successful manner, the different conditions just enumerated must be strictly fulfilled, and we have, therefore, only to ascertain what circumstances prevent the fulfilment of these conditions, in order to discover the contra-indications of lithotrity.

Experience has abundantly established the circumstances most favorable for the performance of lithotrity. Whenever the calculus is small, the bladder healthy, the urinary organs free from irritation, and the general health not deteriorated, the patient presents the most favorable conditions; and nearly all surgeons are now agreed that lithotrity is unquestionably indicated. But such cases are not very frequent. Patients are seldom willing to undergo what they consider a dangerous operation, until the urinary organs, and subsequently the general health, have more or less suffered. According to the nature and degree of these sufferings will the operation be indicated: and it is therefore necessary to point out the conditions which render it unsuitable in any given case.

Serious obstacles to the introduction and mechanical action of the instruments may be regarded as contra-indications, or the latter may be of a more general nature. The contra-indications arising from any mechanical obstacle to the introduction of the lithotrite have been already noticed; we must suppose that the urethra is sufficiently free to admit the passage of the instruments into the bladder, or can be made so by dilatation or other operation.

Omitting these, the nature and mode of operation of which are sufficiently obvious, it may be stated that the principal points on which the applicability of lithotrity, in any given case, depends, are first, the nature of the calculus; the condition of the urinary organs; and, third, the health of the patient, together with some circumstances of a general nature, as age, sex, etc.

As regards the calculus itself, the choice of the operation to be selected will be influenced by the number of calculi, their size, form, position, and density.

A single calculus is more suited to lithotrity than several calculi; this holds good with respect to lithotomy also; but the mere number of foreign bodies taken in the abstract is not a bar to the performance of the operation. M. Civiale has operated with success in a case where the bladder contained forty calculi. What we are to be guided by is the time which may be required for crushing and getting rid of them; and hence the number of calculi is no great obstacle provided they be at the same time small and soft.

The size of the calculus is an important element to be taken into account in any operation employed for its removal. A large stone always creates more or less difficulty, and the question will arise how far it can be crushed with safety to the patient. To determine this question, we must look chiefly to the nature of the calculus, and the condition of the bladder. When the stone is large, *i. e.*, when it measures over two inches in its long diameter, and, at the same time, dense, lithotrity is contra-indicated; if the foreign body be friable, as phosphatic deposits usually are, and if the bladder be not strongly contracted over the calculus, lithotrity is not contra-indicated. The larger the stone, the greater is the necessity that the bladder should be tolerably healthy; yet these two conditions do not often co-exist.

It may be a question how far, with our improved instruments, mere density of the calculus may ever prove so great an obstacle as to constitute a contra-indication. Mulberry calculi are, however, sometimes exceedingly hard; they cannot be reduced into fragments without a considerable number of operations; and in dealing with a calculus of this kind, exceeding an inch or an inch and a half in diameter, the possible consequences of such repeated manipulations in the bladder would indicate that lithotomy is the preferable operation.

The position of the calculus may be unfavorable to the operation of crushing. Thus, a small calculus may become entangled between the meshes of a columnar bladder, and elude all our efforts to seize it: the foreign body may be inclosed in a cyst, either wholly or partially; it may be hooked up above the pubes; attached to the surface of the bladder by fibrous tissue; or finally embraced in a separate compartment, which seems to be produced by irregular action of the muscular fibres, analogous to the hour-glass contraction of the uterus. In the greater part of these cases, lithotrity is contra-indicated, for the simple reason that the stone must be caught before it can be crushed; yet, as the cutting operation is almost equally inapplicable, some attempt may be made to dislodge the stone from its abnormal position. Should this be effected, the difficulty is removed, but otherwise an operation is contra-indicated.

The condition of the urinary organs is the next point to be considered in reference to the indications of lithotrity. Health and disease take, of course, opposite sides in the scale; but it is a delicate matter to determine, in all cases, what morbid conditions of the genito-urinary organs are incompatible with the safe and successful application of lithotrity. We have to avoid rashness on the one side; and, on the other, restricting within too narrow bounds the useful sphere of the operation.

The state of the prostate first demands attention; it may influence the operation and its results, as well as the simple introduction of the instruments. Enlargement of the prostate, unless it be considerable, may be overcome as a mechanical obstacle; but it produces many other unfavorable effects.

Enlargement of the middle lobe, especially, diminishes the capacity of the bladder, deepens its floor, and tends to conceal the stone in a cavity behind the enlarged gland. These circumstances impede the action of the instrument in seizing and crushing the foreign body; they also impede the ready discharge of the fragments—an objection of great importance, when we consider how often enlargement of the prostate is attended by an irritable or diseased state of the urinary reservoir. Enlargement of the prostate is, therefore, an unfavorable circumstance in all cases; but not an absolute contra-indication. The precise nature and extent of the obstacles it may create must be determined by careful examination; and then we proceed to consider how far this unfavorable element may influence other conditions.

Thus, when the enlargement is moderate, the operation is not forbidden, provided the stone be small and the urinary organs not very sensitive. On the other hand, when the enlargement is considerable, we should not have recourse to lithotrity, if there be any reason to suppose, from the size and hardness of the stone, etc., that the operation will require a number of sittings.

Any serious disorder of the urinary organs, co-existing with enlargement of the prostate, should also be regarded as a contra-indication.

Chronic inflammation or catarrh of the bladder is of very frequent occurrence in calculous patients, and it is therefore of importance to examine how far such a condition may influence our choice of the operation to be selected. Simple calculous catarrh ot the bladder, instead of being aggravated by lithotrity, is almost always alleviated by comminution of the foreign body. This opinion has been fully confirmed by the experience of Sir B. Brodie.

In a paper read before the Medical and Chirurgical Society, on March 13th, 1855, Sir B. Brodie expressed his opinion "that inflammation of the mucous membrane of the bladder, induced by calculus, and indicated by great irritability of the organ and the copious secretion of mucus, does not form an absolute objection to the operation, although it is doubtless a reason for proceeding with more caution, for, on the contrary, it often happens, under such circumstances, that the crushing of the calculus is followed by an alleviation of all the bad symptoms." The question may arise as to whether lithotrity is to be preferred to lithotomy for phosphatic calculi, in cases of enlarged prostate, with the bladder in a state of more or less atony and chronic inflammation, and the urine exhibiting an ammoniacal reaction in spite of the constant use of the catheter. Conditions such as these are doubtless extremely unfavorable for operation by lithotrity. The calculus may certainly be readily crushed and an immense quantity of débris removed, either by the natural efforts of the patient or by the aid of injections, but, owing to the state of the urine, fresh deposits of phosphates take place with such rapidity that the amount of calcareous matter contained in the bladder becomes little, if at all, diminished. Each fragment serves as a nucleus for fresh additions, the accumulation of which is facilitated by the tenacious mucus secreted by the bladder. In such a case, also, even if the removal of all fragments be successfully accomplished, it by no means follows that the patient's symptoms will be relieved, for these depend not only upon the presence of the calculus, but, in a still greater degree, upon the state of the bladder, a condition which has given rise to the formation of the stone, and is not due to its presence. Under these circumstances, it would seem more expedient to select lithotomy as the operation to be performed. Not only would there be far less chance of relapse, but the bladder would be placed for a time in a state of rest highly favorable to the subsidence of the chronic inflammation. It has been proposed to perform the lateral operation for the relief of severe chronic cystitis even in the absence of a calculus, and in the opinion of the editor of this work, a similar operation is preferable to lithotrity where a calculus co-exists with the condition above described.

There is another danger connected with purulent catarrh of the bladder; a purulent secretion from any part of the genito-urinary system may, under the influence of an operation, become a determining cause of purulent infection of the blood.

Another condition of the bladder, which contra-indicates lithotrity, is hypertrophy, with diminution of the cavity and extreme irritability. It is unnecessary to point out how dangerous any manipulations would be in an organ thus affected.

Small fibrous tumors about the neck of the bladder may impede the operation without absolutely contra-indicating it; but vascular or cancerous tumors in the body of the bladder are to be regarded as positive contra-indications.

As the expulsion of the detritus must be chiefly effected by the bladder, paralysis of that organ has been mentioned by nearly all writers as a condition which renders lithotrity inapplicable. This observation does not accord with present experience. These cases require great care; the sittings must be short, and the bladder must be injected from day to day with a large catheter in order to promote the discharge of the detritus. The following is a short account of a case of this kind. A gentleman, æt. 64, with complete paralysis of the bladder, had a large calculus, and an affection of the spine and kidney; the former indicated by diminution of power of the lower extremities; and the latter by occasional severe pains in the lumbar regions and down the thigh, with swelling of the testicles. The calculus had not produced much distress about the bladder, but fearing the consequences, it was thought desirable to get rid of it. So much care, however, was required in this case, that the sittings were extended over a period of two years. If the object of medical science be the prolongation of life under all conditions, the delay in this case was the only means to secure the object. In another case of the same kind the paralysis of the bladder was complete. The stone was crushed without difficulty, and the detritus removed by washing out the bladder in the way already described. The operation was perfectly successful. Cases of this kind would now be dealt with by Dr. Bigelow's method.

The bladder is sometimes partially paralyzed, a condition which also requires care. In these cases the catheter must be regularly used, and the bladder washed out daily, as in those cases in which the organ has entirely lost its power.

Much more frequently, however, it happens that the bladder of calculous patients is in a condition directly opposite to the one just noticed. The constant irritation kept up by the foreign body produces increased sensibility of the urinary passages and bladder. In some cases the highly sensitive state of the urinary apparatus depends on nervous influence, and may be controlled by proper treatment; but lithotrity is contra-indicated whenever the increased sensibility depends on organic disease, or is connected with increased contractility and hypertrophy of the bladder. Disappointment and failure will often result if we attempt to crush the stone in cases of this kind, and the most prudent course will be to abstain.

With regard to renal affections as a complication of calculus, and as influencing our choice of an operation, it is, unfortunately, not always possible to decide as to the presence of disease of the kidney in these cases, or to determine its nature and extent. A common form of kidney-disease connected with calculus is chronic pyelitis, with dilatation of the ureter and of the pelvis of the kidney, and sacculation of the organ itself. The symptoms may be completely masked by those of the calculus. Examination of the urine does little to aid the diagnosis. It is impossible to determine the exact source of the pus, and the epithelium-cells from the various portions of the urinary tract, even if distinguishable in the mass of sediment, too closely resemble each other to allow of a positive assertion being made as to their place of origin. If, however, even a few tube-casts be discoverable, the nature of the case is sufficiently evident.

Disseminated abscess is another form of kidney-affection which sometimes complicates calculus, though the symptoms, as in the case of pyelitis, may never be sufficiently marked to enable an accurate diagnosis to be made. Those which are most prominent, viz., pain in the loins, intermittent febrile movements, sickness, loss of appetite, diarrhœa, etc., have nothing peculiarly distinctive about them, and there must always be a doubt as to the source of the pus and blood found in the urine. Should, however, fragments of the renal tissues be discoverable (as is sometimes the case) in the sediment, the existence and nature of the disease may be safely predicated. The various forms of Bright's disease are more or less easily diagnosed, though when the tube-casts are few in number it may be very difficult to discover them among the masses of pus and epithelial débris discharged from the bladder. The constitutional symptoms, however, taken together with the history of the case, will sufficiently indicate the nature of the affection.

When any one of these complications is known, or suspected to exist in a patient suffering from calculus in the bladder, the question will arise, first as to the advisability of operating, and secondly as to the nature of the operation to be performed. If the stone is small, the renal disease in all probability not far advanced, and the bladder not very irritable, there is a fair chance that the calculus may be removed without aggravating the mischief in the kidney, and for these cases lithotrity is the preferable operation. Even in more severe instances, where the stone is large and the renal disease in all probability far advanced, lithotrity, especially if Bigelow's operation be adopted, is generally to be preferred to lithotomy, as less likely to be attended with fatal results. In a third class of cases, the symptoms may be such as to indicate that neither operation can be attempted with the least prospect of success, and for these palliative treatment is all that can be recommended. No absolute rule can, however, be laid down. It is sometimes necessary to operate in order to mitigate suffering in cases where a successful issue appears almost impossible.

The indications derived from age may finally be noticed. In early editions of this work the opinion was stated that all cases of calculus in children under twelve or fifteen years of age should be treated by lithotomy. The following are the chief objections against the propriety of applying lithotrity to the young subject. The genito-urinary organs are not fully developed before the age of puberty. The diameter of the urethra is small; the antero-posterior diameter of the bladder is short; the bladder of children is irritable, and the neck of this organ is very dilatable, a circumstance which favors impaction of the fragments; lastly, the indocility of the child is a great obstacle.

On the other hand, it is well-known that lithotomy is highly successful in children, not more than 1 in 15 or 20 dying after the operation when performed on subjects under fourteen years of age.

Many of these reasons have only a certain degree of force, and do not constitute insurmountable objections. If the organs be small, our instrument-makers now construct small lithotrites of sufficient force. The irritability of the bladder in children does not arise from a state of disease. Indocility may be overcome by the use of chloroform. Lithotomy is, undoubtedly, most successful in children; but the success is in a great measure connected with the healthy condition of the urinary organs in the young subject; and hence it may be fairly argued that if lithotrity diminishes the mortality in adults, it should have a still greater influence in reducing the comparative mortality when applied to children, because the number of deaths in adults, after crushing the stone, is greatly increased by the state of the urinary organs at that period of life. Sir H. Thompson[1] thinks that if the stone is of such a size that it can be properly crushed at a single sitting, it may be treated by lithotrity, but that the advantages of this operation are very doubtful where more than two or three sittings are necessary. He cites the experience of Dr. Guersant, of Paris, surgeon to the Children's Hospital, who reports 21 cases of children on whom he practised lithotrity with six deaths as a result. Besides these, three others were subsequently subjected to lithotomy, with a fatal result in all cases. Dr. Klien,[2] of Moscow, has furnished us with some statistics of lithotrity in children. Sixty-two patients aged 1 to 15, were operated on, with a mortality of six. In 24 of the cases, one sitting was sufficient, and in several of these, the stone measured eleven lines in diameter. The number of fatal cases (1 in 10.33) shows a rate of mortality much higher than that which follows the operation of lithotomy for similar ages. As far therefore as statistics go, lithotomy would appear to be a much safer operation than lithotrity for patients under fifteen years of age.

In connection with the operation of lithotrity a few words must be said on the recurrence of calculus after operation. In Sir H. Thompson's account of 204 cases, there are 18 (1 in 11) in which the calculus reappeared. It is certain that such an occurrence is more apt to take place after lithotrity than after lithotomy, and it is interesting and important to compare the two operations in this respect. The Norwich tables of Mr. Crosse show twelve cases of relapse after 704 operations of lithotomy, or 1 in 58 cases. According to returns received by M. Civiale from the whole of France the proportion of relapses after lithotomy was 1 in 74 cases, whereas among 548 cases submitted by himself to lithotrity the proportion of relapses was 1 in 10. This latter result is in accord with Sir H. Thompson's experience.

The recurrence of calculus after operation may be due to any one of three causes:—1, a second calculus may descend from the kidneys;—2,

---

[1] Practical Lithotomy and Lithotrity, p. 178.
[2] Die Steinkrankheit in Russland, Langenbeck's Archiv, vol. vi., p. 80.

one or more fragments of the original calculus may have been left in the bladder after lithotrity, and become the nuclei of secondary formations ; —3, the bladder may have been completely freed, and the relapse may depend upon the renewed deposition of phosphatic matter, and the continued secretion of viscid mucus.

When the relapse is due to the first cause, an interval of shorter or longer duration will exist between the operation and the recurrence of the symptoms, and during this time the patient will remain free from urinary troubles. The symptoms of the descent of a renal calculus will form the starting-point of the new affection. A relapse of this kind is not peculiar to lithotrity; it is equally liable to occur after lithotomy.

In a second class of cases the relapse is due to retention of fragments after lithotrity. In such a case the patient has never been completely freed from the symptoms of calculus. For a time he may complain less than usual, but the painful sensations inevitably connected with the presence of a foreign body in the bladder are never entirely absent. The calculi of relapse are often multiple, not single, and these secondary formations, more or less rapidly developed on fragments of the original stone, are usually of the phosphatic kind. The fact that relapses of this kind occasionally occur shows the necessity for careful examination of the bladder before dismissing a patient as cured by lithotrity. Such an examination is best conducted by means of a lithotrite.

In the last class of cases, the recurrence of calculus is due to the continued operation of the same causes which led to the development of the original formation, and these causes may be briefly described as the condition of chronic cystitis and the changes produced in the urine by this affection. The bladder may have been completely freed from the original concretion, but the urine remaining more or less ammoniacal, phosphates are constantly deposited, while their agglutination is directly promoted by the presence of the viscid, tenacious mucus which the bladder secretes. Relapses due to this cause may occur after either operation, and they are the more difficult to treat, inasmuch as the cause is in great measure irremovable. Something, however, may be done by regularly emptying the bladder, and by using injections, with or without a little dilute nitric acid, in order to remove deposits and to improve the condition of the mucous membrane. In a few rare cases, the cause of relapse is the presence of an encysted calculus. A portion of the surface of a stone of this kind is in contact with the urine, and upon this surface a deposit, generally phosphatic, takes place. This goes on increasing until the mass becomes attached to the original calculus by a narrow neck, which may give way and thus the new growth becomes free in the bladder. The symptoms of calculus then become urgent and subside after operation. In the course of time a new formation takes place, and the symptoms recur as before. Such a condition may be suspected during life, but can scarcely be diagnosed with absolute certainty.

The statistics of lithotrity will be noticed in a subsequent chapter with those of lithotomy.

CHAPTER XXIII.

LITHOLAPAXY,[1] OR RAPID LITHOTRITY WITH EVACUATION OF THE FRAGMENTS.

It was stated in the foregoing chapter that to remove débris, by injecting and washing out the bladder after crushing the stone, was formerly the rule, but that the practice had fallen into desuetude. The desire to rid the bladder of its abnormal contents had also led operators to adopt another method, viz. to withdraw fragments of the stone by means of a scoop, or between the blades of the lithotrite. In not a few instances this procedure gave rise to serious accidents, such as laceration of the urethra, impaction of fragments, abscess, infiltration of urine, etc., and it is now universally abandoned. The evacuating catheters and injecting bottles, which were in use up to a recent date, sufficed to effect the removal only of dust, or at most of small fragments which, under ordinary circumstances, would escape with the urine. When these instruments were had recourse to, the larger fragments remained, and it may be doubted whether the mucous membrane was not more liable to suffer from their presence than would have been the case, had its surface been protected by the dust and fine débris. The unsatisfactory results of all attempts to remove the broken-up particles led surgeons generally to believe that, except in cases where the bladder is unable to empty itself, it was better to allow all débris to be spontaneously evacuated.

In 1878, Professor Bigelow,[2] Surgeon of the Massachusetts General Hospital, promulgated the details of a new method, practised by him, for the removal of a calculus from the bladder. His plan may be briefly described to consist in reduction (by crushing) of the calculus to fragments small enough to pass through an evacuating tube of large calibre, and continuous removal of all fragments and débris by means of an aspirator. Dr. Bigelow's experience has led him to believe that "it is better to protract the operation indefinitely in point of time, if thus the whole stone can be removed without serious injury to the bladder," and that "if the bladder can be completely emptied of detritus, we have as little to apprehend from the fatigue of the organ, consequent upon such manipulation, as from the alternative of residual fragments and further operations." Dr. Bigelow's proposition includes the use of a large and powerful lithotrite, and of an evacuating apparatus consisting of an evacuating tube connected with an elastic bottle or aspirator.

The elastic bulb or bottle has a capacity of ten fluid ounces. To its lower extremity is attached, by means of a bayonet-joint, a glass receptacle for the débris. Its upper end is continuous with a piece of elastic tubing, varying in length from six inches to two feet, and attached to

---

[1] From λίθος a stone, and λάπαξις evacuation. The term has been invented by Professor Bigelow.
[2] In the American Journal of the Medical Sciences, Philadelphia, Jan., 1878. The communication has since been published in a separate form, with additions, under the title of Litholapaxy, or Rapid Lithotrity with Evacuation.

the evacuating catheter. In Sir H. Thompson's modification of this portion of the apparatus, the catheter is attached directly to a short tube passing obliquely into the metallic canal, which connects the lower extremity of the elastic bottle with the glass receiver. The tube in question is furnished with a stop-cock, and the upper part of the bottle terminates in a funnel and tap through which water is poured into the instrument. The glass receptacle is attached to the bulb by means of a bayonet-joint. The introduction of fluid into the bottle is much facilitated by the addition of the funnel and tap. It has been recommended

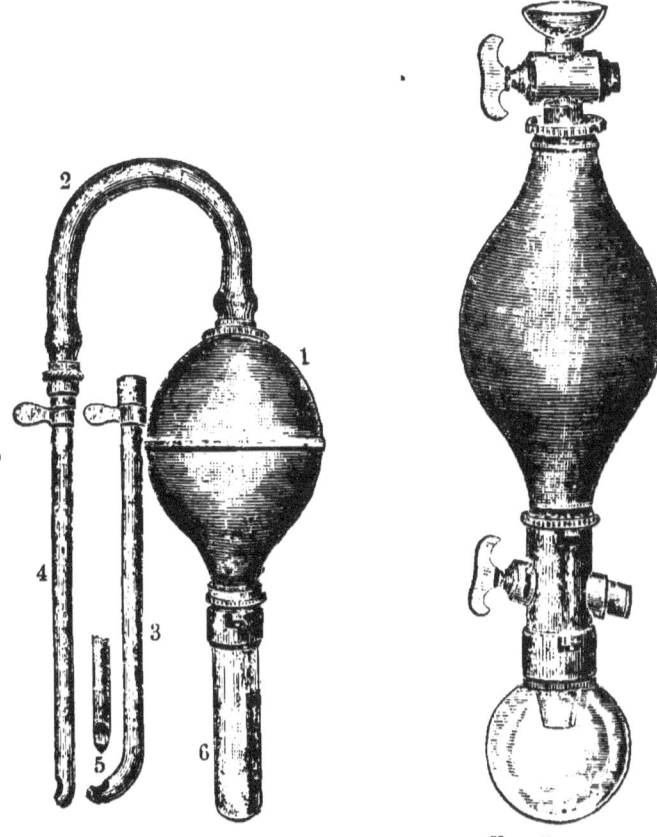

FIG. 7.   FIG. 8.

FIG. 7.—Bigelow's evacuating apparatus: (1) The elastic bulb; (2) India-rubber tubing; (3) Curved evacuating tube; (4) Straight tube; (5) Front view of orifice of tube; (6) Glass receptacle for débris. (From Dr. Bigelow's work.)
FIG. 8.—Sir H. Thompson's modification of Bigelow's aspirator.

that a piece of rubber-tubing should be interposed between the evacuating catheter and the tube connecting it with the elastic bottle, so that any movement of the latter may not be communicated to the evacuating catheter while in the bladder. The bottle, however, may be used to aspirate fragments without moving the catheter directly attached to it, and the shorter the route to be traversed by the fragments the more speedily will their evacuation be accomplished.

The evacuating tubes are of two kinds, one being straight, and the other slightly curved. They are of large calibre, much larger than the full-sized catheters in ordinary use. As a matter of course, the larger the tube admitted by the urethra the more readily will the fragments be removed. Dr. Bigelow points out that the comparatively small calibre of the evacuating catheters attached to Clover's instrument is a bar to their efficiency. He uses evacuating tubes of thin silver, of sizes 27, 28, 29, 30, and 31, Charrière or French scale. No. 27 corresponds to No. 16 English scale, and No. 30 to No. 18. Judging from present experience it would seem that evacuating tubes of the first-mentioned size will generally be found sufficient. Fragments pass more readily through the straight tube than through the curved one, and the former can be as easily passed as the latter, provided that certain precautions are taken. These will be subsequently alluded to. The extremity of the tube is

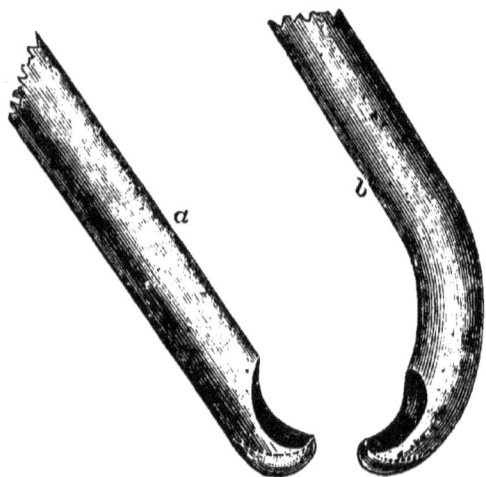

FIG. 9.—Evacuating tubes, with unguiform extremity. *a*, Straight tube. *b*, Curved tube. The dotted lines show the false floor of the extremity. The tubes are of size 31, Charrière.

rendered unguiform by the removal of a small portion of the side walls of the orifice, which is just above the extremity, and slightly oval in form. The edge of the orifice is "thickened and rounded so as to slide smoothly through the urethra; any rim inside the orifice should be masked by a false floor; but the calibre should be nowhere contracted. If a couple of inches of the end of such a tube be bent, it may be inverted after introduction, and will bury itself in the floor of the bladder, which it depresses, while the orifice loooks forward and is unobstructed." The straight tube is, however, to be preferred; it is less liable to lodge fragments, and can be more readily cleared should any obstruction take place. Dr. Bigelow also points out that it is easy to know exactly where the extremity of such a tube lies during an operation.

With regard to the instrument to be employed for crushing the stone,

Dr. Bigelow uses a very powerful lithotrite with a non-fenestrated female blade somewhat longer and wider than usual, and having a small slot or opening above the angle of junction with the shaft, in order to allow of the escape of débris. The male blade has a "series of alternate triangular notches by whose inclined planes the detritus escapes laterally." At the end of the sliding shaft a ball is substituted for the wheel in common use, while the lock is opened and closed by a revolving cylinder-handle. A lithotrite of this kind is adapted for dealing with very large stones, but for ordinary calculi, composed of uric acid or phosphates, and measuring less than an inch and a half in the longest diameter, the lithotrites described in the previous chapter will be found sufficient.[1]

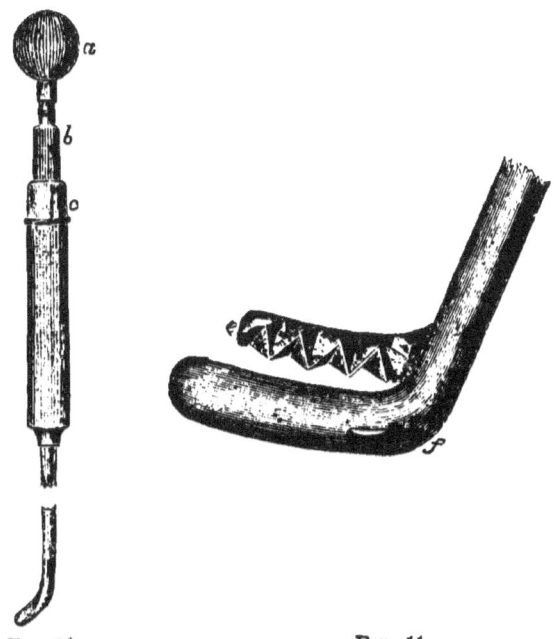

FIG. 10.   FIG. 11.

FIG. 10.—Bigelow's lithotrite (reduced). *a.* Ball which turns the screw. *b.* Revolving cylinder-handle attached to the screw-guard, which also revolves. This guard consists of two square or T-shaped rods, which slide through notches in the cap of the lock. By their revolution the cylinder-handle turns the cap and operates on the lock. *c.* Cap of the lock, which, by its revolution, wedges up the screws.

FIG. 11.—Blades of lithotrite. *e.* Male blade, presenting alternate triangular notches. *f.* Slot in female blade.

They are easier of introduction, and the necessary movements can be executed with greater rapidity. The stone need not be reduced to powder, but only to fragments small enough to pass through the evacuating tube. The fenestrated lithotrite works better than the plain instrument.

---

[1] The writer's experience in this respect is confirmed by that of Dr. Keyes (Surgeon to the Bellevue and Charity Hospitals, New York), in an article on Bigelow's operation, American. Jour. of Med. Sciences, April, 1880. Dr. Keyes advocates the employment of a fenestrated instrument.

The former does not pulverize the calculus, but reduces it to small fragments, which are not repeatedly caught by the instrument, but are sufficiently reduced in size to pass readily through the tube. Besides this advantage, the fenestrated lithotrite is readily freed from débris by simply screwing the male blade home, and thus the repeated withdrawals and reintroductions necessary when using the plain instrument are dispensed with.

The operation is performed in the following manner: the lithotrite having been introduced, the stone is caught and crushed freely several times in succession, the patient being under the influence of ether. If the calculus be a large one, the withdrawal of the lithotrite after several crushings and the introduction of a second instrument will be found advantageous. When the calculus has been sufficiently crushed, the lithotrite is withdrawn and the evacuating-tube introduced, the meatus urethræ being, if necessary, previously incised with a bistourie cachée. There need be no hesitation about enlarging the orifice of the urethra should it be too small to admit the tube. Difficulties, due to various causes, may be experienced in the introduction of the tube, and great care must be taken to avoid injuring the urethra. The obstacles most commonly met with are caused by (1) rigidity of the prostatic portion of the urethra, due to enlargement of the prostate, (2) a generally flaccid condition of the canal, (3) adhesion of pulverulent débris to the lining membrane.[1] Under all circumstances, it is most important to avoid using force. The best method of dealing with these obstacles is to pass the tube as far as it will go, and then attach the bottle and gently inject a little water. The fluid, finding its way into the bladder, dilates the urethra, and the tube readily follows. This latter having been fairly introduced into the bladder, the aspirator, filled with tepid water, is then attached to it, and, by grasping the bulb in the right hand, a few ounces of fluid are injected into the bladder. When the pressure is taken off, the elastic bulb expands, and a current takes place from the bladder, the fluid carrying with it débris and fragments small enough to pass through the evacuating tube, whence they fall into the glass receiver. The process is repeated several times until particles are no longer seen to pass into the receiver. Dr. Bigelow recommends that, as long as the fragments are numerous, the operator should endeavor to separate and float them by the injection, so that they may enter the tube singly and without obstructing it. For this purpose the extremity of the tube should be kept above the floor of the bladder. After the removal of the smaller débris, " the tube may be made to indent the floor, so as to gather instead of separating the fragments; and, as a final measure, the tube should be raised toward the perpendicular, in order to carry the orifice nearer the prostate."

When the fragments are passing freely, a distinct rattle is heard as they strike against the walls of the tube. Should this latter become obstructed, fragments will cease to appear in the reservoir, and the elastic bulb will either not expand at all when compression is taken off, or else do so very slowly and imperfectly. Obstruction may be caused by a fragment too large to pass through the tube, becoming wedged in the aperture. To get rid of it, the extremity of the tube should be raised above the floor of the bladder and the bulb suddenly compressed. If this plan

---

[1] For instances in which these obstacles were experienced and overcome, see reports of cases by Walter J. Coulson, Lancet, Jan. 31st and July 3d, 1880.

fails, a whalebone rod may be passed down the tube. The same method may be adopted for dealing with a fragment impacted within the tube itself, or the latter may be withdrawn and a fresh one introduced. Sometimes the opening of the tube becomes occluded by the wall of the bladder. This source of obstruction is easily recognized by the peculiar sensation communicated to the fingers in contact with the tube, and by the dull sound which is heard. The position of the extremity of the tube should be changed, or, as Dr. Bigelow recommends, more water may be advantageously introduced, so as to distend the bladder. When fragments cease to pass, and yet, from the sound heard as the bulb is allowed to expand, it is obvious that the bladder contains one or more fragments too large to escape through the tube, the latter must be withdrawn and recourse had again to the lithotrite, after which the aspirator is used as before. In making the repeated crushings, it is, of course; in the highest degree necessary to avoid pinching or lacerating the mucous membrane of the bladder. This accident is likely to occur if the lithotrite be pressed against the base of the bladder so as to produce a hollow into which the stone may fall, and it may be guarded against by adopting Civiale's method, in which contact with the vesical walls is as far as possible avoided. In a fatal case, recorded by Dr. Weir,[1] many minute lacerations of the floor of the bladder were caused by the sharp fragments being pressed against the mucous membrane by the female blade of the instrument.

If, owing to its size, the stone cannot be completely crushed at a single sitting, or if, from any cause, the patient cannot be kept for a sufficient length of time under the influence of ether, a few days may be allowed to elapse before repeating the operation. Should, however, any amount of cystitis have supervened, it is not necessary to wait until this has subsided. On the contrary, the operation should be repeated as quickly as possible. When the calculus is small, i. e., under an inch in its longest diameter, and comparatively soft, a sitting of twenty minutes or half an hour's duration will probably suffice for its complete evacuation.[2] In dealing with larger stones, the operation is necessarily a more protracted one. In a case,[3] mentioned in Dr. Bigelow's work, the patient's bladder contained two calculi; one so large that it was barely possible to lock the lithotrite. The first sitting lasted an hour and a half, the second, three hours, and the third, three hours and three-quarters. The total weight of fragments removed, when dried, amounted to four ounces and fifty-two grains. The patient was discharged well a fortnight from the date of the first operation. The composition of the stone is not mentioned, but, from the history of the case, it was probably phosphatic. In a case under the writer's own care,[4] in which three sittings were necessary, owing to the inability of the patient to sustain a prolonged administration of ether, the first lasted for forty-seven minutes, the second thirty minutes, and the third thirty-five minutes. Fragments weighing 335 grains were removed. In another case,[5] 560 grains were removed at two sittings occupying fifty and forty-two minutes respectively. In proof that hard calculi can be satisfactorily dealt with by this method, Dr. Bigelow has reported a case[6] where the stone was composed of oxalate of lime, and measured an inch and five-eighths in its longest diam-

---

[1] American Journal of Medical Sciences, Jan., 1880, p. 134.
[2] Lancet, Nov. 8th, 1879, July 3d, 1880.
[3] Bigelow, loc. cit., p. 29, Dr. C. B. Porter's case.   [4] Lancet, July 3d 1880.
[5] Lancet, Jan. 31st, 1880.   [6] Lancet, May 17th, 1879.

eter. It was entirely removed in one sitting lasting eighty-one minutes; the fragments weighed 302 grains. Dr. Keyes thinks that, where there is no complication, five grains of dried stone is a fair average for each minute of work.

With regard to the success of the operation, it may be confidently asserted that Dr. Bigelow's expectations have been abundantly fulfilled. Numerous cases have been published in which the relief afforded was immediate and complete, the patient being freed from the cause of his troubles without any local or constitutional symptoms as an after-result. Up to February, 1880,[1] a hundred and seven operations have been recorded as performed in this country and the United States, with only six deaths, all of which occurred in patients who were far from healthy. It is probable that a few fatal cases have not been reported, but, making a reasonable allowance for these, the result must be pronounced satisfactory, especially when it is borne in mind that the operation has several times been performed on patients in a condition unfitted in all respects for any operative procedure. The comparative harmlessness of long sittings (shown to some extent by Heurteloup's operations) has been clearly proved by Bigelow and those who have adopted his method, while the fact that cystitis occurring after ordinary lithotrity is best treated by freely crushing the remaining fragments, would seem to indicate that this complication may best be guarded against by removing from the bladder those sources of irritation to which the inflammation is largely due. As a matter of course, the operation should be performed with the utmost care and gentleness, and it should not be attempted by those who have not had considerable experience of lithotrity. The presence of sharp angular fragments is more harmful than careful though protracted manipulation with properly constructed instruments, but several cases have terminated fatally in consequence of injuries to the bladder or urethra, caused by the lithotrite or evacuating tube. The advantages of thoroughly clearing the bladder are that further sittings are rendered unnecessary, the risk of irritation from sharp fragments being left behind is obviated, and the chance that small particles may remain, and serve as nuclei for fresh concretions or induce chronic cystitis and phosphatic deposit, is reduced to a minimum. The method has the further advantage of enlarging the range within which lithotrity is applicable, and its success encourages the surgeon to deal with much larger stones than those heretofore subjected to crushing. What the limits of the operation, as determined by the size of the calculus, really are, further experience is required to decide, but the prospect of greatly diminishing the number of cases hitherto subjected to lithotomy on account of the dimensions of the calculus—a class in which the mortality is especially great—seems almost certain to be realized.

---

[1] American Journal Med. Sciences, April, 1880.

CHAPTER XXIV.

LITHOTOMY.

THERE are three different parts at which the bladder may be opened for the purpose of extracting calculi.
I. At the anterior surface above the pubes, by what is called the high operation.
II. At the inferior and posterior part, through the rectum, by the recto-vesical operation.
III. Near the neck through the prostate, by the perineal section and lateral operation, or various other methods.

The High Operation.—In order to perform this operation, the bladder must rise above the superior edge of the pubes, which, when empty, it does not reach. Hence, it is necessary either to distend the bladder by injection; to wait until sufficient urine has accumulated to produce the desired effect; or to elevate the anterior and superior part of the viscus by means of a sound, the point of which is made to glide from below upwards, against the posterior surface of the pubes. It is obvious, therefore, that the operation is impracticable in all cases where the bladder does not readily admit of distention. Should the presence of the stone have produced much irritability of the parietes, a sufficient quantity of fluid cannot be injected into this cavity.

This method of operating has rarely been adopted in this country, though it has been practised occasionally in France; and judging from its results in the hands of M. Souberbielle, with a very fair amount of success, in the worst description of cases, viz., those in which the stone is of large size.

Mr. Carpue having witnessed many of M. Souberbielle's cases, endeavored to re-introduce this operation into England.

Professor Gibson, of Philadelphia, performed it in the case of an old gentleman who was affected with great enlargement of the prostate, but the patient died soon afterwards from peritonitis, consequent upon urinary effusion. Dr. Carpenter, of Lancaster, Pennsylvania, repeated the operation soon after, with a more fortunate result. Dr. George M'Lellan, of Philadelphia, performed the operation several times; but the results did not encourage others to follow his example. It would seem to be an easy and a simple operation; but such is not the case, and the objections which induced Cheselden, Douglas, and others, to relinquish it, are based upon good grounds. It is true there is no danger of hæmorrhage; that neither the rectum, nor the seminal ducts, can be wounded; but then the bladder cannot always be fully distended, and there is great risk that the peritoneum may be injured. Such an accident is immediately followed by the protrusion of some folds of intestine, and probably by acute peritonitis; or the bladder being opened, and the peritoneum being wounded, urine may escape into this cavity, and produce fatal results.

The force sometimes necessary to distend the bladder to a sufficient extent for the high operation, may be illustrated by two fatal cases of Cheselden's, in which the bladder was burst from injecting too much water.

The operation is most easily performed on young subjects, in whom the pelvis is small, and the bladder still to be enumerated among the abdominal viscera. But these are the very patients upon whom the lateral operation may be performed with the greatest prospect of success. A patient may be laboring under such extensive disease of the bladder and kidney as to render almost any operation improper; Mr. Coulson, however, in the course of his long experience, never met with a case in which a better chance of recovery seemed probable from opening the organ above the pubes. Even in the case of a large stone, when fears may justly be entertained that there may be insufficient space for its extraction entire under the arch of the pubes and through the perineum, he did not consider the high operation advisable, on the grounds that the bladder in almost every instance of this kind is thickened, contracted, and incapable of distention enough behind the pubes, and no instrument introduced through urethra would materially assist in elevating the anterior wall of the bladder to the line of incision.

This operation, however, being decided upon, the patient should be placed in the recumbent posture, with the shoulders slightly elevated, and the feet resting on a chair. The parts having been shaved, and the boundary of the distended bladder being clearly ascertained, the surgeon should make an incision from three to four inches in length along the mesial line, commencing about half an inch above the symphysis pubis. The aponeurosis of the obliquus externus abdominis having been exposed, a small opening is made into the suprapubic space, which is filled with fatty cellular tissue, and bounded above by the reflexion of the peritoneum from the urachus to the posterior surface of the abdominal walls. A grooved director having been introduced, the linea alba and fascia transversalis are divided to the requisite extent. Hæmorrhage having been arrested, the bladder is sought for; it feels like a soft fluctuating tumor at the bottom of this deep wound. The external incision being kept open by means of retractors, the organ may be secured by a hook, or tenaculum given to an assistant, to prevent it, when opened, collapsing beyond the reach of the operator. A small opening is then made into its lower and front part, the knife being introduced with its edge directed upwards. The instrument is then cautiously carried upwards for about two inches, its course being guided by the fore and middle fingers of the left hand introduced into the vesical cavity. It is safer for this last part of the proceeding to use a knife with a rounded or a probe-pointed extremity. Or the incision may be made in an opposite direction, viz., from above downwards towards the neck of the bladder. The forceps is now introduced, and the calculus extracted. A gum-elastic catheter, or a silver tube should be left in the wound, for the purpose of preventing infiltration of urine into the cellular tissue of the pelvis and of the lower part of the abdomen.

Professor Bruns, of Tübingen, recommends that the wound in the bladder should be closed by sutures, one end of the threads being cut off close to the knot and the other brought through the wound, the edges of which are to be kept in contact by means of pin-sutures. An elastic catheter, retained in the bladder, provides for the ready escape of urine. In the case of a child thus treated, the sutures were removed on the sixth day, and the catheter two days afterwards. There was then some amount of suppuration about the wound, but no urine had ever escaped from it.[1] The operation in question appears to be favorably regarded by several German surgeons at the present day. Professor Josef Podrazki, of Vienna, the writer of the article on "Diseases of the Bladder," in von Pitha und Billroth's "Handbuch der Chirurgie," thinks that there are many advantages connected with the high operation. The ease with which it can be accomplished, the small number of assistants required, the slight risk of hæmorrhage and the facilities which it offers for the removal of large stones, even if encysted, supply, in his opinion, sufficiently cogent reasons for a more frequent adoption of this operation in preference to others. The operation might also be performed with antiseptic precautions.

When, however, there is a great accumulation of fat in the suprapubic region, the external incisions must be both long and deep; under such circumstances, the bladder, lying at a distance from the surface, cannot be satisfactorily explored, and the extraction of a large calculus would be almost, if not quite, impossible.

In a case operated upon by Pye, fatal hæmorrhage ensued. "In the old method of performing this operation," says Carpue, "hæmorrhage must be common, as I have been sent for to examine three persons who died shortly after the operation, and in whom, on inspection, I found the cause of death to be hæmorrhage from the puncture of a branch of the circumflexus ilii, which anastomosed with the epigastric."

The high operation should, it would appear, be reserved for those cases, extremely few in number, in which the pelvis is so deformed that the proper incisions cannot well be made in the perineal space. It might also be the best means at our disposal in cases where the calculus is very large,[2] or where the perineum is so diseased as to forbid any incisions being made through it. After exposing the bladder, it may be found impossible to remove the stone without a further operation. "I once saw," observes Mr. Bransby Cooper, "the high operation performed by Sir Everard Home, and a most difficult operation it was: after he had cut down through the parietes of the abdomen, he was obliged to make an opening in the perineum, and pass an instrument into the bladder from below to enable him to open the bladder above the pubes. After much difficulty, a calculus was removed, but the patient died, in consequence of this complicated operation, a few hours after its completion. A friend of mine," Mr. Cooper adds, "some years since was present at the performance of the high operation of lithotomy, by M. Civiale. In making the first incision through the parietes of the abdomen, he wounded the peritoneum which led to the protrusion of the small intestines. He returned these into the abdomen and continued the operation, and ultimately succeeded in extracting a stone. The patient was put to bed, and M. Civiale, on subsequent consideration, dreading the liability of extravasation of urine into the peritoneum, determined on laying open the bladder from the perineum. He effected this, and so completely suc-

---

[1] Von Pitha u. Billroth, Chirurgie, Bd. iii., 2. Abth., 8. Liefg., s. 111.
[2] Dr. Uytterhoeven, of Brussels, removed, by this operation, from the bladder of a man aged 39, an enormous calculus, weighing 40½ ounces and measuring 16¼ inches in one direction and 12½ in the other. The patient survived the operation eight days. See Erichsen, Science and Art of Surgery, vol. ii., p. 805, and Roberts on Urinary Diseases, p. 281. A cast of the stone is in Mr. Erichsen's possession.

ceeded in preventing the result he feared, that the patient ultimately recovered." Dr. Humphry, surgeon to Addenbrooke's Hospital, Cambridge, has published a successful case of the high operation, and given the results from the practice of various surgeons, which show the mortality to be 1 in 3.35. This result closely coincides with that obtained by Dr. C. W. Dulles,[1] of Philadelphia, who has collected 465 cases of this operation, with 330 recoveries and 135 deaths. The table (given at p. 338), shows that for calculi above two ounces in weight the results of the suprapubic operation are more favorable than those of the lateral method. It must also be remembered, on instituting any comparison between the two methods, that the former has been often adopted under circumstances which would make any operation especially hazardous.

The *Recto-Vesical Operation*, claimed by Sanson as his own, has been recommended on the ground of being free from the danger of hæmorrhage, and of affording a ready passage to the instruments necessary to seize and extract the stone. Sanson divided the sphincter ani and the lower part of the rectum by a vertical incision corresponding to the raphé of the perineum. Continuing the dissection, he exposed the bulb and membranous portion of the urethra, the prostate, and lower wall of the bladder. The prostate is then to be divided upon a staff introduced into the bladder through the urethra, or the knife may be passed behind the prostate, through the wall of the bladder, into the groove of the staff, and the base of the bladder divided.

This operation may also be performed by introducing a sharp-pointed straight bistoury, for about an inch into the rectum, with the blade flat upon the palmar surface of the left index finger, a staff with a central groove having been previously passed into the bladder. With the right hand the operator then turns the edge of the knife upwards, and cuts into the groove of the staff just in front of the prostate. He then withdraws the knife, cutting, as he does so, through the rectum and soft parts in the median line of the perineum for about an inch. Guided by the left forefinger, the bistoury, with its edge downwards, is again placed in the groove of the staff and pushed onwards in the median line, dividing freely the prostate and neck of the bladder. The finger is then passed into the bladder, the staff removed and the calculus extracted by means of the forceps. This method was much practised by the late Mr. Lloyd, of St. Bartholomew's, but it is now very rarely adopted.

The objections to this operation are great. When the cavities of the rectum and bladder have been thrown into one, it is by no means certain that the communication will close. Mr. Coulson witnessed several cases in which a recto-vesical fistula remained for years, the result of an incised wound. The division of the prostate exposes the vasa deferentia to danger, and the wound of the neck of the bladder may be complicated with a wound of the vesiculæ seminales, producing atrophy of the testicles. The recovery from such an operation must necessarily be tedious, and the whole proceeding is even more objectionable than the operation above the pubes.[2]

But it has already been discarded; and cannot be recommended even

---

[1] Gross, loc. cit., p. 206.
[2] In a case in which the late Mr. Southam, of Manchester, performed this operation for the removal of a large calculus, the communication between the bladder and rectum appeared to be closed 4½ months after the operation. It soon, however, re-opened and could never afterwards be completely closed. See Med.-Chir. Trans., vol. xlii., p. 427, and Roberts on Urinary Diseases, p. 459.

in those cases where the calculus is impacted in the walls of the bladder. Although the danger of hæmorrhage is stated to be inconsiderable, arteries of some size may be wounded by so deep an incision. In 185 cases of this operation the mortality was 38, or 1 in 4.86.

The Lateral Operation.—The lateral operation has superseded every other method of extracting calculi from the bladder. Mr. Coulson performed it on patients of all ages, from eighteen months to above eighty years. Mr. Cline operated upon a man of eighty-two, Sir Astley Cooper upon one of eighty-six. Dr. Physick removed, at one operation, from Chief Justice Marshall, of the United States, a man of very advanced age, upwards of a thousand small calculi, varying from the volume of a partridge-shot to that of a bean. Mr. Attenburrow operated successfully upon a man of eighty-five. Mr. Keate performed the lateral operation upon an infant of twelve months; Mr. South upon one of twenty months. Civiale refers to a child that was cut at ten weeks.

In very young subjects the bladder is rather an abdominal than a pelvic viscus; but the perineum is devoid of fat, the parts are very thin, and the bladder can be readily reached by the finger. Cases of lithotomy usually do well in young children, and collapse, so common in middle age, does not occur.

The only instruments usually required in the operation of lithotomy, are a staff, a scalpel, the button-pointed bistoury, a scoop and forceps. The gorget which was formerly employed is now nearly abandoned, and the bistourie cachée is used in France, but seldom in this country.

Before proceeding to perform lithotomy we should be careful to prepare the patient for the operation; in fact, the most successful lithotomists are those who are most attentive to this point. If the patient's health be much impaired, and the bowels are relaxed, a condition which is not unfrequent in children, or if the urine be alkaline, we must, before the operation, lessen by an anodyne the irritability of the bladder and bowels, and improve the patient's health, as far as possible, by medicine and attention to diet. A dose of castor oil should be administered on the morning of the previous day, and a few hours before the operation an enema should be given, so as to make sure that the bowels are thoroughly emptied. If the injection has not come away when the surgeon arrives to perform the operation, he should urge the patient, if an adult, to go to stool; if a child, it will generally happen that the injection, if it has not passed off, will do so when the staff is introduced. It is highly necessary that before the operation is undertaken, the rectum should be free from all fecal matter, for if distended, the space between it and the urethra becomes much narrowed and there would be great risk of wounding the intestine as well as increased difficulty in extracting the stone. On no account should the surgeon undertake the operation till the injection has come away.

The perineum having been shaved, and everything being in readiness, the patient should be placed evenly on his back, and supported by pillows upon a table of convenient height, about two feet and a half from the ground. Chloroform should now be administered. A full-sized staff, with a deep groove, is then to be passed into the bladder, and search made for the stone, which must be distinctly felt with the staff. The patient is to be drawn to the edge of the table, with the buttocks projecting slightly over the edge; the foot and hand on each side are next to be secured together by means of bandages, or the leather anklets and wristbands devised by Mr. Prichard, of Bristol.

The staff should possess a curve of sufficient length to enable the point to enter fairly into the bladder; and the groove should be deep, situated either in the middle or a little to the left side, and cease about a quarter of an inch from the extremity of the instrument. The staff is

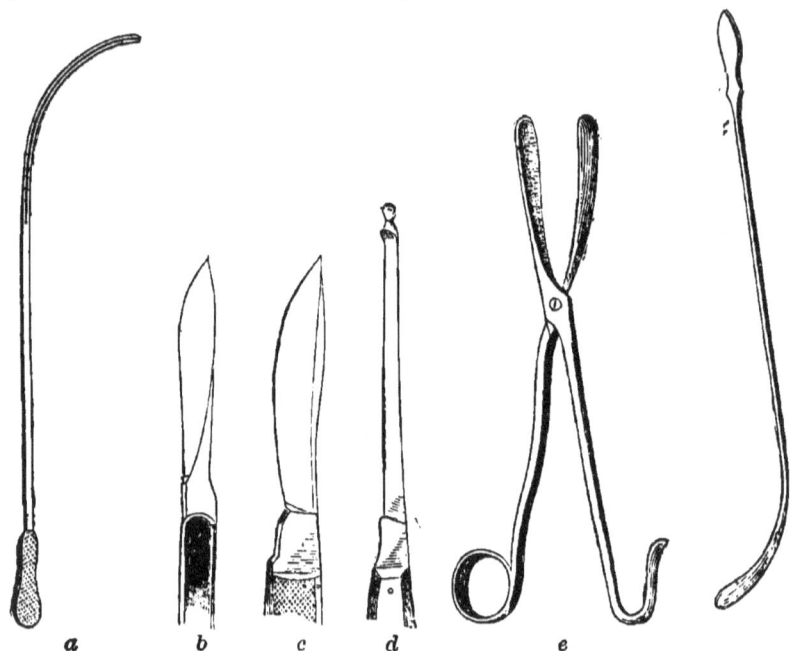

Fig. 12.—Modern instruments. *a.* Staff. *b.* Liston's knife. *c.* Coulson's knife. *d.* Coulson's straight button-pointed bistoury. *e.* Forceps. *f.* Searcher.

given to the care of an assistant, with the directions that the handle, which should be broad and rough so as to be easily and firmly grasped, be held quite perpendicular, and immovable from the first. An assistant

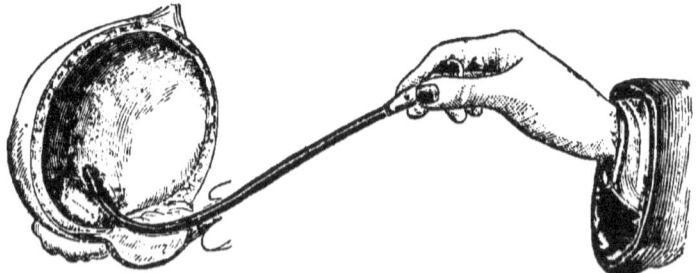

Fig. 13.—Bladder showing the calculus struck with the staff.

on each side takes charge of the lower limb, drawing the thigh well aside, so as to expose the perineum to the fullest extent.

Sitting upon a low chair, the operator commences the first incision, three or four lines to the left of the raphé, and about two fingers' breadth above the anus, and terminates it midway between this opening and the tuber ischii. The knife is held somewhat like a pen, between the thumb and two next fingers of the right hand, and with one or two strokes the surgeon divides the skin, fat, and cellular layer comprised between the left erector penis and accelerator urinæ, together with the fibres of the

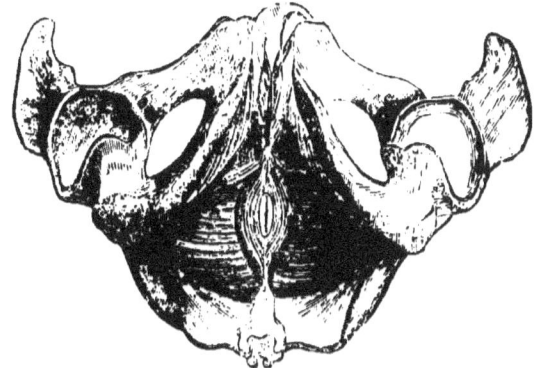

FIG. 14.—Muscles of the perineum. (From Wilson's "Anatomy.")

transversus perinei muscle. With the forefinger of the left hand the scalpel is followed till the staff is felt, and then, pressing the rectum inwards and backwards, the membranous portion of the urethra is opened so that the nail of the left forefinger can be introduced into the groove of the instrument.

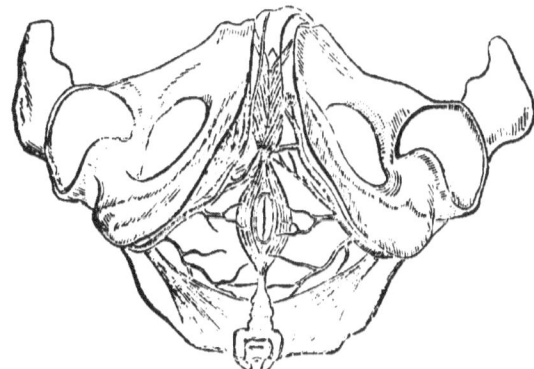

FIG. 15.—Vessels of the perineum. On the right side the superficial arteries are seen, and on the left the deep. (From Wilson's "Anatomy.")

When the surgeon has cut into the groove, there is no occasion to alter the position of the staff, and the forefinger of the left hand is quite at his disposal for protecting the rectum and guiding the knife. The operation can be more rapidly and safely performed in this way than by taking the staff into the surgeon's hand. If the groove be directly in front of the instrument, the latter must be slightly turned towards the left side.

Having made these incisions with the common scalpel, the surgeon then takes a long straight knife with a knob at its point, and holds it in the same manner as the scalpel; he then guides it, the cutting edge being directed downwards, with his left forefinger, into and along the groove of the staff until it reaches the bladder, dividing in its course some of the anterior fibres of the levator ani, and the lateral and lower part of

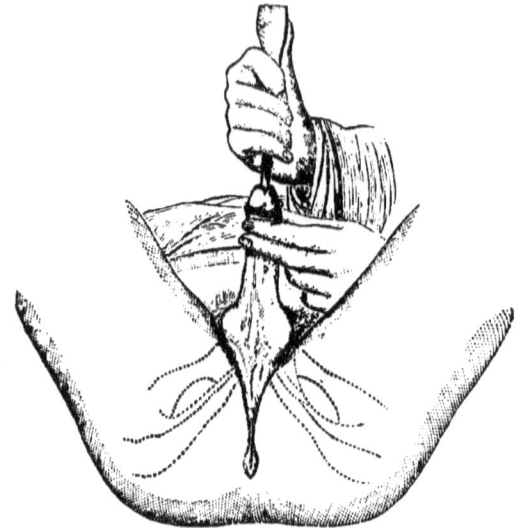

Fig. 16.—Position of the patient.

the prostate, the capsule of which must be left entire. He then withdraws the knife nearly in the direction in which he introduced it, and inserts his left forefinger into the bladder. The operator being satisfied that the opening into the bladder is sufficiently large, directs the staff to

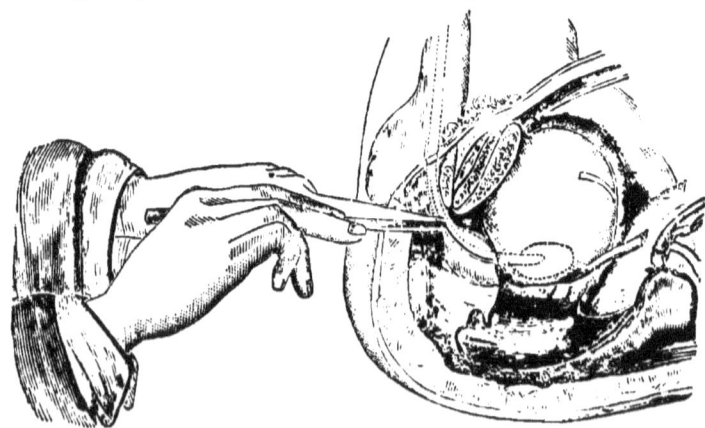

Fig. 17.—Section of the prostate with button-pointed bistoury.

be withdrawn, and introduces a pair of broad-bladed forceps along the left forefinger. The stone is then seized and extracted. In doing this the operator will often find it convenient to use both hands to the forceps, *i.e.* taking a handle of the instrument in each hand. It is scarcely possible to say too much as to the caution necessary in the extraction of a large stone. We are to withdraw it gradually from the bladder, endeavoring to dilate the parts through which it is to pass, instead of tearing them; and it is astonishing to what an extent this gradual dilatation may be accomplished in the hands of a prudent surgeon. The manipulations will be much facilitated by a free incision. After the stone has been removed, the bladder should be searched to see whether it contains any other concretion.

In making the last incision some operators cut onwards, along the groove of the staff, with the same knife which served to make the first incision. It is, however, not always easy to keep the point of the knife perfectly in contact with the groove of the curved staff. The use of the probe-pointed bistoury obviates the possibility of a perforation of the bladder should the knife leave the groove in the staff.

Certain additional precautions are necessary in performing the operation on young subjects. Before puberty the bladder lies somewhat high in the pelvis, and it is therefore especially necessary, in making the deep incision, to depress the handle of the knife so as to keep the point directed upwards and in close contact with the staff. If the knife be pushed forwards in too horizontal a direction, it will probably leave the staff and pass into the cellular tissue between the bladder and rectum. The loose and yielding nature of the parts renders this accident especially likely to happen unless the knife be held as directed.

The incision into the urethra should be free enough to admit the finger without difficulty, for if not, there will be great danger of tearing through the urethra and separating it from the bladder, in the attempt to reach the latter organ. The position of the bladder should be borne in mind in passing the finger and in introducing the forceps. Unless great care be taken, there is much risk of pushing the prostate gland, which is very small in young subjects, inwards and upwards.[1] The tissue of the membranous portion of the urethra is also very delicate, and the tube itself very narrow and easily lacerable. The finger therefore of the operator should be kept close to the knife as the bladder is entered.

Various modifications have been made in the shape and form of the staff used for the guidance of the knife in lithotomy. The late Mr. Aston Key strongly advocated a straight staff, and this form of instrument is still adopted by the surgeons at Guy's Hospital, and certain advantages are claimed for it. It is, however, admitted by Mr. Poland that the straight staff is more apt to be tilted out of the bladder than the curved one, that it is not so easy of introduction, and that, when it is used, the membranous portion of the urethra is placed at a considerable depth from the perineum. It is necessary to make a free external opening, and before entering the membranous portion of the urethra, with the knife, to protect the bulb with the forefinger passed deeply through the loose cellular tissue in the triangular space between the muscles. Keeping the point of the knife in the groove of the staff, the operator takes the handle of the latter instrument in his left hand, and lowers it until the staff makes an angle with the horizon of about 30 degrees, keeping his right hand fixed. With a simultaneous movement

---

[1] For a description of the difficulties of cutting into the bladder in children, and cautions against the too free use of the finger, see Sir W. Ferguson's Lectures, Lancet, 1864, vol. ii., p. 1; also a lecture by Dr. Humphry, Lancet, 1864, vol. i., p. 460.

of both hands, the groove of the staff and the edge of the knife are to be turned obliquely towards the patient's left side. The knife is now passed along the groove and the incision made through the prostate, and if desirable the wound can be enlarged as the knife is withdrawn. The consentaneous action of the hands in making the deep incision, and diminished risk of the knife slipping out of the wound are the main advantages claimed for the straight staff, but as stated above, it is much more difficult to find in the perineum.

Another form of staff is the rectangular one, which was introduced some years ago by Dr. A. Buchanan, of Glasgow. The employment of this staff, however, is not the only distinctive feature of Dr. Buchanan's operation, which is to all intents and purposes a central one. The terminal portion of the staff in question, three inches in length, is directed at a right angle to the shaft, and has a deep groove on its left side. The angle is intended to project prominently at the most posterior part of the perineum, while the short arm lies horizontally under the level of the bulb of the urethra, an assistant pressing the instrument firmly downwards to maintain it steadily in that position. The forefinger of the left hand is introduced into the rectum, and the operation is then performed by a single stroke of a double-edged lithotomy knife, which as it penetrates in the mesial line towards the bladder, passes through the middle of the upper half of the sphincter ani, the membranous part of the urethra and the apex of the prostate, and as it is withdrawn divides the body of the gland, the levator ani, and the skin of the perineum; the direction of the knife as it comes out being first horizontally outwards and then downwards in a curved direction, describing a curved line equal to about one-fourth of a circle round the upper and left side of the rectum. The external wound is about an inch and a quarter in length. It is claimed for this operation that only the left lateral lobe of the prostate is divided, and that the bulb and rectum are safe. Dr. Buchanan also asserts that it is easier of execution than the lateral operation, and that there is less risk of hæmorrhage and urinary infiltration. The operation, however, has but few advocates at the present time.

Having thus described the lateral operation, it will be convenient to follow the method adopted in the chapter on lithotrity, and consider the obstacles to the performance of the operation, and the accidents attending it.

Lithotomy consists of four parts or steps, viz., the introduction of a grooved staff to guide the knife; the external incision, the internal incisions, and, finally, the extraction of the stone. Anything which impedes the performance of these successive steps becomes an obstacle, from which arise most of the accidents of lithotomy.

Obstacles to the introduction of the staff and first incision are few, and do not require any extended notice. An unusual depth of the perineum may cause some impediment during the second step of the operation, or the internal incisions may be somewhat impeded by the circumstance that, in children, the bladder is apt to recede before the point of the knife. Some difficulty may exist in making the point of the knife, unless a button-headed bistoury be used, to glide regularly and easily along the groove of the staff into the bladder. In more than one case the knife has slipped out of the groove. Accidents of this kind, however, seldom occur at the present day. We usually get into the bladder with ease, the great difficulty in many cases is to extract the stone. Numerous obstacles may oppose this step of the operation; they are chiefly connected with certain conditions of the parts, or with the nature, position, and size of the calculus.

The foreign body is extracted with a particular kind of forceps, which is passed through the incisions into the bladder, expanded, closed over the stone, and then withdrawn in the direction of the external wound. During each of these steps the operator may encounter unexpected obstacles.

Enlargement of the prostate presents a greater or less obstacle to the easy performance of lithotomy. If indurated at the same time, it will impede the action of the cutting instrument; but it constitutes a more considerable obstacle to the seizure of the stone. It places the calculus beyond the reach of the finger, and increases the distance of the vesical cavity from the external surface of the wound; besides this, the enlarged gland, rising above the floor of the bladder, leaves a cavity behind it in which the calculus lodges, and may easily escape the forceps. In such a case, extraction will be facilitated by using forceps which have a considerable curve, and also by introducing the finger into the rectum, so as to tilt up the base of the bladder.

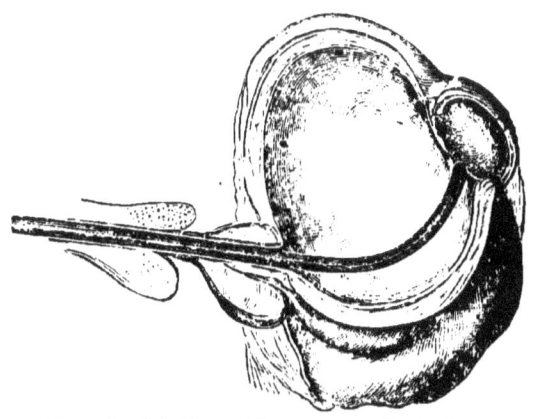

FIG. 18.—Bladder with an encysted calculus.

Several morbid conditions of the bladder may impede the surgeon in his attempts to seize the stone. Most of these have been already noticed in the chapter on lithotrity.

Calculi, when encysted, adherent to the walls of the bladder, embraced in an abnormal contraction, or placed in an unfavorable position, will prove a source of embarrassment to the lithotomist as well as the lithotritist. Every surgeon, who has operated frequently for stone, must have met with cases illustrating these facts. A case of this kind occurred to Mr. Coulson at St. Mary's Hospital. The patient, a boy aged 11, had suffered since birth from incontinence of urine, a symptom which became more distressing during the six months preceding his admission. The child manifested no signs of pain; but there was a small quantity of blood and mucus mingled with the urine; frequency of micturition; and occasionally the water suddenly stopped.

The parents stated that some time previously he had been tormented with pain, and was continually pulling the prepuce; but this symptom passed away, and for a few weeks before admission he had apparently not suffered at all. On being sounded, a stone was detected, evidently not resting on the base of the bladder, but lodged in the superior fundus, closely adjacent to the anterior wall, for it was only when the handle of the sound was depressed, and its point brought forward, that the stone could be struck.

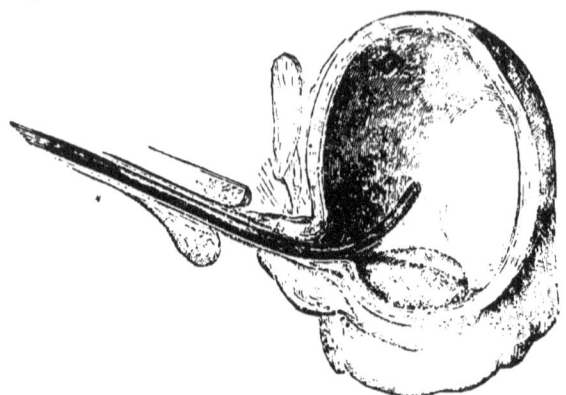

FIG. 19.—Calculus behind prostate.

The bladder was penetrated by a beaked knife, but on introducing the finger the stone could not immediately be felt, being situated quite anteriorly, and, in all probability, adherent to the front wall of the bladder; some difficulty would have been experienced in extracting the calculus if the pelvis of the child had not been raised before the introduction of the forceps. By adopting this measure the stone was readily seized, and the operation completed within a minute. There was

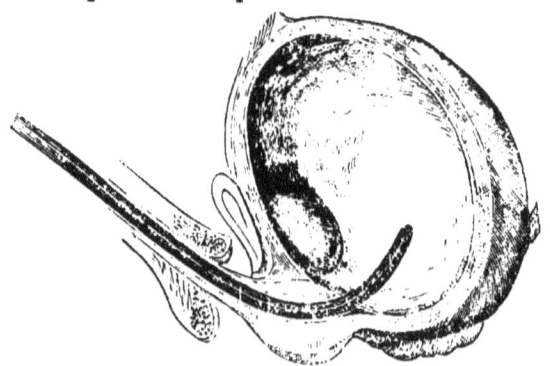

FIG. 20.—Bladder with calculus behind the pubes.

scarcely any blood lost. The subsequent progress of the patient was good; he was able, in a fortnight, to void his urine by the natural passage, and soon got well.

These cases of adherent or encysted stone are sometimes accompanied by a peculiar circumstance, which is worthy of attention.

From the calculus being fixed in a particular spot, and probably from its being thus prevented from irritating the neck of the bladder, the rational symptoms of the disease are sometimes greatly modified, or may even disappear altogether. The manner in which calculi may thus become adherent to the walls of the bladder, is well illustrated by a case reported in the "Transactions of the Pathological Society,"[1] by Mr. Shaw. "Midway between the orifices of the ureters, there was an oval spot, of the size of a sixpence, where the mucous and muscular coats were absent, and a quarter of an inch above the orifice of the left ureter there was a rough patch on the surface, somewhat smaller. A calculus had been adherent to each of these points. On one side of the larger calculus a circular spot, about half an inch in diameter, was covered with a thick flocculent layer of fibrous tissue, and on closely inspecting the mode of union between the fibrous tissue and calculus, it was seen to be effected by the fibres dipping into and being incorporated with the calcareous substance. It was so firm that it resisted separation by tearing with the forceps."

The same volume contains an interesting case of encysted stone, which would have been equally inaccessible to lithotomy as to lithotrity. A small calculus had been crushed and removed, but the symptoms of stone soon returned with severity. The man was most carefully examined under chloroform, yet no stone or fragment could be detected. He died, worn out with pain and debility. The mucous membrane of the blader was highly congested; the organ itself thickened and contracted; while a moderate-sized calculus was detained in the superior fundus by an apparently permanent hour-glass contraction of the muscular fibres.

If it were possible to detect, during life, the impaction of a stone from hour-glass contraction, as above noticed, recourse to the high operation might be, perhaps, justifiable; it is the only operation which affords the least chance of relief.

When the stone is partially encysted, attempts have been made to liberate it from the cyst. Callot succeeded in one case by changing the position of his patient, and surgeons are sometimes fortunate enough to grasp and dislodge it with the forceps. In one case, Sir B. Brodie succeeded in dilating the neck of the sac with a probe-pointed bistoury, after which he separated the calculus from the sac with his finger, and extracted it. When the cyst is within reach, we may attempt to dilate the neck, and dislodge the stone with the finger or scoop; but if that cannot be done, the operation should be abandoned. Division of the neck of the sac with a cutting instrument should not be attempted.

A source of difficulty in the seizure of the stone and its extraction not unfrequently arises from the forcible contractions of the bladder upon it, which the introduction of the forceps, and attempts to find the stone, tend to provoke. This impediment sometimes considerably prolongs the operation, and even occasionally prevents its completion; the danger being increased in proportion to the length of time occupied in the attempts. In these cases, by waiting a little, the contractions of the bladder will often cease, and permit the removal of the stone with facility: whereas on the other hand, the repeated introduction of instruments, and the fruitless attempts to seize the stone, when the surgeon is anxious to perform the operation quickly, without due regard to the difficulties he may encounter, will, in many cases, excite still more the contractility of the bladder, and thus prevent the desired object. Mr. Crosse mentions, in his work, a case where three-quarters of an hour was uselessly employed in attempts with the forceps to find a stone, of the size of a pigeon's egg,

---

[1] Vol. vi., p. 250.

which came out spontaneously some time after they had been discontinued.

If the stone breaks under the forceps, as will probably happen when there is a fusible calculus, we must endeavor to remove as much of the calculous matter as is practicable with the scoop at the time of the operation, and then well wash out the bladder. We must not, however, be content with washing out the bladder at the time of the operation, but should repeat this proceeding before the wound in the perineum closes. In a patient of Mr. Coulson's, 56 years of age, in whom the stone was like a mass of mortar, more than an hour was occupied in removing with the scoop the calculous matter; but a great deal was still left behind. It was thought unadvisable to protract the operation, and the patient was removed to bed. At the end of a fortnight a large catheter was introduced through the urethra, and the bladder well washed out, a great deal of calculous matter escaping through the wound in the perineum. This was repeated every other day for three weeks; in fact, until no more débris escaped. The patient entirely recovered. Dr. Snow administered the chloroform at the operation, and at each washing out of the bladder.

The size of the calculus exercises a most important influence on the results of lithotomy. Difficult as may be the extraction of an impacted calculus, still more so is that of one so large as to require for its passage incisions longer than those usually made. A stone of from four to eight ounces in weight cannot be safely removed without the greatest coolness and skill. If the size of the stone be such that no dilatation of the wound as first made will admit of its extraction, it may be crushed; or the right lobe of the prostate may be divided, or a combination of both these proceedings may be adopted. If the blades be sufficiently strong, a well-made forceps will break most urinary concretions; but a variety of crushing forceps of different lengths, some straight, others curved, with blades flat, and broad, or narrow, spoon-shaped, rough, or cutting, should always be at hand. Mr. H. Earle paid particular attention to the subject of extracting large calculi, and has communicated to the profession some interesting matter in the eleventh volume of the "Transactions of the Royal Medical and Chirurgical Society."

Cheselden withdrew a calculus of twelve ounces, and succeeded in curing his patient; Mr. Hamer, of Norwich, removed by lateral section a stone of the weight of fifteen ounces, its diameter being four inches and three quarters in one direction, and three and a half in the other.

In 1818, Mr. Charles Mayo, of Winchester, operated upon a man aged 28, and extracted a stone of fourteen ounces and two drachms avoirdupois. It measured eight inches and a half in its smaller circumference, by rather more than ten in the largest, and broke into several large pieces in the attempt to extract it.

Ambrose Paré relates the case of a confectioner, who was cut in 1570 by John Collot, where the stone weighed nine ounces, and was three inches and a half in diameter. In an instance described by Folet, a calculus weighing ten ounces, and measuring nearly four inches in diameter, was extracted, and the patient had recovered from the immediate effects of his wound, when an abscess formed in the kidney from the presence of a concretion, and terminated fatally on the tenth day after the operation.

The influence which the size of the calculus has upon the mortality after the operation of lithotomy, is a point which has of late years attracted considerable attention. The cases related above are exceptional. It is well known that the extraction of a very large stone is generally fatal to the patient; but we have scarcely any data from which we can deduce the relative influence exerted by stones of different sizes. The late Mr. Crosse was the first writer who viewed this subject as one of interest and importance, and who had materials at his disposal from which other than either purely individual, or extremely general inferences could be drawn. From the valuable Norwich data, Mr. Crosse[1] has presented us with the following table of 704 calculi, arranged according to the weight, with the result of the operations performed for their removal.

Weight	Number of cases	Cured	Died	Ratio of mortality
1 oz. and under	529	482	47	1 in 11.25
1 " to 2 oz.	119	101	18	1 " 6.61
2 " " 3 "	35	19	16	1 " 2.18
3 " " 4 "	11	4	7	1 " 1.57
4 " " 5 "	5	2	3	1 " 1.66
5 " " 6 "	2	2	0	0 to 2
6 " " 7 "	2	0	2	2 " 2

The information which this table conveys is extremely important. With a stone one ounce in weight and under, it is obvious that the chances of a favorable issue are nearly twice as great as with a stone between one and two ounces in weight, and nearly six times as great as when the stone weighs between two and three ounces. The mortality with stones of every weight over an ounce is nearly 1 in 3¾; and with those over two ounces it is something more than 1 in 2. Of the 55 instances in which the stone weighed two ounces and upwards, 27 recovered, and 28 died. The chance which a patient has of recovery after lithotomy can therefore be calculated beforehand, and independently of every other consideration, from the ascertained dimensions or weight of the stone. Doubtless, some considerable portion of the greater comparative success that attends lithotomy in early life is due to the generally smaller size of the stones that are then extracted.

A certain moderate size of stone, however, would seem rather to be advantageous than otherwise; the most favorable results are not associated with the smallest stones, as is apparent from the particulars conveyed by the following table taken from Mr. Crosse's work:—

Weight	Number of cases	Cured	Died	Ratio of mortality
1 dr. and under	134	122	12	1 in 11.17
1 " to 2 drs.	111	101	10	1 " 11.10
2 " " 3 "	95	90	5	1 " 19
3 " " 4 "	68	60	8	1 " 8.50
4 " " 5 "	29	28	1	1 " 29
5 " " 6 "	38	35	3	1 " 12.66
6 " " 7 "	24	21	3	1 " 8
7 " " 8 "	30	25	5	1 " 6
Total	529	482	47	1 in 11.23

With stones under two drachms in weight the mortality is therefore greater than with those that are between two and three drachms, and

---

[1] Op. cit., p. 162.

particularly with those between four and five drachms: the chances of recovery with this last weight of stone are actually about two and a half times greater than with a stone that weighs less than a drachm. The difficulty of finding a stone of this latter size may have something to do with the result experienced.

Measurements of calculi would be far more important elements in forming estimates of this kind than weight: it is the size of the stone that determines the extent of outlet necessary to give it passage, and this extent of outlet it is that influences the mortality: large stones often weigh less than might be inferred from their size; small ones, again, are often remarkably solid and heavy.

From what has been said, it is evident that for the extraction of calculi beyond a certain size, the lateral operation must be modified; in other words, when the surgeon finds he cannot extract the stone through the ordinary incisions, without employing a dangerous degree of force, he must employ other expedients. The operation may be abandoned for that above the pubes or through the rectum; the stone may be broken in the bladder; or the opposite side of the prostate may be divided, so as to obtain more space for the extraction of the calculus. Mr. Coulson preferred this latter method whenever it could be employed with any prospect of success, and had recourse to it with favorable results in two cases, where the calculi were large; but we must not forget that it does not enable us to gain much beyond half an inch for the required space. The plan which he adopted is the following: After the division of the left portion of the prostate, and the withdrawal of the staff, he introduced his left forefinger as far as he could into the wound, carrying along it the button-headed bistoury, and dividing the right side of the gland in the same manner as the left.

Breaking up the calculus with a forceps introduced through the wound is a very old operation. Even Franco, and after him Le Cat, attempted to facilitate the process by drilling the stone, and thus performed perineal lithotrity. There are, however, many strong objections to this method, but it may be adopted, if division of the right side of the prostate does not furnish sufficient room. Civiale's instrument, consisting of a fair of forceps furnished with a drill, enables the operator to fix the stone, and then perforate it and break it into fragments.

Having thus noticed the principal obstacles to the performance of lithotomy, we are prepared to appreciate many of the accidents which attend or follow the operation.

That these are numerous and severe may be gathered from the general fact that about one-fifth or one-sixth of all patients cut for stone die. The accidents may occur during the operation or after it. The former are few; indeed, the only one likely to occur is injury to the rectum, an accident of no great importance, and not always to be attributed to the surgeon. The chief accidents occur sooner or later after the operation, and may be regarded as the causes of death. They are hæmorrhage, infiltration of urine, suppuration or abscess, cystitis, peritonitis, purulent infection, and renal disease.

The shock of the operation was formerly an occasional cause of death, but since the introduction of chloroform, this accident has lost much of its gravity, and seldom gives rise to a result immediately fatal.

The parts concerned in the lateral operation are abundantly supplied with blood-vessels, and we might hence conclude that hæmorrage must be a frequent accident. Present experience, however, would lead to the inference that dangerous hæmorrhage does not occur frequently, and Mr. Liston has stated that he had only one case of severe bleeding in 100 operations. On the other hand, the accident would seem at one time to have been more frequent in France. Boyer considered hæmorrhage as one of the most frequent accidents of lithotomy, and M. Begin calculated that 1 out of every 4 deaths, after the lateral operation, was occasioned by hæmorrhage.

This accident may occur during the operation or after it, and the bleeding may be venous or arterial. Hæmorrhage during the operation may arise from division of the larger arteries of the perineum and bulb, or of the trunk of the internal pudic itself. When the external incision is carried too far downwards and outwards towards the ramus of the ischium, the trunk of the pudic artery may, it is said, be divided; but this must be an extremely rare accident. The artery lies so close to the bone, and is so well protected by dense fascia, that it can scarcely be wounded, unless there be some considerable deviation from the normal course and distribution of the vessel.

The researches of modern anatomists show that the arteries of the perineum are subject to several important variations, and it seems probable that most cases of fatal primary hæmorrhage may be traced to the cause now mentioned. The most important variation occurs when the artery of the bulb arises farther behind than usual, and crosses the front of the ischio-rectal fossa; in such a case, it would almost necessarily be divided. A greater quantity of blood than usual may appear on division of some of the superficial vessels, but, in the normal state of things, these vessels do not bleed profusely, and can be easily secured.

Venous hæmorrhage may occur during or soon after the operation, from division of the bulb, or, in old persons more commonly, from division of the venous plexus which surrounds the neck of the bladder. Sir B. Brodie lost a patient within a few hours from this cause, having been foiled in all his efforts to arrest the bleeding.

Secondary hæmorrhage may be, in some sort, a continuation of the bleeding which took place during the operation, the divided vessels continuing to pour out blood in small quantities. More usually it commences about the fifth or sixth day. The discharge of blood is often moderate, yet it continues obstinately, sometimes bringing the patient to the brink of the grave. This form of hæmorrhage is due to separation of the plugs of coagulum from divided arteries, or of the sloughs from contused vessels, or, in cases of sloughing, to implication of the vessels.

In the treatment of this accident, general principles must be followed. If considerable hæmorrhage arise during the operation from any large vessel, it should be tied, if possible, but it is not easy to get at the artery of the bulb, and still less so at the internal pudic. When ligature is impracticable, the latter vessel should be compressed against the ascending ramus of the ischium.

Secondary hæmorrhage, as above mentioned, may be either venous or arterial. In the former case, the wound may be carefully plugged, yet this plan is sometimes inefficacious, and often produces great irritation, besides preventing the escape of urine. To obviate these difficulties, Mr. Hilton contrived an apparatus which consists of an ordinary tube (such as those employed after lithotomy by Liston and others), to which is attached, by means of a ring, fixed about an inch and a half from its extremity, a conical bag of soft muslin or linen. One end is passed up through the wound into the bladder, and the other end is left projecting externally, the cavity of the bag being filled with sponge, so as to exert pressure on the surrounding parts. In this way the wound may be most effectually plugged, while the free escape of urine is provided for. When it is thought safe, or becomes necessary to remove the pressure, the sponges may be taken out, one by one, without the least pain to the patient, as they are prevented, by the intervening layer of muslin, from adhering to the sides of the wound.

In a case for which it was devised, the expedient answered admirably. As had been expected, there occurred after the extraction a rather free oozing of venous blood, which was, however, completely arrested by the plan described.[1] Another convenient contrivance for the same purpose is an india-rubber ball connected with a tube. This is passed into the wound, and inflated, and in this way pressure is kept up on the bleeding surface.

In cases of secondary arterial hæmorrhage, the most certain remedy would be to take up the vessels, but this can seldom, if ever, be done. The surgeon must have recourse to plugging the wound, pressure on the trunk of the main vessel, cold, and the usual modes of arresting hæmorrhage. The hæmorrhage sometimes takes place into the bladder, which then becomes distended and prominent, while the patient exhibits all the symptoms which attend loss of blood. Under such circumstances, the bladder should be washed out with cold water, the wound carefully examined, and the hæmorrhage arrested by ligature or plugging, the pelvis meanwhile being kept raised.

The most frequent cause of death after lithotomy is infiltration of urine into the cellular tissue of the pelvis, and this accident is intimately connected with the methods adopted for division of the prostate.

The direction which should be given to this incision, and its extent, are points of the greatest importance to the success of lithotomy.

Cheselden is supposed to have divided, in his operation, as little of the prostate as possible.

The first account we have of Cheselden's[2] operation, is in his Appendix to the fourth edition of the "Anatomy of the Human Body," which appeared in the year 1730. In this appendix, after describing the first stage of the operation, he goes on to say: "I then feel for the staff, and cut upon it the length of the prostate gland, straight on to the bladder, holding down the gut all the while," etc., etc. In the Appendix to the fifth edition of "Anatomy," published ten years later, viz., 1740, the description is the same, till we pass the words, "I then feel for the staff," when he proceeds: "holding down the gut all the while with one or two fingers of my left hand, I cut upon it in that part of the urethra which lies beyond the corpora cavernosa urethræ, and in the prostate gland, cutting from below upwards, to avoid wounding the gut, and then passing the gorget very carefully in the groove of the staff into the bladder, bear the point of the gorget hard against the staff, observing all the while that they do not separate and let the gorget slip to the outside of the bladder," etc. There is evidently a considerable difference between these two operations. In the first, the prostate is cut in its whole length, and the knife is passed along the groove of the staff from before backwards

---

[1] Medical Times, May 19th, 1855, p. 490.
[2] Willis, On the Treatment of Stone in the Bladder by Medical and Mechanical Means, 1842.

into the bladder. In the second and later operation, the knife is used to open the membranous and the commencement of the prostatic portion of the urethra, the incision being made from behind forwards, and the operation is completed by means of the gorget. Much controversy has, however, taken place as to the exact method which Cheselden finally adopted. Dr. Douglas,[1] who published, during the life-time of Cheselden, an account of the operation, described the internal incision as beginning on the internal or vesical side of the prostate. Dr. Willis, however, characterizes this description as confused and inaccurate, and it is highly improbable that Cheselden ever divided the prostate gland in the manner described by Dr. Douglas.

Mr. Martineau, one of the most successful lithotomists of modern times, divided only a part of the prostate. Mr. Crosse,[2] who witnessed for many years Mr. Martineau's public practice, says, "that he (Mr. Martineau) seldom, if ever, divided the prostate through its entire depth, is the opinion I have formed from observation of many operations by him, and which is supported by the only dissection I had the opportunity of making; still, enough was cut to allow the blunt gorget to enter the bladder, and then the operator invariably adopted a proceeding which forms no important part of the operation; the staff being withdrawn, he was accustomed to introduce his left forefinger (which was particularly long and large) upon the concavity of the gorget, into the bladder, forcibly dilating the opening, and using the finger as a powerful but safe instrument for rendering the neck of the bladder ample to admit the forceps. The force and determination with which the finger was thus used, dilating, if not lacerating, the remaining undivided portion of the prostate gland and the neck of the bladder, I always regarded as a peculiar and intrinsic part of Mr. Martineau's method of operating. The opening into the bladder being thus effected, partly by cutting and partly by dilating, the forceps usually entered readily, and in the use of these, rapidly carrying them to different parts of the bladder if required, or more frequently seizing at once the stone (previously felt, and its situation ascertained by the finger), Mr. Martineau possessed a degree of freedom and dexterity unparalled within my observation, and surpassed most surely by very few among those who have become conspicuous in this branch of operative surgery."

Mr. Martineau, however, expressly stated his conviction that it was undesirable to use much force in extracting a large stone. In cases where any difficulty was experienced, his practice was to give the handles of the forceps to an assistant, and use the knife to cut the part forming the stricture, and, rather than lacerate, to repeat this enlargement of the internal wound two or three times.

The dangers of completely dividing the prostate have doubtless been much over-exaggerated. It was supposed that if the knife were carried beyond the boundaries of this gland, the urine would escape into the loose cellular tissue of the pelvis, and produce inflammation, suppuration, and sloughing of the cellular tissue. Pelvic cellulitis is, however, far more likely to be induced by bruising or laceration of the neck of the bladder and adjoining parts, in the attempt to extract a large stone through an incision too small for the purpose. This complication is especially liable to occur in patients suffering from chronic disease of the bladder or kidneys,

---

[1] See Postscript to his History of the Lateral Operation for the Stone, p. 31.
[2] On the Formation, Constituents, and Extraction of the Urinary Calculus.

and is a frequent cause of death in such cases. The question arises whether for the extraction of a large calculus, it is safer to make free incisions into the prostate and to carry the knife even beyond its limits, or to be satisfied with a smaller incision and to run the risk of producing extensive laceration. The former alternative would appear to be the less mischievous. In children the incision must of necessity be carried quite through the prostate, for, owing to the smallness of the gland, it would be impossible otherwise to reach the bladder. Notwithstanding the free incision, and the consequent escape of urine into the cellular tissue, children never suffer from cellulitis, which, when it occurs in the adult, is generally attributed to these causes. As a matter of course, however, the incision through the prostate should always be as limited as is consistent with the ready extraction of the calculus. The symptoms of pelvic cellulitis are heat and pain at the neck of the bladder, with pain behind the pubes; the face expresses anxiety; the pulse is small and weak; the tongue brown, furred, and dry; the wound glazed, the discharge fetid. Prostration rapidly comes on, and the patient dies in a few days after the operation. The treatment is obvious, though too often of little avail; a free outlet should be made for the extravasated and infiltrated urine by enlarging the external incision; the patient's strength should be supported by beer, wine, brandy, beef-tea, and whatever nutriment he can take. A gum-elastic catheter should be retained in the bladder, that the urine may pass off as soon as it enters this organ. Abstraction of blood is inadmissible, the patient's strength being too much reduced; but hot fomentations may be applied to the hypogastric region.

Acute inflammation of the bladder sometimes occurs after lithotomy, but Boyer's statement that it gives rise to three-fourths of the deaths after that operation is quite contrary to our present experience. This may have been so formerly; it is assuredly not the case now. Under ordinary circumstances we have little to apprehend from cystitis. This complication is not likely to occur unless the bladder has been previously inflamed, or injured by the forceps during the extraction of a calculus in a difficult operation. Extension of the inflammation to the kidneys or peritoneum constitutes the main danger.

Peritonitis rarely occurs in adults, but it is occasionally set up by extension of inflammation from the mucous membrane. Sometimes, also, the peritoneum becomes inflamed in cases of pelvic cellulitis—a complication which much increases the gravity of the case. The attack usually proves fatal on the third or fourth day. There is not always much room for treatment, the patient being considerably reduced by the operation. The application of leeches, and the administration of calomel and opium so as to affect the gums, are the usual remedies; but opiates here are of great avail when given in full doses, and continued so as to produce a decided and prolonged effect.

Mr. Crosse, whose experience was large, says, "that as a separate malady, unaccompanied by urinary infiltration and diffuse reticular inflammation, peritonitis is, according to my experience, not frequent; it is known by the usual signs, and commences in the vicinity of the bladder, where the pain is first felt spreading thence over the peritoneal surfaces; the most active antiphlogistic treatment, by bleeding, low diet, fomentation, and counter-irritation, are required, and, if promptly and early practised, will arrest the disease. In children, who are rarely sufferers from diffuse reticular inflammation and urinary infiltration, peritonitis occurs, and yields to active treatment, particularly leeching freely the abdomen; but in the aged, urinary infiltration and reticular inflammation and suppuration are frequent in comparison with unmixed peritonitis; the surgeon seldom has to treat the latter, and when the combined diseased action is present, I must repeat that general bleeding and purging sink the powers without arresting the malady, and should be very guardedly undertaken by the surgeon." Children are especially liable to peritonitis after operations on the bladder. Even sounding for stone has been followed by fatal peritonitis in young subjects.

Inflammation of the prostatic veins, followed by purulent infection of the blood and death, has been invariably described by French authors amongst the causes of a fatal termination in lithotomy. It is said to be much less frequent in this country. When it occurs, it is probably due to the bruising or laceration the parts have undergone in the attempts to extract a large calculus through an insufficient opening.

In connection with pyæmia as a cause of death, it is important to notice a point which has been too much overlooked. Purulent infection may arise from other causes besides phlebitis. The presence of pus in any part of the genito-urinary system is sufficient to determine the affection under the influence of operations performed on the urethra or bladder. Mr. Coulson saw a case in which it arose from ulceration with secretion of pus in the urethra, and likewise from the pus contained in a cyst in the bladder. Abscess of the prostate has also been a determining cause of the infection after lithotomy. These facts prove that we must not be content with examining the venous system alone, but should direct our attention to all the sources whence the introduction of pus into the blood may take place.

With regard to diseases of the kidney as a complication of stone in the bladder, and as a cause of death after lithotomy, to avoid repetition the reader is referred to the statements in the chapter on Lithotrity (p. 289). When renal disease is known or suspected to exist, lithotomy is never to be preferred to lithotrity, provided that the conditions of the case are such as to justify the surgeon in believing that the stone can be removed by the method invented by Dr. Bigelow. If, for example, the urethra be of good calibre, the prostate not very rigid, and the state of the bladder such that the instruments can be used with facility, even in dealing with a very large stone, the presence of renal disease is an additional reason for preferring rapid lithotrity to lithotomy. If, however, these favorable conditions do not exist, and lithotomy is performed on a patient known to be suffering from renal disease, the prognosis is of the gravest character. Rigors, followed by vomiting, are the first indications that the mischief has become aggravated. The treatment consists in applying hot fomentations to the loins; but death within forty-eight hours from uræmia is the ordinary termination.

The after-treatment in cases of lithotomy is simple, provided no accident has occurred during the operation. The patient is placed on his back in bed, with the knees slightly raised, and is kept as quiet as possible. The surface of the wound does not require any application; it will be sufficient to keep the parts clean and dry by removing the napkins, sponges, etc., whenever they are wetted by the urine.

It is a matter of great importance that the urine should escape freely through the wound soon after the operation. To insure this object some surgeons are in the habit of passing a gum-elastic tube through the wound into the bladder, and securing it by tapes round the waist; the tube is kept in for about twenty-four hours, if the patient be a child; from forty to fifty hours if an adult. Mr. Liston was a strong advocate for this practice, which presents some advantages, but it is not absolutely necessary, and the presence of the tube may excite irritation of the bladder, which is always to be avoided: the use of the tube is therefore not to be recommended.

Retention of urine sometimes occurs after this operation. When this is the case immediate relief should be afforded by passing a gum-elastic catheter, or the finger, through the wound as far as the neck of the bladder. Warm fomentations and opiates will assist in relieving any pain or irritation to which the retention may have given rise.

For the first three or four days the diet must be light, consisting of beef-tea, milk, arrowroot, etc., with plenty of any bland fluid to drink. About the fourth or fifth day the wound begins to contract, and then some urine begins to escape through the natural passages. In one case on which Mr. Coulson operated, a considerable proportion of the urine passed through the urethra on the morning after the operation. As the wound gradually contracts more urine is discharged by the urethra, until at last the whole takes the natural course.

From this time the external wound begins to close, and the cure is completed in from fifteen to forty days. As the case progresses favorably the diet must be gradually improved, and it is well to remember that children and aged persons will require support at an earlier period than others.

The Transverse or Bilateral Operation may be viewed as a modification of the lateral operation. It was first described by Chaussier, and performed, in 1824, by Baron Dupuytren.

The patient is placed in the same position as for the lateral operation, an assistant holding the staff in an exactly vertical direction. The surgeon keeps the integuments tense with the fingers of his left hand, and, with a double-edged knife, makes a semilunar incision in front of the anus. This incision commences on the right side, between the anus and right ischium, ascends toward the raphé, and terminates on the left side, on a level with the point whence it set out. The middle point of the semilunar incision should traverse the raphé, about ten lines in front of the anus. It involves the skin, the superficial perineal fascia, and the anterior fibres of the external sphincter ani. The left forefinger is now passed into the wound, and guides the knife during a second incision, which lays bare the membranous portion of the urethra; the nail now guides the point of the knife into the groove of the staff at this part, and the membranous portion is opened transversely to the extent of two or three lines, in order to avoid any danger of wounding the rectum. The extremity of the double lithotome is next introduced into the groove of the staff through the small incision alluded to, with its convexity turned towards the rectum; and once fixed in the groove, it is pushed on into the bladder. The staff is now withdrawn, and the lithotome turned so as to present the concavity downwards; the blades are opened, and the instrument withdrawn in a perfectly horizontal direction. The parts divided in this second stage of the operation are, the membranous portion of the urethra along the median line, the prostate and the neck of the bladder on both sides in a nearly transverse direction, and to an extent proportioned to the separation of the blades of the lithotome.

Dupuytren attributed many advantages to this method of operating, yet it is not likely that it will ever replace the lateral operation.

Tested by results, we find that Dupuytren lost 9 male patients out of

38 on whom he operated, or 1 in 4.22; of 47 operations performed by Sanson, Roux, and other French surgeons, 10 proved fatal.

Professor Eve, of Nashville University, United States, sent Mr. Coulson an account of 22 cases in which he performed the bilateral operation; only 4 patients died, or 1 in 5.50.

Allarton's Method.—The Marian method, or the "Apparatus Major" as it has sometimes been called, consisted in making a nearly vertical incision into the spongy or membranous portion of the urethra, on a grooved staff, in introducing certain instruments into the bladder through the incision, and then forcibly dilating the parts, until a sufficient opening was obtained for the extraction of the stone. In 1727, Dr. Douglas suggested that the urethra might be opened in the perineum, and dilated so as to extract any stone without cutting the bladder or lacerating the structures concerned in the operation. The subject was again taken up, in 1796, by Deschamps, who proposed a mixed method, resembling the lateral operation in the external incisions, and the Marian method in dilatation of the prostate and neck of the bladder; he suggested that the integuments, etc., should be divided obliquely until the knife reached the membranous portion of the urethra; this was opened and the prostate slightly nicked, after which the parts were dilated in a very slow and gradual manner. It is curious to find the celebrated lithotomist exclaiming, with prophetic spirit, "Who knows but at some future time surgeons may not be tempted to return to this method?" Mr. Allarton has the credit of yielding to the temptation and fulfilling the prophecy.

His operation he thus describes: "I introduce a grooved staff, of he usual size, and confide it to an assistant, with directions to keep it perpendicular and hooked up against the pubes; I then introduce the index finger of my left hand into the rectum, placing its extremity in contact with the staff, as it occupies the prostate, and press it firmly against the staff so as to steady it; then, with a sharp-pointed, straight knife, with a tolerably long and rough handle, *I pierce the perineum* in the middle line, about half an inch above the anus, or at such a distance as may appear necessary to avoid dividing the fibres of the sphincter. I carry the knife steadily and firmly on till it strikes the groove of the staff, the deep sphincter lying between the knife and the directing finger, which latter enables me to judge of the distance as the knife passes along. Having struck the groove of the staff, I move the point of the knife along the groove towards the bladder a few lines, and then withdraw it, cutting upwards, so as to leave an external incision of from three-quarters of an inch to an inch and a half, according to the presumed size of the stone. The escape of urine indicates the entrance to the urethra. I then introduce a long, ball-pointed probe, or wire, through the external opening into the groove of the staff, and slide it into the bladder to sufficient depth to insure its safe lodgment in that viscus, and withdraw the staff. I then well grease the index finger of the left hand, and pass it along the probe, with a semi-rotatory motion, through the prostate into the bladder; and when the stone is free it comes at once into contact with the finger, and, *if of moderate size, passes readily into the wound on withdrawing the finger*, the patient having power to strain upon, and thereby facilitate the extraction of the stone; this last-mentioned power being one of the great advantages of this operation." The author then goes on to observe that, "the incision being strictly in the median line, no muscles are divided, and no gaping-open wound is left, as in the lateral section. The integrity of the bladder being preserved, and no chloroform given, the patient himself helps to expel the stone."

This operation is suitable only for cases in which the stone is known to be small. Its advantages consist in the smaller risk of hæmorrhage and the less danger of wounding the pelvic fascia. On the other hand, it affords very little room for manipulation with the forceps, and unless the stone is very small, the prostate and neck of the bladder must necessarily be subjected to a dangerous amount of dilatation and bruising. Statistics are decidedly unfavorable to this operation as compared with the lateral method. The cases, occurring in adults, for which it is suitable, are those which can be far more safely treated by lithotrity, and with regard to children, the lateral operation is easier of execution and yields better results.

CHAPTER XXV.

LITHOTOMY AND LITHOTRITY IN THE FEMALE.

The operations performed for the removal of calculi from the female bladder require a separate notice, and a chapter will therefore be devoted to them.

Women are not very subject to stone in the bladder; at least, not so much so as men.

Several reasons may be assigned for this difference. Women are more regular in their habits, and not addicted to those excesses which, as has been shown, by increasing the proportion of nitrogenous elements in the body, promote the deposit of uric acid. For the same reason they are less subject to gout. It is also evident that the shortness and dilatability of the urethra, its peculiar form, the absence of a prostate, and several other anatomical circumstances facilitate the passage of small calculi through the female urethra and prevent the development of the complaint. Foreign bodies, however, introduced into the bladder, are a somewhat frequent cause of calculus in the female.

Admitting, with all writers, the comparative rarity of calculus in the female, it is probable that it exists oftener than is generally supposed. The following are the principal statistical facts which are available in illustration of this point. Mr. South informs us that in 146 cases operated on at St. Thomas's Hospital from 1822 to 1845, 144 were males, and only 2 females.[1]

This gives a proportion of 1 female to 72 males, which is greatly below the average. In the Norwich collection of 704 calculi, 669 were taken from males, and 35 from females, or from 1 female to 19 males.[2]

This agrees closely with the proportion observed at the Hotel-Dieu, from 1808 to 1830; 284 calculous patients were operated on, of whom 17 were females, giving a proportion of 1 to 16 males.[3]

From the various tables collected by M. Civiale, it appears that the proportions in Italy, out of 1,104 operations, were as 1 female to 18 males; and in France, out of 2,834 operations, as 1 to 22 males. These various statistics agree sufficiently to enable us to conclude that the proportion of females to males, at least amongst those who are submitted to operations, is about 1 to 20.

The symptoms of stone in the female resemble those in the male, but owing to the close connection and sympathy between the bladder and uterus, they are often referred to the latter organ, and the patient is apt to complain of "bearing-down" pains. Incontinence of urine is also another prominent symptom. Large and rough stones are likely to cause acute cystitis or even ulceration of the bladder, and in some cases a vesico-vaginal fistula has been the result. Out of 204 cases collected by Bouqué,[4] of this latter affection, 6 were said to have been caused by a

---

[1] Chelius's Surgery, vol. ii., p. 635 (table).
[2] Crosse, Treatise on Urinary Calculi, etc. London, 1835.
[3] Civiale, Traité d'Affection Calculeuse. Paris, 1838, p. 504.
[4] Du traitement des fistules urogénitales. Paris, 1875, p. 84.

calculus in the bladder. Spontaneous expulsion through the urethra not unfrequently takes place, and calculi of large size, several ounces in weight and more than an inch in diameter, have been thus passed. Stone in the bladder of a parturient woman may become a serious obstacle to the passage of a child's head, and numerous cases of this kind have been placed on record.

The diagnosis of calculus in the female is for the most part unattended with difficulties. The examination is made by means of a sound passed into the bladder, and a finger in the vagina. Cystocele sometimes co-exists with calculus, of which it may be either the effect or the cause, most commonly the latter, and particularly when this complication exists the calculus will be readily felt through the vagina. In any doubtful case, the finger should be passed through the urethra into the bladder, which can thus be thoroughly explored. The examination is conducted in this manner. The patient is put under chloroform and placed in the lithotomy position; a director is passed into the urethra, and over it the forefinger, well oiled, is introduced with a slight rotatory movement. Weiss' urethral dilator or Simon's specula may, if necessary, be used to dilate the urethra, previous to the introduction of the finger. The situation, size, and other particulars with regard to the stone, can be thus ascertained.

The methods proposed and practised for the extraction of calculi from the female bladder are somewhat numerous, but many of them have now fallen into desuetude. Those now practised may be classed under the heads of urethral dilatation and extraction, lithotrity, and various cutting operations.

1. *Dilatation of the urethra and extraction.*—This method has been employed from a very early period, and numerous appliances have been invented for the purpose. In the sixteenth century Franco applied dilatation to cases of small calculi. His successors, however, went further. In the seventeenth century Jannot practised dilatation alone, but the method did not prevail long. Experience soon demonstrated the inconvenience arising from excessive dilatation of the neck of the bladder and surgeons returned to the only sound practice, viz., that of confining this method to small, or at most moderately-sized calculi. The dilatation may be effected either slowly by means of sponge-tents, or rapidly by means of Weiss' dilator or Simon's specula. Rapid dilatation is, however, greatly to be preferred to the more tedious process, inasmuch as it is far less likely to be followed by incontinence of urine. Mr. Bryant, who has strongly advocated this method, states that in children a stone three-quarters of an inch in diameter, and in adults once inch, may be removed through the urethra after rapid dilatation without fear of subsequent incontinence; the patient being under the influence of chloroform. Even larger calculi have been successfully removed by rapid dilatation without subsequent mischief. Mr. Poland[1] has reported a case in which he thus removed a calculus measuring, with the lithotomy forceps applied, nearly two inches in diameter. For these cases, however, as will be presently mentioned, lithotrity is the safer proceeding.

Rapid dilatation may be performed as follows : The patient being in the lithotomy position, and fully under the influence of chloroform, Weiss's dilator is introduced and expanded rapidly and sufficiently to admit of the introduction of the forefinger of the left hand. The posi-

---

[1] Holmes's System of Surgery, vol. iv., p. 1088.

tion of the stone having been ascertained, the forceps is introduced as the finger is withdrawn, and the stone is then seized and removed. The dilatation may also be accomplished by means of the finger alone, or by means of Simon's specula, a description of which has been given in the chapter on "The Methods of Examining the Bladder." It is claimed for these specula, that by their means dilatation is accomplished more perfectly and accurately than by Weiss's dilator or any similar instrument. Professor Simon,[1] however, recommended in addition, that before the introduction of the specula, the margin of the urethral orifice should be slightly notched by two lateral incisions each one-fourth of a centimeter in depth, in the upper margin, and another of half a centimeter in depth through the urethro-vaginal septum. These incisions are effected with a pair of scissors, and, being made in the narrowest and most unyielding part of the urethra, they prevent laceration and facilitate the introduction of the finger or instruments. The specula are 7 in number, the largest measures 2 centimeters in diameter, the smallest three-quarters of a centimeter in the same direction. The latter is to be first introduced, and to be followed in rapid succession by the others. When the largest is withdrawn, the urethra will be found sufficiently dilated to admit the finger without difficulty. Professor Simon asserts that dilatation to this extent may always be accomplished without any danger of producing incontinence, and in extreme cases that dilatation may be increased to a circumference of $6\frac{1}{2}$ to 7 centimeters. In girls from 11 to 15 a circumference of 4.7 to 5.6 centimeters is the limit of dilatation, and for patients between 15 and 20 a circumference of 6.3 centimeters must not be exceeded. The forefinger will be enabled to reach farther into the bladder, if the middle finger is at the same time passed into the vagina; and with the other hand placed above the pubes, the superior fundus of the bladder should be pressed down upon the exploring finger.

2. *Lithotrity.*—If the calculus is too large to be safely extracted through the urethra, lithotrity may be undertaken, provided that the bladder is neither very irritable nor extensively diseased, and the stone is not too large.

It is curious to observe the phases through which lithotrity has passed in its application to females. When this method of operating was first introduced, the enemies of lithotrity pretended that it was applicable only to the female. In one of the early editions of M. Velpeau's "Operative Surgery" the following passage occurs:—"Lithotrity is performed on the female with infinitely greater ease than on the male, and it is almost free from danger; the female urethra being wide, short, and dilatable, without any curvature, without prostate, or seminal orifices, is admirably fitted to give passage to the instruments, and it is not necessary to reduce the calculus into such small fragments." Yet after having such a favorable opinion as this, M. Velpeau subsequently declared that lithotrity is not applicable to female patients. He would exclude women as well as children from the benefit of the operation. In England there existed some prejudice also against the application of lithotrity for the cure of calculus in the female; but it has now passed away, and, even with the opinions formerly held, the only objection of any real force was that drawn from the supposition that the female bladder is incapable of retaining the fluid injected into it previous to the operation.

---

[1] Volkmann's Sammlung Klinischer Vorträge, No. 88, p. 652.

The conditions of simplicity which distinguish the female urinary apparatus from the male, render the introduction of instruments so easy that it is unnecessary to enter into any examination of the obstacles which may present themselves during this part of the operation. Still it may be well to remark that the orifice of the meatus urinarius is narrow in some females, so much so that the operator may experience some difficulty in getting the point of the lithotrite into it. This obstacle, however, is of little importance. The orifice of the meatus can be easily dilated by means of bougies; and no great inconvenience can arise from nicking it with the bistoury or scissors if judged necessary. In some cases the urethra is more or less contracted along its whole length, and in a very irritable condition from sympathy with the disordered bladder. Although such a condition occurs much less frequently than in the male, and never reaches such a degree of intensity as when depending on disease of the prostate, or of the neck of the bladder in males, it still requires attention, and should be combated by the ordinary means already described.

The female bladder is less liable to become seriously diseased than the male from the effects of calculous irritation, and the obstacles arising from enlargement or disease of the prostate are, of course, entirely absent in the female. However, the depression just behind the neck of the bladder, caused in the male by tumefaction of the middle lobe of the prostate, sometimes exists in the female also, though from a different cause. Owing to relaxation of the wall of the vagina the base of the bladder becomes depressed, and in an aged patient this condition may be so developed as to form a kind of sac in which the calculus or its fragments are apt to lodge. In addition to this, as M. Civiale observes, the uterus sometimes presses against the posterior and inferior walls of the organ, dividing its floor into two pouches in which the fragments may likewise become engaged. It is well to bear these circumstances in mind, although any obstacle which they can create is far from having the same importance as those produced by enlargement of the prostate in the male. The canal of the urethra is so short and straight in the female that almost every portion of the bladder can be explored without any great difficulty, and the conditions of each case ascertained with a degree of precision which renders the task of the operator comparatively light.

The operative proceeding is the same for the female as for the male, and the indications relative to disease of the urinary organs are also the same for both sexes. It is unnecessary therefore, to repeat what has been already said on this latter point; but it may be remarked that lithotrity has been successfully applied to cases of stone in the female bladder under circumstances which would certainly have contra-indicated the operation in the male. There are two principal reasons for this— in the first place, the operation can be terminated, *cæteris paribus*, much more rapidly in the female than in the male; it is easier to seize the calculus, it is easier to crush the calculus into fragments. All these advantages render it much less likely that the operation will be followed by irritation of the urinary organs, or by any dangerous accidents. In the second place, the fragments of the calculus are much less liable to be retained in the bladder or to become impacted in the urethra; and it has been already shown that many of the evil consequences of an operation in the male depend on the irritation excited by these two accidents.

In the female they are rare. There is no enlarged prostate to impede the free evacuation of the detritus; no curved canal to arrest their discharge, while the great dilatability of the urethra in the female renders it less necessary to break up the calculus into very small fragments, and thus contributes to render the operation less protracted in difficult cases.

The main objection brought forward in this country against the expediency of applying lithotrity to the female was derived from the alleged difficulty of making the female bladder retain the fluid injected into it.

It would certainly appear from many cases on record that some operators have found it impossible to make the female bladder retain the fluid which they injected; and hence, adopting the false method of reasoning from particulars to universals, they rejected lithotrity altogether for the female. It may happen that the bladder of a female patient is small, contracted, excessively irritable, and incapable of retaining even a small quantity of fluid. This occurs in the female as well as in the male, though probably less frequently, and when it does occur lithotrity is contra-indicated. But this is no reason for converting the exception into a rule. Under ordinary circumstances and with modern instruments, preliminary injections are unnecessary; the bladder will usually retain sufficient urine for the operator's purpose if the patient be kept at rest for some days before the operation. The rapidity with which the operation may be completed in the female renders it unnecessary to have the bladder distended with fluid.

The same instruments may be employed for the female as for the male, or they may be slightly modified. There is no necessity for having the female lithotrite as long as the male instrument, and it is probable that the facility of working may be increased by having the extremity more scoop-shaped than in the lithotrites commonly used. In favorable cases all the fragments may be removed at one sitting. Lithotrity is particularly applicable for calculi occurring in young girls, and numerous successful cases have been placed on record.[1] With a finger in the vagina, or in the rectum in the case of children, the calculus may easily be guided towards the lithotrite. It is not necessary to pulverize it, but only to break it up into such fragments as will pass by the urethra. After the operation, the bladder should be thoroughly washed out with warm water, so as to remove as much of the débris as possible. Bigelow's method is especially suitable for effecting this latter object.

3. *Lithotomy.*—Some form of operation by incision is the third method by which calculi may be removed from the female bladder. This method is the oldest, and the urethra and neck of the bladder have been divided in every possible direction. The objection, however, to all operations of this kind consists in the frequency with which they are followed by incontinence of urine. Vaginal lithotomy is the operation which is now most commonly practised, the urethra being left intact, and the wound being subsequently closed by means of stitches. This operation is adapted for those cases in which the calculus is too large to be extracted *per urethram*, and the bladder is too irritable to admit of lithotrity. For the performance of the operation, chloroform having been administered, the patient should be placed in the lithotomy position, with a Sims's speculum in the vagina, and a director should be introduced into the bladder. An incision is then to be carried from behind forward through the anterior wall of the vagina. The incision should be as free

---

[1] Fergusson, Lancet, Oct. 11th, 1862. Walsham, St. Bart. Hosp. Rep., vol. xi., p. 129.

as possible in order that the stone may be extracted without laceration or bruising; it should not, however, involve the urethra. The calculus having been removed, the wound should be closed by means of wire sutures, as in the operation for vesico-vaginal fistula. The after-treatment is for the most part simple; a catheter should be introduced from time to time, in order to spare the patient the necessity of making efforts to pass her urine.[1] If the stone be very large, Professor Simon[2] recommended that a T-shaped incision should be made through the vesico-vaginal septum, the transverse cut being three centimeters in length, and from a quarter to half a centimeter in front of the anterior lip of the os uteri, and, at right angles to this, the longitudinal incision, two centimeters long, directed towards the urethra.

A calculus in the bladder of a parturient woman has been known to become a serious cause of obstruction during labor. Twenty-three instances of calculus occurring during pregnancy have been collected by Hugenberger.[3] In four of these, the calculi were removed by operation during the early months of pregnancy. One occurred to Mr. Henry Thomas, and is reported in the *Lancet*, 1839, Vol. I., No. 21. The calculus measured 1½ in. by 1 in., and weighed 6 drachms, and was removed during the 4th month of pregnancy by the vestibular incision. The wound healed in 32 days, but the patient gave birth to a dead child in the 7th month. In another case, which occurred to Hugenberger, the calculus was removed by vesico-vaginal lithotomy during the 8th month. Symptoms of pyæmia occurred after the sutures were removed; premature delivery took place on the 23rd day, and death 16 days afterwards. In seven of the cases spontaneous delivery took place, but in the majority of these, serious mischief ensued. A case of this kind is reported by Smellie.[4] The stone weighed between 5 and 6 oz., and was forced out of the bladder under the influence of the labor-pains. Permanent incontinence of urine ensued. In a case which Mr. Coulson[5] saw with Mr. Shillitoe, of Hitchin, the calculus had formed around a foreign body introduced into the bladder. When labor came on, a hard substance was found on the superior surface of the vagina, and the woman confessed that she had passed a hair-pin into her bladder six years previously. Owing to its size, the tumor obstructed the progress of the child's head; it was, however, raised above the pubes, and delivery progressed favorably. About two months afterwards lithotrity was performed, and fragments, weighing 2½ ozs. were removed. Among them was an ordinary double hair-pin measuring 2½ in. In a few cases, the stone has been removed during the progress of delivery. In one which occurred to M. Monod,[6] it was found impossible to raise the stone above the pubes. Vesico-vaginal lithotomy was performed, and delivery subsequently accomplished by means of the forceps. The opening in the vesico-vaginal septum healed spontaneously in about three weeks. A stone in the bladder of a parturient woman has necessitated craniotomy. Mr. Erichsen[7] mentions the case of a patient from whom he removed, by vaginal lithotomy, a calculus measuring 8 in. by 6. The stone had so impeded

---

[1] For a full account of this operation see Walsham, St. Bart. Hosp. Rep., vol. xi. Aveling, Obstet. Soc. Trans., vol. v.
[2] Volkmann's Sammlung Klinischer Vorträge, No. 88, s. 660.
[3] Quoted by Prof. Winckel, loc. cit., p. 201.
[4] Treatise on Midwifery, New Syd. Soc. edition, vol. ii., p. 100.
[5] Med. Times and Gazette, 1858, vol. i., p. 21.
[6] Med. Times and Gazette, 1858, vol. i., p. 356.
[7] Science and Art of Surgery, vol. ii., p. 837.

the descent of the child's head during labor that craniotomy had been rendered necessary.

If a calculus be detected in the bladder of a pregnant woman, it should be removed by one of the methods described at the commencement of this chapter. If detected only after the commencement of labor, and if large enough to obstruct the passage of the child's head, an attempt should be made to push it back into the bladder above the brim of the pelvis. If this attempt be unsuccessful, vesico-vaginal lithotomy, as in M. Monod's case, is the only alternative.

CHAPTER XXVI.

STATISTICS OF LITHOTRITY AND LITHOTOMY.

THE earliest statistics of lithotrity are those which were published by M. Civiale and Sir Benjamin Brodie.

It is well known that the statements of M. Civiale were submitted to severe criticism, both in this country and in France; but it must be borne in mind that the identical statistics of M. Civiale, first reported on by M. Doubois, were again submitted for examination to a committee of the Institute, in 1835; that MM. Larrey and Doubois were members of this committee, as they had been of the former one, and that they permitted M. Civiale to state, without contradiction, that he had lost only 6 in 257 cases. This statement was printed under the authority of the Academy of Sciences, and M. Civiale was subsequently elected a member of the Institute of France.

The following is a tabular view of M. Civiale's results:

1821 to 1846. Calculous patients	No. of operations	Deaths	Proportion.
848	581	14	1 in 41.50

The ages in 512 cases were as follows:

Age.	Cases.
1 to 20 years	25
21 " 40 "	80
41 " 50 "	124
51 " 60 "	44
61 " 80 "	234
81 " 90 "	5
Total	512

This table shows a large proportion of aged patients; the mortality is 1 in 41½, a result which no lithotomist can pretend to approach.

M. Civiale's statistics embrace two periods, one from 1824 to 1836, during which 305 operations were performed, and the number of deaths was 7. During the second period, from 1836 to 1846, 276 operations were performed, and the number of deaths was also 7. M. Civiale did not attempt, however, to conceal the fact that 10 other patients, on whom lithotrity had been commenced, died. These patients were not included in the general table, because the operation had not been persevered in, the cases after a few trials having been found unsuited for lithotrity. As a matter of course these cases should have been comprised in the general table, and if this had been done we should have had a total of 591 operations and 24 deaths, or 1 in 24; a result with which the most sanguine advocates of the method may be content.

M. Civiale's Necker statistics yield a result which is more in harmony with the experience of surgeons in general:

Necker Hospital, Paris.

Years	No. of cases	Cured	Died	References
1836-42	78	73	5	'Traité de la Lithotritie,' Paris. 1847, p 567

Death-rate, 1 in 15.6, or 6.4 per cent.

The Necker Hospital statistics have never been contested by any one, for the simple reason that the hospital books were not kept by M. Civiale himself. They were kept by persons who were far from being friendly to one who had been placed and maintained at the Necker, without having undergone a *concours*. They were deposited at the end of each year at the "Bureau Central," and M. Civiale has repeatedly appealed in his writings to the head of that bureau in confirmation of his results without receiving a contradiction.

Sir Benjamin Brodie also published a statistical account of all his operations, but without entering so much into details as M. Civiale.

The results may be expressed in the following simple form:

No. of Cases	Deaths	Proportion
115	9	in 12.77

In five of the cases death resulted directly from the operation, and in four it was attributable to the co-existence of some other disease brought into activity by the shock of the operation.

In three of the five cases first mentioned, death arose from urinous abscesses; in one, from cystitis; and in one, from fever and general irritation. In three out of the four cases where death resulted indirectly, the co-existing disease, which subsequently proved fatal, was seated in the kidney. In the fourth case the patient was cut off by diarrhœa not positively connected with the operation, but the case was included in the list by Sir Benjamin Brodie, apparently to prevent all cavil.

The mortality was 1 in 12¾. Yet in all fairness the last case should be deducted, leaving the mortality as 1 in 14.37 of the cases operated on. Inasmuch, however, as several of the patients were operated upon more than once—eight operations were performed upon one individual in the course of as many years—the number given in the table cannot be held to refer to distinct operations in estimating the proportion of fatal to successful cases.

Other statistics, though more or less incomplete, were published in France by MM. Leroy d'Etiolles, Heurteloup, and Amussat. Of those given by the first named, we find 11 deaths in 116 cases, while M. Heurteloup asserted that he lost only 1 patient in 38. Of the first 112 operations performed in Italy, 9 proved fatal.

Professor Campanella stated that 10 patients were treated by lithotrity in the hospital of Loretto between 1834 and 1839, and all recovered. Seven of them were above 40 years of age.[1]

In more recent times statistics of lithotrity have been published by Sir W. Fergusson,[2] Professor Keith,[3] of Aberdeen, and Sir Henry Thomp-

---

[1] Repertorio di Piemonte, Aug, 1839. [3] Brit. Med. Journal, March 20th, 1869.
[2] Lectures on the Progress of Anatomy and Surgery, p. 220.

son.[1] Fergusson's statistics included 109 cases with a mortality of 12; Keith lost 7 out of 129 patients. Sir H. Thompson's statistics, published in 1870, refer to 195 cases, in only 12 (i.e. 6.15 per cent.) of which a fatal result followed. These cases were consecutive in point of time (1863-70); all were adults, only 3 were below thirty years of age—the youngest was 22, the oldest 84, and 51 of the patients were 70 years of age and upwards. Classifying 184 of the cases according to the nature of the stone—in 122 instances this was found to consist of uric acid and the urates, in 40 of phosphates, in 4 of oxalate of lime, in 16 the calculus was mixed, 1 was pure phosphate of lime, and 1 was cystine. A second operation for recurrence of the stone was performed for 13 of the 184 cases, and these thirteen are recorded twice. In another communication to the same society, Sir H. Thompson has lately given the results of a more extended experience,[2] comprising 500 cases. Of these, 422 were operated on by lithotrity and 78 by lithotomy. The number of individuals operated upon was 420, several of the lithotrity patients being operated upon twice, and a few of them three times. The mortality in the lithotrity cases was 32, i.e. 1 in 13, or 7.5 per cent.; in the lithotomy cases 29, i.e. 1 in 2¾, or 37.2 per cent.; and, in both sets of cases combined, 12.2 per cent., or one death in 8¼ cases.

General Mortality after Lithotomy.

Locality	No. of operations	Males	Females	Deaths	Proportion of deaths to cases
Luneville Hospital	1,492	1,433	59	141	1 in 10.58
*Hotel Dieu, 1808 to 1830	100	95	5	28	1 " 3.57
*La Charité, 1806 to 1831	70	—	—	35	1 " 2
*Beaujon, *La Pitié, *Maison De Santé	56	53	3	18	1 " 3.11
*Ten Departments of France	110	—	—	24	1 " 4.58
Private patients in Paris in 10 years:					
Dupuytren's table	356	—	—	61	1 " 5.19
*Austria	133	—	—	25	1 " 5.32
*Bavaria	136	—	—	28	1 " 4.85
*Lombardy	1,044	—	—	217	1 " 4.81
Naples	308	298	10	47	1 " 6.80
*Wurtemberg	120	120	—	7	1 " 17.14
*Bohemia	36	—	—	4	1 " 9
*Dalmatia	40	—	—	4	1 " 10
*Roman States	33	—	—	3	1 " 11
*Sardinia	21	—	—	6	1 " 3.50
*Sweden	36	—	—	5	1 " 7.20
*Denmark	35	—	—	12	1 " 2.91
*Cork Infirmary	15	—	—	0	
St. Thomas's Hospital	144	—	—	15	1 " 9.60
Bristol Infirmary	354	347	7	79	1 " 4.48
Leeds Infirmary	197	—	—	28	1 " 7
Norwich Infirmary	704	669	35	93	1 " 7.57
Radcliffe Infirmary, Oxford	101	95	6	13	1 " 7.76
St. Mary's, Moscow	411	—	—	42	1 " 9.78
Pennsylvania Hospital	83	—	—	10	1 " 8.30
Dupuytren's bilateral operation	42	38	4	9	1 " 4.56
Cheselden	213	—	—	20	1 " 10.65
Liston	115	—	—	16	1 " 7.18
Total	6,505	3,148	129	990	1 in 6.56

[2] In a paper read before the Royal Medical and Chirurgical Society, March 12th, 1878. See also Practical Lithotomy and Lithotrity, by the same author, 3rd ed., p. 208. [1] Med.-Chir. Trans., vol. liii., p. 127.

The statistics of lithotomy are much more complete, for here we are aided by numerous hospital returns and registers. They illustrate various interesting questions connected with the subject of calculous disease in general.

The general mortality of lithotomy may be determined, as closely as the materials permit, from authentic returns published by a few practitioners and from several public institutions. The preceding table was constructed by Mr. Coulson.

From the above table it would appear that the general mortality of lithotomy is 1 in 6.55, or about 2 in every 13 cases. The returns from the Luneville Hospital were published by Castara; those from Moscow by Dr. Roos, of St. Petersburg; those marked with an asterisk by M. Civiale, in his work on calculous disorders.

The principal causes of death after lithotomy have been explained in a preceding chapter, where it was asserted that the probability of a fatal result increases in proportion to the size of the stone and the age of the patient. It is interesting to test this by statistics. The following table is composed from the returns furnished by Castara, Mr. Smith, of Bristol, Mr. Crosse, Cheselden, Dupuytren, and Mr. South:

Age: Years	No. of cases	Deaths	Proportion
1 to 10	1,466	112	1 to 13.08
11 " 20	731	71	1 " 10.28
21 " 30	205	31	1 " 6.61
31 " 40	141	24	1 " 5.83
41 " 50	123	27	1 " 4.50
51 " 60	161	44	1 " 3.65
61 " 70	126	39	1 " 3 23
71 " 80	19	7	1 " 2.71
Total . .	2,972	355	1 to 8.37

The above table illustrates very clearly the influence of age on the mortality of lithotomy. Below 10 years of age the mortality is 1 in 13, and gradually augments with each decennial period to 1 in 10, 1 in 6, 1 in 5, 1 in 4, 1 in 3¾, 1 in 3¼, 1 in 2¾.

The results of a table of 1,827 cases, collected by Sir H. Thompson,[1] differ slightly from those just given as regards some of the periods. The average mortality, however, 1 in 7.97, corresponds very closely with that shown in the above table. More recent statistics, however, show that the average mortality in persons under 20 years of age is now considerably less than that yielded by the table,[2] while in after-life the mortality is higher. As an explanation of this latter difference it must be remembered that, as a general rule, only unfavorable cases—e.g. those in which the stone is large—are nowadays relegated to lithotomy, the mortality of which operation must necessarily be increased if favorable cases are universally dealt with in another manner.

Notwithstanding the favorable results obtained at the Luneville Hospital (1 in 10) and by Dupuytren (1 in 18) on children below 10 years of age, those of Cheselden and the surgeons of St. Thomas's Hospital stand

---

[1] Quoted by Mr. Poland in Holmes's System of Surgery, vol. iv., p. 1061.
[2] See tables on page 337, which show a remarkable similarity in the average rates of mortality in cases under 16 years of age. In the four hospitals the mortality varied from 4.5 to 5 per cent.

out in extraordinary relief. Cheselden lost only 1 patient out of 35 under 10 years of age; and on analyzing the cases published by Mr. South, in his translation of Chelius, it appears that during a period of 23 years at St. Thomas's Hospital, the mortality at the same period of life was only 1 in 58. This, perhaps, is the most brilliant success of which modern surgery can boast.

St. Thomas's Table.
144 cases in all; in 19 no age given—1 in 9.6.

Age	Cases	Deaths	Proportion
1 to 10	58	1	1 in 58
11 " 20	26	3	1 " 8.86
21 " 30	7	1	1 " 7
31 " 40	6	1	1 " 6
41 " 50	4	1	1 " 4
51 " 60	9	2	1 " 4.25
61 " 70	9	3	1 " 3
71 " 81	5	2	1 " 2.50
Total	125	14	1 in 9

Age not given for one death.

It is generally agreed that the size of the calculus exercises a marked influence on the result of lithotomy. The size cannot be correctly judged from its weight, though an approximate estimate may be formed. The only statistical returns which we possess on this point are due to the late Mr. Crosse, who weighed all the calculi in the Norwich collection, and having compared the weights with the registered results of each case, drew up the following tables:

Influence of weight of Calculus on results of Lithotomy.
FROM MR. CROSSE.

Weight	Cases	Deaths	Proportion
Up to 1 oz.	529	47	1 in 11.25
1 oz. to 2 "	119	18	1 " 6.61
2 " 3 "	35	16	1 " 2.18
3 " 4 "	11	7	1 " 1.57
4 " 5 "	5	3	1 " 1.66
5 " 6 "	2	0	0 " 2
6 " 7 "	2	2	2 " 2
Total	703	93	1 in 7.56

Weights of Calculi in 100 Fatal Cases.
FROM MR. CROSSE.

Weights.	Deaths.
Up to 1 oz.	50
1 " 2 "	22
2 " 3 "	17
3 " 4 "	5
4 " 5 "	3
5 " 6 "	0
6 " 7 "	2
Above 7 "	1

STATISTICS OF LITHOTRITY AND LITHOTOMY. 335

The above returns are highly instructive. They show how rare is the occurrence of a calculus weighing more than 4 or 5 ozs.; and, on the other hand, they indicate the influence which the size of the stone exercises on the result of the operation, the mortality increasing in nearly the same ratio as the weight.

In order to arrive at definite conclusions with regard to the relative frequency with which the two operations, lithotomy and lithotrity, are performed in London, and likewise as to the results obtained, the following tables have been compiled from the Reports of four large London Hospitals, and of St. Peter's Hospital for Stone:

St. Thomas's Hospital.

Year	Lithotomy						Lithotrity			References.
	Age under 16			Age over 16						
	Number of cases	Cured	Died	Number of cases	Cured	Died	Number of cases	Cured	Died	
1869	2	2	—	—	—	—	—	—	—	Vol. i. (new series), p. 679
1870	3	2	1	—	—	—	1	1	—	Vol. ii. p. 338
1871	1	1	—	1	1	—	1	1	—	Vol. iii. p. 331
1872	3	3	—	1	—	1	1	—	1	Vol. iv. p. 324
1873	1	1	—	1	—	1	—	—	—	Vol. v. pp 427 and 432
1874	1	1	—	5	4	1	2	1	1	Vol. vi. p. 339
1875	3	3	—	2	2	—	—	—	—	Vol. vii. p 360
1876	3	3	—	4	3	1	—	—	—	Vol. viii. p. 542
1877	3	3	—	—	—	—	—	—	—	Vol. viii. p. 608
1878	2	2	—	4	2	2	2	—	2	Vol. ix. p. 326
Total	22	21	1	18	12	6	7	3	4	

(a) Death-rate of children after lithotomy, 1 in 22, or 4.5 per cent.
(b) Death-rate of adults after lithotomy, 1 in 3, or 33.3 per cent.
(c) Death-rate of adults after lithotrity, 4 in 7, or 57 per cent.
(d) Proportion of lithotomy to lithotrity, adults, 18 to 7, or 2.5 to 1.

Guy's Hospital.

Year	Lithotomy; ages not given			Lithotrity			References.
	Number of cases	Cured	Died	Number of cases	Cured	Died	
1870	9	7	2	1	1	—	Vol. xvii. p. 519
1871	7	6	1	—	—	—	Vol. xviii. p. 455
1872	15	11	4	4	3	1	Vol. xviii. p. 481
1873	7	4	2	—	—	—	Vol. xix. p. 533. One lithotomy case unrelieved
1874	8	6	2	3	3	—	Vol. xx. p. 544
1875	8	6	2	1	1	—	Vol. xxi. p. 444
1876	9	5	4	—	—	—	Vol. xxii. p. 492
1877	8	5	3	3	2	—	Vol. xxiii. p. 396. One lithotrity case unrelieved
1878	8	7	1	2	2	—	Vol. xxiv. p. 472
Total	79	57	21	14	12	1	

(a) General death-rate after lithotomy, 1 in 3.7 or 26.5 per cent.
(b) Death-rate of adults after lithotrity, 1 in 13, or 7.6 per cent.

St. Bartholomew's Hospital.

Year	Lithotomy						Lithotrity			References
	Age under 16			Age over 16						
	Number of cases	Cured	Died	Number of cases	Cured	Died	Number of cases	Cured	Died	
1869	3	3	—	—	—	—	1	1	—	Vol. vi p. 54. Statistics
1870	10	8	2	2	2	—	2	2	—	Vol. vii. pp. 28, 63 and 70
1871	7	7	—	2	1	1	—	—	—	Vol. viii. pp. 74 and 82
1872	9	8	1	6	4	2	4	2	2	Vol. ix. pp 67 and 77
1873	4	4	—	4	3	1	—	—	—	Vol. x. p. 67
1874	7	7	—	5	1	3	—	—	—	Vol. xi. p. 77. One not relieved
1875	8	8	—	3	1	2	2	2	—	Vol. xii pp. 66 and 71
1876	4	4	—	1	1	—	4	3	1	Vol. xiii. pp. 108, 109
1877	6	6	—	2	2	—	—	—	—	Vol. xiv. p. 106
1878	2	2	—	1	1	—	—	—	—	Vol. xv. p. 113
Total	60	57	3	26	16	9	13	10	3	

(a) Death-rate of children after lithotomy, 1 in 20, or 5 per cent.
(b) Death-rate of adults after lithotomy, 1 in 2.8, or 34.6 per cent.
(c) Death-rate of adults after lithotrity. 1 in 4.3, or 23 per cent.
(d) Proportion of lithotomy to lithotrity operations (adults), 2 to 1.

St. George's Hospital.

Year	Lithotomy						Lithotrity			References
	Age under 16			Age over 16						
	Number of cases	Cured	Died	Number of cases	Cured	Died	Number of cases	Cured	Died	
1869	2	2	—	2	2	—	4	3	1	Vol. v. p. 306
1870	4	4	—	—	—	—	4	3	1	Vol vi. p. 381
1871	2	2	—	1	1	—	1	1	—	Vol. vi. p. 383
1872	4	3	1	1	—	1	—	—	—	Vol. vii. p. 306. In the fatal case (child) supra-pubic operation contracted pelvis
1873	—	—	—	1	1	—	1	1	—	Vol. vii. p. 318
1874	1	1	—	1	—	1	—	—	—	Vol. viii. p. 445. Median in fatal case
1875	2	2	—	1	1	—	3	3	—	Vol. viii. pp. 502-3
1876	2	2	—	3	—	3	—	—	—	Vol. viii. pp. 502-3
1877	1	1	—	—	—	—	4	2	1	Vol. ix pp. 297-8. One lithotrity patient left hospital before completion of treatment
1878	2	2	—	4	3	1	1	1	1	Vol. ix. p. 372
Total	20	19	1	14	8	6	18	14	3	

(a) Death-rate of children after lithotomy, 1 in 20, or 5 per cent.
(b) Death-rate of adults after lithotomy. 1 in 2.3, or 42.8 per cent.
(c) Death-rate of adults after lithotrity, 1 in 6, or 16.6 per cent.
(d) Proportion of lithotomy to lithotrity operations (adults), 7 to 9.

St. Peter's Hospital.

October to October	Lithotomy						Lithotrity			References
	Age under 16			Age over 16						
	Number of cases	Cured	Died	Number of Cases	Cured	Died	Number of Cases	Cured	Relieved	Died	
1870-71	3	3	—	3	3	—	12	7	4	1	Hosp. Reg.
1871-72	—	—	—	3	1	2	7	4	1	2	"
1872-73	4	4	—	3	2	1	5	5	—	—	"
1873-74	2	2	—	5	3	2	8	6	1	1	"
1874-75	3	3	—	—	—	—	15	10	4	1	"
1875-76	1	1	—	5	4	1	11	7	—	3	"
1876-77	2	2	—	4	1	3	14	13	1	—	Hosp. Reg. (One ab-
1877-78	—	—	—	4	3	1	12	10	—	2	" [sconded)
1878-79	6	5	1	1	—	1	12	11	1	—	"
Total	21	20	1	28	17	11	96	73	12	10	

(a) Death-rate of children after lithotomy, 1 in 21, or 4.7 per cent.
(b) Death-rate of adults after lithotomy, 1 in 2.55, or 39.2 per cent.
(c) Death-rate of adults after lithotrity, 1 in 9.6, or 10.3 per cent.
(d) Proportion of lithotomy to lithotrity operations (adults), 1 to 3.4.

General Table.

Lithotomy	St. Thomas's		St. Bartholomew's		Guy's	St. George's		St. Peter's	
	Under 16	Over 16	Under 16	Over 16	All ages	Under 16	Over 16	Under 16	Over 16
Cured	21	12	57	16	57	19	8	20	17
Died	1	6	3	9	21	1	6	1	11
Total	22	18	60	26	78	20	14	21	28
Death rate	1 in 22, or 4.5 per cent.	1 in 3, or 33.3 per cent.	1 in 20, or 5 per cent.	1 in 2.8, or 34.6 per cent.	1 in 3.7, or 26.9 per cent.	1 in 20, or 5 per cent.	1 in 2.3, or 42.8 per cent.	1 in 21, or 4.7 per cent.	1 in 2.55, or 39.2 per cent.

Lithotrity	St. Thomas's	St. Bartholomew's	Guy's	St. George's	St. Peter's
Cured	3	10	12	14	73
Died	4	3	1	3	10
Received	—	—	—	—	12
Total	7	13	13	17	95
Death-rate	1 in 1.75, or 57 per cent.	1 in 4.3, or 23 per cent.	1 in 13, or 7.6 per cent.	1 in 6, or 16.6 per cent.	1 in 9.6, or 10.3 per cent.

General Totals.

THE FOUR HOSPITALS.

| Lithotomy | . . . | 238 cases: | deaths 47, or 19.7 per cent. |
| Lithotrity | . . . | 50 cases: | deaths 11, or 22 per cent. |

288   58
General death-rate, 20.1 per cent.

St. Peter's

| Lithotomy | . . . | 49 cases: | deaths 12, or 24.4 per cent. |
| Lithotrity | . . . | 95 cases: | deaths 10, or 10.3 per cent. |

144   22
General death-rate, 15.2 per cent.

The following tables may be interesting as an attempt to illustrate the mortality of lithotomy according to the different methods employed:—

Apparatus Major.[1]

Locality	Cases	Deaths	Proportion
La Charité, 1720–1727 .	208	71	1 in 2.83
Hôtel Dieu, 1720–1727 .	604	184	1 " 3.27
La Charité, 1731–1735 .	71	32	1 " 2.21
Total . . .	883	287	1 in 3.08

High Operation.[2]

Ages	Cases	Deaths	Proportion
1 to 20 years	38	9	1 in 4.22
20 " 50 "	8	3	1 " 2.66
50 " 70 "	18	5	1 " 3.6
70 " 80 "	19	6	1 " 3.16
Age not indicated	21	8	1 " 2.62
Total .	104	31	1 in 3.35

The results of a table of 465 cases of this operation, collected by Dr. C. W. Dulles, of Philadelphia,[3] closely coincide with those above given. Of these cases 330 recovered and 135, or 1 in 3.44, died. The following table, taken from Dr. Gross's work, shows that, as compared with the lateral method, the supra-pubic operation yields better results when the calculus is over two ounces in weight.

Table showing the mortality, with calculi of the same weight.

	Lateral operation (Mr. Crosse's table)				Supra-pubic operation			
Weight of calculus	Re-covered	Died	Total	Death ratio	Re-covered	Died	Total	Death ratio
Under 1 oz.	482	47	529	1 to 11.25	11	3	14	1 to 4.66
1 oz. to 2 ozs.	101	18	119	1 " 6.61	17	4	21	1 " 5.25
2 " 3 "	19	16	35	1 " 2.18	10	4	14	1 " 3.50
3 " 4 "	4	7	11	1 " 1.57	13	6	19	1 " 3.16
4 " 5 "	2	3	5	1 " 1.66	9	7	16	1 " 2.28
5 " 6 "	2	—	2	0 " 2.00	7	4	11	1 " 2.75
6 " 7 "	—	2	2	1 " 1.00	1	1	2	1 " 2.00

[1] Morand, Traité de la Taille.
[2] Dr. Humphry, Addenbrooke's Hospital, Cambridge.
[3] Gross, loc. cit., p. 206.

The Bilateral Operation.[1]

Ages	Cases	Deaths	Proportion
1 to 10 years	15	1 ⎫	
11 " 20 "	3	0 ⎬	1 in 18
20 " 50 "	3	1 ⎫	
50 " 70 "	2	2 ⎬	3 in 5
Total . .	23	4	1 in 5.75

The Recto-Vesical Operation.[2]

Ages	Cases	Cures	Recoveries with fistula	Deaths	Proportion of deaths
Under 5 years	12	8	1	3 ⎫	
From 5 to under 10	13	9	2	2 ⎬	1 in 5
" 10 " 20	16	13	1	2	1 " 8
" 20 " 30	5	3	1	1	1 " 5
" 30 " 40	5	4	—	1	1 " 5
" 40 " 50	3	2	—	1	1 " 3
" 50 " 60	9	4	2	3	1 " 3
" 60 " 70	6	6	—	—	—
" 70 and over	9	4	2	3	1 in 3
Age not stated	5	3	2	—	—
Total . . .	83	56	11	16	1 in 5.18

Median Lithotomy.[3]

Operators	Cases	Recoveries	Deaths	Proportion
American Surgeons . . .	205	196	9	1 in 22.77
Reyer of Cairo . . .	56	47	9	1 " 6.22
Norfolk and Norwich Hospital .	64	51	13	1 " 4.92
Pemberton of Birmingham .	25	24	1	1 " 25
Total	350	318	32	1 in 10.93

Surgeon Major Harris's[4] statistics of 300 operations (one bilateral, the remainder lateral) are especially interesting from the fact that they were performed in India, where lithotrity is almost unknown. Of these 300, 191 were performed on children, under 16, the mortality was 1 in 27.29 cases, while in the adults the mortality was 1 in 6.05.

[1] Report from Professor Eve, of Nashville University.
[2] This table is taken from Mr. Poland's article in Holmes's System of Surgery. It is based upan Prof. König's statistics.
[3] Gross, loc. cit., p. 290.
[4] Lithotomy and Extraction of Stone, p. 241.

CHAPTER XXVII.

THE SOLVENT TREATMENT OF CALCULUS.[1]

FROM a very early period in the history of surgery, attempts have been made to remove renal and vesical calculi by means of internal remedies administered by the mouth, as well as, in the case of the latter, by injecting chemical agents into the bladder. No great success, however, could be anticipated from such experiments so long as the chemical properties of the different varieties of stone were imperfectly investigated, and the means were inadequate for determining which variety was present. It was of little use to administer the so-called specific whilst the influence of food and medicine on the composition of the urine was scarcely if at all known. Certain quack nostrums, have, however, had a wide reputation, and some are extensively employed even at the present day. Numerous cures have been attributed to them, and probably in some few cases, not without reason. It cannot be denied that relief of symptoms of bladder-irritation has sometimes followed, but more commonly the only result has been serious damage to the health.

Various more or less complicated and barbarous preparations, all of them containing some proportion of alkali or alkaline earth, together with certain vegetable decoctions, are the remedies which, since very early times, have been recommended for the cure or relief of calculus. As a proof of the reputation which at least one of these compounds obtained, the fact is worthy of mention that in 1739 the English Government paid no less than 5,000*l*. to a Mrs. Joanna Stephens for disclosing the composition of a remedy which was supposed to dissolve stone in the bladder. Many cases were recorded in which it was stated to have worked marvels, not only by dissolving the stone, but by removing various unpleasant and distressing symptoms connected with renal and vesical diseases. The remedy in question was found to consist of Castile soap, calcined egg-shells and snail shells, and the decoctions of certain herbs.

About twenty years after the purchase of Mrs. Stephens' secret, Dr. Whytt, of Edinburgh, published an "Essay on the Virtues of Lime-water and Soap in the Cure of Stone." He recorded a few remarkable cases in which the remedy appeared to produce great relief, and among them the following: "David Miller had long been the subject of stone in the bladder, which had caused him the most excruciating sufferings. After trying a great variety of medicines, he took Castile soap to the amount of an ounce and a half daily, washed down with three pints of lime-water. He soon experienced great relief, and, after passing several fragments of stone, was considered cured. When he died, at the end of eleven years, no stone was discovered."

Fourcroy and Vauquelin first laid down rules for the selection of

---

[1] Several of the paragraphs in this and the following chapter have been taken from the editor's work on "Stone in the Bladder, with special reference to its Prevention, Early Symptoms," etc.

those remedies which might exert the most favorable chemical action. The uric acid calculus was to be dissolved by the diluted alkalies; the phosphatic, by the diluted hydrochloric acid; the oxalate of lime, by the diluted nitric acid.

Experience having demonstrated the good effects to be obtained from alkalies, certain mineral waters known to be possessed of alkaline properties were next tried as solvents for stone, and it seems undeniable that in several persons in whose bladders the presence of calculi had been ascertained by sounding, the symptoms were relieved or removed while using the water internally.

M. Chevallier was the first who instituted direct experiments on the powers of the alkaline water of Vichy, to dissolve and disintegrate urinary calculi. The following are the particulars of some of his trials:

1. A quantity of uric acid gravel was subjected to the action of the Vichy water, kept at a temperature of 97° F. The concretions were speedily destroyed, the uric acid being entirely dissolved and nothing remaining suspended in the fluid but a few loose flocculi of animal matter.

2. The half of a uric acid calculus, weighing 1 ounce, 1 drachm, 36¼ grains, was placed in a little bag of wire muslin, and subjected to the action of the Vichy water' during 151 hours; dried carefully and weighed after this, the calculus was found reduced to 2 drachms, 52 grains, so that in less than a week it had actually lost 6 drachms, 47 grains, or more than two-thirds of its original weight.

3. In another experiment, five calculi, one of phosphate of lime, weighing 1 drachm, 18 grains, a second of uric acid, weiging 1 drachm, 8 grains, a third of uric acid, weighing 25 grains, a fourth and fifth, fragments of phosphatic calculi, and weiging 29 and 13 grains respectively, were inclosed together in a bag of fine muslin, and exposed for six days to a constant stream of Vichy water of the temperature of about 98° F. When the bag was examined, it was found completeley empty; the calculi of uric acid had been dissolved, those of the phosphates disintegrated, and their particles washed through the meshes of the muslin.

Another experimenter, M. Petit, asserted (contrary to the opinion usually entertained) that, on the whole, calculi of the triple phosphate of ammonia and magnesia lost more under the action of the Vichy water —i.e. of a solution of bicarbonate of soda in water, supersaturated with carbonic acid—than those of the uric acid. For example, five specimens of uric acid calculi, weighing together a little more than 7 oz. and 416 grs., after remaining, on an average, 27 days exposed to the action of the water, lost very nearly 53 per cent. of their original weight; but five specimens of the ammonio-magnesian phosphatic calculus, which together weighed rather more than 6 oz., 14 grs., and only remained under the action of the water, one with another, for the space of 23 days, lost 60 per cent. of their original weight.

---

[1] There are seven principal springs which pour out these waters, the product of each differing but little from that of the rest. The composition is as follows:

Water	from 991.9	to 994.9
Carbonic acid gas	" 1.3	" 0.93
Carbonate of soda	" 5.5	" 4.7
Carbonate of lime	" 0.6	" 0.3
Carbonate of magnesia	" 0.9	" 0.6
Chloride of sodium	" 0.47	" 0.52
Sulphate of soda	" 0.47	" 0.27
Silica	" 0.04	" 0.04

M. Petit, like M. Chevallier, found that the Vichy water had extremely little influence on calculi composed of oxalate and phosphate of lime. This water, however, would dissolve cystine and xanthic oxide calculi with at least as much readiness as it disintegrated and dissolved those of lithic acid and the triple phosphate. Water surcharged with carbonic acid, and holding a little bicarbonate of soda in solution, was consequently believed to be a solvent for all the more common calculi, with the exception of those composed of oxalate of lime.

M. Petit's conclusions, however, as to the solubility of the triple phosphate in an alkaline water were not borne out by subsequent experience. The deposit or calculus in question is formed in alkaline urine, and the administration of an alkali would only tend to expedite the formation. The good effects of Vichy water in cases of uric acid have, however, been placed beyond dispute by a number of well authenticated cases. One, quoted by Dr. Roberts,[1] from Dr. Petit's treatise, may be cited as an example. "M. de L., 51 years of age, was sounded by Leroy d'Etiolles, who found a stone in the bladder. This he believed to be not large and suitable for crushing. The patient, however, went to Vichy, and drank the first day 7 or 8 glasses of the waters. The next day he took 15 glasses, and the urine, which was previously very acid, became constantly and strongly alkaline. In a few days he took 22 and 24 glasses. The symptoms, which were before severe, now subsided more and more, and, after seventeen days of treatment, he voided a smooth, uric acid concretion, which bore evident traces of dissolution. From this moment he continued wholly free from symptoms and was able to take violent equestrian exercise without the least inconvenience."

Among other litholytic remedies which have been used internally may be mentioned borax and phosphate of soda. These were recommended by the late Mr. Ure,[2] who published several cases in which marked relief was obtained and partial dissolution of the calculus took place as a result of treatment. In one case, that of a gentleman advanced in years, with well-marked symptoms of calculus, owing to the enfeebled state of the patient's constitution, it was deemed unadvisable to resort to any surgical operation. Mr. Ure, believing the calculus to be composed of uric acid, prescribed phosphate of soda to be given twice or thrice daily in infusion of uva ursi, with a little tincture of henbane. This treatment was persevered in for some months, with decided relief to the calculous symptoms. Eventually the patient died under a gradual decline of the powers of life. A post-mortem examination was made by Dr. Hodgkin and Mr. Ure. On opening the cavity of the urinary bladder, a fragment of a calculus was discovered. This weighed 10 grains only. It consisted of uric acid. It bore evident marks of extensive disintegration of its concentric layers, the surface being irregularly eroded, resembling the appearance of a calculus that has been exposed to the action of a chemical solvent.

More recently numerous experiments have been conducted, by which the question of the solution of stone has been placed on a thoroughly scientific basis. The solvent action of various substances on all the forms of calculi, under conditions resembling as nearly as possible those met with in the body, has been experimentally ascertained. The microscope has enabled us to say almost positively what variety of stone exists in

[1] Med.-Chir. Trans., vol. xlviii., p. 111, and Urinary and Renal Diseases, p. 301.
[2] Provincial Med. Journal, vol. i., 1842, p. 116.

any given case, and experiment and observation have taught us how to induce certain desired conditions of the urine. To Dr. Roberts, of Manchester, the credit is due of having systematized a plan of treatment whereby, under certain favorable circumstances, calculi may be at least partially dissolved without injury to the general health.

The two propositions with which Dr. Roberts commenced his investigations were: (1) that solutions of the alkaline carbonates exercise a solvent action upon uric acid; and (2) that the urine can be rendered alkaline from alkaline carbonates by the administration of certain salts by the mouth. Starting from these data,[1] Dr. Roberts next investigated the solvent powers of carbonate of potash and carbonate of soda, and proved that the former is decidedly more effective than the latter. Solutions containing from 40 to 60 grs. of carbonate to the imperial pint were found to possess the greatest solvent power, and 2 pints of such a solution allowed to flow over a calculus of uric acid produced a daily dissolution of 17.1 per cent. The best way of alkalizing the urine was the next subject for inquiry, and this was found to consist in giving repeated doses of the acetate or citrate of potash. If 40 or 50 grs. of either of these salts are dissolved in 3 or 4 oz. of water, and administered every three hours, the degree of alkalescence produced in the urine will correspond to the maximum solvent power of solution of carbonate of potash. The urine thus rendered alkaline may become turbid from deposition of amorphous phosphate of lime, but under these circumstances solution is not checked. If, however, the urine becomes ammoniacal, the mixed phosphates will be deposited on the surface of the stone, and the solvent power of the alkalized urine entirely ceases.

Calculi composed either of uric acid, cystine, or urostealith are alone amenable to the action of solvent remedies taken internally. Oxalate of lime resists any solvent which can be introduced into the bladder by any method, and the phosphate of lime and the ammoniaco-magnesian phosphate require an acid which can be introduced only by injection. But for the removal of even a small uric acid concretion, a considerable time is required; a period of five to six weeks is calculated from experiment, and during the whole of this time the urine must be kept constantly alkaline. Under the most favorable conditions this prolonged administration of alkalies, and the consequent diuresis, must be attended with waste of tissue and loss of strength. These inconveniences may, however, be reduced to a minimum, by employing the acetate or citrate of potash as above described.[2]

When the stone is larger, not only must the time required be much longer, but there will also be the greater probability that one or more of the layers are composed of oxalate of lime, or that the calculus has received a coating of phosphates, on which the alkalies have no effect. For any chance of success, therefore, in attempting solution of calculus, the urine must be acid, the stone small and composed of uric acid, and without a coating of other ingredients. The application of the treatment is therefore confined within very narrow limits, and the cases in which benefit may be expected are precisely those which are promptly and successfully dealt with by means of lithotrity. For renal concretions, however, the large majority of which are composed of uric acid, the treatment by alkaline solvents is especially indicated.

---

[1] Urinary and Renal Diseases, p. 293.
[2] For evidence on this point see Dr. Roberts' treatise, p. 309.

According to Dr. Roberts, cystine is even more suitable for the application of the alkaline solvent treatment than uric acid. In two of his experiments in which a cystine calculus was exposed to a solution of carbonate of potash, dissolution took place at the rate of 20 per cent. of the weight of the stone in 24 hours. It does not appear, however, that any experiments of this kind have been conducted upon a patient the subject of a cystine calculus.

The solvent treatment has been successfully adopted in a case recorded by Heller, in which the calculus was composed of urostealith. Two drachms of carbonate of soda were administered daily, and with an increased flow of urine the patient voided some soft concretions, always accompanied by fine sand or gravel of ammoniaco-magnesian phosphate. Some of the fragments were as large as half a hazel nut; they were concavo-convex in shape, and evidently part of a large stone. The patient was discharged cured in 14 days. When admitted, the urine was found to be neutral in reaction and to contain no uric acid. The urostealith began to appear in the urine very shortly after the alkaline treatment was commenced. The cure is said to have been complete.[1]

In addition to the method of dissolving calculi by the internal administration of alkalies, numerous attempts have been made to produce solution by injecting solvents into the bladder. The substances that have been used for this purpose are lime-water, alkaline solutions, acid solutions, and a solution of the nitro-saccharate of lead.

Dr. Hales appears to have been the first who endeavored to prove by experiments that fluids injected into the bladder by the urethra were capable of diminishing the size of a vesical calculus. He invented a double catheter, which he used for washing out the bladder, and compared the fluid injected with that expelled after contact with the stone.

Dr. William Butler, of Edinburgh, instituted in 1752 a series of experiments with lime-water, and soon after a case occurred in the practice of Dr. Rutherford, which was reported to be attended with complete success.

Solutions of caustic potash and of carbonate of soda have also been used as injections for a similar purpose, and a few cases have been reported in which success is said to have been obtained.[2] Dr. Roberts,[3] however, has instituted numerous experiments in order to test the solvent power of various alkaline solutions directly applied to the calculus. For this purpose he placed a section of a uric acid stone in a ten-ounce phial, and passed over it at blood heat a current of the solvent as large as the capacity of the urethra might be supposed to permit. The substances used in solution were—carbonate of potash, carbonate of lithia, borax, borax with liquor sodæ, phosphate of soda, lime-water, caustic soda, and caustic potash. As a general result, however, of all these experiments it was found that dissolution progressed only at the rate of one or two grains per hour. Such a result, as Dr. Roberts points out, clearly shows that nothing is to be expected from injecting similar solutions into the bladder.

For the treatment of phosphatic calculi, acid solutions have been used as injections with satisfactory results. Sir Benjamin Brodie employed a weak solution of nitric acid, composed of $2\frac{1}{2}$ minims of strong acid to each ounce of distilled water. A double catheter was used, and

---
[1] Heller's Archiv, ii., 1845, p. 2.
[2] Pereira's Materia Medica, 3d edition, vol. i., p. 261.
[3] Med.-Chir. Transactions, vol. xlviii.

the solution was injected into the bladder for from 15 to 30 minutes every two or three days. A phosphatic calculus was thus dissolved, with relief to all the symptoms. More recently the same method was tried by the late Mr. Southam, of Manchester, for dissolving the fragments of a phosphatic stone after repeated operations by lithotrity. In this case the calculus could be easily crushed, but fresh phosphatic concretions continued to form as fast as the old ones were broken up, and in order to get rid of them an injection containing two drachms of dilute nitric acid to a pint of water was used every second or third day. In the course of a short time the old fragments were completely dissolved, and the formation of fresh deposits ceased.' It is probable that, under similar circumstances, the operation of lithotrity might be considerably aided by injections of the kind, and especially for those patients who are unable completely to empty the bladder.

Other substances which have been used for a similar purpose are the nitro-saccharate and acetate of lead. These were suggested by Dr. Hoskins,[2] of Guernsey. When a solution of acetate of lead comes into contact with a phosphatic calculus, double decomposition ensues, and results in the formation of phosphate of lead and acetate of lime and magnesia. Dr. Hoskins believed that the principle of double chemical decomposition was especially adapted for the disintegration of vesical calculi, and that the agent employed was more energetic as regards the concretion, and less injurious to the organ in which it is contained than any of the uncombined solvents.

As the salts in the urine tend to decompose the solution, and lessen its effects on the concretion, the bladder should be evacuated, and washed out with tepid water before the solvent is introduced. A double current caoutchouc catheter is the best for this purpose, as it enables a continuous stream to be employed; and, on account of its flexibility, it is less liable to irritate the urethra. From 4 to 8 fluid ounces of the solution may be thrown into the bladder at a time, and renewed every 10 or 15 minutes, as often as may be deemed proper. By renewing the liquid at short intervals, much greater effect on the calculus is insured than when the injection is allowed to remain a long time; for the precipitate formed by the decomposition soon envelops the stone, and puts a stop to further action until a fresh surface is exposed. Exercise during the retention of the injection increases its effects. Some slight reaction may be produced by the first introduction of this or any other fluid into the bladder; when such is the case, the operation should be remitted for a day or two, and cautiously renewed. The injection may be either warm or cold, as may be most agreeable to the sensations of the patient. Warmth favors the decomposition of the calculus.

As an aid to lithotrity, in cases of phosphatic calculus, a solution of the acetate of lead is doubtless worthy of trial. From half a grain to a grain may be dissolved in each ounce of distilled water with two or three drops of acetic acid. Super-acidulation is necessary on many accounts; it secures perfect solution, increases the decomposing activity of the liquid, and prevents the formation of any carbonate of lead.

The attempts made to dissolve calculi by the aid of electricity have not resulted in the discovery of any method by which this agent can be employed with any prospect of success. Prevost and Dumas in France,

---

[1] Dr. Roberts, Med.-Chir. Transactions, vol. xlviii.
[2] London Journal of Medicine, Oct, 1851.

and Sir W. B. O'Shaughnessy of Calcutta, proposed to electrolyze urinary calculi in the bladder. For this purpose, however, an electric current of very great intensity would be required to be used for a considerable time, and it is obvious that such a procedure could not be carried out in the bladder. Bonnet suggested that a solution of nitrate of potash should be injected into the bladder, and the calculus subjected to the action of electricity in this liquid. The decomposition of the salt would yield nitric acid and potash, and the former of these would dissolve the phosphates, while the latter would act in a similar manner upon the uric acid and urates. Dr. Bence Jones showed by numerous experiments that lithic and phosphatic calculi could be thus dissolved, out of the body, in a solution of this kind, but it appears to be impossible to apply the method in the case of a living patient.

From the above account of the attempts that have been made to dissolve calculi by remedies administered by the mouth or injected into the bladder, the conclusion appears to be inevitable that the treatment in question, however valuable as an auxiliary to operative measures, cannot by itself suffice for the cure of the large majority of cases of vesical calculus. As a general rule, patients do not come under treatment until the stone has attained a considerable size, and this in itself is an obstacle, apparently insuperable, to the successful carrying out of the treatment by solvents. If, on the other hand, the patient is seen at an early period of the complaint, and the calculus is small and presumably consists of uric acid, the continuous administration for several weeks of a solution of citrate of potash might so diminish the size of the calculus as to enable it to pass through the urethra and thus render an operation unnecessary, while if it failed in this respect it would in no way prejudice the chances of recovery. As an aid to lithotrity in cases of uric acid calculus, the solvent treatment by alkalies is particularly applicable, provided always that the urine remains free from any trace of ammoniacal decomposition. In like manner in cases of phosphatic calculi, the injection of water acidulated with nitric acid may occasionally prove a very useful adjunct to lithotrity.

CHAPTER XXVIII.

THE PREVENTIVE TREATMENT OF CALCULOUS DISEASE.

It will have been seen from the account given in a previous chapter of the different kinds of stone, and of the mode in which they are produced, that calculus is not a sudden formation. Usually the condition of urine to which it is due comes on gradually, and exists for some time before the actual deposition of solid matter takes place within the urinary passages, giving rise to symptoms or to changes in the appearance of the urine which are sufficiently marked to attract the attention of the patient. It is here the opportunity is afforded of preventive treatment. Very commonly, also, one or more small calculi escape, or there are indications of the presence of a stone in the kidney or ureter, and preventive treatment includes the attempt to dislodge such calculi before they attain too great a size. In these cases the solvent treatment is most useful, and at the same time most likely to prove successful. The stone, of course, cannot be reached by instruments before it arrives in the bladder; there is, further, every probability that while in the kidney it is both small and of uniform composition, and it is moreover so situated that there will be a constant flow of urine over its surface, so that the solvent will be continually renewed.

The premonitory symptoms indicating the tendency to the formation of calculus are found in the sensations of the patient, and in the appearances presented by the urine. The sensations are due to the presence in the urine of some abnormal constituent, which irritates the kidney and the mucous membrane of the urinary passages. They rarely amount to positive pain, and are often unheeded or attributed to other causes. Sometimes they are altogether absent. There is frequently a dull aching in the loins, or a sense of weight and fatigue, caused probably by congestion of the kidneys, but very commonly supposed to be the result of fatigue.

Another symptom of the presence of uric acid or of oxalates in the urine is a sensation of heat in the bladder. This gives rise to a more frequent desire to pass water than is habitual with the individual, and sometimes micturition is accompanied by a feeling of heat, which may amount even to scalding, along the urethra.

Any habitual or frequent pain in the loins, more especially when not to be accounted for by over-exertion, any uneasiness about the bladder, or heat along the urethra, or any unusual frequency in micturition or urgency in the desire should direct attention to the state of the urine, an examination of which will very commonly reveal the cause of the symptoms.

The appearances presented will be some of those already described, but in the case of oxalate of lime no change will be noticed by the patient, or visible to the naked eye. In other cases it frequently happens that the urine is perfectly clear when passed, but throws down subsequently, on cooling, a more or less copious precipitate of urates. This occurs from time to time to every one, but it should not be habitual. When it is so, the urine should be examined with the microscope. Most probably, as soon as the deposit has fallen, crystals of uric acid will be found, which, on subsequent examination at intervals of twelve hours, will be seen to have increased in number and size. The rapidity with which these crystals form, and the proportion in which they exist, will measure the tendency to deposit in the urinary passages, and the necessity for treatment. In more urgent cases there will be found, after the urine has stood for a short time, a red crystalline sediment of uric acid, with or without a superincumbent stratum of lithates, and these crystals are often found firmly adhering to the bottom and sides of the vessel. It sometimes happens that crystals of uric acid or oxalate of lime are voided with the urine, and are precipitated before the latter cools. This is a symptom of serious import, inasmuch as it shows that a similar precipitation may take place within some part of the urinary tract, and give rise to the formation of a calculus.

In cases where there is danger of the formation of a phosphatic stone, the symptoms will already have been so urgent that a special description of them is necessary here.

The case of children requires special mention. Statistics show that calculus occurs as often before ten as during the whole life afterwards. We cannot get from children an account of their sensations, and any abnormal appearance in their secretions is likely to escape notice. Usually, however, there are indications which may call attention to the state of the urine, even before calculus is formed. The mucous membrane of the child's bladder seems to be particularly sensitive to the presence of abnormal constituents in the urine. This condition is shown by frequency of micturition, sometimes accompanied with pain, and the urine is expelled with unusual force. There may also be a slight purulent discharge from the urethra, with soreness at the meatus. Nocturnal incontinence is another common consequence of the irritant properties of urine containing uric acid in excess or oxalate of lime, and strangury has been met with which could be attributed to no other cause.

In children the urine is naturally abundant, paler than in the adult, and less liable to deposits of urates from accidental causes. But all the deposits met with in the adult may occur, and the urates may be present in such quantity as to give the urine a turbid appearance when passed. It is only in children that calculi composed exclusively of urates are found. As the urates are paler the urine may have a milky aspect. Phosphates would also give a similar appearance. The distinction is readily made out by simple chemical tests or by the microscope.

The treatment to be adopted in order to prevent the formation of concretions or calculi will vary according to the nature of the deposit, but in all the same principles are to be applied. These are:—

1. To diminish the amount of the abnormal constituent in the urine; or to remove, so to speak, the diathesis.

2. To prevent the calculous material from being precipitated.

3. To keep the urine in a dilute state, by prescribing abundance of fluids, and to flush the urinary system from time to time.

To attain the first of these objects, strict attention must be paid to diet, exercise, and the secretions generally. It must be remembered that the constituents of the urine have a double source in food and tissue, and to keep these constituents in a normal condition, perfect digestion and thorough metamorphosis are required.

When uric acid is the deposit to be apprehended, the diet must be carefully regulated. It must be simple, digestible, and above all, moderate. Animal food must be taken in limited amount, and it must be divided among the different meals, so that at no time shall a large excess of highly nitrogenized material be taken into the blood. Farinaceous food must be substituted for meat, according as the state of the digestive organs permits. Under a strictly vegetable diet the urine becomes alkaline, and uric acid calculus is an impossibility; but in many instances even a moderate preponderance of vegetables causes dyspepsia. Perfect digestion is a condition absolutely necessary to success in preventing the formation of uric acid. Ale and porter should be excluded from the list of beverages, so also the stronger wines, port and sherry. The wines best suited for these cases are hock, burgundy, claret. Only one kind of wine should be taken at a meal, and that in moderate amount; the particular kind will be determined by the peculiarities of the case. If stimulants are required, brandy, with soda-water or effervescing lithia-water, taken weak, forms an excellent drink.

Exercise is of the greatest importance. It should be moderate and habitual, not violent and with long intervals of inaction. It introduces abundance of oxygen by the quickened respiration, increases the metamorphosis of the tissues, and by this consumption of material makes a demand for the nitrogenized constituents of food, which, by their imperfect oxidation, give rise to uric acid. The patient, therefore, should take a daily walk or ride, sufficient to promote a healthy action of the skin without inducing undue fatigue.

The necessity for attending to the secretions and excretions is obvious. The abundant secretions poured into the alimentary canal have not only an important action on the food with which they are mixed, but are themselves in part resorbed. They have, therefore, a double influence on the ultimate results of metamorphosis; first, by their action on the first stage of assimilation, and again, as containing matter both nitrogenized and non-nitrogenized, in a state of retrograde change intermediate between the food or tissues and the excretions. Any deficiency or perversion in them must affect the composition of the excretions. It is necessary, also, that the different excretory organs should be in healthy action. Excretions retained in the blood retard and vitiate metamorphosis, just as the products of combustion, prevented from escaping, will extinguish fire.

The skin, therefore, and the bowels must be attended to. Flannel frictions, and an occasional Turkish bath, will meet the first indication. The bowels should be kept quite open. In selecting aperients, salines, except in cases where the patient is well nourished and plethoric, should be sparingly resorted to. The best purgatives will be those which at the same time act on the liver. The important part this organ takes in the process of assimilation by the changes it effects in the blood passing through it, as well as by the action of the bile on the food in the intestines, would lead us to expect that any derangement of its functions would react on the condition of the urine, and such is found to be markedly the case in organic disease of the liver. This must always be borne in mind; any hepatic derangement should be corrected, and even where no indication of this exists, great benefit will often follow the use of medicines believed to increase the activity of this organ.

It is a work of time materially to diminish the amount of uric acid in the urine, and until this is effected, it is necessary to prevent it from being set free and precipitated—the second of the great indications. This is accomplished by the administration of alkalies; but it is not a matter of indifference either which of the alkalies should be selected or what preparation employed. Ammonia is useless, as it possesses little or no power of affecting the acidity of the urine. Soda, potash, and lithia do this readily; but the uric acid salts with these bases differ in solubility, the urate of potash being the most soluble. Abundant evidence in proof of this statement will be found in an excellent paper by Dr. William Roberts,[1] of Manchester, on the "Solvent Treatment of Urinary Calculi," in which he details the results of the most carefully conducted experiments with solutions of soda, potash, and lithia on uric acid.

"The potash salt was found sensibly to excel the soda salt as a solvent for uric acid." And in a foot-note he says: "Some experiments were also made with carbonate of lithia, which has been much vaunted in recent times as a solvent for uric acid. Its power was found much inferior to that of the carbonates of potash and soda. Its reputation seems to have been gained through its comparative insolubility. Only weak solutions of it could be employed, and these were compared with solutions of potash and soda, which were much too strong for effective dissolution."

Potash is, for all practical purposes therefore, the most useful of the alkalies, and it may be given in the form of liquor potassæ, or of the bicarbonate, or of a salt of some vegetable acid—the effect, however, of each of these being very different. When the potash or soda salts of vegetable acids, citrates, tartrates, acetates, etc., are given, the acid, by a slight process of oxidation, becomes changed into carbonic acid, and the carbonate of soda or potash appears in the urine, rendering it alkaline. In this form the potash will prevent the precipitation of uric acid, but will do little or nothing more. It does not affect the chemical processes going on in the tissues, and it may be taken for a considerable time without producing any lowering effect on the system. Liquor potassæ, on the other hand, does not so speedily or persistently render the urine alkaline, but it exerts a powerful influence on the metamorphosis of the tissues. It appears in the urine in combination with sulphuric or phosphoric acid, increasing the absolute amount of these acids in the secretion. They are formed at the expense of protein compounds, the oxidation of which has been determined by the potash, just as the same alkali oxidizes organic matter out of the body, and gives rise to acids by which it is neutralized. This liquor potassæ, therefore, is a powerful agent, and may be most useful where semi-effete material accumulates in the blood, and oxidation requires to be accelerated, or most injurious where the waste is already excessive. It should never be employed merely for the purpose of rendering the urine alkaline. The bicarbonate is intermediate in its effects. It has not the alterative power of the liquor potassæ, but it has a greater eliminant action than the citrate, acetate, etc.

Combining these different indications, the following is the plan of medicinal treatment, the results of which, with regulation of the diet, have been most satisfactory.

Usually a minute dose of pil. hydrarg., gr. $\frac{1}{8}$th to $\frac{1}{4}$th, is given twice or three times a day with extract of taraxacum in the form of a pill, and, if the bowels do not act regularly, pil. rhæi co. added or substituted for

---

[1] Med.-Chir. Trans., vol. xlviii.

taraxacum till this is the case. The use of mercury in minute doses, as favoring the action of other diuretics, is an advantage, in addition to its influence on the liver and the general nutritive processes. At the same time, 15 to 30 grains of the bicarbonate of potash with infus. calumbæ, or other vegetable tonic, is given three times a day. This often improves the digestion; but, when such is not the case, or when the bicarbonate does not seem to answer, or is not required, the citrate of potash, with a bitter infusion, may be substituted. Quinine and iron, or strychnia and iron, will often be found useful, and may be given with the citrate of potash, as ferri et quiniæ cit., or ferri et strychniæ cit. The liquor potassæ is very rarely required. To any of these, taraxacum in the form of the extract of the British Pharmacopœia, or of the fluid extract, may be added with advantage, but by far the most useful form is the infusion, dandelion-tea, in considerable quantity. This treatment will not only reduce the amount of uric acid in the urine, but will often result in remarkable improvement of the general health where this has previously suffered. In some instances of obstinate dyspepsia and general weakness the mineral acids may be required; and, when it is found that they relieve these conditions, they may be given for some time. They do not greatly increase the acidity of the urine, and will not cause the precipitation of uric acid, while, by their influence on the digestion, they may prevent its formation.

Very commonly, however, patients who are passing uric acid in considerable quantity, or in whom a calculus is actually in process of formation, are otherwise in excellent health. In such cases, very little beyond strict attention to diet will be required, with a plentiful use of water as drink; and the precipitation of the uric acid must be prevented by the administration of the citrate or acetate of potash. Dr. Roberts has shown that the urine may be kept persistently alkaline by this means for many months, without injurious effect either on the system or on the urinary passages. To effect the solution of a uric acid calculus he administers 40 to 60 grains of one of these salts every three hours in three or four ounces of water; but a minor degree of alkalescence will suffice to prevent the uric acid from being deposited.

When oxalate of lime occurs as an habitual deposit, the treatment cannot be laid down with the same exactness. It is not probable that food containing oxalates is taken in such quantity as to give rise to danger of calculus, but substances known to contain them should be avoided. The chemical derivation of oxalic acid from either saccharine or nitrogenous substances gives no indication as to the diet. The omission of sugar from the food and the limitation of starch have, however, been recommended on hypothetical grounds. Addition of lime-water to urine will cause deposit of oxalate of lime when previously absent; and administration of salts of lime has been followed by the appearance of oxalates in the urine. On these grounds, the removal of lime from the water used as drink has been prescribed. It has also been ascertained by experiment that the addition of any mineral acid to urine retards precipitation of oxalate of lime, and for this reason the administration of mineral acids has been recommended. Oxalate of lime, however, is often associated with urates and free uric acid; and a better reason for giving the mineral acids is that experience has shown them to have a special influence on the peculiar dyspepsia sometimes accompanying oxaluria. There are thus no sufficient indications for any special line of treatment, and we have to fall back on general principles. It must be borne in mind that oxalate of lime is usually associated with excess of urea, indicating waste of food or tissue, this furnishing an important clue to the treatment.

All known causes of waste must be avoided—such as over-work, especially mental anxiety; and excesses, especially sexual. Obstinate spermatorrhœa is very frequently accompanied by oxalate of lime in the urine. Rest, relaxation, change of air and scene, will often effect much good. As to food, no absolute rule can be laid down. The quantity and kind must be regulated not so much by the chemical composition as by the digestibility; it must, however, be of a nourishing character. It will usually be found necessary to limit the amount. Saccharine matters should be taken sparingly, or, if necessary, prohibited altogether. If there is pain in the epigastrium soon after eating, and especially if accompanied with acidity and flatulence, vegetables, sugar, and beer should not be taken. If none of these inconveniences are experienced, vegetables and ripe fruit may be used in moderation. As regards stimulants, brandy and water, or brandy and soda-water, a dry sherry, or better than these a good Burgundy, are most suitable. The action of the skin must be favored, and the bowels must be kept gently open. The most important matter, however, is to render digestion easy and perfect. This will sometimes be found very difficult. The mineral acids, as recommended by Dr. Golding Bird, are often very useful, especially the dilute hydrochloric or nitro-hydrochloric acid, with inf. calumbæ. In other cases, bismuth or alkalies with ammonia are of great service. The good effects of small doses of pil. hydrarg. are often very evident in these cases, as well as in deposit of uric acid, and infusion of taraxacum rarely fails to do good. When the digestion and the general health are improving, the amount of oxalate of lime deposited usually diminishes. Its continued presence, under these circumstances, need not excite apprehensions of the formation of calculus when it takes the form of octahedra. These are rarely found formed in the urinary passages, or in urine newly passed. But when dumb-bells are met with, as they are formed even in the tubules of the kidney, every measure which experience or even hypothesis has suggested for preventing the formation of oxalate of lime in the urine should be tried. Rigorous diet, water free from lime as drink, mineral acids as medicines, with regular exercise and frequent baths, till the threatening deposit ceases to appear, must be insisted on. From time to time, also, the urinary organs should be flushed. Large quantities of pure water should be drunk on an empty stomach, or some diuretic infusion with citrate or acetate of potash may be substituted. Nothing will answer better than the dandelion-tea already mentioned, weaker and in greater quantity than would be ordered for habitual use.

When there is reason to suppose that a calculus exists, the same stringent measures should be adopted, even when only the octahedral form of crystal is present. The presence of a foreign body will determine precipitation which would not otherwise occur; and a calculus may increase in size when there would be no danger of a new formation.

Little need be said respecting cystine. It is exceedingly rare, and when met with may be kept in solution by alkalies. Little is known experimentally as to the prevention of its formation. Its composition suggests some relation with the bile, and may direct attention to the functions of the liver. The fact that cystine is commonly met with in scrofulous individuals may lead us to employ cod liver oil. When cystine calculi are formed, and are of small size, they may be treated by the administration of alkali, or by acid injections into the bladder. Cystine calculi are even more favorable for the alkaline solvent treatment than uric acid, not only from their greater solubility in alkaline urine, but from the fact that they are usually unmixed, being composed of pure cystine for the most part.

When the urine is phosphatic, the danger of the formation of a calculus will vary according to circumstances. If the condition be due to the precipitation of phosphate of lime by fixed alkali there is little fear of such a result. It is when the urea has undergone ammoniacal decomposition that a calculus is liable to be formed. If inflammation of the mucous membrane of the bladder has not preceded this change, it will certainly be set up by the irritant action of the ammonia, and the mucus poured out offers the condition required for the formation of stone by glueing together the precipitated mixed phosphates.

When the urine is retained in the bladder by some mechanical obstacle, as enlarged prostate or stricture, or by paralysis of the muscular coat, the danger is proportionably increased, as in paraplegia or in atony from general or local debility. When, on the other hand, there is spasm and frequent micturition, it is less. The one all-important point of treatment is to see that the bladder is completely emptied by introduction of the catheter twice a day, and to wash it out with warm water simply, or with water to which a small proportion of dilute nitric acid has been added.

Preventive measures are demanded with especial urgency when one or more calculi have been passed, or when there is reason to suspect the existence of a small calculus in any part of the urinary tract. It is rare that calculi are formed singly. The same cause is in operation in every part of each kidney, and if precipitation takes place at one point it will probably do so at others. A nucleus once formed, growth is a matter of certainty. Sometimes all the calculi will make their way out; sometimes all will be detained in the kidney, or one or more may be arrested at any part of the urinary apparatus. The results are various—suppurative inflammation of the kidney from direct irritation of one or more calculi; expansion of the pelvis and inflammation and ulceration of its mucous membrane, with atrophy of kidney or peri-renal abscess; suppression of urine by impaction of calculus in the ureter; renal colic, from difficult passage of stone along the ureter. These do not come within the scope of this work. The point here to be considered is the detection of stone while so small that there is a chance of expelling it, and the best means of effecting this object.

Except in the case of the passage of a small stone along the ureter, symptoms experienced by the patient give little aid. A calculus of some size lodged in the pelvis will often give rise to pain sufficiently characteristic to warrant a diagnosis; but for the most part careful watching of the urine from day to day, and under different circumstances, is required. This usually affords grounds for diagnosis of the existence and seat of a calculus.

However small a calculus may be, violent exercise, prolonged jolting in riding of driving, will cause it to bruise the part with which it is in contact, and give rise to bleeding; and, long before blood becomes evident to the naked eye, the corpuscles may be detected by the microscope. Blood-corpuscles appearing constantly after exercise, disappearing on rest, associated with some precipitate in the urine of a nature to form calculus, and with no evidence, local or general, of other disease to account for it, furnish a strong presumption of calculus. But before a calculus can rupture capillaries, it must abrade the mucous membrane and disturb the epithelium; and these cells, with the blood, not only furnish strong corroborative signs, but indicate the exact seat of the calculi. The cells from the calyces are irregularly spheroidal and small; those from the pelvis larger, but still irregularly spheroidal in form. The epithelium of the ureter is conical, that from the bladder between spheroidal and squamous in character, readily distinguished by the large size of its cells.

A patient in whom there is reason to suspect the existence of calculus should be made to take exercise, at first gentle, a simple walk or drive; and this giving no results, exertion of a more vigorous kind, horse-exercise and the like. The urine passed an hour or two afterwards should be allowed to stand for a short time, and should then be examined for blood-corpuscles and epithelial cells, and compared with urine passed before the exercise. A single experiment must not be relied on, the results being sometimes negative for a time, though afterwards conclusive. When blood-corpuscles are met with they usually diminish gradually in number, and sometimes disappear. In other cases they are never altogether absent, and the effect of exercise is to increase the number of epithelial cells without greatly affecting the blood-corpuscles.

When a calculus has been lodged for a length of time in the pelvis there is often little or no blood, and the epithelium is replaced by a constant, but not considerable, discharge of pus-corpuscles. These cases often require watching for some time; all the urine passed should be submitted to the medical attendant, and most commonly there will eventually be detected small scales, detached from the exterior of the calculus, which, by their crystalline appearance under the microscope and behavior with reagents, will afford some clue as to the character of the stone.

A stage in the early history of calculus requires special mention— that is, the passage of the stone along the ureter. This generally occasions some pain, but not always of a severe character. When, however, the stone is so large as to traverse the canal with difficulty, the suffering occasioned is most acute. There is great pain in the region of the kidney on one side, never altogether absent, but coming on in paroxysms. It shoots upwards and forwards, but more especially downwards, along the course of the ureter and down the groin to the testicle of that side. The testicle, also, is retracted. More or less local tenderness accompanies the pain, and usually there is violent sympathetic vomiting. The treatment will consist of diluents, baths, opium, or subcutaneous injections of morphia, or the inhalation of chloroform so as to produce partial anæsthesia. The occurrence of such an attack, more or less acute, will be an important fact in the history of a case in which stone is suspected to exist.

When positive evidence of the presence of a calculus is obtained, no time should be lost in endeavoring to bring it away. This brings us to the third point of preventive treatment. Fluids should be taken in very large quantities, either distilled water or rain-water, weak linseed or dandelion-tea; or, if the stone is uric acid, weak solutions of acetate or citrate of potash. The use of the more stimulant diuretics, such as juniper, turpentine, etc., is not unattended with danger, and should be avoided. While these diluents are being given, the urine should be retained as long as possible, so that the bladder, the ureters, and their renal expansions may be distended. At the same time warm baths should be taken, with the view of inducing a general relaxation of the muscular structures, and the dislodgment of the stone may be favored by movement, walking, riding, etc.

Should there be any reason to suspect constriction of any portion of the urethra, a full-sized instrument should be passed, and if any stricture is detected it would be best treated, if possible, by rapid dilatation.

When from the symptoms detailed, there is reason to believe that a concretion has traversed the ureter from the kidney to the bladder, the patient should be directed to watch closely all the urine excreted. After the calculus has reached the bladder, there may be a complete cessation of all the acute symptoms, until the stone has reached a certain size, when the symptoms characteristic of vesical calculus will show themselves. In other cases the distressing symptoms will be present from the first.

If, after the lapse of a few days, no calculus has been passed, no further time should be lost in bringing it away. This may be sometimes accomplished by washing out the bladder by means of an aspirator and a full-sized injecting catheter. If, however, the concretion is too large to pass through the tube, it should be broken up by means of the lithotrite. After this has been accomplished, the aspirator and evacuating tube may be employed to remove all fragments from the bladder.

CHAPTER XXIX.

ACUTE AND CHRONIC INFLAMMATION OF THE PROSTATE GLAND.

THE prostate gland is subject to inflammation which may be either acute or chronic; the affection in either case being generally associated with, and dependent upon, a similar condition of the urethra or bladder.

Gonorrhœal inflammation is the most frequent cause of acute prostatitis, another, but less common cause, is severe stricture of the urethra. In this latter case the obstacles presented to the free flow of urine produce inflammation of the parts posterior to the stricture, in a degree corresponding to the degree and extent of the constriction. The prostate may also become inflamed during the course of cystitis, and as the result of the presence of a calculus in the bladder and of the lodgment of calculi in the prostatic urethra. Other local causes, such as mechanical violence of various forms—*e.g.* rough catheterization, the use of strong injections, and the application of caustics—may give rise to inflammation of the prostate.

There are other more or less indirect causes of prostatitis. Some amount of inflammation of the gland may be set up in gouty and rheumatic subjects by exposure to cold and damp, as by sitting on moist ground. The influence of horse-exercise, to which an attack of prostatitis has sometimes been attributed, may be doubted, and the same may be said with regard to sedentary habits. Inordinate sexual intercourse, especially when urethritis exists, may be an occasional cause of the affection, and, under similar circumstances, too nutritious or highly-seasoned viands, and the excessive use of vinous and spirituous liquors, may be credited with setting up prostatic inflammation. Prostatitis has sometimes been attributed to the use of purgatives and of certain medicines, such as copaiba or cubebs, which have a specific action on the urethral mucous membrane. The evidence, however, of such a causation is at least doubtful. An overdose of cantharides may, however, produce cystitis, and in such a case the inflammation may easily extend to the prostate.

The incipient symptoms of acute prostatitis are a feeling of weight and uneasiness in the perineum, soon followed by an increased desire to expel the urine, while at the same time the patient is unable to satisfy the calls in a complete manner. These symptoms soon become more acute. The pain in the perineum increases and extends in various directions, shooting down the thighs, to the lumbar region, or often extending towards the anus or along the urethra. The pain is particularly increased by pressure on the perineum.

There is frequent micturition and scalding in making water, and the pain is increased as the bladder contracts to expel the last drops of urine. Evacuations from the bowels also cause great uneasiness, and there often remains a sensation as if the rectum was not completely emptied, giving rise to distressing tenesmus. If the finger be introduced into the rectum, the gut is felt to be hot, and the prostate seems a smooth, round, and hard body, projecting downward on the bowel, and exceedingly painful to the patient under pressure of the finger. All the symptoms are aggravated by movement.

The inflammatory action, if not checked, shortly extends to the neck of the bladder and not unfrequently to its whole inner lining—total retention of urine frequently succeeds; intense febrile symptoms supervene, accompanied with delirium; and in this stage the patient may die if the bladder be not relieved, and the inflammation not subdued.

If a catheter or sound be introduced into the bladder, it passes without any difficulty or causing uneasiness as far as the membranous part of the urethra, but its passage through the prostatic portion is attended with most acute pain, and gives rise to severe spasmodic contraction.

If the inflammation has continued for seven or eight days, and rigors then occur, with increase of febrile symptoms, quickened pulse, hot skin, furred tongue, etc., as well as with greater sense of fulness and tension in the perineum, more frequent calls for, and greater difficulty of, micturating, it is most likely that suppuration is taking place. Complete retention of urine is now likely to occur. If at this period we examine the prostate through the rectum, we shall no longer find it hard and resisting, but, on the contrary, it will be found to resemble the bladder, so that inexperienced persons would most likely be led into error; the examination per rectum, as well as the discharge of feces, causes great pain; and there is constant tenesmus, with a burning sensation in the part.

The fibrous investment of the prostate, softened by inflammation and distended by pus, bursts, and allows the escape of a creamy and sanguineous matter into the urethra, the rectum, or the bladder; then the tumor diminishes, the urethra becomes free, the urine flows in a large stream, and, as the bladder empties itself, the patient feels his sufferings decrease. If, as commonly happens, the abscesses make their way into the urethra, they may involve and obliterate the openings of the ejaculatory ducts. In very rare cases the abscess has burst into the peritoneal cavity at the posterior part of the prostate, and thus caused fatal inflammation. The occurrence is marked by acute pain in the pelvic region, small pulse, rigors, and rapid prostration of strength.

If the abscess opens into the urethra, a very copious purulent discharge instantly manifests itself from this passage; the urinary discharge is accompanied, but more frequently preceded or followed, by the evacuation of a large quantity of purulent matter. After the last drops of urine are voided there is a stinging pain, which lasts for a few minutes or for a much longer period, according to circumstances. There is in some cases more or less blood mixed with the discharge, and sometimes considerable hæmorrhage takes place. There is also pain in the glans penis, as if a burning coal were applied to it.

Authors relate cases in which a single abscess in the prostate contained half an ounce or more of pus—such a case is mentioned by Petit; but more commonly the abscesses are of smaller size, and so numerous that after death the prostate presents a sieve-like appearance from their number and minuteness. When the abscess is of long standing, the pus is contained in smooth, organized cysts, and is extremely fetid. In other cases the fluid is of a light straw-color, and of a cream-like consistence; with it is mixed some of the débris of the gland, and occasionally blood. The surrounding textures are softened, infiltrated by serum, and of reddish hue. They may be almost disorganized, and converted into a substance not unlike wet tow.

When, through active and appropriate treatment, acute inflammation of the prostate terminates in resolution, all the symptoms above indicated diminish; at the same time the liquid secreted by the follicles of the gland augments in quantity, and, mixing with the urine in the form of a whitish or grayish viscid matter, settles at the bottom of the vessel without adhering to its sides, having very much the appearance of imperfectly elaborated pus. This matter diminishes in its turn, and finally disappears as the affected parts return to their natural condition.

Chronic inflammation sometimes remains as the result of the acute form, or the process may be slow and gradual from the commencement, and may be due to exposure to cold or damp, mechanical injury, or venereal excesses. The chief symptoms are: irritability of bladder, evinced by frequency of micturition, the act being accompanied and followed by increase of pain, a sense of weight and uneasiness in the perineum and rectum, loss of power in propelling the urine, which remains cloudy and sometimes contains a little blood, and usually a gleety discharge from the urethra. All these symptoms are aggravated by exercise, and they closely simulate those of calculus.

Suppuration sometimes takes place in the connective tissue external to the prostate gland. The symptoms are the same as those of prostatic abscess, but are less defined, and the pus is more apt to make its way towards the ischio-rectal fossa. An abscess of this kind may co-exist with suppuration in the substance of the gland.

Acute prostatitis demands the employment of the most active antiphlogistic treatment.

If the attack is severe, blood may be taken by cupping from the loins or perineum. Leeches, moreover, may be applied above the pubes, particularly when pain is experienced there, or to the perineum. The French practitioners recommend the application of leeches on the rectal surface of the prostate by means of a speculum introduced into the anus.

After depletion, a warm hip bath, followed by hot fomentations to the hypogastrium and perineum, will aid in relieving the painful symptoms. The bowels should be kept open by means of saline purgatives. To subdue the pain and irritability of the bladder, morphia may be given internally or administered in the form of a suppository. At the same time, the acidity of the urine should be diminished by scruple doses of the citrate of potash every few hours. The patient must be kept in bed, and the diet must be strictly confined to articles of a bland, non-stimulating character. The hot hip-bath may be repeated once or twice daily. Should retention of urine supervene, and not yield to the hot bath or a dose of morphia, a gum-elastic catheter, warmed and well-oiled, should be passed into the bladder and the urine withdrawn. It better not to allow the catheter to remain in the bladder. When all active symptoms have subsided, and the patient is able to leave his room, he should be directed to abstain for some time from beer, coffee, and spirits, and to avoid, as much as possible, exposure to cold or damp, or any sudden change of temperature.

If an abscess forms, it is desirable that the matter should find its way neither into the rectum, nor behind the bladder, but should pass either into the urethra or to the surface, in order that it may be freely and readily discharged. Should it pass into the rectum, a troublesome fistula may result, and should suppuration invade the cellular tissue between the bladder and rectum, it may give rise to a fatal pyæmia or peritonitis. We must therefore not hesitate, when fluctuation can be felt in the swelling in the rectum or perineum, to open the abscess by puncture in the former position, and by free incision should the matter be finding its way towards the surface of the perineum.

Should the disease have anticipated the operator, and the abscess have opened in the perineum or into the rectum, nothing can be done beyond maintaining the general health. A fistulous opening may remain for some little time, but usually closes spontaneously. Should the abscess have burst into the urethra, or at the neck of the bladder, little is required in the way of treatment. Most commonly the matter gradually diminishes in quantity, and eventually ceases. In other instances, the discharge disappears for a time, and then shows itself afresh, and this may occur several times before the patient completely recovers.

In these cases, ulceration of the surface of that portion of the prostate which projects into the bladder sometimes occurs. In this state, as even in that of mere enlargement, the gland is liable to bleed. This symptom usually subsides spontaneously; while it lasts the patient should be kept entirely in the recumbent position.

When the inflammation assumes the chronic form, counter-irritation to the perineum is the most effectual method of treatment. This may be accomplished by applying the liquor epispasticus of the pharmacopœia to a small spot on each side of the raphé. Small doses of iodide of potassium may also be given, and the bowels should be kept open by means of saline purgatives, with the view of preventing congestion of the pelvic veins. The patient's general health should also be attended to. If, as is usually the case, there be indications of debility, tonics, such as quinine and iron, with full diet and change of air, will materially aid in expediting his recovery. All sources of excitement, and exposure to cold or damp, must, of course, be carefully avoided.

Abscesses in the neighborhood of the prostate require the same treatment as those which occur in the gland itself. As soon as there are any symptoms of matter forming in the perineum, a free incision should be made. It should be remembered that collections of pus, of this kind, may give rise to retention of urine, which will be relieved by free incisions.

CHAPTER XXX.

CHRONIC ENLARGEMENT OF THE PROSTATE GLAND.

THE most common affection of the prostate is chronic hypertrophy, which may involve the whole, of only part of the organ, producing changes in the neck of the bladder, and in the urethra, which interfere with the power of the patient to expel the urine. This change was formerly considered to ensue from chronic inflammation, the opinion being that frequently an attack of acute inflammation, partially subdued, and not totally extinguished by treatment, led to a state of engorgement, which either spontaneously, or through some accidental excitement, resumed that action known, in its abated form, as chronic inflammation.

However, if we carefully examine the enlarged gland, we fail to detect any of those changes commonly recognized as inflammatory, such as the effusion and organization of lymph, by which the component parts become glued together and rendered indistinct from one another. There is also, as a general rule, an absence of any history of an acute attack.

Although the change which has taken place in the substance of the gland resembles the mere effect of increased nutrition more than any other, yet the gland often presents appearances which cannot be referred to simple hypertrophy.

The prostate gland is composed of plain muscular tissue, fibrous tissue, and glandular substance proper, the muscular element being the most abundant of the three constituents. The hypertrophy may consist in an enlargement of the whole gland—that is, of both lateral lobes and of the median portion. This is the most common form of hypertrophy. Less frequently, the enlargement mainly affects the median portion of the gland, or, while all portions are enlarged, the right or left lobe is found to predominate. It very rarely happens that the lateral lobes alone are affected, the median portion remaining unaltered.[1]

In addition to the enlargement due to hypertrophy of its elements, the size of the prostate is often increased by the development of one or more circumscribed tumors, which are either imbedded in the interior of the substance of the gland, or project from it in various directions. These tumors are inclosed in capsules of connective tissue. On section they are found to be composed of muscular, fibrous, and glandular elements in varying proportions; their structure, therefore, closely resembles that of the gland itself. Rokitansky first pointed out that these tumors, though often so closely invested and isolated by connective tissue that they can be enucleated, are composed of elements precisely similar to those of the gland with which they are connected. This observation has been confirmed by Sir James Paget,[2] who compares the manner in which these tumors lie imbedded in the enlarged prostate to that in which mammary glandular tumors sometimes lie isolated in a generally enlarged breast. Billroth, however, states that the so-called hypertrophy of the

---

[1] Sir H. Thompson, Diseases of the Prostate, p. 80.
[2] Surgical Pathology, p. 509.

prostate is never connected with the formation of new gland-tissue, but only with dilatation of the acini and epithelial hyperplasia—" the so frequently observed enlargement of these glands depends essentially upon diffused or nodular formation of myoma."[1] Dr. Gross, also, is of this opinion.[2] The growths in question closely resemble the fibroid tumors of the uterus, and the resemblance is a point of considerable interest, inasmuch as the homological relations between the prostate (or, strictly speaking, the vesicula prostatica) and the uterus are extremely close. Sir James Paget points out that the relations of these new-formed portions of prostate gland are very intimate to tumors, and also to general hypertrophy of glands. These tumors are rarely found in the upper part of the prostate; their most common seat is the thicker part of the base of the gland at the neck of the bladder, behind the urethra, the orifice of which they seriously obstruct.

The changes in the form and size of the gland vary considerably in different cases. In some, it is only slightly enlarged, in others it may attain the size of a man's fist. Bartholin relates a case in which the gland is said to have equalled in size a man's head; but this statement is probably loose and exaggerated. Mr. Cadge,[3] however, has met with an instance in which the prostate weighed twenty ounces, and measured five inches in length, four inches in width, and three and a half in depth.

In a great many cases the gland is indurated as well as hypertrophied; and hence the error into which the older surgeons so frequently fell of believing that the prostate is often the seat of schirrhous disease. When the induration is partial, it depends upon the presence of fibrous tumors. On the other hand, although an increase of density may be considered as the ordinary condition of an hypertrophied prostate, the latter may be softer than usual when the ducts are dilated and filled with secretion, and the veins more developed than in the healthy state. The fibrous capsule of the gland is often thickened, but in many cases it has been found reduced to a mere film from excessive development of the subjacent parts.

Enlargement of the prostate influences very materially the length, diameter, and direction of the canal of the urethra. The increase is chiefly in the long axis of the organ, so that the lateral lobes become oval and elongated, and larger in the middle than at the extremities. They may project inwards to such a degree as to touch one another, but more commonly there is a median groove or gutter left between them. The normal length of the prostatic urethra is about an inch and a quarter, but, in cases of hypertrophy, this measurement is sometimes more than doubled. When the enlargement affects the glandular substance in its entire circumference, the antero-posterior diameter of the urethra becomes increased, and the lateral or transverse diameter greatly diminished, so that the canal is reduced to a mere longitudinal slit; if one lobe is more enlarged than the other, the urethra deviates to the opposite side. This lateral bend or obliquity in the passage will increase the difficulty in passing the urine and introducing the catheter. In some cases the prostatic portion of the urethra has been found dilated into a kind of sinus, capable of containing some ounces of urine.

Sir Everard Home observed that the left lobe of the gland increases

---

[1] Billroth, Surgical Pathology and Therapeutics, New Syd. Soc. Translation, vol. ii., p. 437.
[2] Gross, Diseases of the Urinary Organs, p. 398.
[3] Trans. Path. Soc., vol. xviii., p. 182.

more rapidly and to a greater extent than the right one; and hence that the urethra will commonly be found to deviate to the right side. Subsequent experience, however, enabled him to modify this opinion. Mr. Coulson noticed that the right lobe was usually more affected than the left, and it is probable that the greater prevalence of the enlargement on the left side, noticed by Sir E. Home, was accidental.

When the enlargement is irregular, and confined to certain portions which project above the surface of the gland, the latter will present a rough, knobbed appearance. These irregular masses may be developed towards the bladder or press backwards on the rectum; and when the base by which they are attached to the body of the gland is at all narrow they have often been mistaken for fibrous tumors.

Enlargement, however, of the middle lobe is most important from the impediment which it offers to the discharge of urine. Commencing in the form of a nipple, under the mucous membrane, which is pushed before it so as to be put on the stretch, it encroaches upon the vesical cavity in the direction backwards from the verumontanum. In its further increase, observes Sir E. Home, it loses the nipple-like appearance, becoming broader from side to side, and forms a traverse fold by pushing forward the membrane connecting it to the lateral lobes, which also become proportionately extended.

It presents, says Rokitansky, the appearance of a rounded tumor, of the size of a bean or hazel nut, which projects into the neck of the bladder; it may increase to the size of a walnut, hen's or duck's egg, or more; and then protrude into the cavity of the bladder in the shape of a smooth or rough, nodulated, slightly lobular, rounded or cordiform, pyramidal, or cylindrical tumor.

Enlargement of the middle lobe may produce an apparent splitting of the urethra at the neck of the bladder, or the neck of the organ may be compressed and narrowed. When carried to any great degree it gives rise to many important changes in the condition of the urethra and bladder. It elongates the prostatic portion of the urethra, and sometimes in such a manner that the surgeon has been unable to pass an ordinary sound through the neck of the bladder. It throws up this latter part towards the pubes, producing a considerable curvature of the urethra, with the convexity of the curve towards the rectum. It increases the depth of the perineum, and presents an obstacle to the performance of lithotomy; it deepens the floor of the bladder, so that a pouch is formed, in which a calculus may lodge.

But the great inconvenience arising from hypertrophy of the middle lobe is the impediment which it offers to the due evacuation of the bladder.

As the tumor and the transverse fold already noticed are situated immediately behind the internal orifice of the urethra, they are pushed before the urine in every attempt that is made to void it, acting like a valve and closing the opening till the cavity of the bladder is very much distended; then the anterior part of the bladder being pushed forward, and the tumor drawn back, in consequence of the posterior part of the bladder being put on the stretch, the valve is opened, so that a certain quantity of water is allowed to escape, but the bladder is not completely emptied. A very small enlargement, however, of the middle lobe, projecting into the bladder, with the transverse membranous fold connecting it to the lateral lobes, is sufficient to produce a complete retention of urine.

The prostatic ducts may be so enlarged as to contain a considerable quantity of fluid, and to present upon post-mortem examination several

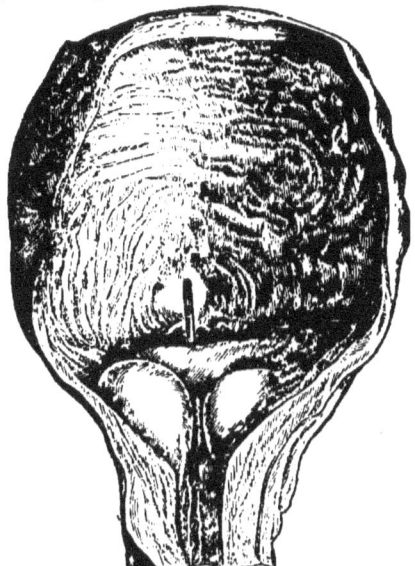

Fig. 21.—Enlargement of the lateral lobes with incipient increase in size of middle lobe—a bar or ridge is formed at the neck of the bladder, through which a false passage has been made.

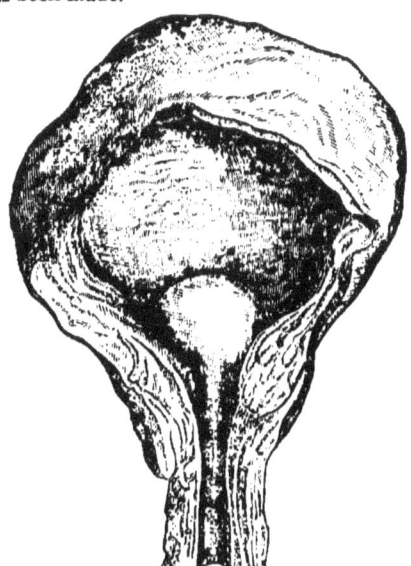

Fig. 22.—Enlargement of middle lobe.

fluctuating prominences. Mr. Coulson recorded a case of an old man who died from retention of urine, in which, upon examining the bladder and its appendages, he found an enlargement of both lateral lobes of the prostate, and several cysts formed of dilated ducts which projected from either side of the generally hardened mass. There was no trace whatever of inflammatory disturbance; the tissues surrounding these structures were perfectly healthy, and separated readily, when dissected, from the neighboring parts to which they were connected by loose cellular tissue. One of the cysts contained as much as half an ounce of fluid.

The morbid conditions of the bladder connected with hypertrophy of the prostate gland depend chiefly on the impediment offered to the free discharge of urine. In the early stages of the disease the accompanying derangement of the bladder appears to be merely functional, and to consist in an increase of irritability; and this is the principal cause of the great and frequent desire, and consequent straining, to make water, when there is only a very small quantity in the bladder. The irritation extends to the muscular coat, so that the bladder does not contain more than half the usual quantity of urine before strong efforts are made to empty it.

As the disease advances, and with it the difficulty of voiding the urine, sometimes amounting to complete retention, inflammation of the mucous coat of the bladder is apt to set in, and extend from the vicinity of the tumor to the other parts of the cavity. In severe cases this inflammation may terminate in ulceration and destruction of the mucous tissue, or even in perforation of the bladder from sloughing.

Hypertrophy of the muscular tissue and a sacculated condition of the bladder are, as is well-known, the frequent effects of any great impediment to the evacuation of the urine.

In cases of chronic enlargement of the prostate, the longitudinal fibres of the bladder are often found to be greatly hypertrophied, and of a red color, passing in bundles over the thickened inner coat, but the cellular tissue connecting them is usually loose and free, unless the part has previously been the seat of inflammation.

The sacculated or encysted state of the bladder has been fully described in another part of this volume. It results from the violent efforts to expel the urine, whereby the mucous coat is protruded through the hypertrophied fibres in the form of a pouch. This is an unfavorable complication, for the lining membrane of these sacs is very liable to take on suppurative inflammation, either from the sojourn of urine within their cavities, or, as Mr. Adams has ingeniously observed, from the mere circumstance of displacement. Dilatation or inflammation of the ureters, and secondary disease of the kidneys, frequently supervene during the last stages of chronic enlargement of the prostate. The effect of retention of urine as a mechanical obstacle is sufficiently obvious; in other cases the kidneys appear to suffer sympathetically. The capsule becomes firmly connected with the cortical substance by chronic inflammation; the effusion of lymph blocks up and compresses the minute vessels in the secreting substance, and the whole organ eventually becomes smaller and firmer than natural. The flow of urine, which during the period of vascular excitement had been increased, is now much diminished, and albumen with blood-discs and other abnormal elements are found mixed with the proper renal secretion.

Vesical calculus is another frequent result and complication of enlargement of the prostate. Owing to the imperfect evacuation of the bladder, the urine is very apt to become alkaline and to deposit phosphates, and, in addition to this source of calculus, should a uric acid concretion descend from the kidney, the prostatic enlargement will in all probability prevent its escape by the urethra.

The irritation reflected from the hypertrophied prostate is apt to produce vascular excitement in the vesiculæ seminales, the walls of which become thickened and blended with the surrounding parts. Hence patients are sometimes troubled with nocturnal emissions, from which even the aged are not always free.

The testicles seldom suffer in these affections, although hydrocele has occasionally been noticed in patients the subjects of prostatic disease.

A variety of causes have been assigned for hypertrophy of the prostate—excessive venery, strictures of the urethra, riding on horseback, gonorrhœa, irritation from vesical calculi, etc. Gonorrhœa is a disease common among the young, but uncommon in those of advanced years, in whom alone enlargement of the prostate is met with. The chronic prostatis, which is an occasional result of gonorrhœal inflammation, is in no way connected with the affection under consideration. Strictures of the urethra are usually classed among the exciting causes of this affection. Sir Everard Home entertained this opinion, but, in his account of the cases which came under his observation, he has not confirmed his statement. Civiale, on the other hand, declares that the prostate is usually healthy in persons affected with permanent stricture; and Mercier thinks that, so far from being a cause of hypertrophy of the organ, stricture of the urethra is calculated to produce atrophy.

In the opinion of some surgeons, hypertrophy of the prostate gland is connected with the declension of the generative powers which takes place after a certain period of life. Facts, however, do not support the theory, that in a state of perfect health the prostate gland becomes enlarged with advancing years. On the contrary, it would seem that if no morbid change is going on, the prostate, like other parts of the body, decreases in size. That the affection occurs chiefly amongst persons beyond fifty years of age is undoubted; but the cause of this tendency has not yet been clearly ascertained.

Sir Henry Thompson's[1] researches show that prostatic hypertrophy is by no means a necessary concomitant of old age, and that the occurrence is not normal but exceptional. The conclusions at which he arrives, after dissecting 164 prostates from individuals of 60 years of age and upwards, are, that while hypertrophy exists in about 34 per cent., it produces marked symptoms in about 15 per cent. of men at and above 60 years of age. He has never seen or heard of a true example of the affection before the age of 54 years.

The symptoms of incipient enlargement of the prostate gland are vague, and easily passed over by the patient. At first, he experiences some uneasiness and occasional pain, though of a slight kind, in the region of the bladder. The uneasiness is dull, and the pain, when aggravated by any excess, may extend along the urethra and down the thighs, or to the back; but it is seldom so acute as to excite much attention. Up to this point the irritation may be considered as confined to the gland itself. As this enlarges, the irritation quickly extends to the bladder, the neck of which begins to resent the encroachment made on it by the gradually augmenting prostate. The patient is now sensible that his calls to make water are more frequent, and he perceives that some time elapses before the urine appears, and that the stream is slower

---

[1] Diseases of the Prostate, p. 138.

and weaker than natural. After some time the difficulty of expelling the urine becomes evident; its flow is irregular, and the last drops, especially, are not evacuated without considerable effort, and their passage is often accompanied by a painful sensation resembling scalding. During the early stage, this difficulty of evacuating the bladder varies much, even in the course of twenty-four hours—at one time the patient passes his water freely; in an hour or two he experiences great difficulty in making water, and endeavors to promote the discharge of the last drops by pressure along the urethra. He is now tormented by the idea that he labors under stricture, or he may attribute his sufferings to internal piles. The urine during this stage is clear, or slightly charged with mucus.

The greater part of the symptoms now mentioned may continue without increase for some time; but, as soon as the gland has become enlarged to a certain size, symptoms evidently dependent on a mechanical obstruction at the neck of the bladder set in. As the gland increases in volume, a portion juts out into the bladder, and some of the urine is retained; the bladder is unable to expel the whole of its contents; when the patient has voided all the urine he possibly can, several ounces may still be left behind. The desire to micturate now increases in frequency. The moment after urine has been passed the patient has an inclination to make more; the pain is more intense, and extends along the penis to the glans; the prostate is the seat of constant soreness, which may be throbbing or pulsative, hot to scalding, or even neuralgic in character. The urine becomes excessively fetid and alkaline, and emits a strong ammoniacal odor; it is also occasionally mixed with pus or blood. It is not difficult to recognize a patient suffering from hypertrophy of the prostate when striving to pass his water; he bends the body forwards; his legs are straddling or wide apart; he strains with all his might, and uses every muscle to overcome the resistance to the flow of urine. During these efforts the fæces sometimes escape involuntarily, and prolapsus of the rectum may occur. The stream of urine is small; a quantity escapes after the patient supposes that the bladder is emptied, and the clothes become soaked, wet, and offensive. The general health soon begins to fail; the tongue is loaded, there is little appetite, the feet are cold, the skin dry, and the bowels irregular. His sleep is broken, for he suffers more at night; he may be obliged to rise every quarter or half-hour to relieve his bladder. Exposure to atmospheric influences aggravates all these symptoms; he becomes at last haggard, thin, weak, almost worn out from the irritation, pain, and watchfulness.

The symptoms, therefore, of incipient enlargement of the prostate are irritability of the bladder, combined with difficulty in expelling the urine, which flows in a slower stream than natural. But when the prostate is once considerably enlarged, the patient is always subject to sudden aggravation of his symptoms, amounting to complete retention upon exposure to cold or indulgence in any excess.

These symptoms are all increased by much standing or walking: rheumatic pains are felt down the thigh and leg, and likewise in the loins; occasionally there is hæmorrhage from the urethra; constipation, dyspepsia, headache, and frequently some scaly eruption on the skin exist.

The hæmorrhage may be due to ulceration, or to rupture of the engorged veins of the prostatic plexus. It is sometimes profuse, and is very likely to occur when the passage of instruments has been attended with difficulty.

Owing to the prostate being intimately connected with the neck of the bladder, every increase in the size of the former must affect the latter, causing some alteration in the exercise of its functions. The loss of power to empty the bladder is apparently the first effect thus produced, and its consequences are very serious. Sometimes these symptoms occur suddenly. The patient awakes with a strong desire to void urine; he finds that he can scarcely pass any, and, on inquiry, it is discovered that exposure to cold or wet, or irregularity in diet or exercise, had previously occurred.

As the complaint progresses, the imperfect evacuation of the bladder causes the organ to become more and more distended; some of the surplus urine dribbles away, or is passed by the efforts of the patient, but a gradually increasing quantity always remains behind. Under these circumstances, the bladder forms a prominent tumor in the hypogastric region, where it can be detected on percussion. The changes which the urine undergoes are due to the chronic cystitis, and the conversion of the urea into carbonate of ammonia. The precipitation of the earthy phosphates is a necessary consequence of the ammoniacal condition, and the growth of calculus is favored by the presence of the deposit and the abundant secretion of viscid, tenacious mucus, which serves to bind the particles together. The mischief does not remain confined to the bladder; pyelitis may occur from extension of the inflammation, and the backward pressure of the urine tends to produce dilatation and sacculation of the kidneys.

As the symptoms of hypertrophy of the prostate chiefly depend on mechanical obstruction to the discharge of urine, it becomes necessary to distinguish them from those of other diseases in which obstruction may exist. These are stricture, urinary calculi, and tumors about the neck of the bladder.

The diagnostic signs of tumors have been described in another chapter. The obstacle arising from enlargement of the prostate has been frequently confounded with the impediment produced by organic stricture, and the numerous preparations contained in our museums show the unfortunate result of such a mistake.

It should be remembered that stricture generally occurs at an earlier period of life than hypertrophy of the prostate. The seat of the obstacle is often much nearer the external orifice of the urethra. When it occurs further back, the impediment to the discharge of urine being nearly the same in both cases, may lead to some obscurity, but a careful examination with the sound and through the rectum, will enable us to detect enlargement of the prostate in one case, and absence of it in the other.

Careful observation of the manner in which the urine flows will also aid the diagnosis. In prostatic hypertrophy, the stream is much diminished in force, but not materially altered in volume, and the flow is scarcely, if at all increased by the efforts of the patient. In cases of stricture, on the other hand, the stream is always smaller than natural, but the patient can, when he chooses, propel it with increased force.

Hypertrophy of the prostate produces symptoms which may by possibility be mistaken for those of stone in the bladder. In the latter affection, however, the flow of urine is suddenly stopped at any period by the calculus falling across the orifice of the urethra; while in the former, the stream of urine having once commenced to flow, continues until the evacuating effort ceases. Again, the pain and uneasiness about the neck of the bladder, in cases of enlarged prostate, are not aggravated when the urine has been drawn off by the catheter, whereas in stone, these symptoms are usually more severe.

Observation of the effect of exercise upon the symptoms also aids in distinguishing the two complaints; in cases of calculus all the symptoms are increased, and hæmaturia not unfrequently occurs as a result of exercise. Another distinguishing characteristic is the fact that, in cases of prostatic hypertrophy, the calls to pass urine are especially frequent during the night, whereas irritability of the bladder, due to calculus, is always less marked when the patient is at rest. When the two affections co-exist, the subjective symptoms of calculus are less severe than usual, owing to the fact that the concretion remains in the depressed and less sensitive portion of the bladder, behind the enlarged prostate, and does not occlude the internal meatus. In a patient with enlarged prostate, the occurrence of hæmaturia after exercise strongly suggests that a calculus complicates his other troubles.

It has been stated by some writers that the fæces are often flattened when the prostate is sufficiently enlarged to press upon the rectum; the same appearance, however, may be observed when the canal of the gut is from any cause contracted. The shape which the fæces assume is also materially modified by the condition of the sphincter ani. But no satisfactory opinion or correct diagnosis can be formed without careful examination of the prostate *per anum;* and the enlarged gland here lies so completely within the reach of the finger, that its size, form, and density can be sufficiently estimated in the greater number of cases. For this purpose, the patient may be placed on his back with the thighs bent and widely separated, or he may support himself upon his knees and elbows. The bowel having been previously emptied by an enema, the surgeon introduces the index finger, well oiled, and runs it upwards along the mesian line, and then directs it from side to side. A space measuring about three inches longitudinally may thus be accurately examined, and enlargement of the prostate always detected. It is also requisite to employ the long catheter, that deviations in the course of the urethra or changes about the neck of the bladder may be detected.

It is at all times a difficult task to attempt the removal of changes which occur in advanced life, and it must be confessed that we posses no certain means of diminishing the size of the hypertrophied prostate. The morbid change, once established, goes on until it interferes, more or less, with the expulsion of the urine, when surgical interference becomes indispensably necessary. Still, these ultimate stages may be retarded, and urgent symptoms may be so far prevented, that the patient may pass the remainder of his days in comparative comfort under suitable treatment. He can, by degrees, learn what habits to pursue, what diet to adopt, and may acquire practical experience in the selection of suitable medicines to act upon the alimentary canal. He should repose as much as convenient in the horizontal posture, avoid taking exercise upon horseback; he should prefer carriage conveyance, and his walks upon foot should not be of a length to excite fatigue. The food should be simple, plain, and nutritious. Wine, brandy, and other stimulants should be avoided. The bowels should be relieved of all accumulation, not by drastic purgatives, such as aloes or colocynth, but by the confection of senna, castor oil, bitartrate of potash, or sulphur. An occasional dose of blue pill, or of calomel, may be administered with advantage. Enemata are, in most cases, especially useful. There seems to be no good reason for disturbing the whole length of the alimentary canal by purgative medicines administered by the mouth, when the end to be desired can be obtained by injecting a pint of warm water or thin gruel, with salt, into the rectum.

A variety of remedies have been employed with the view of diminishing the hypertrophy, or of preventing further enlargement. Local depletion, counter-irritation, iodine, mercury, hemlock, and hydrochlorate of ammonia, have all been tried, but without any good results. It is true that cures have from time to time been reported, but with regard to these cases, it seems certain that the enlargement was due to chronic inflammation or congestion. Iodine was very warmly recommended some years ago by Mr. Stafford. He administered it internally and in the form of suppositories, and also applied a weak ointment of iodide of potassium to the urethral surface of the prostate. He asserted that this treatment proved very successful, not only in some cases where the enlargement was of inflammatory origin, but also in others of well-marked senile hypertrophy. In the hands of several other surgeons, however, Mr. Stafford's method of treatment has not been attended with success. Recently, Professor Heine,[1] of Innsbruck, has treated several cases of hypertrophied prostate by injecting solutions of iodine and iodide of potassium into the gland itself. Considerable relief to all symptoms is reported to have followed in four cases. Two patients died; in one of these an ulcer was found between the rectum and prostate, in the other, the fatal result was in no way due to the operation.

In the treatment of hypertrophy of the prostate, the main indication to be fulfilled is the prevention of accumulation of urine. An examination of the bladder will show that more or less urine always remains behind, notwithstanding the patient's repeated and violent efforts at micturition, and this residual urine, by its decomposition, leads to cystitis, whilst its constantly increasing quantity dilates the bladder and gradually abolishes its expulsive power. The bladder must, therefore, be emptied by means of the catheter, which should be introduced at stated intervals, or as often as the exigencies of the case require. A full-sized gum-elastic catheter, used without the stylet, or a French catheter " à boule," is the best form of instrument for the majority of such cases, and the patient can generally be taught to pass it for himself. When a silver catheter is necessary, one with a long shaft and large curve should be selected, in order that it may pass the more readily over the obstruction at the neck of the bladder. An over-curved gum-elastic instrument, prepared as described at page 22, will also be found useful. A straight, flexible catheter, with a short, abrupt curve, "cathétère à coudé," is another very convenient form of instrument.

A winding passage is sometimes formed through the prostate by alterations in the shape of its canal, the channel being thrown to one side, frequently to the right—a circumstance to be recollected on passing the catheter when obstruction is felt at the neck of the bladder. It obviously is one of the cases in which an elastic catheter will sometimes pass easily without a stylet, though with difficulty, or not at all, when the stylet is retained.

For difficult cases, various forms of instruments and methods of using them have been suggested. The late Mr. Hey, of Leeds, pointed out a plan which has occasionally succeeded. The elastic catheter, mounted on its stylet, is to be passed as far as it will go without difficulty; then

---

[1] Langenbeck's Archiv, bd. xv., s. 88, and bd. xvi., s. 79.

the stylet is to be partially withdrawn, a manœuvre which tilts up the end of the catheter, and may enable it to pass over the enlarged portion of the gland. The vulcanized india-rubber catheter has the advantage of softness and pliancy, and, if allowed to remain in the bladder, it causes less irritation than other instruments. When the parts are very irritable, Dr. Gideon Mantell has suggested that the catheter should be smeared with an ointment containing five grains of acetate of morphia to the ounce; he has related an interesting case of a large and inflamed prostate in which this application was of great service.

The question may arise as to whether frequent introductions of the catheter are more desirable than allowing it to remain, and, as a general rule, it may be stated that the best plan is to withdraw the catheter after each emptying of the bladder. If, however, retention has occurred and much difficulty has been experienced in introducing the catheter, the latter may be allowed to remain, and, under such circumstances, an elastic instrument or the vulcanized india-rubber one should be selected. The catheter may be kept in its place by fastening a piece of twine to its neck, and attaching the ends of the thread to the penis by means of adhesive plaster. By removal of the plug, the urine can be allowed to flow through the catheter at certain intervals.

When the catheter is introduced for the relief of retention the patient should invariably be in the recumbent position. The bladder may contain several pints of urine, and the rapid withdrawal of so large a quantity when the patient is erect may give rise to fatal syncope. Sir Henry Thompson[1] refers to a case of the kind in which six pints of urine were withdrawn from the bladder of an old man, standing erect against a wall; fatal syncope took place immediately the urine ceased to flow. Another precaution to be adopted is not to withdraw all the urine on first commencing the use of the catheter for patients with enlarged prostate. In these cases the bladder has been for some time accustomed always to contain more or less urine, and if completely emptied by artificial means, a condition of excessive irritability is very apt to be induced. A few ounces of urine should therefore always be left in the bladder, and this quantity may be reduced by degrees until finally all the urine is withdrawn. In this way the bladder becomes gradually accustomed to the state of contractility.

Hæmorrhage from the prostate, whether arising from ulceration, from the accidental rupture of vessels, or from injury by the catheter, may be treated by opium and gallic acid, or the acetate of lead with opium, in doses of from one to two grains of the former to one of the latter, every six hours, or by the infusion of roses with sulphuric acid and alum. In plethoric patients, or those subject to active hæmorrhage, such remedies should be preceded by local bleeding, injections of cold or iced water, etc. The catheter should, if possible, be avoided during hæmorrhage; or, if irritation and useless efforts to void urine render it necessary, it should be allowed to remain in order to keep the parts at rest. The bladder, if filled by coagula, should be left to empty itself unless absolute retention supervenes.

Should chronic cystitis, a very frequent complication, also exist, it must be treated in the manner described in Chapter VII. Washing out the bladder, so as to remove decomposing urine, is particularly requisite.

Hitherto we have been going on the supposition that, though there

---

[1] Diseases of the Prostate, p. 242.

be an obstruction to the evacuation of the urine, it is to be surmounted by appropriate remedies and the employment of the catheter. But it happens at times, in this as in other affections of the bladder and urethra, that all our endeavors to relieve the distended bladder fail; and, in order to save the patient's life, our sole remaining resource is that of puncturing this viscus, and thus drawing off the urine. The spots usually selected for this operation are either above the pubes or through the rectum.

A description of the manner of performing these operations will be found in Chapter XVI. If the enlargement of the prostate be so considerable that the bladder cannot be distinctly felt by the finger introduced into the rectum, the puncture above the pubes is the operation which must be selected. When, however, fluctuation can be detected by the finger, the puncture by the rectum is an operation unattended with difficulty and devoid of danger.

For the relief of patients suffering from advanced prostatic enlargement, and who require the almost hourly use of the catheter, it has been proposed to establish a permanent outlet for the urine above the symphysis pubis. A hollow sound, having a strongly marked curve, and closed by a bulbous-ended stylet, is introduced into the bladder, and so directed that its end projects just above the symphysis. A small incision is then made above the pubes, through the skin and tissues, till the bladder is reached. An opening, just large enough to expose the end of the sound, is then made in the bladder. The stylet having been withdrawn, a piece of elastic tubing, which passes through and is attached to a small silver plate, is introduced into the open end of the sound, which is caused to project at the bottom of the wound. The sound is then withdrawn, and the piece of tubing left in the bladder. The silver plate is kept in position by means of tapes and plaster. Sir H. Thompson,[1] who has suggested this plan, has performed the operation five times for the relief of patients suffering from advanced prostatic disease and cancer of the bladder. The operation has for its object the mitigation of suffering in hopeless cases, and for these it would seem well worthy of trial.

With regard to the consequences of retention of urine from chronic enlargement of the prostate gland, it would appear that extravasation of urine from rupture of the bladder is a very rare termination. Mr. S. Cooper[2] states that he never witnessed such a result. Sir Astley Cooper met with but one instance of this accident; and the reason of its rarity is given by Desault, who remarked, long ago, that in every case of retention, where the urethra is free from obstruction, the urine, after distending the bladder to a certain point, generally dribbles away through that canal, and the patient may live a good while in this condition. But when the retention depends upon any stoppage in the urethra, the urine then does not partially escape; the distention increases, and, if relief be not speedily afforded, the urethra gives way behind the stricture of obstruction.

---

[1] Diseases of the Urinary Organs, p. 284.
[2] Surgical Dictionary, p. 110.

CHAPTER XXXI.

MALIGNANT DISEASE OF THE PROSTATE GLAND—TUBERCLE OF THE PROSTATE.

THE prostate is rarely the seat of malignant disease, which, however, when it does occur in the gland, is usually of the encephaloid variety; very few cases of schirrhus have been met with. Cancer of the prostate is usually of primary origin, but, once established, the disease spreads to neighboring organs, as the rectum and bladder. Secondary cancer is rare; the disease originated in the prostate in 39 out of 45 cases collected by M. Maurice Joly.[1] With regard to age, the patients are either children or persons forty years old and upwards. Dr. Picard states that among children cancer invades the prostate more frequently than any other organ, the eye excepted. There is no evidence that any previous affection of the prostate exerts any influence on the development of malignant disease, but a deposit of encephaloid has been known to take place in a prostate gland previously the seat of hypertrophy. Sir Henry Thompson has reported a case of this kind.[2]

The rarity of schirrhus of the prostate is shown by the fact that very few cases have been reported, and that, with regard to the majority of these, the accounts given point rather to encephaloid than to the other form of cancer. Sir H. Thompson goes so far as to state that there are only two unequivocal cases of schirrhus of the prostate on record, one reported by Mr. Adams and the other by himself.[3] Two other cases recorded by Messrs. Travers[4] and Howship[5] respectively, Sir H. Thompson considers to be examples of encephaloid, while with regard to another, which occurred to Mr. Cock,[6] the evidence of the schirrhous nature of the disease is at least very doubtful. In Mr. Adams's case the gland was only twice its usual size; the left lobe was occupied by a carcinomatous tubercle; the right was healthy. The adjacent glands were enlarged, and similarly diseased, but no other abdominal viscera were affected. An experienced microscopist pronounced the tumor to be "true schirrhus in every particular."

On the other hand, undoubted cases of encephaloid cancer of the prostate have been described by Mr. Langstaff,[7] Mr. Howship,[8] Mr. Stafford,[9] Mr. Solly,[10] Sir H. Thompson,[11] M. Mercier,[12] M. Civiale,[13] Sir. W. Ferguson,[14] Mr. Moore,[15] Mr. Haynes Walton,[16] and Mr. Simon.[17] Many

---

[1] Du cancer de la Prostate. Arch. gén. de Méd., Mai, 1869. Quoted by Dr. Henri Picard, Traité des Maladies de la Prostate, p. 188.
[2] Diseases of the Prostate, p. 269.
[3] Anatomy and Diseases of the Prostate Gland, 2d edit., p. 149.
[4] Med.-Chir. Trans., vol. xvii., p. 346.   [5] Ibidem, vol. xix., p. 35.
[6] Quoted in Adams on Diseases of the Prostate Gland, p. 147.
[7] Med.-Chir. Trans., vol. viii., p. 279.   [8] Ibidem, vol. xix.
[9] Ibidem, vol. xxii., p. 218.   [10] Trans. Path Soc., 1860–51, p. 130.
[11] Ibidem, vol. v., p. 204.   [12] Mal. des Voies urinaires, p. 169.
[13] Traité Pratique, vol. ii., p. 343.
[14] Lancet. vol. i., 1853, p. 473.   [15] Med. Times, 1851, vol. ii., p. 174.
[16] Path. Soc. Trans., vol. ii., p. 287.   [17] Lancet, 1850, vol. i., p. 291.

of the cases occurred in children, in whom the disease runs a very rapid course, proving fatal in from three to nine months. In adult and advanced age the duration varies from nine months to several years. The disease is often accompanied by similar growths in other parts of the body.

The size of the tumor varies—in several of the cases it was reported to be of the size of an orange: in Mr. Haynes Walton's case the tumor entirely filled the true pelvis; in M. Mercier's the tumor was as large as an ostrich's egg. In one of Mr. Simon's cases the disease affected only the mucous surface of the prostate, and no tumor could be felt in the rectum. Among children, cases are described in which the morbid growth reached the size of a hen's egg.

The morbid growth is seldom contained within the proper capsule of the organ. Sometimes, however, the tumor is thus circumscribed, and forms a tolerably rounded mass, occupying more or less space in the pelvis, and encroaching upon the bladder and rectum. When the capsule ceases to confine the tumor, the latter assumes an irregular shape, and projects in the form of sprouting fungous masses in various directions. In this way the cavity of the bladder may become much encroached upon, while the expulsion of its contents is rendered difficult, owing to the obstruction caused by the tumor. When the urethra becomes involved, a large cavity outside the bladder is sometimes formed, into which the catheter may penetrate. If the disease has continued for any length of time, the lymphatic glands in the neighborhood are always implicated. The tumor is for the most part soft and spongy in texture; its color varies from pale red to dark brown. Some portions are deeply pigmented or melanotic in hue, owing to copious extravasation of blood. Other portions contain collections of sanious matter. Some of the original elements of the gland may be distinguishable in the tumor; it would appear that the gland-follicles are first involved in the disease.

The symptoms of cancer of the prostate are often obscure, and they are likely, for some time at least, to be confounded with those of simple tumor or ordinary enlargement. Their course, however, is ordinarily much more rapid than that of the latter affection.

The tumor, according to its size, position, etc., will create more or less impediment to the discharge of urine, and will cause irritability of the bladder; but these symptoms may arise from a variety of causes. Some difference in the symptoms will also occur, according as the tumor projects or not into the cavity of the bladder. According to Mr. Adams the irritability of the bladder is greater than under ordinary hypertrophy, and the evacuation of the organ is less impeded. The pain after micturition is, however, much more severe, and hæmorrhage from the urethra frequently takes place.

As the disease advances, the pain in the perineum, rectum, and about the neck of the bladder becomes more severe, and symptoms of constitutional cachexia rapidly manifest themselves. The hæmorrhage is often profuse, the blood being usually voided pure and unmixed with urine. If, however, a fungous growth projects into the bladder, hæmorrhage may arise from this source, and in this case the blood will be more or less mixed with the urine.

The local symptoms are not always more characteristic than the general. On examining the prostate through the rectum it is easy to detect enlargement, or perhaps discover the presence of a tumor, but the nature of the enlargement has still to be made out.

In a case of schirrhous prostate, the excessive hardness and nodulated feel of the tumor would assist us, when joined to other signs, in forming a diagnosis, yet it must be remembered that the gland often contains calculous concretions. Doubt must always exist, unless in addition to the local signs there are other symptoms which indicate the existence of a cancerous affection.

Thus, malignant tumors may be developed in the lymphatic glands of the groin, or in other parts of the body, or the patient may have the appearance of laboring under malignant disease. Under such circumstances, if a tumor be discovered in the prostate, giving rise to pain, hæmorrhage, and a highly irritable state of the bladder, its cancerous nature may be reasonably predicted.

The urine should of course be examined in all cases; cells of a suspicious character may, perhaps, be detected in it; but we cannot expect this appearance to occur so long as the tumor is entirely confined to the gland. Moreover, as there is no specific "cancer cell," and the epithelial cells of the urinary passages exhibit great varieties of form and size, much reliance cannot be placed in this means of diagnosis.

The following cases illustrate some of the features presented by malignant disease of the prostate gland.

Mr. Stafford has related a case in a child only 5 years of age. The gland was equal in size to the largest walnut, its form was somewhat globular, as is often the case with the enlarged prostate in advanced life. There advanced from the prostate into the bladder, immediately behind the orifice of the urethra, a rounded nipple-like projection, nearly equal in size to a small hazel nut, and exactly resembling, both in its appearance and situation, the projection of the gland usually ascribed to an enlargement of its third lobe. On making an incision through one lateral lobe of the prostate, the cut surface exhibited none of the natural texture of the gland: it was decidedly encephaloid in color, consistence and texture, and one part of the cut surface exhibited so dark a color as to present the appearance of there being melanotic deposit mixed with the encephaloid matter.

It may be necessary to observe that the prostate gland, in a child of 5 years of age, is about the size of a small hazel nut; consequently its increase of volume, in this case, was great; and even its third lobe alone was as large as the whole gland itself in its normal condition at that period of life.

The bladder was contracted to about the size of a turkey's egg, and found to contain about an ounce of urine mingled with purulent matter. The mucous membrane of the bladder was somewhat thickened; in other respects this organ was sound.

Mr. Solly relates a similar case in a child 3 years old, in whom the prostate was the size of a hen's egg. The tumor presented a cream color on section, and a lobulated condition, with a fibrous tissue surrounding the lobules, giving it a gristly feel. It was of the same nature as the one described by Mr. Stafford, and contained innumerable cells.

Mr. Moore, of the Middlesex Hospital, has recorded a remarkable case of prostatic cancer, attended with peculiar circumstances.[1] The patient, a man aged 53, was admitted into the Middlesex Hospital with a large pulsating tumor in the left groin. The tumor lay precisely in the situation of the external iliac artery, and resembled an aneurism, but

---

[1] Medical Times, vol. ii., 1851, p. 174.

was partly lobulated. It was considered to be aneurismal, and the common iliac artery was tied.

The patient survived the operation 43 hours, and died of peritonitis. The tumor was the largest of several masses of enlarged glands affected with encephaloid cancer. It lay against the ilium, and was inclosed in thick fascia, which had given way at the soft prominences felt during life. It was somewhat tightly girt in almost a circle of arteries, the external iliac being stretched and flattened along its upper surface and beneath the fascia, the internal iliac lying in contact with its posterior part, and the obturator artery beneath it completed the circle, except where the tumor and pubes intervened between it and the external iliac trunk. Moreover, the large branches of the internal iliac passed through the tumor. There were small secondary cancerous deposits in the kidneys, and an encephaloid growth sprang from the back of the prostate gland (which was otherwise healthy) and extended upwards between the bladder and rectum. This appeared to be the primary growth. Some urine, pressed out through the urethra after death, was examined by the microscope, and was found copiously pervaded with cancer-elements. That which had been observed during life, being clear, was not examined microscopically.

Mr. Haynes Walton has also related an interesting case of malignant disease of the prostate, the symptoms of which were for some time obscure.

The patient, a gentleman aged 59, had enjoyed good health, when he was suddenly seized with feverishness, pains in the loins, and difficult micturition.

These symptoms were soon followed by pain in the right lumbar region, constipation, tenesmus, pain in the rectum just before making water, and shooting pains on the inside of the thighs and along the urethra to the glans while water was being passed. This state continued, with intervals of ease, for six months, when he came under Mr. Walton's care. A slight enlargement of the prostate was now detected; the urethra was found somewhat pushed forwards by an apparent enlargement of the third lobe. The catheter was passed, and the urine was drawn off, at first five or six times daily, then two or three times.

About a month after Mr. Walton's first visit, the desire to pass water became most urgent, and on introducing the finger into the rectum, the prostate was found to be greatly enlarged; there was constant and severe pain at the neck of the bladder. Soon afterwards the patient was obliged to keep his bed, and the prostate became so enlarged that it was impossible to introduce the finger between it and the sacrum. The patient now gradually sank, and died comatose, about six weeks after Mr. Walton's first visit.

After death the true pelvis was found filled by a morbid growth, which proved to be the prostate gland converted into a malignant tumor. No trace of its natural tissue remained. On the anterior part of the diseased mass was a cavity with irregularly ulcerated walls, containing unhealthy pus; there was no trace of urethra, but the bladder opened directly into it. The pelvic lympathic glands on the right side were carcinomatous.[1]

Other cases, besides the above, illustrate the fact that malignant disease of the prostate may exist for some time without giving rise to any characteristic symptoms. This is more likely to occur when the medul-

---

[1] Trans. of Path. Society, vol. ii., p. 287.

lary matter is infiltrated through the substance of the gland, as in the following example:—

The patient, a man 60 years of age, had been under the care of Sir Henry Thompson for frequent and difficult micturition, evidently arising from enlarged prostate gland. His symptoms were much relieved by medicine and the use of the prostatic catheter. He continued in this state for two years, suffering occasionally from retention of urine, but otherwise experiencing little pain or inconvenience. After a lapse of two years he was attacked by paraplegia, for which he was admitted into the Marylebone Infirmary, on December 9th, 1853. At first he seemed to improve slightly, but early in February, 1854, the urine was passed with increased frequency, became bloody, and contained much phosphatic deposit. He died on the 23rd of the same month, in a state of great emaciation.

The prostate was found enlarged to the size of a small orange, being nearly equally increased in all directions. On dividing the right lobe, its structure was found softened, and of a pink color, from congestion; the fibres of the prostatic tissue were seen to be separated from each other by soft, almost creamy, matter infiltrated amongst them. The juice, when mixed with water and placed under the microscope, exhibited cancer-elements. Cancerous matter was also found adherent to the dura mater of the cord, and in a lymphatic gland near the bladder. Beneath the mucous membrane of the neck of the bladder were five or six small masses, varying in size from a pea to a horse-bean, and firmly fixed to the muscular coat, but unconnected with each other or the prostate. These small tumors likewise contained cancerous matter.[1]

The treatment of malignant disease of the prostate is, of course, merely palliative. The pain and irritability must be relieved by full doses of opium, while retention of urine is obviated by a cautious use of the catheter.

Tubercular Disease of the Prostate.

Tuberculous deposit occasionally takes place in the substance of the prostate gland, in the same manner and under the same circumstances, as in other tissues. It is never found in the prostate alone, other parts of the genito-urinary organs being always similarly affected. The kidneys, testicles, and vesiculæ seminales are the organs in which the disease usually co-exists, and in the majority of the cases on record the lungs are also reported to have been the seat of tubercle.

The tuberculous deposit appears to commence in the proper glandular tissue of the prostate, in the form of minute gray points. These increase in size and number, become confluent, and form masses as large as a pea or bean. Softening and disintegration subsequently take place, and one or more abscesses result. The whole of the gland may become excavated into an irregular cavity, bounded by the capsule and containing a mixture of tubercle and pus. Unless the contents be speedily evacuated by puncture from the perineum, there is great danger of the abscess bursting into the urethra or rectum; and even under favorable circumstances, it not uncommonly happens that this communication takes place from the encroachment of the tuberculous masses upon the parts bounding the purulent collection and their subsequent softening. Under such circumstances the verumontanum and seminal ducts are destroyed.

---

[1] Trans. of Path. Soc., vol. v., p. 205.

The deposit, however, sometimes passes through other changes. After becoming caseous, it may undergo calcification.[1] In the vesiculæ seminales the tuberculous matter is deposited in a continuous layer, which produces ulceration of the mucous membrane. Shreds of ulcerated tissue, infiltrated with tuberculous matter and floating in pus, are often found within their convoluted tubes. The disorganization may extend through their coats, and cause perforation of the bladder, when the contents of this latter organ will escape into the cavity of the abdomen, and may cause death by peritonitis.

It has been already remarked that tuberculous deposit in the kidneys is frequently present in these cases. Small masses in the cortical substance, gradually increasing in size, become confluent and softened, as in the prostate, and an irregular cavity is formed containing a mixture of pus and tubercle. If the patient live long enough this abscess may burst internally; but it is generally combined with extensive disorganization of other important organs.

The symptoms of tubercular disease of the prostate are extremely obscure, and a diagnosis can be formed only from the general condition of the patient. So long as the tubercular matter is small in quantity and concrete, the patient experiences no inconvenience; the change gives rise to no symptoms, and therefore cannot be detected. The earliest signs are those of chronic enlargement of the gland, which proceeds gradually and with little or no pain. When the enlargement has proceeded to such a degree as to excite any irritability of the bladder, or oppose any impediment to the free discharge of the urine, it may be detected by the finger introduced into the anus; but then will arise the question, What is the nature of the tumor or enlargement with which we have to deal? This cannot be determined with certainty at an early stage; but there will be strong reason to suspect its tuberculous origin, when the tumor has been of slow formation, and unattended with pain; and when the patient presents signs of the deposition of tubercles in other parts of the body. The occasional hæmaturia, and the presence of pus in the urine may give rise to the suspicion of calculus, the absence of which, however, will be demonstrated by sounding.

The treatment, in cases of this kind, must be mainly directed to that state of the constitution in which the tubercular deposit takes place. Local remedies will be of little or no avail. When abcess forms, it must be treated in the ordinary manner.

---

[1] Picard, loc. cit., p. 209.

CHAPTER XXXII.

PROSTATIC CONCRETIONS AND CALCULI.

The calculi so often found in the prostate gland, although not properly ranked with urinary calculi, must not be passed without notice. These calculi are composed of phosphate, with a little carbonate, of lime and animal matter. They are found in the cells of the prostate, are usually small, and often rounded and numerous, but become irregular as they increase in size. Larger calculi, sometimes of considerable bulk, are occasionally found in abscesses and cysts in the prostate; these are smooth and sometimes highly polished, resembling porcelain in appearance. They are sometimes laminated with radiating striæ, and sometimes compact. Occasionally they consist of the neutral phosphate, but far more frequently of the bone phosphate. These calculi may be analyzed by dissolving them in hydrochloric acid, and adding liquid ammonia, until all the phosphate of lime is precipitated, filtering and precipitating the remainder of the lime, which existed as carbonate, by oxalate of ammonia.

The distinct calculi of the prostate gland are to be distinguished from the minute concretions which form in its follicles. The first have been long known, and have been described by many pathologists. The latter have been more recently brought under the notice of pathologists, and their microscopic appearances have been described by Dr. H. Jones, Mr. Quekett, and Sir H. Thompson. These concretions occur in minute follicles of the gland, and the average diameter of the follicles does not, as it appears, exceed the hundredth part of an inch. The color of these minute bodies is deep yellow or red; sometimes, however, they are pale and colorless. Being so very minute, they are hardly distinguishable from the texture in which they are deposited. Dr. Handfield Jones thus describes their mode of formation: "They arise in a large oval vesicle of a single wall of homogeneous membrane. This is occupied by a colorless, finely mottled substance, in the centre of which a nuclear corpuscle sometimes occurs. Their mean diameter is about $\frac{1}{1000}$th of an inch. In those of larger size the envelope is still seen, but the contained amorphous matter is beginning to be arranged in layers concentric to the envelope. In the further stage the vesicle measures $\frac{1}{500}$th of an inch or more, showing concentric layers which are more developed on one side than on another, like so many repetitions of the original envelope, the intervals between the layers being occupied by a finely mottled deep yellow or red substance. There is a central cavity corresponding with the external contour in its form, which is triangular with rounded angles, or quadrilateral. From this normal appearance these bodies present numerous variations in form and internal arrangement, and appear to occupy an intermediate position between organic growths and inorganic concretions; they are related to the former by their vesicular origin and by their growth, which appears to take place chiefly by the dilatation of the vesicle and successive deposits in its interior; to the latter by their shape, their tendency to become infiltrated with earthy matter, and to pass into the condition of a dead amorphous mass of a deep yellow, red, or even almost black. The chemical composition varies probably with their different stages of development, at first consisting of little else but animal matter, then acquiring, especially when in a state of degeneration, calcareous salts, stated by Dr. Prout to be phosphate with a little carbonate of lime. The coloring matter is unaffected by ether, liquor potassæ, and muriatic acid."[1]

According to Mr. Quekett, these concretions are met with so frequently that they might seem to be a part of the constituents of the gland or its secretion. From his account it appears that the concretions commence by a deposit of earthy matter in the secreting cells of the gland; that they increase in size either by aggregation or by deposition in the form of concentric layers, and thus in the former case that they take the moulds of the follicles, and in the latter that they exhibit on section the usual appearance of an uric acid calculus. Mr. Quekett says that the parietes of the adjacent cells in which a deposition has taken place are in contact and may be destroyed, so that on the addition of dilute muriatic acid, by which the earthy matter is dissolved, a many-celled cavity is left. Owing to the peculiar mode in which the calculi mould themselves to the follicles, they not unfrequently present on the external surface an appearance resembling that of mulberry calculus. Mr. John Adams, in his able work on the 'Prostate Gland,' remarks, that the opinion of Prout, as to the deposition of earthy salts being the result of a deranged action in the mucous membrane is thus fully borne out. "I have examined," he says, "these concretions, with Mr. Quekett, in a prostate removed from a young man who had died of phthisis; they were exceedingly numerous, especially in the middle lobe. The middle lobe was much larger than either lateral lobe."

Sir Henry Thompson,[2] who has minutely investigated the subject of the formation of these concretions, points out that yellow granules similar to those occurring in the prostate are constantly to be found in the fluid of the vesiculæ seminales, and he is inclined to believe that they are the natural product of gland-secretion. Virchow thinks that the organic matter of these concretions is derived from a peculiar insoluble protein substance mixed with the semen. Their origin would therefore appear to be somewhat uncertain, but they doubtless form the starting-points of prostatic calculi. The irritation they cause leads to the secretion of earthy matter, consisting of phosphate and carbonate of lime, which becomes deposited upon the nucleus of organic matter and forms a minute calculus. This increases in size by further deposition of phosphate of lime.

The calculi thus formed are usually round and firm, somewhat transparent, having a pearly aspect, so as in some measure to resemble grains of pearl barley. They have a brown coating which they seem to derive from the secretion of the gland. They gradually increase in size, and, coming into contact with each other in the adjacent follicles, become facetted or articulated; their smoothness is so great that they put on the appearance of porcelain. In the progress of their development they cause absorption of the gland, so that it, or a part of it, may become reduced

---

[1] Medical Gazette, August 20th, 1847.
[2] Diseases of the Prostate, p. 318.

to a sac, often composed of many cells, containing 50 or 60 calculi. When the prostate gland has passed into this state, it is felt through the rectum as if it were a bag full of marbles. At other times, a single cavity is found in one lobe, containing one large calculus. The smaller stones often escape through dilated ducts into the urethra, and may be extracted by the urethral forceps. The section of these calculi generally shows a radiated or laminated structure; in other cases the structure is compact.

The analysis of these concretions shows the presence of phosphoric acid and lime. Their chemical composition, according to Dr. Wollaston, is phosphate of lime 84.5, carbonate of lime 0.5, animal matter 15.0. Mr. Adams, in the work just quoted, regards the combination of the lime and phosphoric acid in such calculi as reducible to two salts, namely, the neutral phosphate and the basic phosphate of lime. Those composed of the neutral phosphate are somewhat fusible before the blow-pipe, and generally exhibit a crystalline structure; on the contrary, those composed of the basic phosphate are altogether infusible before the blow-pipe. It should be remembered, however, that if the calculus under examination has passed into the bladder or urinary passages, it may have acquired a new coating on its surface. And thus when calculi formed in the prostate have passed into the bladder, that which was composed of basic phosphate of lime, and therefore wholly infusible, may have become fusible by a coating of triple phosphate. It appears also that when exposed in the prostatic portion of the urethra, such calculi often become coated with uric acid.

It is sometimes observed that the urine becomes milky from the great abundance of phosphates secreted in the prostate, and finally mingled with the urine. This milkiness is to be distinguished from that which might arise from a deposition of phosphates from the urine, and the diagnosis will be assisted by noticing the symptoms of irritation of the prostate gland and neck of the bladder, which cannot but accompany such a state of the prostatic secretion.

Calculi derived from a different source may be found in the prostate gland. Thus a small urinary calculus, or a fragment produced by lithotrity may, after escape from the bladder, become impacted in the prostatic urethra and remain lodged in the gland, or a vesical calculus may send a prolongation along the urethra into the gland just as a prostatic calculus proper has been known to extend backwards into the bladder.

The symptoms produced by calculi in the prostate gland will depend on their size, number and situation. The prostate itself may be hypertrophied; if the calculi have attained any size, the tissue of the organ is absorbed, and it is converted into a number of sacs, in which the foreign bodies are lodged.

Very minute prostatic calculi give rise to no symptoms, and their presence is not suspected during life. When larger and projecting into the cavity of the urethra, they are generally attended with more or less irritation about the neck of the bladder. The patient complains of a sense of uneasiness in this region, and has frequent desire to void his urine, or some difficulty in passing the urine is often experienced. The pain and uneasiness are occasionally increased by violent exercise, as Mr. Wilson observes, but the same effect would result from stone in the bladder. In some cases, these calculi excite acute inflammation of the gland, followed by abscess.

The general symptoms are seldom of a nature to determine, with any degree of certainty, the presence of calculi in the prostate; recourse must be had to examination, but even this is not always satisfactory. Unless the foreign bodies project into the urethra, or can be felt through the rectum, instrumental examination can be of little avail. In the former case, when a sound is introduced, the peculiar grating feel of the metallic instrument cannot be mistaken, but even here the calculus may be pushed into the cavity of the prostate, or into the bladder, during the first examination, and thus escape subsequent examination. If a wax bougie of large size be used, it surface may perhaps be more or less indented by passing over the concretion, if the latter projects into the urethra. When numerous or large concretions have caused absorption of the tissue of the gland, they can often be detected by the finger introduced into the rectum.

The treatment is to be regulated according to the circumstances of each individual case. Unless the calculi produce some irritation, they should not be meddled with. Should the symptoms be so urgent as to render an operation expedient, an attempt must be made to extract them by means of Weiss's forceps. This has often been done with success. When the calculi are too large to admit of being extracted whole, the urethral litholabe may be tried. M. Civiale has succeeded in detaching and crushing the calculi with this instrument in three cases. It has been recommended to push back the stone into the bladder; this proceeding would render an operation necessary at some subsequent period, and, considering the ease with which a stone may be crushed, it is the best course which can be adopted.

When the calculi are so impacted as to elude the grasp of the forceps, while the patient's sufferings render an operation necessary, there is no recourse left but to cut down on the foreign body in the median line of the perineum, and remove it. However, if the calculus formed a considerable projection into the rectum, it may be questioned whether the recto-prostatic operation would not be preferable.

INDEX.

Abnormities of the bladder, 31
Abscess between coats of bladder, 95, 97
   of prostrate, 357
Absorption by mucus membrane of bladder, 10
Accidents of lithotomy, 314
   of lithotrity, 278
Acid injections for chronic cystitis and phosphatic deposits, 91, 284, 353
   nitro-hydrochloric, for oxaluria, 352
   uric, amount of, excreted, 221
      calculus, 203
      deposit of, prevention, 348
      origin of, 221
Acidity of urine as a cause of stone, 222
Adams, Mr., on prostatic calculi, 379
   on scirrhus of the prostate, 372
After-treatment in cases of lithotomy, 319
   in cases of lithotrity, 275
Age, as affecting frequency of calculus, 231
   lithotomy, 303
   lithotrity, 289
Alkaline remedies for uric acid deposit, 350
   irritable bladder produced by, 160
Alkalinity of urine from ammoniacal decomposition, 86
Allarton's operation for stone, 321
Alternating calculi, 215
Ammonia, urate of, calculus, 206
Ammoniaco-magnesian phosphate calculus, 211
Amorphous urates, composition of, 206
Analysis of calculi, method of, 201
Anatomy of bladder, 1
Animal food as affecting composition of calculus, 234
Animals, urinary calculi in, 242
Arrest of development of bladder, 33
Ash, Dr., on bilobular bladder, 105
Atony of bladder, 165
Atrophy of bladder, 32
Auscultation for detecting calculus, 252

Banks, Mr., on urethral fever, 25
Barnes, Mr., elastic bottle of, for vesical fistula, 144

Bartels, Dr., on wounds of the bladder, 54
Beatty, Dr., case by, of eversion of bladder, 43
Beaumont, Mr., needle of, for vesical fistula, 148
Beketow, Dr., on prevalence of calculus in Russia, 238
Belladonna for nocturnal enuresis, 175
Bennett, Dr. Risdon, case by, of cystine calculi, 216
Bilateral operation, 320
Billroth, Prof., case by, of myoma of bladder, successfully removed, 118
   on irritable bladder, its possible cause in some cases, 158, 174
   on prostatic hypertrophy, 360
   on vesical catarrh as a symptom of kidney disease 84
Bladder, diseases and disorders of:
   abnormities of, 31
   absence of, 32
   fissures of, 41
   imperforate, 32
   multiple, 32
   extroversion, 32
      cases, 35, 36
      causes of, 33
      complications, 35
      treatment, 37-40
   pervious urachus, 40
   atony (see Paralysis)
   eversion of, 42
   fistulæ of, 138
      recto-vesical, 154
      vesico-vaginal, 139
         appearances, 141
         causes, 139
         classification of, 139
         diagnosis and symptoms, 142
         prognosis, 143
         treatment, 143
            after-treatment, 148
            caustics, 144
            cautery, 144
            Dr. Bouqué's method, 145
            Dupuytren's method, 146
            Lallemand's method, 146

Bladder, vesico-vaginal fistula, treatment;
    Mr. B. Brown's operation, 150
    Mr. Lawson Tait's operation, 152
    needles, 147
    palliative treatment, 143
    plugs to close opening, 143
    radical treatment, 144
    results, 143
    spontaneous closure, 142
    sutures, 149
    various knives, 146
  foreign bodies in, 261
    bone, fragments of, 263
    bougies introduced, 261 [263
    bullets and other projectiles, cases by Mr. Bryant, Mr. Cock, Mr. Coulson, Mr. Callaway, Mr. Hutchinson, Mr. Toogood, and Mr. Walton, 261–263, 265
    Civiale on extraction of, 264
    hairs, 265
    hydatids, 265
    instruments required, 264
    symptoms of 263
    treatment, 263
  gangrene of, 78
  gunshot wounds of, 45
  hernia and displacements of, 42
    cases by Dr. Murphy, Mr. Croft, Dr. Beatty, Mr. Lowe, Mr. Cross, and Dr. Thompson, 42–44
    symptoms of, 44
    treatment, 44
  inguinal cystocele, 45
    causes, 45
    complications, 48
    illustrative cases, 46–48
    symptoms, 48
    treatment, 49
  hysterical affections of, 162, 182
  hypertrophy of, 99
    causes of this state, 99, 100
    columnar and fasciculated bladder, 7, 99
    morbid appearances, 99, 100
    sacculated bladder, 101
    "stammering bladder," 100
    hypertrophy of, symptoms, 101
    treatment, 101
    vesical cyst, Drs. Murchison and Warren's cases, 104
  inflammation (acute) of mucous membrane of, 69
  inflammation (chronic) 85
  inflammation of peritoneal coat of, 107
  inflammation of walls of, 95
  injections for chronic cystitis and phosphatic deposits, 91, 284, 353

Bladder, irritability of, 158
    Billroth's views, 158
    causes of, 158
    symptoms of, 158
    treatment, 163
  mucous membrane of, acute inflammation of, 69
    appearances of urine in, 71
    causes of, 69
    diagnosis of, 78
    diphtheritic inflammation, cases of, 77
    morbid appearances, 73
    prognosis, 81
    symptoms of, 70
    treatment, 78
  mucous membrane, chronic inflammation of, 85
    causes of, 85
    changes in the urine, 86
    morbid appearances, 101
    symptoms, 85
    treatment, 89
  croupous inflammation, 70, 77
  paralysis of, 165
    causes of, 165
    differences between atony and paralysis, 165
    over-distention as a cause of atony, 167
    morbid appearances, 169
    symptoms, 168
    treatment, 170
  peritoneal coat of, inflammation of, 107
    morbid appearances, 107
    symptoms, 107
    treatment, 109
  prolapsus vesicæ, 49 [51
    Mr. B. Brown's operation. Prof. Stolz's operation, 51
    symptoms, 50
    treatment, 50
  rupture of, 57
  spasm of, 162
    causes, 162
    symptoms, 163
    treatment, 164
  tubercle of, 121
    diagnosis, 135
    morbid appearances, 121
    Mr. Prescott's Hewett and Dr. Wilks's cases, 121
    Mr. Thomas Smith's case, symptoms, 122, 135 [122
    treatment, 135
  tumors, malignant, 123
    cases by Mr. Coulson, 130–133
    case by Dr. Lankester, 129
    diagnosis, 128
    encephaloid, 123
    epithelioma, 124
    cases by Sir H. Thompson and Dr. Fagge, 124

Bladder, tumors, malignant:
  scirrhus, 123
  symptoms, 124
  treatment, 137
 tumors, non-malignant, 111
  cystic growths, 114
  fibrous tumors and fibro-my-
    omata, 118
    cases by Dr. Winckel and
      Profs. Billroth and
      Volkmann, 118, 119
  mucous polypi, 111
    Mr. Crosse's case, 111
    Mr. Savory's case, 113
  papilloma or villous tumor,
    sarcoma, 121   [115
  symptoms of, 133
  treatment of, 135
  tumors in female bladder, 134
  Ultzmann's cases of papillo-
    ma, 116
  "villous cancer," 116
  villous growths, 115
 ulceration of, 74
  cases of, 81-83
  Dr. Prout's views, 83
  Professor Billroth on, 84
 wounds and injuries of, 53
Blood in urine (see Hæmaturia)
Bone-earth calculi, 211
Bougies, fragments of, in bladder, 262
Bouqué's statistics of vesical fistula,
  140
Brander, Dr., operation by, of tapping
  bladder through pubic symphysis,
  181
Brodie, Sir B., on chloroform in operation of lithotrity, 268
  on encysted calculus,
    255
  statistics of lithotrity,
    331
Brown, Mr. Baker, operation by, for
  prolapsus vesi-
    cæ, 51
  operation by, for
    vesico - vaginal
    fistula, 150
Brüns, Prof., on closing wound after
  supra-pubic operation for stone, 300
Bryant, Mr., analysis by, of 230 cases
  of lithotomy, 232
  case by, of removal of
    stiletto from female
    bladder, 264
  on incision through peri-
    neum for rupture of
    bladder, 67
Buchu, for irritable bladder, 89

Cadge, Mr., on prevalence of stone in
  Norfolk, 235
Calculi, adherent to bladder, cases of,
  310
  animal matter in, 199
  causes of 221
25

Calculi, chemical classification, 203
  colors of, 194
  composition of nuclei, 195
  conditions of, as affecting li-
    thotrity, 278, 286
  diagnosis of, 243, 250
  difficulty of detecting stone,
    encysted, 254, 309   [252
  forms of, 194
  fracture during lithotomy, 312
  fracture, spontaneous, 248
  in animals, 242
  internal structure of, 195
  method of examining, 201
  nucleus of, 196
  numbers of, 195
  prostatic, 378
  rate of increase in size of, 248
  size and weight of, 195
  size of, as affecting results of
    lithotomy, 312
  solution of, in the bladder, 340
    Chevallier's experiments,
      341
    Dr. Roberts's experiments,
    Heller's case, 344   [343
    Sir B. Brodie and Mr.
      Southam's cases, 344
    Petit's experiments, 341
  species of:
    alternating calculi, 215
    ammoniaco - magnesian
      phosphate, 211
    carbonate of lime, 214
    cystine, 216
    fibrinous, 218
    fusible, 212
    oxalate of lime, 208
    phosphate of lime, 210
    silicious concretions, 218
    urate of ammonia, 206
    uric acid, 203
    urostealith, 219
    xanthine, 207
  symptoms and diagnosis of,
    243, 249
  symptoms simulating, 258
Calculous disease in China, 239
  in Egypt, 239
  in Finland, 238
  in India, 239
  in Russia, 238
  in United States, 237
  in various counties of
    England, 235, 236
Callaway, Mr., case by, of bougie in
  bladder, 262
Cancer of bladder, 110
  of prostate, 372
Cancer-elements in urine, 129, 187
Carbonate of lime calculus (see Calculus)
Carter, Dr. Vandyke on calculus in
  India, 197, 239
Catarrh (see Chronic Inflammation of
  Mucous Membrane of Bladder), 85

Catheters, 20, 21
    English and French measurements, 21
    introduction of, 21
    kinds of, 21, 22
    prostatic, 22
Catheterism, accidents sometimes attending, 23
    precautions regarding 21
Caustics, treatment of vesical fistulæ by, 144
Cautery, treatment of vesical fistulæ by, 144
Cheselden's lateral operation for stone, 316
Chevallier on solution of calculi, 341
Children, incontinence of urine in, 174
    lithotomy in, 307
    lithotrity in, 290
    position of bladder in, 1
    prostatic cancer in, 372
Chloral, for nocturnal enuresis, 176
Chloroform in lithotrity, 268
Cider-producing counties, rarity of calculus in, 236
Civiale on extraction of foreign bodies from bladder, 264
    on injections for catarrh of bladder, 91
    on injections for paralysis of bladder, 171
    on method of performing lithotrity, 272
    on removal of pedunculated growths from bladder, 136
    on removal of prostatic calculi, 381
    statistics of lithotrity, 330
Clark, Mr. Le Gros, on tolerance of presence of urine by peritoneum, 66
Clement, Mr., case by, of hernia of bladder, 46
Climate, influence of, in production of calculi, 236
Clubbe, Mr., case by, showing hereditary tendency to calculus, 241
Cock, Mr., case by, of bougie in the bladder, 262
    on puncture of bladder through rectum, 179
Concretions, prostatic, 378
    urinary (see Calculi)
Copaiba for vesical catarrh, 89
    injections of, into bladder, Devergie on, 93
Croft, Mr., case by, of eversion of bladder, 43
Crosse, Mr., on eversion of the bladder, 42
    on peritonitis after lithotomy, 318
    on polypus of bladder, 111
    tables by, concerning size of calculi in fatal cases, etc., 313
Curling, Mr., on worms in bladder, 265

Cystine calculus (see Calculus)
Cystitis (see Bladder, inflammation of)
Cystitis, acute, after lithotomy, 318
    after lithotrity, 280
Cystocele, inguinal, 45
    vaginal, 49
Cysts of bladder, an obstacle to lithotomy, 309
    cases by Drs. Murchison and Warren, 104
    cases by Dr. Wilks and Mr. Solly, 103
    effects of, in concealing a calculus, 254
    effects upon contents of bladder, 103
    impediments to lithotrity, 278
    treatment, 105
    with hypertrophy, 101

Deficiency of bladder, 31
Demarquay on absorption by mucous membrane of bladder, 11
Dentition in children a cause of irritable bladder, 160
Desault, catheter of, for vesical fistulæ, 143
Development of bladder, arrest of, 33
Dickinson, Dr., case by, of malignant disease of the bladder without marked symptoms, 129
Diet as affecting the production of calculi, 234
    milk, for cystitis, 80, 93
Dilatation of urethra for examination of bladder in female, 27
    for extraction of stone in female, 324
Dilators for female urethra, Simon's, 28, 325
    Weiss's, 324
Diphtheria of the bladder, 77
Displacements of bladder, 42
Distention of bladder a cause of paralysis, 167
Dulles, Dr., statistics of supra pubic lithotomy, 338
Dupuytren, case by, of laceration of the bladder, 62
Dyspepsia as a cause of uric acid deposits, 222

Electricity for paralysis of the bladder, 171
    employed for solution of calculi, 345
Elliotson, Dr., case by, of abscess between bladder and symphysis pubis, 108
Endoscope, use of, 26
Enlargement, chronic, of the prostate, 360
Enuresis, 173   [174
Ergot of rye for paralysis of bladder,

INDEX. 387

Erichsen, Mr., case by, of extroversion of the bladder, 35
   case by, of sacculated bladder, 114
Estlander, Dr., on rarity of calculus in Finland, 238
Evacuating tubes, Dr. Bigelow's, 294
Eve, Professor, on the bilateral operation, 321
Eversion of bladder, 42
Examination of bladder and prostate, method of, 19
Extravasation of urine from wound of bladder, 53, 59
Extroversion of bladder, 32

Fagge, Dr. Hilton, case by, of epithelioma of the bladder, 124
Fasciculated bladder, 99
Fayrer, Sir J., on calculus in India, 239
Females, calculi in, 233, 323
   incontinence of urine in, 173
   method of examining bladder, 27
   lithotomy and lithotrity in, 218 [323
Fibrinous calculus,
Fibrinuria, 188
Fissures of the bladder, 33, 41
Fistulæ of the bladder, 138
Fleming, Mr., case by, of laceration of bladder with very slight hæmaturia, 59
   case by, of xanthine calculus, 208
   on rarity of operations for stone in Ireland, 236
   on supra-pubic puncture for retention, 179
Fletcher, Mr., case by, of fatal peritonitis after sounding, 252
Fleury, M., case by, of absence of bladder, 31
Food, influence of, in production of calculus, 234
   nitrogenized, source of uric acid, 222
Forceps for extracting foreign bodies, 264
   the crushing, 312
   use of in lithotomy, 306
Foreign bodies, forceps for extracting, 264
   in the bladder, 261
   producing calculus, 263
Forms of calculi, 194
Fourcroy and Vauquelin on the solution of stone, 340
Fragments of stone, discharge of, after lithotrity, 275
   Dr. Bigelow's catheters for evacuating, 292, 294
   lodgment of, in urethra, after lithotrity, 281

Fuller, Dr., case by, of rupture of dermoid cyst of ovary into bladder, 114
Fusible calculus, 212

Gland, prostate (see Prostate)
Gout a cause of irritable bladder and stone, 161, 226
Gross, Dr., on abnormities of the bladder, 31, 39
   on calculi in the United States, 237
   on neuralgia of the bladder, 157
   on size of prostate in early life, 18
Growth of calculi, 248
Grünfeld, Dr., on the use of the endoscope, 26
Gunshot wounds of bladder, 55
Guthrie, Mr., on gunshot wounds of the bladder, 56
   on the sphincter of the bladder, 96
Gutta-percha bougies, fragments of in the bladder, 262

Hæmaturia, 184
Hæmorrhage, a sympton of calculus, 244
   after lithotomy, 314
   after lithotrity, 279
Hæmorrhage from the bladder, 187
   from the kidneys, 190
   from the prostate, 185
   from the urethra, 186
   from the urinary passages in general, 190
   in wounds of bladder, 53, 59
   symptomatic of tumors, especially of villous growths, 127, 188
   treatment of, 191
Hair discharged from bladder, 265
Harris, Dr., lithotomy in India by, 339
   on prevalence of stone in India, 239
Harrison, Dr., on laceration of bladder, 63
Haskins, Dr., on internal structure of calculi, 200
Hauff, case by, of rupture of bladder, 64
Heath, Mr., case by, of multiple cystine calculi, 216
   on dilatation of the female urethra, 28
Heine, Prof., on treatment of hypertrophied prostate, 369
Heller, case by, of urostealith calculus treated by solvents, 344
   classification of calculi, 226
Hemp-seed calculi, 210
Hereditary predisposition to calculous disease, 216, 241
Hernia of bladder, 45

Hernia of bladder, symptoms of, 48
  treatment of, 49
Hewett, Mr. Prescott, cases by, of laceration of the bladder, 61
  case of tuberculosis of the bladder, 121
Hey, Mr., case by, of rupture of the bladder, 64
  catheter for prostatic enlargement, 369
Hicks, Dr. B., case by, of villous tumor of bladder, 116
High operation for stone, 299
Hird, Mr., case by, of laceration of the bladder, 60
Holland, calculous disorders in, 236
Holmes, Mr., case by, of rupture of the bladder, 58
  operation for extroversion of the bladder, 37
Home, Sir E., on prostatic hypertrophy, 361
Hoskins, Dr., on solution of calculi, 345
Hospitals, London, statistics of operations for stone at five, 335
Howard, Dr., case by, of eversion of bladder through a vaginal fistula, 45
Huber, case by, of kidney disease with bladder symptoms, 122
Hutchinson, Mr. C., case by, of bougie broken in bladder, 262
  calculi in sailors, 240
Hydatids in bladder, 265
Hymen, imperforate, causing retention of urine, 183
Hypertrophy of bladder, 99
  of prostate, 360
Hysteria, a cause of irritable bladder, 162
  retention of urine in, 182

Incontinence of urine, 173
  after dilatation of the urethra for extraction of stone, 173, 324
  in children, 174
  in females, 173
  from paralysis of bladder, 173
Indigo, renal calculus containing, 220
Infarctions of uric acid in kidneys of infants, 223
Infection, purulent, of the blood after lithotomy, 319
  after lithotrity, 288
Infiltration of urine after lithotomy, 316
  after wounds of bladder, 53, 59
Inflammation of bladder, diagnosis of, from irritability, 84
  acute of the mucous membrane, 69
  after lithotomy, 318
  after lithotrity, 280
  causes of, 69
  diagnosis of, 78
  morbid appearances, 73

Inflammation, acute, of the mucous membrane of bladder, symptoms of, 70
  treatment of 78
  chronic, of mucous membrane of bladder, 85
  after lithotrity, 280, 287
  causes of, 85
  influence of, on results of lithotrity, 288
  morbid appearances, 88
  symptoms of, 85
  treatment of, 89
  of peritoneal coat of bladder, 107
  after lithotomy, 107, 318
  morbid appearances, 107
  symptoms of, 107
  treatment of, 107
  of walls of bladder, 95
  causes of, 95
  morbid appearances, 97
  symptoms of, 95
  termination in abcess, 97
  treatment of, 98
  acute, of prostate, causes, of, 356
  formation of abscess, 357
  symptoms of, 356
  treatment of, 358
  chronic, of prostate, causes of, 358
  symptoms, 358
  treatment, 359
  of cellular tissue of pelvis after injuries to the bladder, 59
  of cellular tissue of pelvis after lithotomy 316, 317
  of peritoneum from injury to bladder, 53, 59
  after lithotomy, 107, 318 [319
  of prostatic veins, after lithotomy,
Injections for vesical catarrh, 91, 370
  in cases of phosphatic calculi, 344 [171
  in cases of vesical paralysis,
Iron, iodide of, for nocturnal enuresis, 174, 176
Irritability of the bladder, 158
  causes of, 158
  treatment of,163

Johnson, Dr. George, on milk for cystis, 80, 93
Jolyet and Alling on absorption by mucous membrane of bladder, 10
Jones, Dr. Bence, on acidity of the urine, 222
Jones, Dr C. H., case by, of cystine calculus, 216
  on prostatic concretions, 378

Kidney, disease of, as affecting lithotomy, 319
  disease of, as affecting lithotrity, 289
  state of, in chronic cystitis, 87
Klien, Dr., on calculus in Russia, 197, 233, 238.

INDEX.   389

Knife for section of prostate, 305
Küss, Prof., on absorption by vesical mucous membrane, 12
 on sphincter of bladder, 14

Laceration of the bladder, 57
Lallemand on treatment of vesico-vaginal fistula by caustics, 146
Lankester, Dr., case by, of cancer of bladder, 129
Lateral operation, the, 303 (see Lithotomy).
Lawrence, Mr., case by, of atony of bladder mistaken for incontinence of urine, 167
Lebert on strychnia as a remedy for paralysis of the bladder, 172
Lime, carbonate of, calculus (see Calculus).
 oxalate of, calculus (see Calculus).
 phosphate of, calculus (see Calculus).
 water charged with, influence in causation of calculus, 235, 242
Liston on hæmorrhage after lithotomy, 315
Lithates, amorphous, composition of, 207
Lithic acid (see Uric acid).
Litholapaxy, or rapid lithotrity, 292
 advantages of, 297
 instruments employed in, 292
 method of operating, 296
 obstacles to, and how to overcome them, 295
 results of, 297
Lithotomy, 299
 Allarton's method, 321
 bilateral operation, 320
 the high operation, 299
 objections against 299
 preference for, by some German surgeons, 300
 the lateral operation, 303
 accidents of, 314
 after-treatment, 319
 instruments required for, 303
 obstacles to, 308
 precautions necessary in operating on children, 307
 the straight staff, use of, 307
 the recto-vesical operation, 302
Lithotomy, and lithotrity in the female, 323
Lithotrites, different kinds of, 269
 Dr. Bigelow's, 295
 measuring stone by means of, 251
 mode of introducing, 271
 mode of seizing and crushing stone, 272
Lithotrity, 267
 accidents of the operation, 278
 bladder, state of, as affecting, 287

Lithotrity, density and position of calculus, 286
 employment of chloroform, 268
 finding and seizing the stone, 272
 fragments impacted in urethra, 281
 hæmorrhage after, 279
 in children, 289
 indications and contra-indications, 285
 instruments employed, 269
 introduction of instruments, 271
 kidney, disease of, as affecting, 289
 obstacles to the operation, 277
 preparatory treatment, 267
 prostate, state of, as affecting, 287
 relapses after, 290
 removal of the fragments, question as to, 283
 size of calculus, as affecting, 278
 urinary organs, condition of, as affecting, 277, 283
 washing out bladder, question as to, 275
Lloyd, Mr., operation by, for extroversion of bladder, 37
Loebisch, method of analyzing calculi, 201
Lowe, Dr., case by, of eversion of bladder, 43, 44

Majendie, M., on food as a cause of calculus, 235
Malignant tumors of bladder, 123
 of prostate, 372
 illustrative cases, 374
Marcet, Dr., on xanthine calculus, 207
Martineau on dividing prostate in lithotomy, 317
Mason, Dr. E., case by, of rupture of bladder successfully treated, 67
Maury, Dr., operation by, for extroversion of bladder, 39
Meckel on the origin and metamorphoses of calculi, 226, 228
Medullary cancer of bladder, 131
Mercier, Dr., on absorption by mucous membrane of bladder, 11, 12
 on anatomy of neck of bladder, 5, 7
Method of analyzing calculi, 201
Microscope, use of, for detection of blood in the urine, 184
 of pus in the urine, 72
 for detection of villous and malignant disease of bladder, 129
Moore, Dr., case by, of urostealith calculus, 219
Moore, Mr., case by, of malignant tumor of prostate, 374
Mortality after lithotomy, statistics of, 332, 335
 after lithotrity, 330, 335
Mucous membrane of bladder, absorption by, 10
 structure of, 10

Mucus in urine, 72, 86
Mulberry calculus, 208
Murchison, Dr., case by, of enormous cyst of the bladder, 104
Murexide test for uric acid, 201
Murphy, Dr., case by, of eversion of the bladder, 42
Muscular coat of bladder, anatomy of, 9
hypertropy of, 99

Neck of bladder, anatomy of, 5, 6
Niemeyer on alkaline fermentation of the urine, 69, 86
on oxaluria as a cause of irritable bladder, 161
on tannin as a remedy for chronic cystitis, 90
Nitrate of silver, application of, to neck of bladder, 92
as an injection for chronic cystitis, 92
Nitric acid, injections of, for chronic cystitis and phosphatic deposits, 91, 344, 346
Nitrogenized food a source of uric acid, 222
Norfolk, prevalence of calculus in, 234, 235, 237
Nucleus of calculi, chemical composition of the, 196
Number of calculi, 195

Old age, paralysis of bladder in, 168
state of prostate in, 364
Orchitis after lithotrity, 283
Ord, Dr. W. M., on calculus containing indigo, 220
on forms of uric acid, 227
on spontaneous fracture of a calculus, 248
O'Shaughnessy, Sir W. B., on solution of calculi by electricity, 346
Oxalate of lime (see Calculus).

Paget, Mr., calculus removed by, from umbilicus, in case of pervious urachus, 40
median operation performed by, but no stone found, 259
Paget, Sir J., on tuberculosis of the bladder, 122
on tumors of the prostate gland, 360
on urinary stammering, 100
Paralysis of bladder, 165
causes of, 166
in old persons, 168
in women, 168
lithotrity in cases of, 283, 288
morbid appearances in, 169
state of urine in, 166
symptoms of, 168
treatment of, 170
Parasite in the bladder causing hæmaturia, 189

Pareira brava for irritable bladder, 161
for chronic cystitis, 89
Peritoneal coat of bladder, inflammation of, 107
symptoms, 107
treatment, 109
Peritonitis after lithotomy, 107, 318
from wounds of bladder, 53, 59
Petit, M., on solution of calculi, 341
Phlebitis after lithotomy, 319
Phosphatic deposits, cause of, 87, 367
treatment of, 91, 344, 346
Piles, paralysis of bladder after operation for, 166
Pitha, Von, on passing bougie for atony of bladder, 171
Plug for hæmorrhage after lithotomy, 315
Podrazki, Prof., on irregular contraction of bladder, causing a difficulty in finding stone, 256
on supra-pubic operation, 301
Poland, Mr., on the use of the straight staff in lithototomy, 307
Pollock, Mr., case by, of cancer of bladder, 132
Polypus of bladder, cases by Messrs. Crosse and Savory, 111, 113
Pregnancy, calculus in the bladder during, 328
retention of urine during, 183
Preparatory treatment in lithotomy, 303
in lithotrity, 267
Prepuce, contracted, in children, a cause of irritable bladder, 160
Preventive treatment of calculus, 347
Prostate gland, abscess of, 357
acute inflammation of, 356
causes of, 356
symptoms of, 356
treatment of, 358
anatomy of, 15
calculi in, 379
character and composition of, 380
symptoms of, 380
treatment, 381
calculus as a complication of enlargement of, 364
chronic enlargement of, 360
age at which it occurs, 365
changes in form and size, 361
complications, 364
diagnosis, 367
effects of, on bladdder and urethra, 361, 364, 367
enlargement of middle lobe, 362
operation for relief in advanced cases, 371
retention of urine in cases of, 178, 371
symptoms, 365
treatment, 368
use of catheter, 369

Prostate gland, concretions in, 378
 encephaloid cancer of, 372
 enlargement of, an obstacle to lithotomy, 309
 enlargement of, effect in diagnosis of stone, 256
 fibrous tumors of, 360
 measurements of, 16
 method of examining, 26
 scirrhus of, 372
 section of, during lateral operation, 306
   Cheselden on, 316
   Martineau's method, 317
   risk in complete division of, 317
 state of, as affecting lithotrity, 277, 287
 structure of, 16
 tubercular disease of, 376
 tumors of, 360, 372
Prout, Dr., on irritable bladder as a symptom of kidney disease, 83
 on parenchymatous cystitis, 95
Pus in urine, characters of, 72
 question as to origin in cases of calculus, 289, 319
Pyæmia after lithotomy, 319
 after lithotrity, 288
Pyelitis caused by cystitis, 87
 lithotomy where suspected to exist, 319
 lithotrity where suspected, 289

Quekett, Mr., on prostatic concretions, 379

Recto-vesical operation for stone, 302
Rectum, examination through, for calculus, 257
Recurrence of calculus after lithotrity, 290
Rees, Dr. Owen, on transformation of uric acid, 196, 229
Retention of urine, 177
 after lithotomy, 320
 lithotrity, 280
 from atony of bladder, 177
   enlargement of prostate, 178
   inflammation of prostate, 178
   paralysis of bladder, 166
   stricture of urethra, 178
 puncture of bladder for, 179
   above the pubes, 179
   by capillary trocar and aspirator, 181
   perineal section, 180
   through the rectum, 179
   through the pubic symphysis, 181
 Voillemier's operation, 181
 of urine, treatment of, 178
   in the female, 182
Rhenish wines, influence of, in preventing calculus, 236

Roberts, Dr., on ammoniacal decomposition of urine, 87
 carbonate of lime calculus, 214
 citrate of potash for solution of uric acid, 284
 composition of mixed urates, 207
 cystine calculus, 217
 detection of blood in urine, 184
 hæmaturia from presence of parasites in the bladder, 189
 hard water as a cause of calculus, 235
 preventive treatment of calculus, 350
 solvent treatment of calculus, 343
 suppression of urine after catheterism, 25, 279
Rokitansky on prostatic hypertrophy, 362
Rupture of bladder, 57
 treatment of, 65
Rutenberg, endoscope for female bladder, 26

Sacculated bladder, 101
Sailors, alleged rarity of calculus among, 240
Scharling, Prof., on animal matter in calculi, 199
Scirrhus of the bladder, 123
 of the prostate, 372
Scoop, use of the, in lithotomy, 312
Scudamore, Sir C., on gout as a cause of irritable bladder, 161
Senfleben, case by, of sarcoma of the bladder, 121
Sex as regards frequency of calculous disease, 233, 323
Silicious calculi (see Calculi)
Simon, Mr., operation by, for extroversion of bladder, 37
Simon, Prof., on catheterism of the ureters, 29
 on dilatation of the female urethra by means of plugs, 29
 on method of examining bladder in the female, 28
 on removal of calculi from female bladder, 325, 328
 on removal of tumors from female bladder, 137
Sloane, Sir Hans, on discharge of hairs from bladder, 265
Smith, Dr. Stephen, on laceration of the bladder, 63
Smith, Mr. Thomas, on a case of tuberculosis of the bladder, 122
Sodium chloride, influence of, in preventing formation of uric acid, 179
Solly, Mr., case by, of malignant disease of the prostate, 374
 case by, of sacculated bladder, 103

Solvent treatment of calculus, 340
Sounding the bladder for calculus, 250
   occasional danger of, 251
   occasional failure in detecting stone, 252
South, Mr., on fistula of bladder after lithotomy, 156
   on statistics of lithotomy at St. Thomas's Hospital, 333
Southam, Mr., case by, of a large calculus, removed by recto-vesical operation, 213
   case by, of spontaneous disruption of calculus, 248
   on acid injections as an aid to lithotrity in cases of phosphatic calculus, 345
Spasm of bladder, 162
Species of urinary calculi, 203   [205
Specific gravity of uric acid calculus,
Spine, injuries of, a cause of paralysis of bladder, 165
Spontaneous closure of vesico-vaginal fistula, 143
   fracture of calculus, 248
Staff, for lithotomy, 304
   the rectangular, 308
   the straight, 307
Stafford, Mr., case by, of malignant tumor of prostate, 374
Statistics of lithotomy, 332
   general table, 332
   influence of age, 333
   influence of size of stone, 334, 335
   of operations for stone at Guy's Hospital, 335
   at St. Bartholomew's Hospital, 336
   at St. George's Hospital, 336
   at St. Peter's Hospital, 337
   at St. Thomas's Hospital, 335
   of the apparatus major, 338
   the bilateral operation, 339
   the high operation, 339
   the lateral operation, 338
   the median operation, 339
   the recto-vesical operation, 339
   300 operations performed in India by Dr. Harris, 339
   Thompson, Sir H., by, 333
   of lithotrity, 330
   by M. Civiale, 330
   by Prof. Keith, 331
   by Sir B. Brodie, 331
   by Sir H. Thompson, 331
   by Sir W. Fergusson, 331
   results of operations for stone at four large London Hospitals and at St. Peter's Hospital, 335, 337
Statistics of wounds and injuries of bladder, 65
Stephens, Mrs., remedy for dissolving stone, 340

Stolz, Prof., operation for prolapsus vesicæ, 51
Stone in the bladder (see Calculus)
Suppression of urine after catheterism, 25
   lithotrity, 279
Suture, treatment of vesical fistulæ by, the button, 150   [145
   the clamp, 149
Syme, Mr., case by, of laceration of bladder, recovery, 62

Tait, Mr. Lawson, operations by, for vesico-vaginal fistula, 152
Taylor, Mr., on case of xanthine calculus, 207
Thompson, Dr., case by, of eversion of bladder, 43
Thompson, Sir H., case by, of epithelioma of bladder, 124
   case by, of fatal syncope after withdrawal of urine, 370
   case by, of malignant disease of prostate, 376
   case by, of villous tumor of bladder, 115
   on absorption by mucous membrane of bladder, 10
   on prostatic concretions, 379
   on prostatc hypertrophy, 365, 371
   on use of endoscope, 26
   statistics, lithotomy and lithotrity, 332, 333
Thorp, Mr., case by, of rupture of bladder successfully treated, 67
Tubercle of bladder, 121
   of prostate, 376
Tumors of bladder, 110
   of prostate, 360, 372

Ulceration of the bladder, 74
   cases of, 81
   distinguished from irritable bladder, 83
   Dr. Prout on, 83
   Prof. Billroth on, 84
   tubercular, 121
Ultzmann, Dr., on blood-corpuscles in urine, 184
   cystine calculus, question of hereditary predisposition, 217
   fibrinuria, 127, 188
   growth of calculi, 248
   kidney-infarctions as a cause of calculus, 224
   on predominance of uric acid as nucleus of calculi, 196
   on primary and secondary stone-formation, 226
   villous tumor of bladder, cases of recovery from, 116
Umbilicus, calculus removed at, 40
   fistulous opening at, 40

Urachus, pervious, 40 [Ins).
Urate of ammonia calculus (see Calculus)
Urea, conversion of, into carbonate of ammonia, 69, 86
Ure, Mr., on solution of calculi, 342
Ureters, catheterism of the, 29
Urethra, dilatation of, in the female for removal of calculi, foreign bodies and tumors, 27, 136, 264, 324
    impaction of fragments in, after lithotrity, 281
    incision into, in lateral lithotomy in children, 307
Urethral fever, 24
Uric acid, a principal constituent of urinary calculi, 203
    calculus (see Calculus),
    composition of, 221
    predominance of, as nucleus of calculi, 196
    prevention of formation and deposit, 348
    solvent treatment of, 342
    sources of, 222
Urinary concretions (see Calculus).
Urine, blood in (see Hæmaturia).
    changes in, in acute cystitis, 71
        in chronic cystitis, 86
    color of, when containing blood, 184
    detection of cancer-elements in, 128
    detection of villous growths in, 115, 116
    fibrine in, 188
    infiltration of, after lithotomy, 316
    retention of (see Retention).
Utero-vesical fistula, 142, 153
Uterus, displacement of, causing retention of urine, 182
Uytterhoeven, case by, of atrophy of the bladder, 32

Vagina, prolapse of bladder into, 49
Varix of the bladder, 189
Vauquelin, on solution of calculus, 340
Vegetable and animal food as affecting the composition of calculi, 234 [325
Velpeau, on lithotrity in the female,

Vesico-vaginal fistula (see Bladder, fistula of). [341
Vichy waters for solution of calculi,
Virchow, Prof., on uric acid infarctions in kidneys of infants, 223
Voillemier's operation for puncture of bladder, 181
Volkmann, Prof., case by, of myoma of the bladder, removed by operation, 119
    method (bimanual), of examining the bladder, 19

Walker, Dr., operation by, for case of rupture of bladder, 67
Walls of bladder, information of, 95
Walter, Dr, operation by, for case of rupture of bladder, 67
Walton, Mr. Haynes, case by, of foreign body in bladder, 261
    case by, of malignant disease of prostate, 375
Warner, Mr., case by, of polypus of bladder, 113
Warren, Dr., case by, of sacculated bladder, 104
Water, influence of, in production of calculus, 235
Wheeler, Mr., on absorption by mucous membrane of bladder, 11
Wiblin, Mr., case by, of extroversion of the bladder, 36
Wilks, Dr., case by, of tubercle of the bladder, 122
    cases by, of villous tumor of bladder, 115
Willett Mr., case by, of rupture of bladder treated by abdominal section, 67
Winckel, Dr., on a case of fibro-myomata of the bladder, 118
    on tubercle of the bladder, 122
    on vesico-vaginal fistula, 139
Wood, Mr. John, operation by, for extroversion of the bladder, 33, 38
Wounds and injuries of the bladder, 53

Xanthine calculus (see Calculi, species of).

Ziemssen, Prof., on induced current for paralysis of the bladder, 171

www.ingramcontent.com/pod-product-compliance
Lightning Source LLC
Chambersburg PA
CBHW030554300426
44111CB00009B/979